Adaptive Signal Processing Algorithms

PRENTICE HALL INFORMATION AND SYSTEM SCIENCES SERIES

Thomas Kailath, Editor

ANDERSON & MOORE	*Optimal Control: Linear Quadratic Methods*
ANDERSON & MOORE	*Optimal Filtering*
ASTROM & WITTENMARK	*Computer-Controlled Systems: Theory and Design, 2/E*
BASSEVILLE & NIKIROV	*Detection of Abrupt Changes: Theory & Application*
BOYD & BARRATT	*Linear Controller Design: Limits of Performance*
DICKINSON	*Systems: Analysis, Design and Computation*
FRIEDLAND	*Advanced Control System Design*
GARDNER	*Statistical Spectral Analysis: A Nonprobabilistic Theory*
GRAY & DAVISSON	*Random Processes: A Mathematical Approach for Engineers*
GREEN & LIMEBEER	*Linear Robust Control*
HAYKIN	*Adaptive Filter Theory*
HAYKIN	*Blind Deconvolution*
JAIN	*Fundamentals of Digital Image Processing*
JOHANSSON	*Modeling and System Identification*
JOHNSON	*Lectures on Adaptive Parameter Estimation*
KAILATH	*Linear Systems*
KUNG	*VLSI Array Processors*
KUNG, WHITEHOUSE & KAILATH, EDS.	*VLSI and Modern Signal Processing*
KWAKERNAAK & SIVAN	*Signals and Systems*
LANDAU	*System Identification and Control Design Using P.I.M. + Software*
LJUNG	*System Identification: Theory for the User*
LJUNG & GLAD	*Modeling of Dynamic Systems*
MACOVSKI	*Medical Imaging Systems*
MOSCA	*Stochastic and Predictive Adaptive Control*
NARENDRA & ANNASWAMY	*Stable Adaptive Systems*
RUGH	*Linear System Theory*
SASTRY & BODSON	*Adaptive Control: Stability, Convergence, and Robustness*
SOLIMAN & SRINATH	*Continuous and Discrete-Time Signals and Systems*
SOLO & KONG	*Adaptive Signal Processing Algorithms: Stability & Performance*
VISWANADHAM & NARAHARI	*Performance Modeling of Automated Manufacturing Systems*
WILLIAMS	*Designing Digital Filters*

Adaptive Signal Processing Algorithms

Stability and Performance

Victor Solo

Macquarie University
Sydney, Australia

Xuan Kong

Northern Illinois University
DeKalb, Illinois

PRENTICE HALL
ENGLEWOOD CLIFFS, NEW JERSEY 07632

Library of Congress Cataloging-in-Publication Data

Solo, Victor
Adaptive signal processing algorithms: stability and performance / Victor Solo, Xuan Kong
p. cm. — (Prentice Hall information and system sciences series)
Includes bibliographical references and index.
ISBN 0-13-501263-5
1. Adaptive signal processing — Mathematics. 2. Adaptive filters — Mathematical models. I. Kong, Xuan. II. Title. III. Series.
TK5102.9.S66 1995
621.382'2—dc20 93-38907
CIP

Acquisitions editor: Linda Ratts
Project manager: Irwin Zucker
Production coordinator: David Dickey
Supplements editor: Alice Dworkin
Cover designer: Jerry Votta
Editorial assistant: Susan Handy

Printed in the United States of America

10 9 8 7 6 5 4 3 2 1

ISBN 0-13-501263-5

Prentice-Hall International (UK) Limited, *London*
Prentice-Hall of Australia Pty. Limited, *Sydney*
Prentice-Hall Canada Inc., *Toronto*
Prentice-Hall Hispanoamericana. S.A., *Mexico*
Prentice-Hall of India Private Limited, *New Delhi*
Prentice-Hall of Japan, Inc., *Tokyo*
Simon & Schuster Asia Pte. Ltd., *Singapore*
Editora Prentice-Hall do Brasil, Ltda., *Rio de Janeiro*

To Padmini, Michael, and Alexander,
and to the memory of Paul.

V.S.

To Ning

X.K.

Contents

Preface

This book is chiefly concerned with the analysis of the behavior (i.e., stability and performance) of adaptive signal processing (ASP) algorithms.

There has been considerable research activity in the theoretical study of ASP (and adaptive control) algorithms in the last decade or so and there are now a number of books devoted to the subject (see the notes at the end of Chapter 1). Most of these books only dwell briefly on performance analysis. Most of those that include stability analysis restrict themselves to the LMS (least mean square) algorithm as outlined below and its close relatives. Our aims on these counts are much more ambitious. We discuss general methods of algorithm construction and general methods of algorithm analysis.

The book is intended for use as a graduate text while providing a bridge to and overlapping with more advanced research monographs. More generally the book should be of interest to practitioners and researchers in signal processing, communication, and control.

The book is divided into three parts, each requiring successively more background. But the first two can certainly be used (and have been used) for a graduate level, first course on ASP.

The first part of the book introduces ASP through its applications, shows how to construct ASP algorithms, and gives a detailed global stability and performance analysis in both deterministic and stochastic settings. The stochastic analysis is done under very restrictive assumptions but requires little background and enables a whole range of issues to be investigated. Further, the conclusions of the stochastic analysis turn out to have wider applicability. Background expected of the student would be an undergraduate course in digital signal processing or control and an undergraduate course in probability.

The second part turns to deterministic stability analysis by means of averaging methods, which are necessary to handle more complicated algorithms. Our hope is that this part will be accessible to typical graduate students in a systems program, especially because the basic results are developed in a very simple way. Background expected of the student would be a course in mathematical analysis at the level of [A6].

The third part treats stochastic stability and performance by averaging methods. Some continuity is provided here with the second part because the arguments used are stochastic versions of the deterministic ones. In this way we hope to make a step towards bringing the deterministic viewpoint (prevalent in the control community) and the stochastic viewpoint (prevalent in the communication and signal processing communities) closer together. Background expected of the reader would be a course in measure-theoretic probability at the level of [B9].

Before turning to a chapter by chapter discussion of the contents, let us point out that each chapter is ended by a set of notes that direct the reader to additional references. In these notes we occasionally sketch a bit of history and occasionally apportion credit to earlier workers. But no attempt has been made to do this in a systematic way and we have instead chosen the references to ease the path of the student.

PART I: FUNDAMENTALS

In Chapter 1 ASP problems are introduced by way of examples. In Chapter 2 relevant offline signal processing results are discussed because they throw considerable light on the behavior of adaptive algorithms. In Chapters 3 and 4 algorithm construction is covered. Chapters 5 and 6 are concerned with algorithm analysis.

An adaptive algorithm is, in general, a nonlinear time varying system, and so the analysis of its behavior is not straightforward and revolves around two issues: (i) stability and (ii) performance. Stability has two parts: convergence and boundedness. In the absence of noise the adaptive lter parameters will typically converge to a limit value. In the presence of noise there is no parameter convergence, although there is typically convergence in distribution. Nevertheless the noise-free convergence behavior has a lot to do with whether there is stochastic boundedness in the noisy case.

Once an adaptive filter is found to be stable, then one looks at its performance as compared, say, to an offline filter design. To put it another way, one seeks to calculate the loss of performance (so called misadjustment) due to the learning or adaptation process.

Chapter 1. Here we introduce ASP through some of its applications: Channel Equalization, Time Delay Estimation, System Identication, and Noise or Interference Cancelling. These examples are pursued in subsequent chapters through analysis and simulation. We identify a common "adaptive signal processing" element in these applications.

Chapter 2. As a reference for performance comparison, as well as for general insight, it is very valuable to understand the behavior of offline methods for the solution of typical ASP problems. In particular the signal processing ideas behind equalization, noise cancelling, and time delay estimation are laid out clearly. In Part II of the book, this link between offline behavior and adaptive behavior is made more precise by means of averaging analysis.

Chapter 3. This chapter is partly intended as review material on iterative optimization. There is some novelty in the discussion of improvements to steepest descent. More important however is the fact that much insight about adaptive algorithms can be obtained by a careful study of their offline iteration equivalents. Again the justication of this heuristic idea is achieved later through averaging analysis when it is shown that the averaged system is the ofine iteration corresponding to the adaptive algorithm.

Chapter 4. Here various algorithms are introduced. A classification scheme is presented that emphasizes properties that distinguish algorithms in important ways. An attempt has been made to give a fairly comprehensive discussion.

Chapter 5. This is the major chapter of Part I. In this chapter, both stability and performance are treated. The reader may be surprised that we devote so much space to the Gaussian white noise case. Especially since the analysis techniques used do not generalize. The point is, however, that the conclusions do. Also, in a white noise setting we are able to cover several issues of importance in ASP. In particular we would like to emphasize our detailed discussion of algorithm behavior in the presence of time varying parameters. In the past a lot of the adaptive signal processing literature has operated in the Gaussian white noise framework, and indeed there is "folklore" that the results thus obtained hold more generally. In fact averaging analysis, covered in Part II, makes precise this "folklore" idea while showing when and how it needs to be altered.

Chapter 6. This chapter concerns global stability analysis of some special algorithms when all signals are deterministic. The stability problem then reduces to the convergence problem. Historically these analyses delivered the first global results, while allowing fairly general signal specifications. The results and methods of this chapter have been especially important in adaptive control theory. Unfortunately, the techniques of this chapter, while important, do not extend beyond a limited class of algorithms. If we add noise into the algorithms of this chapter, then the question of performance arises. Although the consequent performance analysis can be done, we have chosen not to include it, so that space is available for the more powerful techniques of Parts II and III.

PART II: DETERMINISTIC AVERAGING

To handle algorithms where the parameters appear nonlinearly, the techniques of Part I are inadequate and we must turn to the powerful averaging methods to investigate stability and performance. Averaging is capable of providing a stability analysis of just about any adaptive algorithm that can be constructed and is certainly robust enough to allow a full treatment of time varying parameters. Furthermore, it has not been well appreciated that averaging also is capable of providing global stability results.

In this part of the book we concentrate on stability by omitting noise from the problem. This also makes it much easier to introduce averaging methods and show how to use them.

Chapter 7. This chapter introduces deterministic averaging for single time scale problems and works through several examples. The development is very simple, being based on a perturbation idea. There are three types of averaging result. The first is a finite time averaging result or "trajectory locking" result, which shows how, over a fixed time interval, the averaged system trajectory "locks" onto the original system trajectory as the gain, $\mu \to 0$. The second type of result is an infinite time averaging result, which shows how, if the averaged system has at least one attracting point and the original algorithm trajectory starts inside the corresponding region of attraction, then it remains very close (uniformly in time) to the averaged system trajectory, if μ is small enough. Note that the averaged system may have attracting points that are not attracting points of the original system. The third type of result is an infinite time averaging stability result, which shows that if the averaged system has an attracting equilibrium point and the original system has the same equilibrium point and is initialized in the region of attraction of the equilibrium point, then this equilibrium point is also an attraction point for the original system.

Chapter 8. Averaging is extended here to more difcult mixed time scale problems. The development remains very straightforward however. The extra analysis needed beyond the single time scale case of Chapter 7 is provided by a stability theorem for slowly time varying linear systems.

PART III: STOCHASTIC AVERAGING

In Part I some stochastic problems were discussed under very special assumptions. In this part of the book we extend the general analysis of Part II to a stochastic setting and address stability as well as performance. The techniques used for analysis bear a strong resemblance to the deterministic methods of Part II; this is not only pedagogically pleasing, it also makes them easier to understand. Nevertheless it must be said that this part of the book requires somewhat more background than do Parts I and II.

Usually the asymptotic analysis of stochastic adaptive algorithms has employed weak convergence methods, but for various reasons we have been able to bypass these methods. This does not mean weak convergence methods are not useful but rather that a considerable analysis is possible without them.

A complete and rigorous development of all the material in this part would lengthen the book considerably and anyway properly belongs in a separate research monograph. For that reason we have developed some material with full rigor while only sketching other areas. This is consistent with our aim of providing an introduction and bridge to the research literature.

Chapter 9. This Chapter parallels Chapter 7 for single time scale systems. In the stochastic setting a first order averaging and a second order averaging analysis is possible. The first order analysis produces results much like deterministic averaging. Indeed the development is, surprisingly, rather similar. The second order analysis describes stochastic fluctuations about the first order averaging path. Second order analysis enables asymptotic computation of performance measures such as mean square tracking error to be made. It hinges on perturbation-based computation of steady state behavior. The possibility of second order averaging makes the treatment of time varying parameters somewhat richer than in the deterministic setting.

Chapter 10. In this chapter, finite time averaging is developed and a second order analysis is sketched for mixed time scale ASP algorithms.

Our attempts at description and rationalization notwithstanding, we deliver our small offering to the reader with the same spirit (but more humility) with which Bunyan offered "Pilgrim's Progress".

"Some said, John, print it; others said, Not so;
Some said, It might do good; others said, No."

ACKNOWLEDGMENTS

We offer thanks to a numbers of colleagues who provided input during preparation of the book. To Bob Bitmead and Wei Ren for valuable detailed comments; to Peter Caines and P.R. Kumar for general encouragement. We also beneted at various times from discussions with Joseph Bentsman, Bruce Hajek, Rick Johnson, P.R. Kumar, Brad Lehman, and Lennart Ljung. Needless to say the remaining weaknesses are entirely ours. We would also like to thank the reviewers Claude S. Lindquist at the University of Miami and Raymond Kwong at the University of Toronto.

The book grew out of lecture notes from courses on Adaptive Signal Processing and Adaptive Control given several times by the first author in the Department of Electrical and Computer Engineering at the Johns Hopkins University. Thanks are due to Nitish Thakor for encouraging his students to attend.

We are also grateful to Yannis Goutsias and Jerry Prince of the Signal and Image Analysis Lab in the ECE Department at Hopkins for facilitating the partly long distance collaboration of the authors.

Acknowledgment is also due to the National Science Foundation who funded research of the first author that led to many of the results in Parts II and III of the book.

The first author owes a special debt to his family who put up with his many absences. The second author thanks his wife Ning for her encouragement, support, and understanding.

Victor Solo

Xuan Kong

List of Symbols

Symbol	Definition
$\tilde{E}$	average for a periodic signal: If ξ_k has period M, $\tilde{E}(\xi_k) = M^{-1} \sum_1^M \xi_k$
e	error signal
s	noise-free primary signal
$\hat{s}$	output of ASP filter
n	noise
x	auxiliary or reference signal
$y = s + n$	noisy primary signal
$\underline{w}$	weight or parameter vector for an FIR filter
$\underline{w}_o$, $\underline{w}_e$	optimal weight
$\underline{\theta}$	parameter vector for an IIR or other non-FIR filter
$\delta\underline{\hat{w}}_k$	$= \underline{\hat{w}}_k - \underline{\hat{w}}_{k-1}$
$\delta\underline{\hat{\theta}}_k$	$= \underline{\hat{\theta}}_k - \underline{\hat{\theta}}_{k-1}$
μ	step size
$O(\mu)$	big order of μ, i.e., $O(\mu)/\mu \to 1$ as $\mu \to 0$
$o(\mu)$	small order of μ, i.e., $o(\mu)/\mu \to 0$ as $\mu \to 0$
$[a]$	largest integer less than or equal to a
$\mathrm{sgn}(a)$	sign of a
$\nabla \underline{f}$	$= \partial \underline{f} / \partial \underline{z}^T$

List of Acronyms

Acronym	Definition
acf	Autocorrelation function
acv	Autocovariance function
AR	Autoregressive
ARMA	Autoregressive, moving average
ASP	Adaptive signal processing
cdf	(cumulative) distribution function
CLT	Central limit theorem
EKF	Extended Kalman filter
EWLS	Exponentially weighted least squares
EWSE	Exponentially weighted squared error
ES	Exponential stability
FIR	Finite impulse response
GD	Gaussian density
IIR	Infinite impulse response
IOS	Input-output stability
IR	Impulse response
LMS	Least mean square
LMS_1	single time scale LMS
LMS_2	mixed time scale LMS
LMS_P	Posterior LMS
MCT	Monotone convergence theorem
MG	Martingale
MGCT	Martingale convergence theorem
MSE	Mean square error
NM	Newton's method
ODE	Ordinary differential equation
PE	Persistent excitation
PLR	Pseudo linear regression
pdf	probability density function
RV	Random variable
SD	Steepest descent
SLLN	Strong law of large numbers
SPR	Strictly positive real
TF	Transfer function
TFCN	Transfer function plus colored noise
w.p.1	with probability one

Adaptive Signal Processing Algorithms

PART I

FUNDAMENTALS

1

Introduction

1.1 MOTIVATING ADAPTIVE ALGORITHMS

We think the most helpful way to introduce adaptive signal processing (ASP) methods, is through application examples. Our descriptions are necessarily simplified, but specific enough to show the structure and utility of adaptive methods. In each case we pause to extract the basic idea in the application, and the common features are then outlined in Section 1.3.

1.2 APPLICATION EXAMPLES

1.2.1 Noise Cancellation—Suppression of maternal ECG component in fetal ECG

Suppose we desire to measure the fetal heart rate during labor or delivery. The fetal heart rate signal can be recorded from a sensor placed in the abdominal region. Clearly, however, this signal will be made noisy or distorted, chiefly by the mother's heartbeat, but also by fetal motion. The idea behind noise cancelling is to take a direct recording of the mother's heartbeat and after judicious filtering of this signal, subtract it off the fetal heart rate signal to get a new, relatively noise-free fetal heart rate signal. Such a procedure is described in [W5] and is illustrated in Figs. 1.1, 1.2, and 1.3.

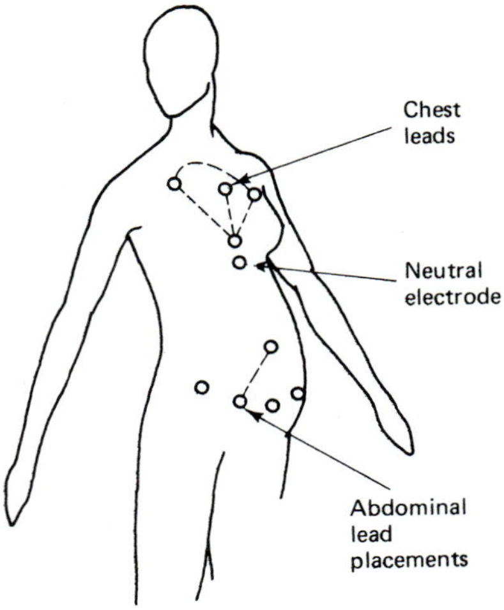

Figure 1.1 Cancelling maternal heartbeat in fetal electrocardiography (ECG): position of leads. From B. Widrow et al, *Adaptive Noise Cancelling: Principles and applications* [W3], © December 1975, IEEE.

The signal processing problem here is the choice of the filter. Since the noise cancellation must be done in real time, an algorithm is required that adjusts the filter weights in real time, being driven by the "error" signal which is the "cleaned" output signal in Fig. 1.2. Thus in Fig. 1.2 we show the filter as programmable or adjustable with the adaptive algorithm feeding the new weight settings. The results (from [W5]) in Fig. 1.3 show clearly the success of the Adaptive Noise Cancelling procedure.

Let us pause to extract the essential features from this problem. We have two inputs: a primary and a reference. The primary signal is of interest but has a noisy interference component which is correlated with the reference signal. The adaptive filter is used to produce an estimate of this interference or noise component which is then subtracted off the primary signal to give what is called either a "cleaned" primary or an error signal. Intuitively, the filter should be chosen to ensure that the error signal and the reference signal

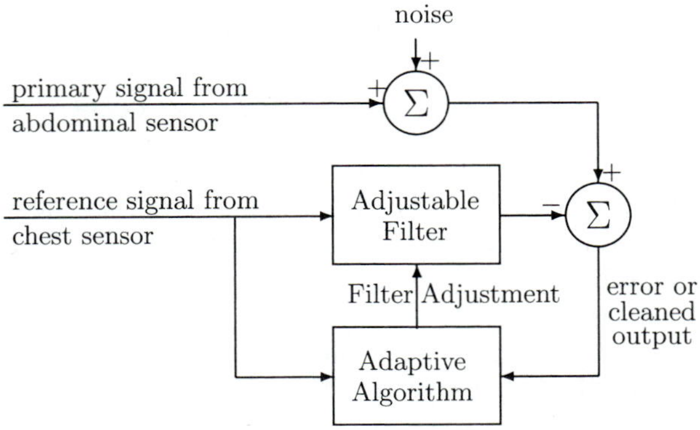

Figure 1.2 Schematic of noise canceller.

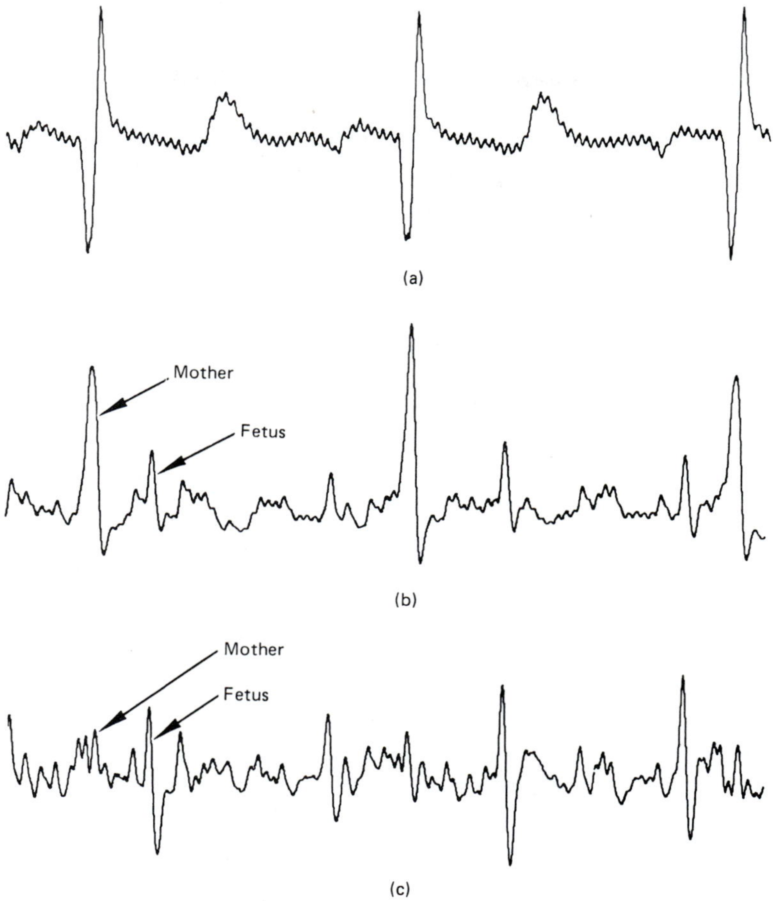

Figure 1.3 Results of fetal ECG experiment (bandwidth, 3–35 Hz; sampling rate, 256 Hz): (a) reference input (chest lead); (b) primary input (abdominal lead); (c) noise-canceller output. From B. Widrow et al, *Adaptive Noise Canceling: Principles and applications* [W3], © December 1975, IEEE.

are uncorrelated (or "decorrelated") at all lags. The Adaptive Algorithm block in Fig. 1.2 is used to adjust or adapt the filter weights or parameters to achieve this decorrelation.

1.2.2 Channel Equalization

Consider the efficient transmission of a signal over a communication channel (e.g., an underground cable, an optical fiber, the atmosphere, a satellite link, etc.). A typical communication system design involves first passing the signal to be transmitted through a whitening filter (to reduce redundancy or correlation) and then transmitting the resultant whitened signal. In a digital system this whitened signal may also be quantized before transmission (in fact optimal performance requires that the whitening and quantization be done jointly). At the receiver, the recorded signal (which ideally will be the same transmitted whitened signal) is passed through the inverse whitening filter and the original signal is thus restored.

In practice, however, the channel will affect the transmitted signal because of (i) channel noise and (ii) channel dispersion (leading to inter-symbol interference) caused, for example, by reflection of the transmitted signal from various objects such as buildings in the transmission path, leading to echoes of the transmitted signal appearing at the receiver. It is therefore necessary to pass the received signal through a so-called equalizing filter to undo the dispersion effect. If we model the channel as a linear time invariant system, the equalizer should be its inverse system. The processing of the various signals is sketched in Fig. 1.4.

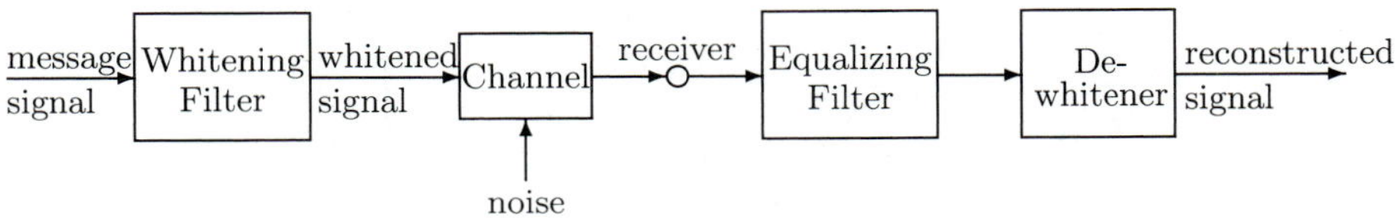

Figure 1.4 Channel equalization.

Channel characteristics typically drift with time so that the equalizing filter weights need to be continually adjusted. This leads to adaptive equalization, which is represented in Figs. 1.5 and 1.6. In Fig. 1.5, the Adaptive Equalizer operates in a training mode, where a pre-recorded training signal is made available at the receiver. In this mode signal transmission must be regularly interrupted to reset the equalizing filter with a brief training session. In Fig. 1.6 so-called Blind Adaptive Equalization is illustrated where no training signal is used. This latter technique is more complicated to use. The primary signal is now the received signal and the reference signal is obtained by delaying the primary signal.

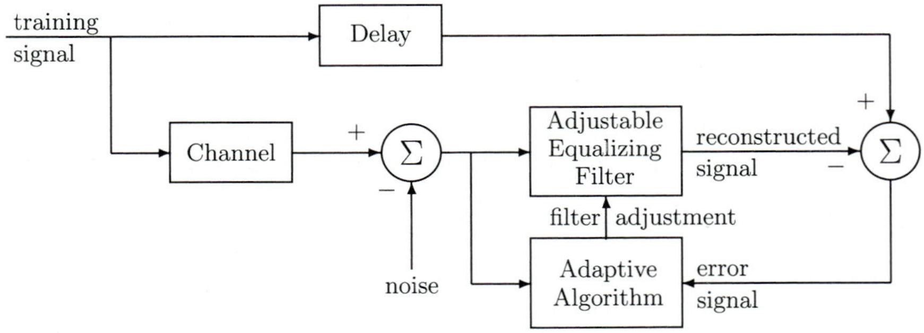

Figure 1.5 Adaptive equalization: training mode.

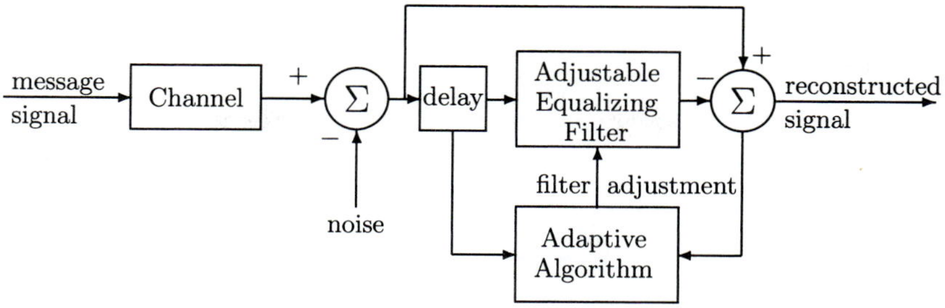

Figure 1.6 Blind adaptive equalization.

1.2.3 System Identification

Suppose we have a linear time invariant communication channel whose impulse response (IR) or transfer function (TF) we wish to know. We may be able to determine it theoretically from physical principles, but for many practical purposes an empirical approach is also suitable. This is the System Identification method where we transmit a known message signal (e.g., a white noise) and cross correlate it with the received signal. If the message signal is a white noise then the transfer function is equal to the cross spectrum between input (= transmitted message signal) and output (= received signal). This is sketched in Fig. 1.7.

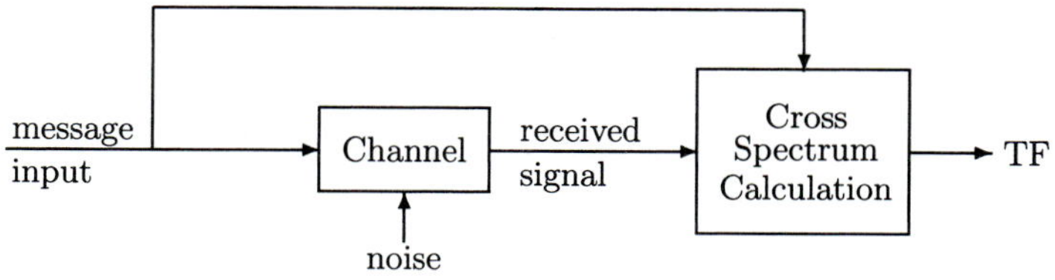

Figure 1.7 Cross spectrum for TF identification.

To do the transfer function identification in real time the approach of Fig. 1.8 is used. The filter parameters are adjusted by the adaptive algorithm, which is driven by the input and error signals, until the error signal is uncorrelated with the input signal at all lags. At that stage the filter transfer function should approximate the channel transfer function. The filter may be regarded as producing an input/output model of the system or channel.

1.2.4 Time Delay Estimation

In the simplest delay estimation problem, signals are received at two spatially separated sensors, one signal being an attenuated and delayed, noisy version of the other. The aim is to estimate the delay Δ. If the distance d between the sensors is known, and the signal speed v is known, then the direction of approach of the the signal front can be determined.

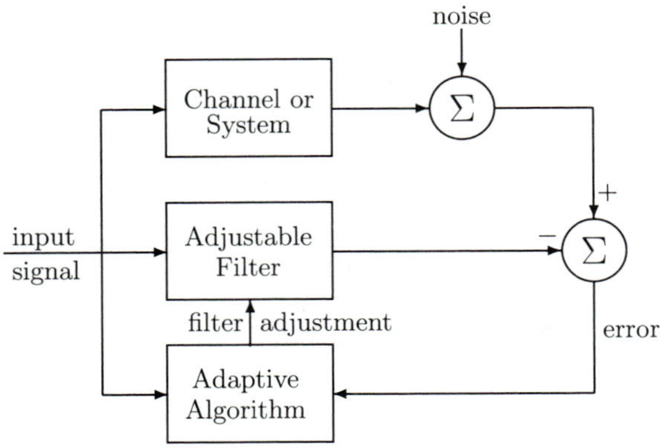

Figure 1.8 Adaptive system identification.

In Fig. 1.9, $\Delta = d \sin\beta / v$. If the delay is time invariant, then it can be determined by cross-correlating the two signals and finding the global maximum of the cross-correlation function. If the delay is time variant (say, because the direction is), then a real time estimation of the delay may be needed and an adaptive algorithm is then appropriate.

1.2.5 Echo Cancellation

Echoes (reflected waves) are generated in long distance telephone communication due to impedance mismatches. Such a mismatch typically occurs at the junction (or hybrid) between the local subscriber loop and the long distance loop. While the junction is designed to prevent any major signal reflection, component variation and aging make it difficult to achieve exact impedance matching.

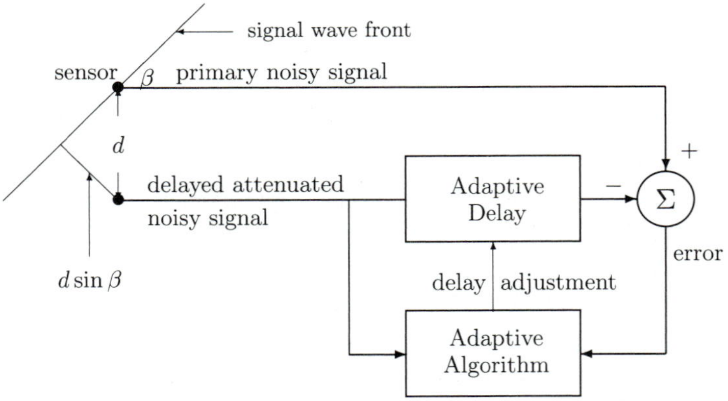

Figure 1.9 Adaptive time delay estimation.

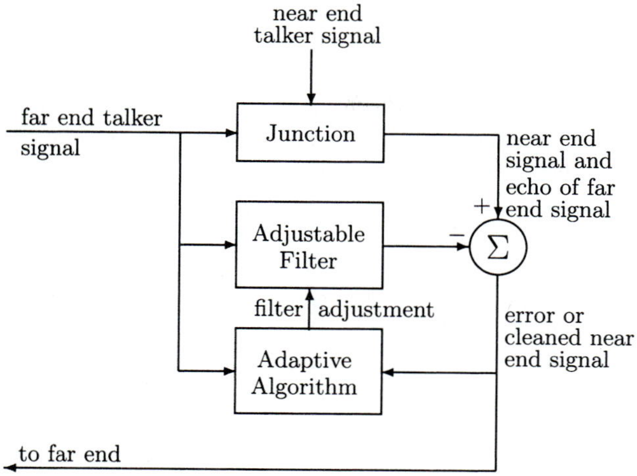

Figure 1.10 Echo cancellation by adaptive prediction.

The idea behind echo cancellation is to predict the echo signal value and thus subtract it out. The mechanism is sketched in Fig. 1.10.

1.2.6 Adaptive Control

Perhaps the most powerful feature of feedback control is that it allows a control system set point to be maintained even if the parameters of the system under control vary. If the parameter variation is severe, however, it may be preferable to adjust the controller parameter settings in such a way as to counteract the effect of the system parameter variation. This is what an adaptive controller attempts to achieve. A typical such adaptive controller is sketched in Fig. 1.11. Driven by an error signal and the reference or set point signal, the Adaptive Algorithm adjusts the controller gains to keep them appropriate to the system as it changes over time.

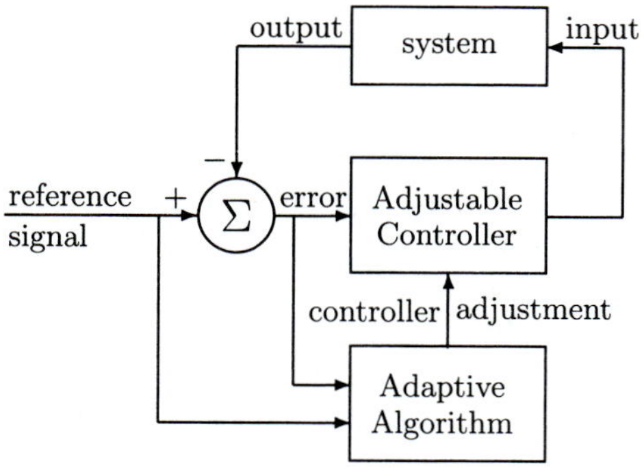

Figure 1.11 Adaptive controller.

1.3 BASIC FEATURES OF ADAPTIVE ALGORITHMS

1.3.1 Structure of Adaptive Element

A perusal of the examples shows that they all have a common feature: an ASP element or subsystem as sketched in Fig. 1.12. Actually the adaptive control example in Fig. 1.11 only fits a subsystem of Fig. 1.12: namely there is an adaptive algorithm fed by a reference signal and an error signal and linked to an adjustable filter. Although we do not treat adaptive control in this book, this variation on Fig. 1.12 is not a troublesome one. The ASP element has two components: first is the adjustable or programmable filter. This could be a finite impulse response (FIR) filter (also called a transversal or tapped delay line filter). It could be an infinite impulse response (IIR) filter (also called recursive, which may be an all-pole or pole-zero filter) or it could be some other type of nonlinear filter. If delay is allowed, then the adjustable filter can be two-sided or noncausal. The adjustable filter has an auxiliary signal input x and also an input from the adaptive algorithm which causes the filter parameters to be altered. The adjustable or programmable filter produces an output $\hat{s}$ which is compared with the primary signal y to generate an error signal e that is used to drive the adaptive algorithm. In general, the presence of the adaptive algorithm makes the ASP element *a nonlinear system or filter*. Note that in prediction problems the reference signal will be a delayed version of the primary signal. This is the case, for example, in some types of blind equalization. So far we have not been explicit about the details of the adaptive algorithm: that is taken up in Chapter 4.

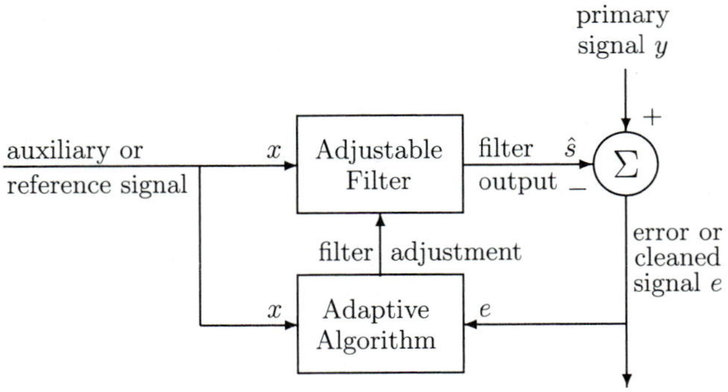

Figure 1.12 Adaptive signal processing element.

1.3.2 Adaptive Element with Noise Whitening

If we look back over all the examples we see that the ever present noise will generally be colored. But the (adjustable) filter as we have presented it does not acknowledge this fact. Presumably by doing so we could increase the quality of the signal processing. It is indeed easy enough to make such a modification and we are led to a modified ASP element with two adjustable filters. The general structure is sketched in Fig. 1.13. The system filter performs the same function as before. The whitening filter is driven by the error signal rather than by an auxiliary signal.

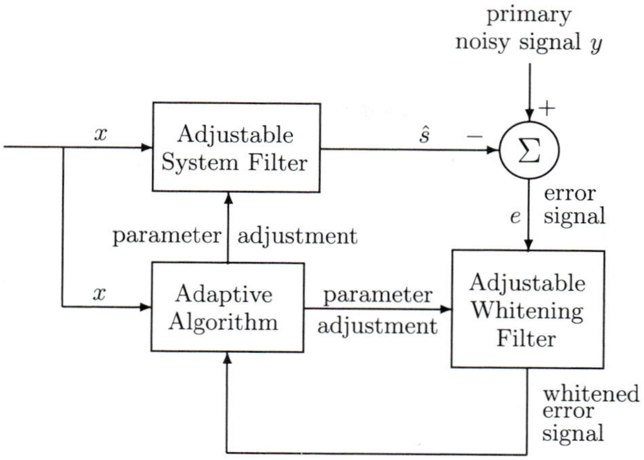

Figure 1.13 ASP element with noise whitening.

1.3.3 Reasons for Adaptation

Again a perusal of the above examples shows several reasons for the use of an adaptive filter rather than a fixed filter.

(i) Filter parameters are unknown and must be estimated or tuned in real time.

(ii) System parameters may vary with time and so need to be tracked.

1.4 NOTES

The fetal ECG example is further described in [W3] and [W5]. Adaptive Equalization is described in [H7] and Blind Equalization in [B7], [D4], and [G8]. System Identification is a huge area; a convenient reference is [L10]. A survey of offline time delay estimation is given by [C3]. There are several books on Adaptive Control: [G9], [J3], [L2], [N2], and [S3]. Echo cancellation is surveyed in [G12] (see also [A2], [B5], and [H10]). Other applications include Beamforming [W5], Speech Coding [G5], and Waveform Coding [J2]. There are a number of texts on ASP: [A2], [B5], [B7], [C7], [H7], [H10], [J3], [M7], and [W5]. Some ASP algorithms are also analyzed in the research monographs [B7] and [K4]. Our block diagram representation of the ASP element as a programmable filter appears in [M7]. Strangely, the noise whitening modification of the ASP element is hard to find outside the Identification area. In these examples and throughout the book we have dealt with a scalar auxiliary signal and a scalar primary signal. The ideas and techniques all extend fairly easily to vector signals.

2

Offline Analysis

2.1 UTILITY OF OFFLINE CALCULATIONS

Considerable insight into the performance of adaptive signal processing algorithms can be gained by studying the associated signal processing problems in an offline setting. The validity of this claim is at least intuitively clear, but the precise reason only becomes fully clear in Part II when we do averaging analysis.

In specifying an adaptive filter or its offline equivalent there are three features of importance:

(i) The filter structure, e.g., FIR filter or IIR filter or nonlinear filter.

(ii) The error criterion used to drive the algorithm, e.g., squared error or some other non-quadratic function of the error signal.

(iii) The adaptation or learning rule, e.g., steepest descent.

With an FIR filter, the error signal is a linear function of the parameters, and if a squared error criterion is used we have a linear-quadratic problem. These problems are the easiest to analyze. With an IIR filter, the error is generally a nonlinear function of the parameters, and such problems are harder to analyze regardless of which error criterion is used. The most common criterion is a squared error, but others are important and will be discussed.

2.2 ONE-SIDED FIR FILTER

Let us consider the Echo Cancellation problem. Fig. 2.1 shows the ASP element in expanded detail (note that q^{-1} is the unit delay operator: see Appendix B). The adjustable filter is shown as a tapped delay line or transversal filter or finite impulse response (FIR) filter with adjustable weights. However, in the present chapter the adaptive algorithm is disconnected and the filter weights are fixed.

The output of the FIR filter is

$$\hat{s}_k(\underline{w}) = \sum_{r=1}^{p} w_r x_{k-r} = \underline{x}_k^T \underline{w} \tag{2.1}$$

where $\underline{w}$ is the weight or parameter vector and $\underline{x}_k$ the stacked auxiliary signal

$$\underline{w} = (w_1 \cdots w_p)^T, \qquad \underline{x}_k = (x_{k-1} \cdots x_{k-p})^T. \tag{2.2}$$

The error signal is

$$e_k(\underline{w}) = y_k - \hat{s}_k(\underline{w}) = y_k - \underline{x}_k^T \underline{w}. \tag{2.3}$$

Here $\underline{x}_k$, y_k are, respectively, the auxiliary or reference signal and the primary signal. For simplicity we assume for the moment that $(\underline{x}_k, y_k)$ are jointly stationary (see Appendix B for definitions) with zero mean.

The success of the filtering will be measured by the mean square error (MSE) namely,

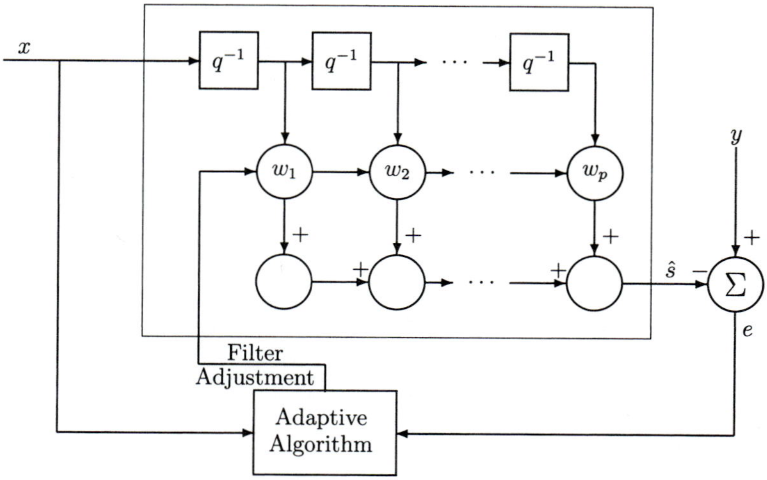

Figure 2.1 FIR adaptive predictor.

$$\begin{aligned}\mathcal{E} = \mathcal{E}(\underline{w}) = E(e_k^2(\underline{w})) &= E(y_k - \underline{x}_k^T \underline{w})^2 \\ &= E(y_k^2) - 2\underline{w}^T E(\underline{x}_k y_k) + \underline{w}^T E(\underline{x}_k \underline{x}_k^T)\underline{w} \\ &= R_y - 2\underline{R}_{xy}^T \underline{w} + \underline{w}^T \underline{R}_x \underline{w}, \end{aligned} \tag{2.4}$$

(see Appendix B for matrix operations with random vectors). Here we have introduced

$$\begin{aligned} R_y &= E(y_k^2) = \text{variance of } y_k, \\ \underline{R}_x &= E(\underline{x}_k \underline{x}_k^T) = \text{covariance matrix of } \underline{x}_k, \\ \underline{R}_{yx} &= E(y_k \underline{x}_k^T) = \text{covariance between } y_k,\ \underline{x}_k = \underline{R}_{xy}^T. \end{aligned}$$

We see that the MSE is quadratic in $\underline{w}$. It is helpful to view the MSE surface geometrically. With two weights we get a bowl shaped surface as in Fig. 2.2. Contours of constant MSE are ellipses as in Fig. 2.3.

The optimal weight is the one that minimizes MSE. In Fig. 2.2 it is the point at the bottom of the bowl. In Fig. 2.3 it is the center of the ellipse. To find the optimal weight $\underline{w}_o$ we solve

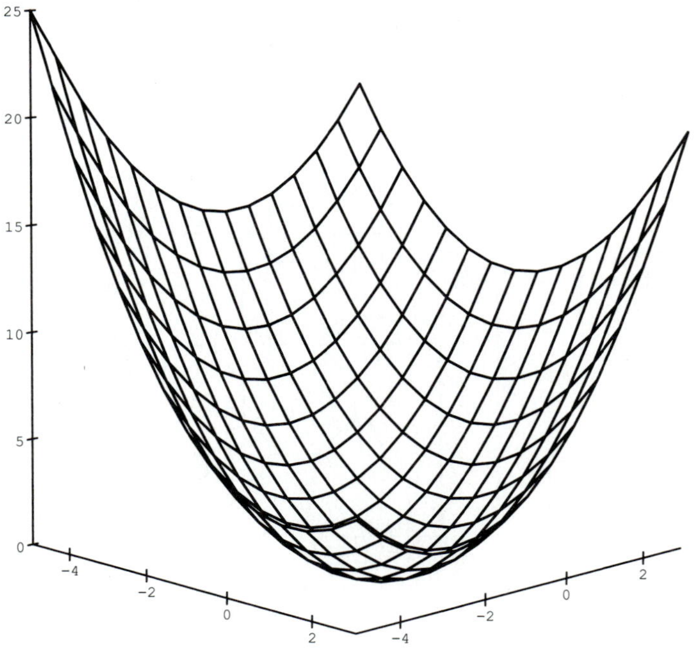

Figure 2.2 Two-dimensional quadratic MSE surface.

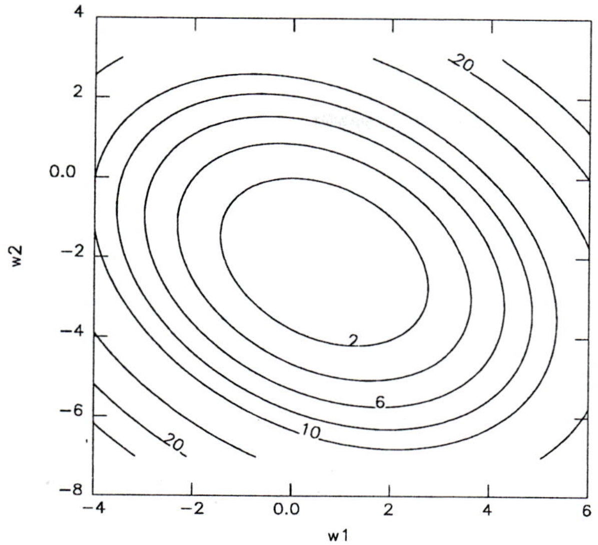

Figure 2.3 Contours of constant MSE.

$$\begin{aligned} & d\mathcal{E}/d\underline{w}_o = \underline{0} \\ \Rightarrow\quad & 2\underline{R}_x\underline{w}_o - 2\underline{R}_{xy} = \underline{0} \\ \Rightarrow\quad & \underline{R}_x\underline{w}_o = \underline{R}_{xy}. \end{aligned} \tag{2.5}$$

(For vector differentiation, see Appendix A.) This is a finite dimensional Wiener-Hopf equation, and provided $\underline{R}_x$ has full rank it has solution

$$\underline{w}_o = \underline{R}_x^{-1}\underline{R}_{xy}. \tag{2.6}$$

The corresponding minimized value of MSE is

$$\mathcal{E}_{min} = R_y - 2\underline{w}_o^T\underline{R}_{xy} + \underline{w}_o^T\underline{R}_x\underline{w}_o$$

$$= R_y - \underline{w}_o^T\underline{R}_x\underline{w}_o \qquad \text{(by (2.5))} \tag{2.7}$$

$$= R_y - \underline{R}_{xy}^T\underline{R}_x^{-1}\underline{R}_{xy}. \tag{2.8}$$

If we rewrite (2.5) as

$$E(\underline{x}_k\underline{x}_k^T\underline{w}_o) = E(\underline{x}_k y_k)$$

then we can reorganize it as

$$E(\epsilon_k\underline{x}_k) = 0 \tag{2.9}$$

where we have introduced ϵ_k, the "optimal" prediction error or output error

$$\epsilon_k = y_k - \underline{x}_k^T\underline{w}_o. \tag{2.10}$$

Equation (2.9) tells us that at the minimum, the error signal is uncorrelated with (or orthogonal to) the vector input signal $\underline{x}_k$. That is, ϵ_k is uncorrelated with the input signal

x_{k-r} at all lags $r = 1 \cdots p$. This characterization of the minimizing weight is called the orthogonality condition and is especially useful from an intuitive point of view.

It is also helpful to rewrite the expression (2.4) for MSE in another way. Use (2.5) in (2.4) and also add and subtract $\underline{w}_o^T \underline{R}_x \underline{w}_o$ in (2.4) to find via (2.7) that

$$\begin{aligned}\mathcal{E} &= R_y - 2\underline{w}_o^T \underline{R}_x \underline{w} + \underline{w}^T \underline{R}_x \underline{w} \\ &= R_y - \underline{w}_o^T \underline{R}_x \underline{w}_o + (\underline{w} - \underline{w}_o)^T \underline{R}_x (\underline{w} - \underline{w}_o) \\ &= \mathcal{E}_{min} + (\underline{w} - \underline{w}_o)^T \underline{R}_x (\underline{w} - \underline{w}_o). \end{aligned} \tag{2.11}$$

This clearly shows $\mathcal{E}$ is quadratic in $\underline{w}$, that its minimum is at $\underline{w}_o$ and has value $\mathcal{E}_{min}$.

The calculations in this section were made assuming $\underline{x}_k$, y_k were stochastic signals. However the same calculations hold if we take $\underline{x}_k$, y_k to be periodic deterministic signals and use the following definition of average. If ξ_k is periodic, with period M we define the average as follows:

$$\mathrm{av}(\xi_k) = \tilde{E}(\xi_k) = M^{-1} \sum_{k=1}^{M} \xi_k. \tag{2.12}$$

Then the reader can rapidly verify that the above development is valid with $\tilde{E}$ replacing E.

To give a firmer idea of the calculations above we look at some examples.

Example 2.1.

Suppose we have two weights, a noise free primary signal, and

$$\text{auxiliary signal} = x_k = \sin(2\pi k f_o)$$
$$\text{primary signal} = s_k = y_k = 2\cos(2\pi (k-1) f_o)$$

where the period of each signal is $1/f_o = M =$ an integer. To compute $\underline{R}_x$ we need the correlations

$$\tilde{E}(x_k^2), \quad \tilde{E}(x_k x_{k-1})$$

while to compute $\underline{R}_{xy}$ we need the correlations

$$\tilde{E}(y_k x_{k-1}), \quad \tilde{E}(y_k x_{k-2}).$$

To obtain these we first calculate the autocovariance of x,

$$\begin{aligned}\tilde{E}(x_k x_{k-n}) &= M^{-1} \sum_{k=1}^{M} \sin(2\pi k/M) \sin(2\pi (k-n)/M) \\ &= \frac{1}{2} \cos(2\pi n/M), \quad n = 0, 1.\end{aligned}$$

To see this, note for instance that

$$\begin{aligned}\tilde{E}(x_k^2) &= M^{-1} \sum_{k=1}^{M} \sin^2(2\pi k/M) \\ &= M^{-1} \sum_{k=1}^{M} (1 - \cos(4\pi k/M))/2 = \frac{1}{2} - 0.\end{aligned}$$

From this we see that

$$\underline{R}_x = \begin{bmatrix} \frac{1}{2} & \frac{1}{2}\cos(2\pi/M) \\ \frac{1}{2}\cos(2\pi/M) & \frac{1}{2} \end{bmatrix}$$

$$\underline{R}_{xy} = \begin{bmatrix} 0 \\ -\sin(2\pi/M) \end{bmatrix}.$$

The performance function is

$$\begin{aligned}\mathcal{E} &= E(y_k^2) + \underline{w}^T \underline{R}_x \underline{w} - 2\underline{R}_{yx}^T \underline{w} \\ &= 2 + \frac{1}{2}\begin{pmatrix} w_1 & w_2 \end{pmatrix}\begin{pmatrix} 1 & \cos\psi \\ \cos\psi & 1 \end{pmatrix}\begin{pmatrix} w_1 \\ w_2 \end{pmatrix} - 2\begin{pmatrix} 0 & -\sin\psi \end{pmatrix}\begin{pmatrix} w_1 \\ w_2 \end{pmatrix}\end{aligned}$$

where $\psi = 2\pi/M$ so that

$$\mathcal{E} = 2 + \frac{1}{2}(w_1^2 + w_2^2) + w_1 w_2 \cos\psi + 2w_2 \sin\psi.$$

A plot of $\mathcal{E}$ is given in Fig. 2.2. Continuing, we find the optimal weight

$$\begin{aligned}d\mathcal{E}/d\underline{w} &= 2\underline{R}_x\underline{w} - 2\underline{R}_{yx} \\ &= \begin{pmatrix} 1 & \cos\psi \\ \cos\psi & 1 \end{pmatrix}\begin{pmatrix} w_1 \\ w_2 \end{pmatrix} - 2\begin{pmatrix} 0 \\ -\sin\psi \end{pmatrix} \\ &= \begin{pmatrix} w_1 + w_2\cos\psi \\ w_1\cos\psi + w_2 + 2\sin\psi \end{pmatrix}.\end{aligned}$$

Finally the optimal or Wiener weight is

$$\begin{aligned}\underline{w}_o &= \underline{R}_x^{-1}\underline{R}_{yx} \\ &= \frac{2}{1-\cos^2\psi}\begin{pmatrix} 1 & -\cos\psi \\ -\cos\psi & 1 \end{pmatrix}\begin{pmatrix} 0 \\ -\sin\psi \end{pmatrix} \\ &= \frac{2}{1-\cos^2\psi}\begin{pmatrix} \cos\psi\sin\psi \\ -\sin\psi \end{pmatrix} \\ &= 2\begin{pmatrix} \cot\psi \\ -\csc\psi \end{pmatrix}.\end{aligned}$$

And the optimal MSE is

$$\begin{aligned}\mathcal{E}_{min} &= E(y_k^2) - \underline{w}_o^T \underline{R}_x \underline{w}_o \\ &= E(y_k^2) - \underline{w}_o^T \underline{R}_{yx} \\ &= 2 - 2\begin{pmatrix} 0 & -\sin\psi \end{pmatrix}\begin{pmatrix} \cot\psi \\ -\csc\psi \end{pmatrix} \\ &= 2 - 2 = 0.\end{aligned}$$

Thus the weights can be adjusted to make $\mathcal{E}$ zero.

Contours of MSE are plotted in Fig. 2.3; the construction of these contours is discussed further in Section 3.2.

2.3 TWO-SIDED FIR FILTER: CHANNEL EQUALIZATION

Here we consider channel equalization both with and without a training signal. The set-up with a training signal is shown in Fig. 2.4. For simplicity, quantization is excluded from consideration.

The training signal s_k is transmitted through the channel becoming dispersed and noisy in the process. The noise-free channel output is denoted ξ_k. The recorded signal x_k is filtered to produce an estimated message signal $\hat{s}_k$ which is compared with the (possibly delayed) training signal to generate an error signal that drives the adaptive algorithm. Again here the dotted line shows the adaptive algorithm disconnected. It is assumed that s_k, n_k are jointly stationary. Also the channel is treated as a linear time invariant system with rational transfer function $H(q^{-1})$ (where q^{-1} is the backshift operator: see Appendix B). We denote the equalization filter by $W(q^{-1})$. The error signal is then

$$\begin{aligned} e_k &= y_k - \hat{s}_{k-\Delta} \\ &= y_k - W(q^{-1})x_k \qquad (2.13) \\ &= y_k - W(q^{-1})(H(q^{-1})s_k + n_k) \\ &= [q^{-\Delta} - W(q^{-1})H(q^{-1})]s_k - W(q^{-1})n_k + m_k. \qquad (2.14) \end{aligned}$$

Before we proceed with a general discussion of equalization, it may be helpful to look at an example.

Example 2.2. Minimum Phase Channel Equalization

With Fig. 2.4 we take the following specifications (see Appendix B for a review of the stochastic models). All signals are zero mean and $\Delta = 0$. We suppose s_k is an AR(1) signal

$$s_k = (1 + \beta q^{-1})^{-1}\epsilon_k, \quad \beta = 0.881 \qquad (2.15)$$

where ϵ_k is a white noise ($\sigma_\epsilon^2 = 0.48$). The noise-free part of the channel is an all pole first order system

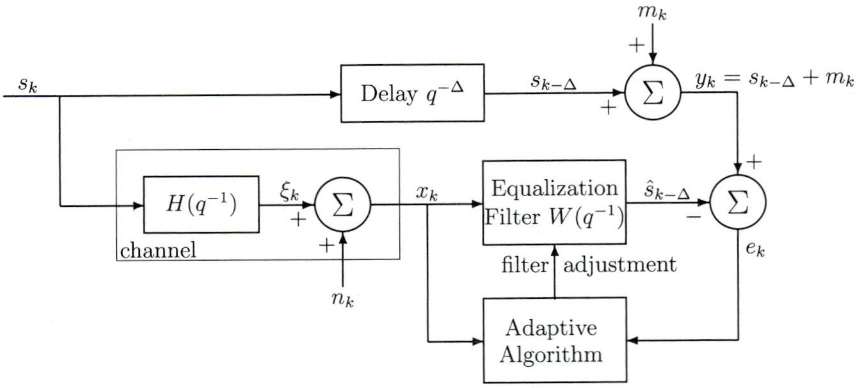

Figure 2.4 Channel equalization in training mode.

$$\xi_k = (1 + \theta q^{-1})^{-1} s_k, \quad \theta = -0.681 \tag{2.16}$$

and the output noise is white with $\sigma_n^2 = 0.1$. The noisy channel output is

$$x_k = \xi_k + n_k. \tag{2.17}$$

Also $m_k = 0$ so that $y_k = s_k$. An FIR equalizing filter with two taps or weights will be used:

$$\hat{s}_k = w_1 x_k + w_2 x_{k-1}. \tag{2.18}$$

As in Section 2.2 the optimal filter weight vector satisfies the Wiener-Hopf equation

$$\underline{w}_o = \underline{R}_x^{-1} \underline{R}_{xy}$$

where

$$\underline{R}_x = E(\underline{x}_k \underline{x}_k^T) = \begin{pmatrix} R_{x0} & R_{x1} \\ R_{x1} & R_{x0} \end{pmatrix}, \quad R_{x\tau} = E(x_k x_{k+\tau})$$

$$\underline{R}_{xy} = E(s_k \underline{x}_k) = \begin{pmatrix} R_{sx,0} \\ R_{sx,1} \end{pmatrix}, \quad R_{sx,\tau} = E(s_k x_{k-\tau}).$$

So we need to calculate these covariances. From (2.17)

$$R_{x0} = \sigma_n^2 + R_{\xi 0}$$
$$R_{x1} = R_{\xi 1}.$$

Next note from (2.15), (2.16) that ξ_k is AR(2). Indeed

$$\xi_k = \left[(1 + \theta q^{-1})(1 + \beta q^{-1})\right]^{-1} \epsilon_k$$
$$= (1 + a_1 q^{-1} + a_2 q^{-2})^{-1} \epsilon_k$$

where $(a_1, a_2) = (\theta + \beta, \theta\beta) = (0.2, -0.6)$. From Appendix B we find variances and covariances as

$$R_{\xi 0} = \frac{(1 + a_2)\sigma_\epsilon^2}{(1 - a_2)(1 + a_2 - a_1)(1 + a_2 + a_1)} = \frac{0.4 \times 0.48}{1.6 \times 0.6 \times 0.2} = 1.0$$

$$R_{\xi 1} = -\frac{a_1}{1 + a_2} R_{\xi 0} = \frac{-0.2}{1 - 0.6} = -0.5.$$

So we find

$$\underline{R}_x = \begin{bmatrix} 1.1 & 0.5 \\ 0.5 & 1.1 \end{bmatrix}.$$

To calculate $\underline{R}_{xy}$ we start with

$$R_{sx,\tau} = E(s_k x_{k-\tau}) = E(s_k(\xi_{k-\tau} + n_{k-\tau})) = E(s_k \xi_{k-\tau}).$$

From (2.16)

$$\xi_k = s_k - \theta \xi_{k-1}$$

$$\Rightarrow \quad E(\xi_k \xi_{k-1}) = E(s_k \xi_{k-1}) - \theta E(\xi_{k-1}^2) \tag{2.19}$$

$$E(\xi_k^2) = E(s_k \xi_k) - \theta E(\xi_{k-1} \xi_k). \tag{2.20}$$

From (2.20) we find

$$E(s_k\xi_k) = R_{\xi 0} + \theta R_{\xi 1} = 1 + (-0.681) \times (-0.5) = 1.341.$$

In (2.19) we find

$$E(s_k\xi_{k-1}) = R_{\xi 1} + \theta R_{\xi 0} = -0.5 - 0.681 \times 1.0 = -1.181$$

so that

$$\underline{R}_{xy} = (1.341,\ -1.181)^T.$$

Finally to calculate the MSE surface we need

$$R_y = R_s = E(s_k^2) = \sigma_\epsilon^2/(1-\beta^2) = 0.48/0.224 = 2.14.$$

Thus the MSE surface is

$$\begin{aligned}
\mathcal{E} &= R_y - 2\underline{R}_{xy}^T\underline{w} + \underline{w}^T\underline{R}_x\underline{w} \\
&= 2.14 + (\,w_1 \quad w_2\,)\begin{pmatrix} 1.1 & -0.5 \\ -0.5 & 1.1 \end{pmatrix}^{-1}\begin{pmatrix} w_1 \\ w_2 \end{pmatrix} \\
&\quad - 2\,(\,1.341 \quad -1.181\,)\begin{pmatrix} w_1 \\ w_2 \end{pmatrix} \\
&= 2.14 - 2.682w_1 + 2.362w_2 - w_1w_2 + 1.1w_1^2 + 1.1w_2^2.
\end{aligned}$$

This is plotted in Fig. 2.5. The optimal filter weights are

$$\begin{aligned}
\underline{w}_o &= \underline{R}_x^{-1}R_{xy} \\
&= \begin{pmatrix} 1.1 & -0.5 \\ -0.5 & 1.1 \end{pmatrix}^{-1}\begin{pmatrix} 1.341 \\ -1.181 \end{pmatrix} \\
&= \begin{pmatrix} 1.146 & -0.521 \\ -0.521 & 1.146 \end{pmatrix}\begin{pmatrix} 1.341 \\ -1.181 \end{pmatrix} \\
&= \begin{pmatrix} 0.921 \\ -0.655 \end{pmatrix}.
\end{aligned}$$

The minimum error is

$$\begin{aligned}
\mathcal{E}_{min} &= E(y_k^2) - \underline{w}_o^T\underline{R}_x\underline{w}_o = E(y_k^2) - \underline{w}_o^T\underline{R}_{xy} \\
&= 2.14 - (\,0.921 \quad -0.655\,)\begin{pmatrix} 1.341 \\ -1.181 \end{pmatrix} \\
&= 0.13.
\end{aligned}$$

This can be compared to $R_y = 2.14$ to see the gain due to equalization. Again see Fig. 2.5 for the position of the optimum.

Now we return to a general discussion of equalization where one usually has many more than two filter weights. Let us for the moment suppose there is no noise and no delay, so $\Delta = 0$. Then (2.14) becomes

$$e_k = \Big(1 - W(q^{-1})H(q^{-1})\Big)\, s_k. \tag{2.21}$$

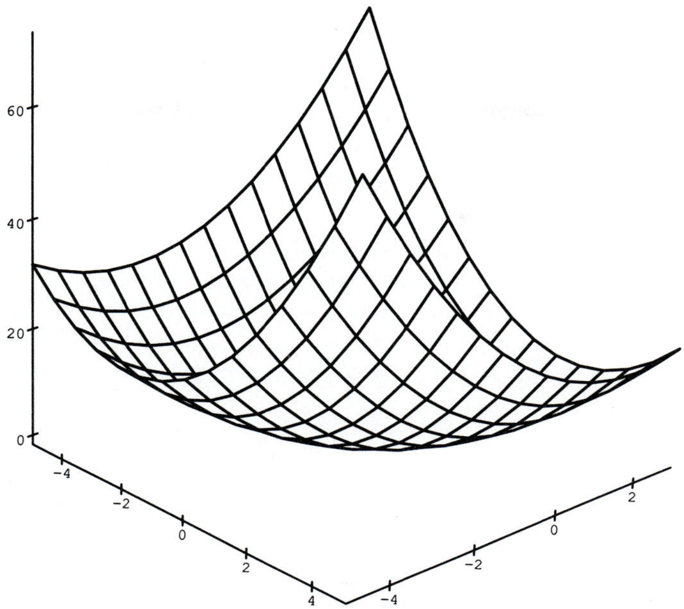

Figure 2.5 MSE surface for Example 2.2.

Clearly this error signal can be nulled (i.e., the MSE can be driven to zero) if we take

$$W(q^{-1}) = \frac{1}{H(q^{-1})}. \tag{2.22}$$

The innocuous looking requirement here, that $W(q^{-1})$ be the inverse filter of $H(q^{-1})$ is not as simple as it seems. The equalization filter must be stable (see Appendix C) so that for (2.22) to work, the inverse of the channel transfer function must be stable. To put it another way, the channel transfer function must be minimum phase (see Appendix C for a definition). However there is no physical reason why this should be true. Indeed it usually will not be, because channel dispersion will usually involve wave reflection and echoing and these phenomena lead to a non-minimum phase channel impulse response. Thus (2.22) cannot be satisfied in general. The way out of this dilemma is to tolerate some delay in processing, thus allowing $W(q^{-1})$ to be a two-sided or noncausal filter. Let us pursue this possibility. We use then an FIR filter with $p = m + r + 1$ taps or weights

$$q^m W(q^{-1}) = \sum_{u=-m}^{r} w_u q^{-u}, \qquad m = \Delta. \tag{2.23}$$

Now the MSE can be written (see Appendix B)

$$\mathcal{E} = E(e_k^2) = \int_{-\pi}^{\pi} F_e(\omega) d\omega / 2\pi$$

where $F_e(\omega)$ is the spectrum of e_k. From (2.13)

$$F_e(\omega) = F_y(\omega) - W(e^{-j\omega})F_{yx}^*(\omega) - W^*(e^{-j\omega})F_{yx}(\omega) + |W(e^{-j\omega})|^2 F_x(\omega)$$

where $F_{yx}(\omega)$ is the cross spectrum between y_k and x_k, etc. (see Appendix B7). Thus

$$\mathcal{E} = \int_{-\pi}^{\pi} (F_y(\omega) - W(e^{-j\omega})F_{yx}^*(\omega) - W^*(e^{-j\omega})F_{yx}(\omega) + |W(e^{-j\omega})|^2 F_x(\omega))d\omega/2\pi. \quad (2.24)$$

This formula is a frequency domain version of (2.4). The aim now is to minimize $\mathcal{E}$ with respect to $W(e^{-j\omega})$. The solution is given as before by solving (2.6). However, we can get an approximate solution which gives considerable insight into the structure of the optimal filter as follows. Let us allow $m = r = \infty$ in (2.23) and then as we did in (2.11) add and subtract $|F_{yx}(\omega)|^2/F_x(\omega)$ to (2.24) to find (much as in (2.11)) that

$$\mathcal{E} = \int_{-\pi}^{\pi} (F_y(\omega) - |F_{yx}(\omega)|^2/F_x(\omega))d\omega/2\pi$$
$$+ \int_{-\pi}^{\pi} |F_{yx}(\omega)/F_x(\omega) - W(e^{-j\omega})|^2 F_x(\omega)d\omega/2\pi. \quad (2.25)$$

The first term does not depend on $W(e^{-j\omega})$ while the second term can be made zero for

$$W_o(e^{-j\omega}) = F_{yx}(\omega)/F_x(\omega). \quad (2.26)$$

The corresponding minimum MSE is then

$$\mathcal{E}_{min} = \int_{-\pi}^{\pi} (F_y(\omega) - |F_{yx}(\omega)|^2/F_x(\omega))d\omega/2\pi. \quad (2.27)$$

Referring to Fig. 2.4, we can re-express (2.26) in terms of the various signal spectra as follows. We have

$$x_k = \xi_k + n_k$$
$$= H(q^{-1})s_k + n_k$$
$$\Rightarrow \quad F_{s_\Delta x}(e^{-j\omega}) = F_{xs_\Delta}(e^{j\omega}) = H(e^{j\omega})F_s(\omega)e^{-j\omega\Delta}$$
$$F_x(\omega) = |H(e^{-j\omega})|^2 F_s(\omega) + F_n(\omega).$$

So when $m_k = 0$ we have $y_k = s_{\Delta k} (= s_{k-\Delta} = q^{-\Delta}s_k)$ so (2.26) becomes

$$W_o(e^{-j\omega}) = \frac{H(e^{j\omega})F_s(\omega)e^{-j\omega\Delta}}{F_n(\omega) + |H(e^{-j\omega})|^2 F_s(\omega)}. \quad (2.28)$$

When there is no channel noise we get

$$W_o(e^{-j\omega}) = e^{-j\omega\Delta}/H(e^{-j\omega}) \quad (2.29)$$

which is a delayed version of (2.22). Again an example will help us understand these formulae.

Example 2.3. Non-minimum Phase Channel Equalization

We suppose no noise ($m_k = n_k = 0$) and consider a channel transfer function as follows:

$$H(q^{-1}) = 0.2 - 0.7q^{-1} + 0.3q^{-2} = 0.2(1 - 3q^{-1})(1 - \frac{1}{2}q^{-1}).$$

Notice that the numerator polynomial of $H(q^{-1})$ has one root of modulus > 1 so that $H(q^{-1})$ is not minimum phase. Equalization of such a transfer function will require a two-sided filter. From (2.29)

$$\begin{aligned} W_o(q^{-1}) &= q^{-\Delta}/H(q^{-1}) \\ &= q^{-\Delta}\left(\frac{6}{1-3q^{-1}} - \frac{1}{1-0.5q^{-1}}\right) \\ &= q^{-\Delta}\left(\frac{-2q}{1-\frac{1}{3}q} - \frac{1}{1-\frac{1}{2}q^{-1}}\right). \end{aligned}$$

If we write

$$q^{\Delta}W_o(q^{-1}) = \sum_{k=-\infty}^{\infty} w_k q^{-k}$$

then we have by inspection that

$$w_k = \begin{cases} -6\left(\frac{1}{3}\right)^{|k|} & k < 0 \\ -\left(\frac{1}{2}\right)^{k} & k \geq 0. \end{cases}$$

This two-sided filter is sketched in Fig. 2.6.

In practice the allowed delay is of course finite, and if the delay is large, then the optimal filter approximates a truncated version of Fig. 2.6 as shown in Fig. 2.7.

2.3.1 Blind Equalization

The setup for the problem of blind equalization is shown in Fig. 2.8. The message signal is transmitted through a noisy dispersive channel. The received signal x_k is filtered to produce an estimated message signal $\hat{s}_k$. From this an error signal e_k is manufactured. There are typically two ways this is done. In one type of equalizer the estimated message signal is passed through a memoryless nonlinearity (e.g., $\hat{s}_k \to |\hat{s}_k|^2$) and this is subtracted from a fixed average value (so e.g., $e_k = |\hat{s}_k|^2 - \text{constant}$). In a decision equalizer the fact that s_k is usually quantized is used. Thus if s_k is binary, then the error signal is $e_k = \hat{s}_k - q(\hat{s}_k)$ where $q(\hat{s}_k) = \pm 1$, depending on which of 1 or -1 $\hat{s}_k$ is closest to. The analysis of decision equalizers is rather difficult and beyond the scope of this book so we will not discuss them further.

To keep the ensuing discussion as simple as possible we assume henceforth that the channel noise $n_k = 0$. The problem then is to design an equalizing filter so as to null (as nearly as possible) the error signal

$$\begin{aligned} e_k &= s_k - W(q^{-1})x_k \\ &= s_k - W(q^{-1})H(q^{-1})s_k. \end{aligned}$$

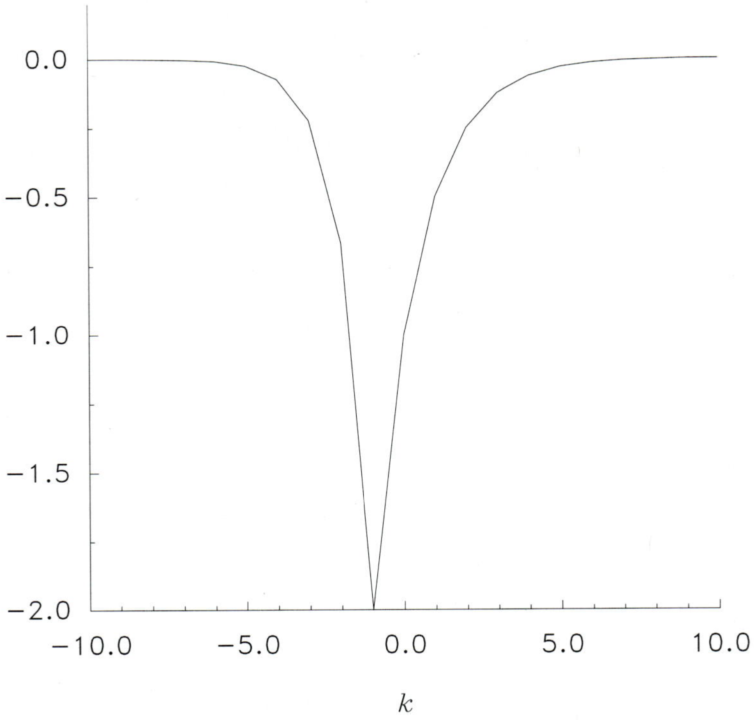

Figure 2.6 Two-sided equalization filter.

We have already seen that it is necessary to allow some signal processing delay; we use two-sided filters to accomplish this. The additional problem here is that s_k is not measured, so that e_k cannot be formed. It follows that we have only statistical properties of x_k (and s_k) to use to design the filter. Since the message signal is compressed before transmission it is reasonable to assume it to be white noise; so

$$s_k = \epsilon_k, \quad \text{a zero-mean white noise whose distribution is known.}$$

To estimate the filter weights, it is natural to consider initially, using second order properties of x_k. However this approach fails as is easily seen from the following example.

Suppose the channel is FIR with two taps so that

$$x_k = h_0\epsilon_k + h_1\epsilon_{k-1}.$$

Then x_k is MA(2) and its second order statistics are fully described by its autocovariance (acv)

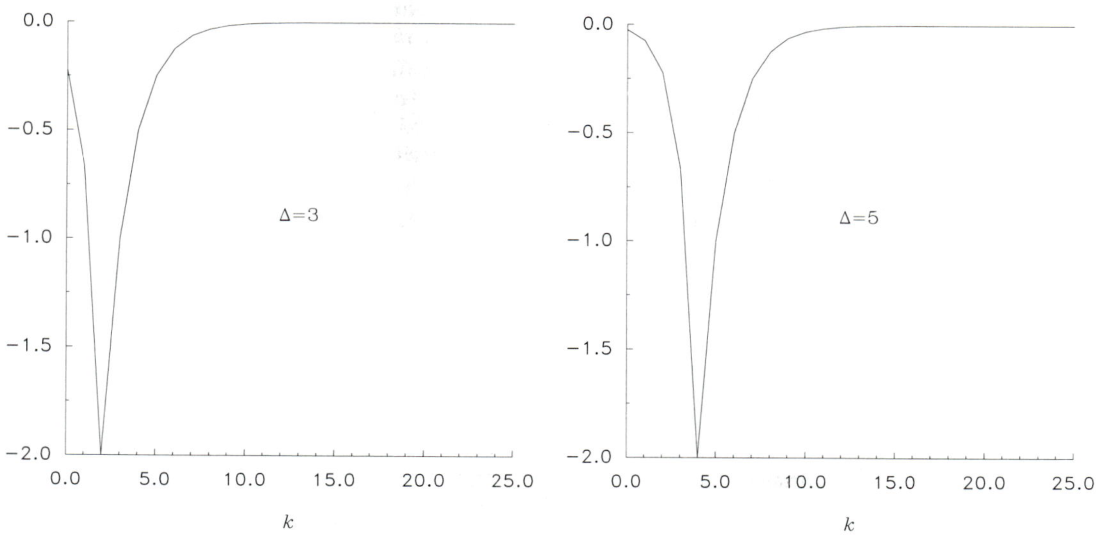

Figure 2.7 Approximate two-sided equalization filter with 25 taps.

$$\gamma_0 = (h_0^2 + h_1^2)\sigma_\epsilon^2, \quad \sigma_\epsilon^2 = E(\epsilon_k^2)$$
$$\gamma_1 = h_0 h_1 \sigma_\epsilon^2$$
$$\gamma_k = 0, \quad k \geq 2.$$

Now consider the alternate channel

$$x_k^* = h_1\epsilon_k + h_0\epsilon_{k-1}.$$

Then x_k^* is also MA(2) and has exactly the same acv as x_k. So the two channels cannot be distinguished on the basis of second order statistics, i.e., the acv.

We are thus led to consider using higher order statistics, a simple idea being to match a higher order moment. Thus consider choosing $\underline{w}$ to minimize the following performance index

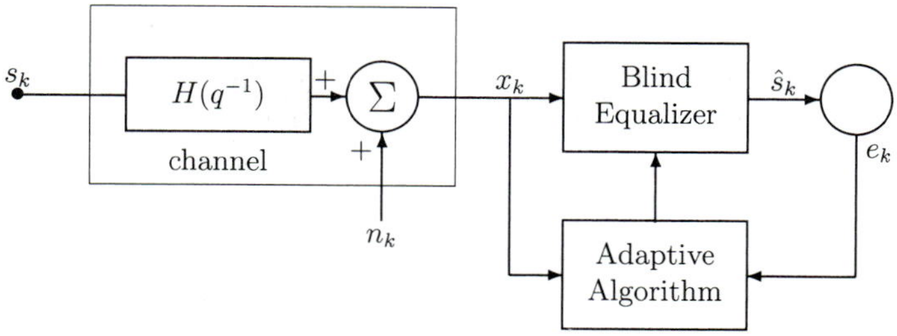

Figure 2.8 Blind channel equalizer.

$$J_p(\underline{w}) = E\left(|\hat{s}_k(\underline{w})|^p - R_p\right)^2 \tag{2.30}$$

$$\hat{s}_k(\underline{w}) = \sum_{s=1}^{p} w_s x_{k-s}$$

where

$$\begin{aligned} R_p &= E|s_k|^{2p}/E|s_k|^p \\ &= E|\epsilon_k|^{2p}/E|\epsilon_k|^p = \mu_{2p}/\mu_p. \end{aligned} \tag{2.31}$$

The reason for this choice of R_p can be seen as follows. From (2.30)

$$dJ_p/d\underline{w} = E\left\{|\hat{s}_k(\underline{w})|^{p-2}(p-2)\underline{x}_k\hat{s}_k(\underline{w})\left[|\hat{s}_k(\underline{w})|^p - R_p\right]\right\}$$

$$\underline{x}_k = (x_{k-1} \cdots x_{k-p})^T.$$

If there is a true weight vector $\underline{w}_e$ which makes $s_k(\underline{w}_e) = s_k = \epsilon_k$, then to ensure $dJ_p/d\underline{w}_e = \underline{0}$ we need

$$E\left[|\epsilon_k|^{p-2}(p-2)\underline{x}_k\epsilon_k(|\epsilon_k|^p - R_p)\right] = \underline{0}$$

$$\Rightarrow \qquad E\left[|\epsilon_k|^p(|\epsilon_k|^p - R_p)\right] = 0$$

since $s_k(\underline{w}_e) = \underline{w}_e^T\underline{x}_k$.

Further investigation of the properties of $J_p(\underline{w})$ as a criterion to be minimized is considerably simplified if we suppose an infinite two-sided equalizing filter is used. This is not practically possible but it does enable considerable conceptual insight to be gained. This leads us to express the problem in terms of new parameters as follows. We have

$$\begin{aligned} \hat{s}_k(\underline{w}) &= \sum_{s=-\infty}^{\infty} w_s x_{k-s} \\ &= W(q^{-1})x_k \\ &= W(q^{-1})H(q^{-1})\epsilon_k \\ &= T(q^{-1})\epsilon_k \end{aligned} \tag{2.32a}$$

where

$$\begin{aligned} T(q^{-1}) &= W(q^{-1})H(q^{-1}) \\ &= \sum_{s=-\infty}^{\infty} t_s q^{-s} \end{aligned}$$

and

$$t_s = \sum_{u=-\infty}^{\infty} w_u h_{s-u}. \tag{2.32b}$$

Now we take $\{t_s\}_{-\infty}^{\infty}$ to be the free parameters. Since a finite delay will not affect the quality of the equalizer, then ideally

$$t_k = \delta_{k-k_0} = \begin{cases} 1, & k = k_0 \\ 0, & k \neq k_0 \end{cases}$$

where k_0 is an arbitrary delay. The criterion (2.30) now becomes

$$J_p(\underline{t}) = E\left(|T(q^{-1})\epsilon_k|^p - R_p\right)^2.$$

To keep the exposition clear, it is convenient to concentrate on the case $p = 2$: little generality is lost in doing this. We now set about expressing $J_2(\underline{t})$ directly in terms of $\{t_s\}$

$$\begin{aligned} J_2(\underline{t}) &= E\left(|\hat{s}_k(\underline{t})|^2 - R_2\right)^2 \\ &= E\left(|\hat{s}_k(\underline{t})|^4\right) - 2E\left(|\hat{s}_k(\underline{t})|^2\right) R_2 + R_2^2. \end{aligned}$$

In Exercise 2.12 the reader is asked to show

$$E|\hat{s}_k(\underline{t})|^4 = (\mu_4 - 3\mu_2^2) \sum_{k=-\infty}^{\infty} |t_k|^4 + 3\mu_2^2 \left(\sum_{k=-\infty}^{\infty} t_k^2\right)^2 \tag{2.33a}$$

so that

$$J_2(\underline{t}) = (\mu_4 - 3\mu_2^2) \sum_{k=-\infty}^{\infty} |t_k|^4 + 3\mu_2^2 \left(\sum_{k=-\infty}^{\infty} t_k^2\right)^2 - 2\mu_4 \sum_{k=-\infty}^{\infty} t_k^2 + \mu_4^2/\mu_2^2. \tag{2.33b}$$

Note that $J_2(\underline{t})$ is non-quadratic in $\{t_k\}$ and so we should expect multiple minima. The stationary points of $J_2(\underline{t})$ are determined by the zeroed gradient

$$\frac{dJ_2}{dt_r} = 0, \quad r = 0, \pm 1, \pm 2, \cdots$$

$$\Rightarrow \quad 4(\mu_4 - 3\mu_2^2)t_r^3 + 12\mu_2^2 E_2 t_r - 4\mu_4 t_r = 0 \tag{2.34a}$$

$$\Rightarrow \quad 4t_r\left[(\mu_4 - 3\mu_2^2)t_r^2 + 3\mu_2^2 E_2 - \mu_4\right] = 0 \tag{2.34b}$$

where

$$E_2 = \sum_{r=-\infty}^{\infty} t_r^2.$$

The gradient is thus zeroed when

$$t_r = 0, \quad \text{or} \tag{2.35a}$$

$$t_r^2 = \alpha^2 \tag{2.35b}$$

where $\alpha^2 = (\mu_4 - 3\mu_2^2 E_2)/(\mu_4 - 3\mu_2^2)$. A non-zero solution requires

$$\mu_4 < 3\mu_2^2 \tag{2.36a}$$

$$\mu_4 \le 3\mu_2^2 E_2 \tag{2.36b}$$

or

$$\mu_4 > 3\mu_2^2 \tag{2.37a}$$

$$\mu_4 \ge 3\mu_2^2 E_2. \tag{2.37b}$$

The reader is asked to show in Exercise 2.13 that (2.37) cannot in fact yield a solution. There clearly are infinitely many solutions to (2.35), which can be described as follows. The solution sets are S_M, $M = 1, 2, \ldots$, where S_M has M of the t_r's non-zero and the other t_r's equal to 0. Thus from (2.35b), S_M has

$$t_r^2 = \alpha_M^2, \quad \text{for } M \text{ values of } r$$

$$\Rightarrow \quad E_{2,M} = M\alpha_M^2.$$

Thus assuming (2.36a), (2.35) gives

$$\alpha_M^2 = \frac{3\mu_2^2 E_{2,M} - \mu_4}{3\mu_2^2 - \mu_4}$$

$$= \frac{3\mu_2^2 M\alpha_M^2 - \mu_4}{3\mu_2^2 - \mu_4}$$

$$\Rightarrow \quad \alpha_M^2 = \frac{\mu_4}{3(M-1)\mu_2^2 + \mu_4} \tag{2.38}$$

$$\Rightarrow \quad E_{2,M} = \frac{M\mu_4}{3(M-1)\mu_2^2 + \mu_4} = \frac{M\mu_4}{(3\mu_2^2 - \mu_4)(M-1) + M\mu_4}. \tag{2.39}$$

From this we see

$$E_{2,M} \le 1, \quad E_{2,1} = 1.$$

It is easy to see that $E_{2,M+1} < E_{2,M}$ and that

$$E_{2,\infty} = \lim_{M\to\infty} E_{2,M} = \frac{\mu_4}{3\mu_2^2}$$

so that (2.36b) is satisfied for all M.

We can now calculate the criterion function. From (2.34), at a stationary point

$$(\mu_4 - 3\mu_2^2) \sum_{r=-\infty}^{\infty} t_r^4 + 3\mu_2^2 E_{2,M}^2 = \mu_4 E_{2,M}.$$

Thus from (2.33b)

$$J_{2,M} = \mu_4(\mu_4/\mu_2^2 - E_{2,M}). \tag{2.40}$$

We conclude that $J_{2,M}$ increases with M and that the global minimum is

$$J_{2,1} = \mu_4 \left(\frac{\mu_4}{\mu_2^2} - 1 \right), \quad \text{at } M = 1.$$

A natural question is how to find a way of minimizing $J_2(\underline{w})$ without getting stuck at a local minimum. This issue is discussed further in Section 3.3.

2.4 TIME DELAY ESTIMATION

The time delay estimation setup is shown in Fig. 2.9 in a little more detail than previously. The true delay is θ_o and it is assumed that all signals are stationary and zero mean. We also assume that s_k, n_{1k}, and n_{2k} are mutually uncorrelated at all lags and allow amplitude attenuation (α) of the reference signal. We also assume that s_k is sampled from a continuous time signal. From Fig. 2.9 the error signal is

$$e_k = e_k(\hat{\theta}) = y_k - x_{k-\hat{\theta}} \tag{2.41}$$

$$= (s_{k-\theta_o} + n_{1k}) - (\alpha s_{k-\hat{\theta}} + n_{2,k-\hat{\theta}})$$

$$= (s_{k-\theta_o} - \alpha s_{k-\hat{\theta}}) + (n_{1k} - n_{2,k-\hat{\theta}}). \tag{2.42}$$

Thus the MSE is on the one hand from (2.41)

$$\mathcal{E} = \mathcal{E}(\hat{\theta}) = E(e_k^2)$$

$$= R_y(0) - 2R_{xy}(\hat{\theta}) + R_x(0) \tag{2.43}$$

where $R_{xy}(\tau)$ is the cross-covariance between x_t, $y_{t+\tau}$. On the other hand the MSE is also, from (2.42),

$$\mathcal{E} = (1 + \alpha^2)R_s(0) - 2\alpha R_s(\theta_o - \hat{\theta}) + R_{n1} + R_{n2} \tag{2.44}$$

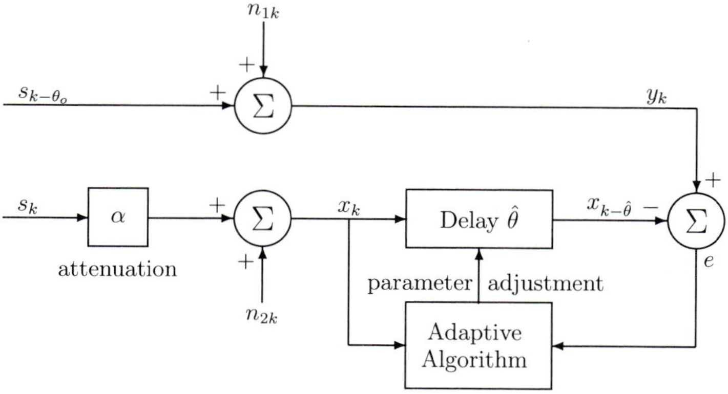

Figure 2.9 Time delay estimation.

where R_{n1} and R_{n2} are noise variances. Expression (2.43) gives $\mathcal{E}$ in terms of covariances that can be empirically determined. Expression (2.44) gives a theoretical connection to the underlying signal autocovariance. It is clear then that minimizing (2.43) with respect to $\hat{\theta}$ is equivalent (via (2.44)) to maximizing $R_s(\theta_o - \hat{\theta})$ over $\hat{\theta}$. This maximum occurs where $\theta_o - \hat{\theta} = 0$, i.e., where $\hat{\theta} = \theta_o =$ true delay. Needless to say $R_s(\tau)$ is not in general a quadratic function of τ.

2.5 TWO-SIDED FIR FILTER: INTERFERENCE CANCELLING

The general interference cancelling setup is shown in Fig. 2.10. A primary signal s_k is contaminated with a noise n_k and recorded in the presence of another noise m_{0k} as y_k. A measurement of n_k is available through a noisy channel of transfer function $H(q^{-1})$ as x_k. This auxiliary signal is passed through the noise cancellation filter $W(q^{-1})$ to produce an estimate of the signal $\hat{s}_k$, which is subtracted from y_k to give a "cleaned" signal e_k. We assume that all signals are jointly stationary with zero means. Let us look at the structure of the error signal. We have from Fig. 2.10 (with $\Delta = 0$) that

$$e_k = y_k - \hat{s}_k = y_k - W(q^{-1})x_k \tag{2.45}$$

$$= s_k + n_k + m_{0k} - W(q^{-1})(m_{1k} + H(q^{-1})n_k)$$

$$\Rightarrow \quad e_k = s_k + n_k - \tilde{n}_k + m_{0k} \tag{2.46}$$

where we have introduced

$$\tilde{n}_k = W(q^{-1})(H(q^{-1})n_k + m_{1k}). \tag{2.47}$$

We may call $n_k - \tilde{n}_k$ the noise or interference cancelling error.

Assume that s_k, n_k, m_{0k}, and m_{1k} are mutually statistically independent with spectra $F_s(\omega)$, $F_n(\omega)$, $F_{m_0}(\omega)$, and $F_{m_1}(\omega)$ respectively; then we see from (2.46) that the MSE is

$$\mathcal{E} = E(e_k^2) = E(s_k^2) + E(n_k - \tilde{n}_k)^2 + E(m_{0k}^2).$$

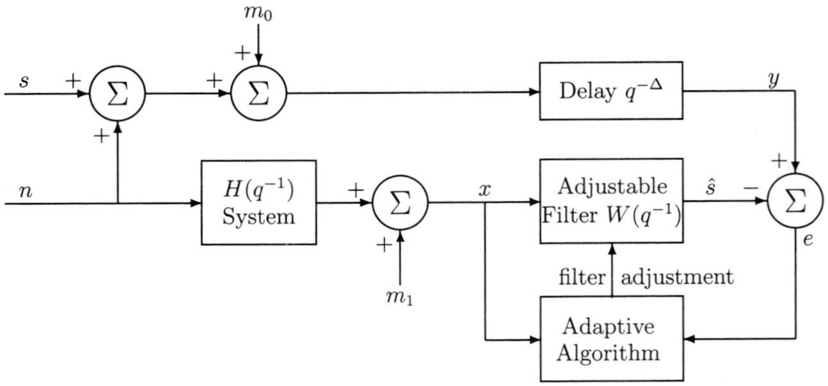

Figure 2.10 Interference cancelling.

Thus, minimizing MSE with respect to $W(q^{-1})$ is equivalent to minimizing the noise cancelling error power $E(n_k - \tilde{n}_k)^2$. From (2.45) we may immediately apply the results of Section 2.3 to see that the optimal noise cancelling filter is given by (2.26), namely

$$W_o(e^{-j\omega}) = F_{yx}(\omega)/F_x(\omega). \tag{2.48}$$

Referring to Fig. 2.10 we can re-express this as follows. We have

$$y_k = s_k + n_k + m_{0k}$$
$$x_k = m_{1k} + H(q^{-1})n_k.$$

Thus

$$F_{yx}(\omega) = H^*(e^{-j\omega})F_n(\omega)$$
$$F_x(\omega) = F_{m_1}(\omega) + |H(e^{-j\omega})|^2 F_n(\omega). \tag{2.49}$$

Thus (2.48) becomes

$$W_o(e^{-j\omega}) = \frac{H^*(e^{-j\omega})F_n(\omega)}{F_{m_1}(\omega) + |H(e^{-j\omega})|^2 F_n(\omega)}. \tag{2.50}$$

Note that in the absence of noise

$$W_o(e^{-j\omega}) = H^{-1}(e^{-j\omega}).$$

The situation here is much the same as with equalization. Since $H(e^{-j\omega})$ will not generally be minimum phase, successful noise cancellation requires a delay in processing thus allowing $W_o(e^{-j\omega})$ to be a two-sided filter.

We can measure the quality of the noise cancelling by comparing signal to noise ratios at the output, before and after cancelling. From (2.48) we have

$$\text{SNR}_{in} = F_s(\omega)/(F_n(\omega) + F_{m_0}(\omega)) \tag{2.51}$$

while from (2.46)

$$\text{SNR}_{out} = F_s(\omega)/(F_{n-\tilde{n}}(\omega) + F_{m_0}(\omega)). \tag{2.52}$$

However, from (2.46), (2.47)

$$n_k - \tilde{n}_k = [1 - W(q^{-1})H(q^{-1})]n_k - W(q^{-1})m_{1k}$$
$$\Rightarrow \quad F_{n-\tilde{n}}(\omega) = |1 - W(e^{-j\omega})H(e^{-j\omega})|^2 F_n(\omega) + |W(e^{-j\omega})|^2 F_{m1}(\omega). \tag{2.53}$$

Let us now suppose we use the optimal noise cancelling filter (2.49). Then we have

$$\begin{aligned} 1 - W_o(e^{-j\omega})H(e^{-j\omega}) &= 1 - |H(e^{-j\omega})|^2 F_n(\omega)/(F_{m_1}(\omega) + |H(e^{-j\omega})|^2 F_n(\omega)) \\ &= F_{m_1}(\omega)/(F_{m_1}(\omega) + |H(e^{-j\omega})|^2 F_n(\omega)) \\ &= F_{m_1}(\omega)/F_x(\omega) \end{aligned}$$

where (2.49) has been used. Thus (2.53) becomes

$$\begin{aligned} F_{n-\tilde{n}}(\omega) &= F_{m_1}^2(\omega)F_n(\omega)/F_x^2(\omega) + |H(e^{-j\omega})|^2 F_n^2(\omega)F_{m_1}(\omega)/F_x^2(\omega) \\ &= F_{m_1}(\omega)F_n(\omega)/F_x(\omega) \\ &= F_n(\omega) - |H(e^{-j\omega})|^2 F_n^2(\omega)/F_x(\omega) \quad \text{(by (2.49))}. \end{aligned}$$

Using this in (2.51), (2.52) we find

$$\begin{aligned} \frac{\mathrm{SNR}_{out}}{\mathrm{SNR}_{in}} &= \frac{F_n(\omega) + F_{m_0}(\omega)}{F_{n-\tilde{n}}(\omega) + F_{m_0}(\omega)} \\ &= \frac{F_n(\omega) + F_{m_0}(\omega)}{F_n(\omega) + F_{m_0}(\omega) - |H(e^{-j\omega})|^2 F_n^2(\omega)/F_x(\omega)} \\ &= \frac{1}{1 - \pi_{prim}(\omega)\pi_{aux}(\omega)} \end{aligned} \tag{2.54}$$

where we have introduced the spectral ratios

$$\pi_{prim}(\omega) = F_n(\omega)/(F_n(\omega) + F_{m_0}(\omega)) \tag{2.55}$$

$$\pi_{aux}(\omega) = |H(e^{-j\omega})|^2 F_n(\omega)/(F_{m_1}(\omega) + |H(e^{-j\omega})|^2 F_n(\omega)). \tag{2.56}$$

Clearly

$$0 \leq \pi_{prim}(\omega),\ \pi_{aux}(\omega) \leq 1$$

so that

$$\mathrm{SNR}_{out}/\mathrm{SNR}_{in} \geq 1.$$

Thus noise cancelling with the optimal filter always improves the signal to noise ratio.

Note that the π-ratios measure primary and auxiliary noise power ratios. They enable us to see that if the major noise component of y_k is due to n_k (i.e., $F_{m_0}(\omega) \ll F_n(\omega)$) and the major component of x_k is also due to n_k (i.e., $F_{m_1}(\omega) \ll |H(e^{-j\omega})|^2 F_n(\omega)$), then both $\pi_{prim}(\omega)$, $\pi_{aux}(\omega)$ are near to 1. And so from (2.54), $\mathrm{SNR}_{out}/\mathrm{SNR}_{in} \gg 1$, so that noise cancelling will work well; otherwise it may not help very much.

2.6 IIR FILTERS

2.6.1 Output Error Filter

Let us return to the echo cancellation problem considered in Section 2.2 and allow the filter to be an IIR filter. So we replace (2.1) with

$$\hat{s}_k(\underline{\theta}) = \frac{b(q^{-1})}{1 + a(q^{-1})} x_k = W_\theta(q^{-1}) x_k \tag{2.57}$$

where

$$b(q^{-1}) = \sum_{t=1}^{m} b_t q^{-t}$$

$$a(q^{-1}) = \sum_{t=1}^{p} a_t q^{-t}.$$

The error signal is

$$e_k(\underline{\theta}) = y_k - \hat{s}_k(\underline{\theta}) \tag{2.58}$$

and we have also introduced the parameter vector

$$\underline{\theta} = (a_1 \cdots a_p,\ b_1 \cdots b_m)^T.$$

Note also the change of notation for the parameter. Thus $\underline{w}$ always relates to an FIR filter while $\underline{\theta}$ always relates to a non-FIR filter. From (2.57) we can write

$$\hat{s}_k(\underline{\theta}) = -\sum_{t=1}^{p} a_t \hat{s}_{k-t}(\underline{\theta}) + \sum_{t=1}^{m} b_t x_{k-t}. \tag{2.59}$$

If we introduce a new signal vector

$$\underline{\phi}_k(\underline{\theta}) = (-\hat{s}_{k-1}(\underline{\theta}) \cdots - \hat{s}_{k-p}(\underline{\theta}),\ x_{k-1} \cdots x_{k-m})^T$$

then we can write (2.59) as

$$\hat{s}_k(\underline{\theta}) = \underline{\phi}_k^T(\underline{\theta})\underline{\theta} \tag{2.60}$$

and this looks like the FIR expression (2.1). We call this a pseudo linear regression. It looks like a linear regression but hides the fact that $\underline{\phi}_k(\underline{\theta})$ also depends on $\underline{\theta}$.

If all relevant signals are stationary, then from (2.57), (2.58) the MSE is given by

$$\begin{aligned}\mathcal{E}(\underline{\theta}) &= E(e_k^2(\underline{\theta})) \\ &= E(y_k - W_\theta(q^{-1})x_k)^2\end{aligned}$$

and exactly as in Section 2.3 we can write this as (see (2.25))

$$\mathcal{E}(\underline{\theta}) = \mathcal{E}_W(\underline{\theta}) + \mathcal{E}_N(\underline{\theta}) \tag{2.61}$$

where

$$\mathcal{E}_W(\underline{\theta}) = \int_{-\pi}^{\pi} |F_{yx}(\omega)/F_x(\omega) - W_\theta(e^{-j\omega})|^2 F_x(\omega)d\omega/2\pi \tag{2.62}$$

$$\mathcal{E}_N(\underline{\theta}) = \int_{-\pi}^{\pi} (F_y(\omega) - |F_{yx}(\omega)|^2/F_x(\omega))d\omega/2\pi. \tag{2.63}$$

These formulae generalize (2.11). Note that the error surface (2.61) is not in general quadratic and may have (many) local minima.

2.6.2 Equation Error Filter

To try to avoid the non-quadratic behavior of $\mathcal{E}_W(\underline{\theta})$ one can use the following idea. Ideally $\hat{s}_k$ is close to y_k, so in (2.59) one is led to replace the lagged values of $\hat{s}_k$ on the right hand side with corresponding lagged values of y_k to obtain the following filter:

$$\hat{s}_k = -\sum_{s=1}^{p} a_s y_{k-s} + \sum_{s=1}^{m} b_s x_{k-s} \tag{2.64}$$

$$= \underline{\psi}_k^T \underline{\theta} \tag{2.65}$$

where now

$$\underline{\psi}_k = (-y_{k-1} \cdots - y_{k-p}, x_{k-1} \cdots x_{k-m})^T. \tag{2.66}$$

Note that $\hat{s}_k$ is now linear in $\underline{\theta}$ and so the MSE will be quadratic in $\underline{\theta}$. Indeed,

$$e_k = y_k - \hat{s}_k$$

$$= y_k - \underline{\psi}_k^T \underline{\theta}$$

$$\Rightarrow \quad \mathcal{E}(\underline{\theta}) = E(e_k^2) = R_y - 2R_{y\psi}^T \underline{\theta} + \underline{\theta}^T R_{\psi} \underline{\theta}.$$

The disadvantage of this equation error method is that if y_k is very noisy, then severe bias is introduced into $\hat{s}_k$. It is also important to note that the equation error based ASP element has an extra input, namely, y_k. This is shown in Fig. 2.11.

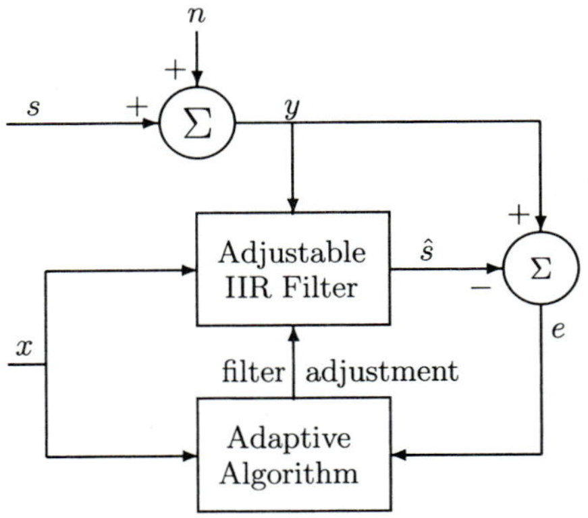

Figure 2.11 ASP element for equation error filter.

2.6.3 Output Error Filter with Noise Whitening

Let us recall the System Identification problem of Fig. 1.8. So far in discussing the system filter we have made no attempt to deal with color in the additive primary signal noise n_k. The natural way to handle this is to build a whitening filter for the error leading to the structure in Fig. 2.12 (which follows from Fig. 1.13).

The whitened error signal becomes, from Fig. 2.12

$$e_k(\underline{\theta}) = W_\alpha^{-1}(q^{-1})(y_k - W_\beta(q^{-1})x_k) \tag{2.67}$$

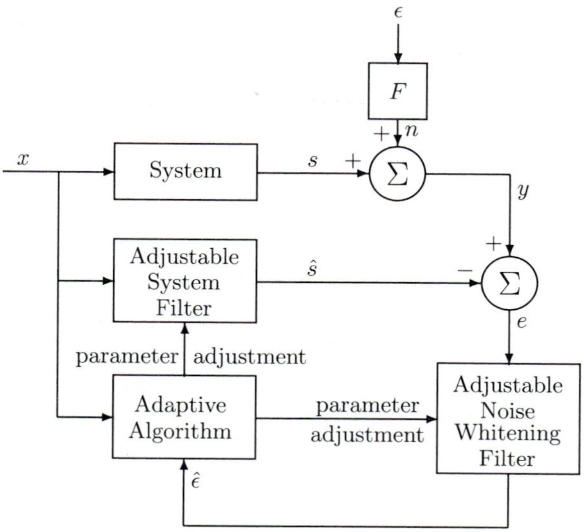

Figure 2.12 Adaptive identification with noise whitening.

where the system filter is

$$W_\beta(q^{-1}) = [1 + a(q^{-1})]^{-1}b(q^{-1}) \tag{2.68}$$

and the noise filter is (an ARMA filter: see Appendix B6(vii))

$$W_\alpha(q^{-1}) = [1 + c(q^{-1})]^{-1}[1 + d(q^{-1})] \tag{2.69}$$

in which $1 + c(q^{-1})$ is a stability polynomial (see Appendix C). Also we have partitioned

$$\underline{\theta} = (\underline{\beta}^T, \underline{\alpha}^T)^T \tag{2.70}$$

with $\underline{\beta}$ denoting system parameters

$$\underline{\beta} = (a_1 \cdots a_p, b_1 \cdots b_m)^T \tag{2.71}$$

and $\underline{\alpha}$ denoting noise parameters

$$\underline{\alpha} = (c_1 \cdots c_r, d_1 \cdots d_r)^T \tag{2.72}$$

and

$$a(q^{-1}) = \sum_{u=1}^{p} a_u q^{-u}, \quad \text{etc.}$$

Now following the method used in Section 2.3 we find from (2.67) that the MSE surface is

$$\mathcal{E}(\underline{\theta}) = E(e_k^2(\underline{\theta})) = \mathcal{E}_S(\underline{\theta}) + \mathcal{E}_N(\underline{\theta})$$

$$\mathcal{E}_S(\underline{\theta}) = \int_{-\pi}^{\pi} |F_{yx}(\omega)/F_x(\omega) - W_\beta(e^{-j\omega})|^2 (F_x(\omega)/|W_\alpha(e^{-j\omega})|^2) d\omega/2\pi \quad (2.73)$$

$$\mathcal{E}_N(\underline{\theta}) = \int_{-\pi}^{\pi} (F_y(\omega) - |F_{yx}(\omega)|^2/F_x(\omega))|W_\alpha(e^{-j\omega})|^{-2} d\omega/2\pi. \quad (2.74)$$

Notice that $\mathcal{E}_N(\underline{\theta})$ depends only on noise parameters $\underline{\alpha}$, while $\mathcal{E}_S(\underline{\theta})$ is a weighted frequency domain norm of the distance between the true transfer function between y_k and x_k and the model transfer function specified by the filter.

2.7 NOTES

Our discussion of channel equalization with a training signal and of interference cancelling borrows from that in [W5]. Blind Equalization is discussed in [B7], [D4], [D5], [D6], [F3], [G8], [J5], and [S5]. The $J_p(\omega)$ criterion in (2.30) is introduced in [G8] and $J_2(\omega)$ is further discussed in [T1]. Time delay estimation is discussed in more general settings in [C3]. IIR filters are discussed in many references such as [G9], and [J3]. The output error filter with noise whitening corresponds to a transfer function plus colored noise (TFCN) model for the relation between y, x, n. This has the virtue of treating system or filter and noise parameters separately. However, an alternative scheme is the ARMAX (autoregressive moving average exogenous) process where filter and noise parameters are mixed as follows:

$$e_k(\underline{\theta}) = [1 + c(q^{-1})]^{-1}[(1 + a(q^{-1}))y_k - b(q^{-1})x_k] \quad (2.75)$$

$$\underline{\theta} = (a_1 \cdots a_p, b_1 \cdots b_m, c_1 \cdots c_r)^T.$$

The reader can construct a block diagram corresponding to Fig. 2.12. The breakup of MSE in (2.25) appears in [S11] and is used by [L9].

EXERCISES

2.1 Consider the adaptive predictor of Fig. 2.1 with

$$\text{auxiliary signal } x_k = \sin(2\pi k f_0) + \cos(2\pi k f_1)$$

$$\text{primary signal } s_k = y_k = \cos(2\pi k f_0)$$

where $1/f_0 = M_0 = 5$; $1/f_1 = M_1 = 3$. Find the optimal weight vector $\underline{w}_o$, the MSE surface $\mathcal{E}(\underline{w})$, the minimum MSE, $\mathcal{E}_{min}$, for (i) 2 weights and (ii) 3 weights. Compare your results and explain the changes you find in $\mathcal{E}_{min}$. Keep an eye on the rank of $\underline{R}_x$.

2.2 Repeat Exercise 2.1 with the following signal specifications:

$$s_k = y_k = \sin(4\pi k f_0)$$

$$x_k = \sin(2\pi k f_0) - \cos(4\pi k f_0)$$

2.3 Repeat Exercise 2.1 (cases (i) and (ii)) with the following signal specifications:

$$s_k = \cos(2\pi k f_0)$$
$$x_k = \sin(2\pi k f_0)$$
$$y_k = s_k + \epsilon_k$$

where ϵ_k is a zero mean, unit variance white noise.

2.4 Repeat Exercise 2.1 (case (i)) with the following signal specifications:

$$x_k \text{ is an AR(1) process}$$
$$x_k = (1 + \alpha q^{-1})^{-1}\delta_k, \quad \alpha = 0.8$$

where δ_k is a zero mean, unit variance white noise. We also assume

$$s_k = (1 + \beta q^{-1})^{-1}\delta_k, \quad \beta = 0.7$$
$$y_k = s_k + \epsilon_k$$
$$\epsilon_k = \sin(2\pi k/10.0).$$

2.5 Repeat Exercise 2.1 (case(ii)) with the following signal specifications:

$$y_k = s_k + \epsilon_k$$
$$s_k = -a_1 y_{k-1} - a_2 y_{k-2} - a_3 y_{k-3}$$

(so that y_k is an AR(3) process); ϵ_k is a Gaussian white noise of mean zero, unit variance, and

$$x_k = y_k.$$

2.6 Rework Example 2.2 with the same signal specifications but a 3 tap filter. Find the optimal weight $\underline{w}_o$ and $\mathcal{E}_{min}$.

2.7 Consider the minimum phase channel equalization problem with the following signal specifications: s_k as in (2.15); ξ_k as in (2.16); $\sigma_\epsilon^2 = 0.48$; $n_k = 0$ so that $x_k = \xi_k$; $y_k = s_k + m_k$ where m_k is a zero mean, unit variance white noise which is independent of ϵ_k in (2.15). Use an equalizing filter with 2 taps. Find the optimal weight vector $\underline{w}_o$ and $\mathcal{E}_{min}$.

2.8 Consider the non-minimum phase channel equalization problem with the following signal specifications:

a white noise message signal s_k (zero mean, unit variance),
an all pole first order channel

$$H(q^{-1}) = (1 + \theta q^{-1})^{-1}, \quad \theta = -0.7$$

a white noise at the filter input n_k (zero mean, unit variance).

Compute the optimal infinite order 2 sided equalizing filter and corresponding error variance $\mathcal{E}_{min}$. Repeat the calculation finding and plotting 25 tap equalizing filters for delays of $\Delta = 0, 1, 2, 4$. Also find and plot the corresponding minimum MSE's.

2.9 Consider the time delay estimation set up of Fig. 2.9 with the following specifications

$$s_k = \sin(2\pi k f_0) + 0.5\cos(2\pi k f_0)$$

where $1/f_0 = M =$ an integer; $\theta_o = delay = 2$. Calculate and plot the MSE, $\mathcal{E}(\hat{\theta})$. All noises are zero.

2.10 For the time delay estimation problem of Fig. 2.41 (with true delay $\theta_o = 2$), calculate and plot the MSE, $\mathcal{E}(\hat{\theta})$, when s_k is an AR(1) process

$$s_k = (1 + \beta q^{-1})^{-1} \epsilon_k, \quad \beta = -0.8$$

and ϵ_k is a zero mean, unit variance white noise. Also suppose $n_{1k} = n_{2k} = 0$.

2.11 Consider the interference cancelling setup of Fig. 2.10 with the following specifications.

m_{1k} is zero mean white noise of variance σ_{m1}^2.
n_k is zero mean white noise of variance σ_n^2.
$H(q^{-1}) = (1 + \beta q^{-1})^{-1}$, $\beta = 0.8$, and $\Delta = 0$.
m_{0k} is zero mean white noise of variance σ_{m0}^2.

$\{m_{1k}, n_k, m_{0k}\}$ are jointly statistically independent. Find an expression for the optimal 2 sided filter weights. Also find frequency domain expressions for π_{prime}, π_{aux}. For $\sigma_n^2 = \sigma_{m0}^2 = \sigma_{m1}^2 = 1.0$, plot $\text{SNR}_{\text{out}}/\text{SNR}_{\text{in}}$ as a function of frequency.

2.12 Suppose $\{\epsilon_k\}$ is a zero mean white noise (i.e., sequence of independent identically distributed random variables) with moments

$$\mu_p = E(\epsilon_k^p), \quad p = 2, 4$$

let $\hat{s}_k$ be a filtered signal

$$\hat{s}_k = \sum_{s=-\infty}^{\infty} t_s \epsilon_{k-s}.$$

Show that (2.33) holds.

2.13 Show that the pair (2.37) do not lead to a viable solution for (2.35).

2.14 Another approach to finding a criterion for Blind Equalization is based on the observation that

$$E\left(|\hat{s}_k(\underline{w})|^2\right) = \mu_2 \sum_{k=-\infty}^{\infty} t_k^2$$

but that if the true equalizer is used then

$$E\left(|\hat{s}_k(\underline{w}_e)|^2\right) = E(\epsilon_k^2) = \mu_2.$$

This suggests imposing the constraint

$$\sum_{k=-\infty}^{\infty} t_k^2 = 1$$

and leads to the following criterion. Choose $\{t_s\}$ to minimize

$$E\left(|s_k(\underline{t})|^4\right) \quad \text{subject to} \quad \sum_{k=-\infty}^{\infty} t_k^2 = 1.$$

Discuss the minimization of this criterion. Show that it allows both cases (2.36) and (2.37).

3

Iterative Minimization

3.1 INTRODUCTION

Most adaptive algorithms are modifications of standard iterative procedures for solving minimization problems in a real time setting.

As a prelude to the development of the adaptive algorithms we therefore consider iterative methods of minimizing an error criterion such as MSE. The two basic iterative methods are

Steepest Descent and its variations

and Newton's method.

We first review the behavior of these procedures in linear quadratic problems where the filter is an FIR filter and the error criterion is quadratic and then look at nonlinear problems such as time delay estimation. Of course linear quadratic problems can be solved exactly so there is no need for iterative solution. But a study of iterative methods is valuable in this simple setting because of the considerable insight into their behavior that can be gained.

3.2 QUADRATIC ERROR SURFACE

Before we begin a study of iterative optimization it is useful to recall a few properties of the quadratic MSE surface. We recall from (2.11) that the MSE (with an FIR filter) can be written

$$\mathcal{E}(\underline{\tilde{w}}) = \mathcal{E} = \mathcal{E}_{min} + \underline{\tilde{w}}^T \underline{R}_x \underline{\tilde{w}} \tag{3.1}$$

where we have introduced the weight estimation error

$$\underline{\tilde{w}} = \underline{w} - \underline{w}_o.$$

Contours of the MSE surface are defined by the equation

$$\begin{aligned} & \mathcal{E}(\underline{\tilde{w}}) = \text{constant} \\ \Rightarrow \quad & \underline{\tilde{w}}^T \underline{R}_x \underline{\tilde{w}} = \text{constant.} \end{aligned}$$

In two dimensions a contour of constant MSE is an ellipse centered at the origin in the $\underline{\tilde{w}} = (\underline{\tilde{w}}_1, \underline{\tilde{w}}_2)$ plane. In higher dimensions a contour is a hyper-ellipse. In the $\underline{\tilde{w}}$ plane the principal axes of the ellipse are normal to the ellipse and so can be found from the gradient

$$d\mathcal{E}/d\underline{\tilde{w}} = 2\underline{R}_x \underline{\tilde{w}}. \tag{3.2}$$

On the other hand any vector through the origin must be of the form $\lambda\underline{\tilde{w}}$ for some λ. So the principal axes (which go through the origin) must satisfy

$$2\underline{R}_x \underline{\tilde{w}} = \lambda \underline{\tilde{w}}.$$

This tells us that the eigenvectors of $\underline{R}_x$ define the principal axes of the MSE surface, while the eigenvalues represent steepness along the principal axes.

If $\underline{R}_x$ is positive definite (and for simplicity suppose it has distinct eigenvalues), then it has an eigenvector decomposition (see Appendix A)

$$\underline{R}_x = \sum_{u=1}^{p} \lambda_u \underline{v}_u \underline{v}_u^T \tag{3.3}$$

where $\underline{v}_u$ are the eigenvectors and λ_u the eigenvalues. Also

$$\begin{aligned} \underline{v}_u^T \underline{v}_r &= \delta_{ur} \\ &= \begin{cases} 1 & u = r \\ 0 & u \neq r. \end{cases} \end{aligned}$$

Thus, (3.1) can be written

$$\begin{aligned} \mathcal{E} &= \mathcal{E}_{min} + \underline{\tilde{w}}^T \sum_{u=1}^{p} \lambda_u \underline{v}_u \underline{v}_u^T \underline{\tilde{w}} \\ &= \mathcal{E}_{min} + \sum_{u=1}^{p} \lambda_u \tilde{v}_u^2 \end{aligned} \tag{3.4}$$

where we have introduced the rotated coordinate

$$\tilde{v}_u = \underline{v}_u^T \underline{\tilde{w}}. \tag{3.5}$$

Note then that the MSE gradient is

$$d\mathcal{E}/d\tilde{v}_u = 2\lambda_u \tilde{v}_u \tag{3.6}$$

so the steepness per unit length is $2\lambda_u$.

An example will clarify these points.

Example 3.1.

Suppose we have the following signal statistics:

$$\underline{R}_x = \begin{pmatrix} 2 & 1 \\ 1 & 2 \end{pmatrix}, \quad \underline{R}_{xy} = \begin{pmatrix} 7 \\ 8 \end{pmatrix}, \quad R_y = 40.$$

The minimizing weight is given from (2.6) as

$$\underline{w}_o = \begin{pmatrix} w_{o1} \\ w_{o2} \end{pmatrix} = \begin{pmatrix} 2 & 1 \\ 1 & 2 \end{pmatrix}^{-1} \begin{pmatrix} 7 \\ 8 \end{pmatrix} = \begin{pmatrix} 2 \\ 3 \end{pmatrix}.$$

Then from (2.8)

$$\begin{aligned} \mathcal{E}_{min} &= R_y - \underline{R}_{xy}^T \underline{R}_x^{-1} \underline{R}_{xy} \\ &= R_y - \underline{R}_{xy}^T \underline{w}_o \\ &= 40 - (7 \quad 8) \begin{pmatrix} 2 \\ 3 \end{pmatrix} \\ &= 2. \end{aligned}$$

Next we find the eigenvalues of $\underline{R}_x$. They satisfy

$$\begin{aligned} |\underline{R}_x - \lambda I| = 0 &= \begin{vmatrix} 2-\lambda & 1 \\ 1 & 2-\lambda \end{vmatrix} \\ &= 4 - 4\lambda + \lambda^2 - 1 = \lambda^2 - 4\lambda + 3 \\ &= (\lambda - 1)(\lambda - 3) \\ \Rightarrow \quad & \lambda_1, \lambda_2 = 1, 3. \end{aligned}$$

To find the eigenvectors we solve for $\underline{a} = (a_1, a_2)^T$

$$\begin{aligned} & \underline{R}_x \underline{a} = \lambda \underline{a} \\ \Rightarrow \quad & \begin{pmatrix} 2 & 1 \\ 1 & 2 \end{pmatrix} \begin{pmatrix} a_1 \\ a_2 \end{pmatrix} = \begin{pmatrix} \lambda a_1 \\ \lambda a_2 \end{pmatrix} \\ \Rightarrow \quad & \begin{aligned} 2a_1 + a_2 &= \lambda a_1 \\ a_1 + 2a_2 &= \lambda a_2 \end{aligned} \\ \Rightarrow \quad & a_2 = (\lambda - 2) a_1 \\ \Rightarrow \quad & (a_1, a_2) = a_1(1, \lambda - 2). \end{aligned}$$

A unit eigenvector satisfies $\underline{a}^T\underline{a} = 1$ which leads to

$$\begin{aligned}1 &= a_1^2(1 + (\lambda - 2)^2)\\ &= a_1^2(\lambda^2 - 4\lambda + 4 + 1) = a_1^2(0 + 2)\\ &= 2a_1^2 \quad \Rightarrow \quad a_1 = \frac{1}{\sqrt{2}}.\end{aligned}$$

So the unit eigenvectors are

$$\frac{1}{\sqrt{2}}(1, -1), \quad \frac{1}{\sqrt{2}}(1, 1).$$

These define the rotated principal axes while $\underline{w}_o$ defines the center of any contour ellipse. The resultant elliptical contours are shown in Fig. 3.1. To label the contours we note that in rotated coordinates (3.4) becomes

$$\begin{aligned}\mathcal{E} &= \mathcal{E}_{min} + \lambda_1\tilde{v}_1^2 + \lambda_2\tilde{v}_2^2\\ &= 2 + \tilde{v}_1^2 + 3\tilde{v}_2^2.\end{aligned}$$

In the original coordinates we find from (2.4) that

$$\mathcal{E} = 40 - 14w_1 - 16w_2 + 2w_1^2 + 2w_2^2 + 2w_1w_2.$$

3.2.1 Steepest Descent

Although we can calculate the minimizing weight vector of the MSE analytically we turn now to look at iterative minimization of the MSE starting with steepest descent. The SD algorithm has the form

$$\delta\underline{w}_k = \underline{w}_k - \underline{w}_{k-1} = -\frac{1}{2}\mu\underline{g}(\underline{w}_{k-1}) \tag{3.7}$$

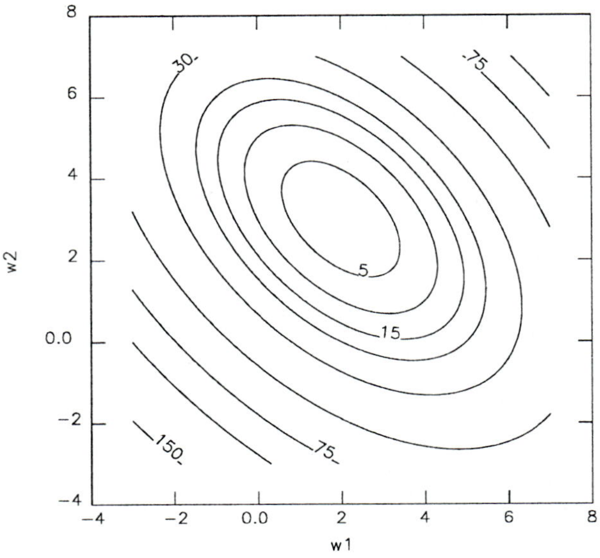

Figure 3.1 Contours of Constant MSE for Example 3.1.

where $\underline{g}(\underline{w})$ is the MSE gradient

$$\underline{g}(\underline{w}) = d\mathcal{E}/d\underline{w}. \tag{3.8}$$

That is, the kth iterate is the $(k-1)^{th}$ iterate plus a correction in the direction of the MSE gradient. The parameter μ is a step size. From (2.4) we find

$$\delta \underline{w}_k = \mu(\underline{R}_{xy} - \underline{R}_x \underline{w}_{k-1}). \tag{3.9}$$

The basic question for any iteration is whether or not it converges. To investigate this (parameter convergence) it is convenient to rewrite the algorithm in so-called error form. Put (2.5) into (3.9) to find that

$$\delta \underline{w}_k = \mu \underline{R}_x (\underline{w}_e - \underline{w}_{k-1}) \tag{3.10}$$

where to avoid confusion we have here denoted the minimizing weight by $\underline{w}_e$ rather than $\underline{w}_o$. If we introduce the weight estimation error

$$\tilde{\underline{w}}_k = \underline{w}_k - \underline{w}_e \tag{3.11}$$

then we can rewrite (3.10) as

$$\delta \tilde{\underline{w}}_k = -\mu \underline{R}_x \tilde{\underline{w}}_{k-1} \tag{3.12}$$

or

$$\tilde{\underline{w}}_k = (I - \mu \underline{R}_x) \tilde{\underline{w}}_{k-1}.$$

This so-called error system is a vector difference equation and its stability or convergence can be analyzed in a number of ways. For the present we will analyze it by diagonalizing the system.

To do this we appeal again to the eigenvector decomposition (3.3) to rewrite (3.12) as

$$\delta \tilde{\underline{w}}_k = -\mu \sum_{u=1}^{p} \lambda_u \underline{v}_u \underline{v}_u^T \tilde{\underline{w}}_{k-1}.$$

Take the inner product of this equation with $\underline{v}_r$ and use the orthogonality of the eigenvectors to find

$$\delta \tilde{v}_{k,r} = -\mu \lambda_r \tilde{v}_{k-1,r}, \quad 1 \le r \le p \tag{3.13}$$

where we have introduced the rotated weight error

$$\tilde{v}_{k,r} = \underline{v}_r^T \tilde{\underline{w}}_k, \quad 1 \le r \le p. \tag{3.14}$$

Of course (3.13) can be written as a set of scalar difference equations

$$\tilde{v}_{k,r} = (1 - \mu \lambda_r) \tilde{v}_{k-1,r}, \quad 1 \le r \le p.$$

Iterating this gives

$$\tilde{v}_{k,r} = (1 - \mu\lambda_r)^k \tilde{v}_{0,r}, \quad 1 \le r \le p. \tag{3.15}$$

For convergence, i.e., in order that $\tilde{v}_{k,r} \to 0$ as $k \to \infty$ for $1 \le r \le p$ we need

$$
\begin{aligned}
& |1 - \mu\lambda_r| < 1, \quad 1 \le r \le p \\
\Rightarrow \quad & -1 < 1 - \mu\lambda_r < 1, \quad 1 \le r \le p \\
\Rightarrow \quad & 0 < \mu\lambda_r < 2, \quad 1 \le r \le p.
\end{aligned}
$$

This is equivalent to the pair of conditions

$$0 < \mu\lambda_{min} < \mu\lambda_{max} < 2 \tag{3.16}$$

where λ_{min}, λ_{max} are respectively the smallest and largest eigenvalues of $\underline{R}_x$. We can say then that convergence occurs if and only if

$$
\begin{aligned}
&\text{(i) } \lambda_{min} > 0 \quad \text{i.e., } \underline{R}_x \text{ has full rank} \\
&\text{and (ii) } \mu\lambda_{max} < 2.
\end{aligned}
\tag{3.17}
$$

For reasons to be made clear later the first condition is called a persistently exciting condition. The second condition puts an amplitude constraint on μ.

From (3.15) the speed of decay or convergence for mode r depends on how close $|1 - \mu\lambda_r|$ is to 1. Thus the overall speed of convergence is controlled by λ_{min} or λ_{max} because

$$
\begin{aligned}
&1 - \mu\lambda_{min} \text{ is closest to } +1 \\
&\text{and } 1 - \mu\lambda_{max} \text{ is closest to } -1.
\end{aligned}
$$

The maximum speed of convergence occurs in general, then, when μ is chosen so that

$$
\begin{aligned}
& 1 - \mu\lambda_{min} = -(1 - \mu\lambda_{max}) \\
\Rightarrow \quad & \mu_{opt} = 2(\lambda_{min} + \lambda_{max})^{-1}.
\end{aligned}
\tag{3.18}
$$

Note that (3.17) holds for this choice of μ. The maximum speed of convergence is then determined by how small is

$$
\begin{aligned}
|1 - \mu_{opt}\lambda_{max}| = |1 - \mu_{opt}\lambda_{min}| &= 1 - 2\lambda_{min}(\lambda_{min} + \lambda_{max})^{-1} \\
&= (\lambda_{max} - \lambda_{min})/(\lambda_{min} + \lambda_{max}) \\
&= (\kappa - 1)/(\kappa + 1)
\end{aligned}
$$

where we have introduced the condition number or convergence spread measure

$$\kappa = \lambda_{max}/\lambda_{min}. \tag{3.19}$$

We can say then that, in general, the speed of convergence is controlled by the eigenvalue spread in the $\underline{R}_x$ matrix. The closer κ is to 1, the faster is the possible convergence speed.

Example 3.2. Continuation of Example 3.1

Fig. 3.2 shows an iteration path on the MSE contour plot.

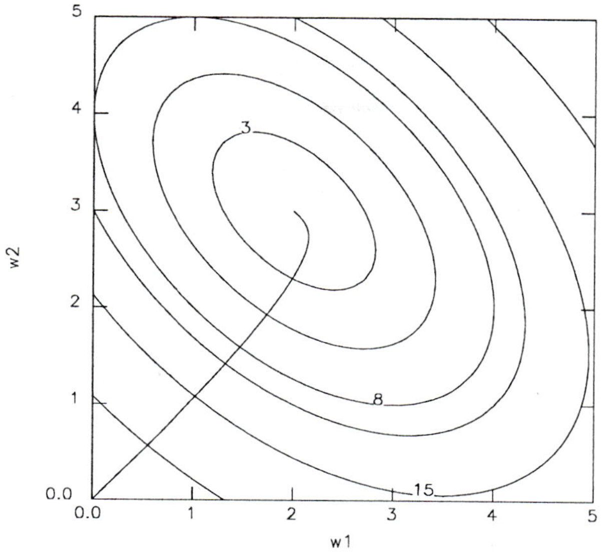

Figure 3.2 Steepest descent iteration path in two-dimensional case.

A convenient way to encapsulate the overall convergence speed of any algorithm is through the learning curve, which is a plot of the MSE at iteration k versus k. From (3.4) the MSE is given by

$$\mathcal{E}_k = \mathcal{E}_{min} + \sum_{u=1}^{p} \lambda_u \tilde{v}_{k,u}^2.$$

Substituting (3.15) gives

$$\mathcal{E}_k = \mathcal{E}_{min} + \sum_{u=1}^{p} \lambda_u (1 - \mu\lambda_u)^{2k} \tilde{v}_{0,u}^2. \tag{3.20}$$

Clearly the ultimate rate of decay or convergence of the learning curve is in general determined by the slowest mode in (3.20), i.e., by λ_{min} or λ_{max} (unless of course the corresponding rotated initial condition happens to be zero). If the step size is chosen according to (3.18) then the convergence speed is determined by κ of (3.19).

Example 3.3.

In Fig. 3.3, the effect of eigenvalue spread on speed of convergence is exhibited in the learning curve (3.20). In each case the optimal (for speed) choice of μ has been made and in each case the minimizing value of $\mathcal{E}$ is $\mathcal{E}_{min} = 2$.

$$\underline{R}_x = \begin{pmatrix} 1 & \rho \\ \rho & 1 \end{pmatrix}, \quad \underline{w}_e = \begin{pmatrix} 1 \\ 2 \end{pmatrix}, \quad R_y = 7 + 4\rho, \quad \underline{R}_{xy} = \begin{pmatrix} 1 + 2\rho \\ 2 + \rho \end{pmatrix}.$$

Also initial conditions are $\underline{w}_0 = \underline{0}$. Then

$$\lambda_{min} = 1 - |\rho|, \quad \lambda_{max} = 1 + |\rho|$$

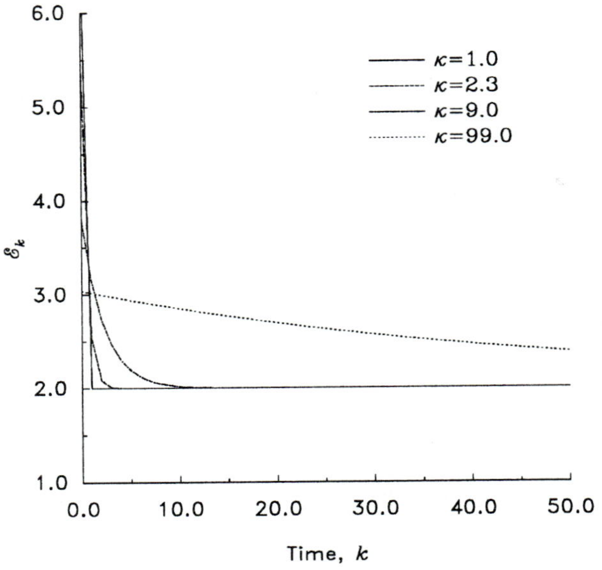

Figure 3.3 Effects of eigenvalue spread on top speed in the learning curve.

so that $\mu_{opt} = 1$ and $\kappa = (1 + |\rho|)/(1 - |\rho|)$. The maximum speed is determined by how small is

$$(\kappa - 1)/(\kappa + 1) = |\rho|.$$

Initial conditions are $v_{01} = -(1 + 2\,\mathrm{sgn}(\rho))/\sqrt{2}$, $v_{02} = -(1 - 2\,\mathrm{sgn}(\rho))/\sqrt{2}$. Eigenvectors are $1/\sqrt{2}(1 \quad \mathrm{sgn}(\rho))$ and $1/\sqrt{2}(1 \quad -\mathrm{sgn}(\rho))$. The learning curve is from (3.20):

$$\mathcal{E}_k = \mathcal{E}_{min} + |\rho|^{2k}(v_{01}^2\lambda_{min} + v_{02}^2\lambda_{max}).$$

One sees in this expression the effect of the initial conditions.

3.2.2 Persistent Excitation

In (3.17) we met conditions for convergence of the SD iteration (3.12). We now investigate their meaning a little further by considering two special cases. Let us suppose $\underline{x}_k$ is deterministic and periodic. Then we claim

$$\begin{aligned}&\underline{R}_x \text{ lacks full rank if and only if there is}\\ &\text{a vector } \underline{\alpha} \text{ which nulls } \underline{x}_k\text{, i.e., } \underline{\alpha}^T\underline{x}_k = 0 \text{, for all } k.\end{aligned} \tag{3.21}$$

Proof. If such an $\underline{\alpha}$ exists then from (2.12)

$$0 = \mathrm{av}(\underline{x}_k^T\underline{\alpha}) = \tilde{E}(\alpha^T\underline{x}_k)^2 = \underline{\alpha}^T\underline{R}_x\underline{\alpha} \quad \Rightarrow \quad \lambda_{min} = 0.$$

If $\underline{R}_x$ lacks full rank then an $\underline{\alpha}$ exists so that $\underline{\alpha}^T \underline{R}_x \underline{\alpha} = 0$. But if M is the period of $\underline{x}_k$ then

$$\sum_{k=1}^{M} (\underline{\alpha}^T \underline{x}_k)^2 = 0 \quad \Rightarrow \quad \underline{\alpha}^T \underline{x}_k = 0, \quad 1 \le k \le M$$

$$\Rightarrow \quad \underline{\alpha}^T \underline{x}_k = 0, \quad \text{for all } k, \text{ by periodicity.}$$

Now we look at two cases.

(i) A one-sided FIR filter where

$$\underline{x}_k = (x_{k-1} \cdots x_{k-p})^T$$

(ii) A noise-free IIR equation error filter setup (see Section 2.6) with

$$\underline{x}_k = (-y_{k-1} \cdots - y_{k-p}, x_{k-1} \cdots x_{k-m})^T \tag{3.22}$$

$$y_k = s_k = [1 + a(q^{-1})]^{-1} b(q^{-1}) x_k \tag{3.23}$$

For case (i) note that the result (3.21) can be rephrased as follows:

> $\underline{R}_x$ lacks full rank if and only if there is a non-zero FIR filter $\alpha(q^{-1}) = \sum_{s=1}^{p} \alpha_s q^{-s}$ which nulls x_k, i.e., $\alpha(q^{-1}) x_k = 0$, for all k.

This leads us to the following definition:

Definition 3.1. We say the deterministic periodic signal x_k is persistently exciting (PE) of order r if there is no FIR filter of order r that nulls (annihilates) x_k.

In Exercise 3.10 the reader is asked to show that

> If $F(q^{-1})$ is a stable filter (see Appendix C) then
> x_k is PE of order r if and only if $x_{Fk} = F^{-1}(q^{-1}) x_k$ (3.24)
> is PE of order r.

Example 3.4.

$x_k = \sin(\omega_o k)$, $\omega_o \neq 0$ is PE of order 2 but not PE of order 3 since $1 - 2q^{-1}\cos(\omega_o) + q^{-2}$ annihilates x_k.

We see that for a deterministic periodic signal to be PE of order $2r$ it must have r sines/cosines of distinct frequencies so that it can excite r distinct modes. This explains the name "persistent excitation."

We turn now to case (ii). The idea here is to work the full rank criterion back to a direct condition on the input signal x_k. We claim the following (noting that two polynomials are co-prime if they have no common factors):

If $b(q^{-1})$, $1+a(q^{-1})$are co-prime then $\underline{R}_x$ has full rank if and only if x_k is PE of order $\geq p+m$.

Proof. If $\underline{R}_x$ lacks full rank there is a vector $\underline{\alpha}$ with $\underline{\alpha}^T\underline{x}_k=0$ for all k. That is, there are FIR filters $\alpha_0(q^{-1})$ of order $p-1$, $\alpha_1(q^{-1})$ of order $m-1$ with

$$-\alpha_0(q^{-1})y_k+\alpha_1(q^{-1})x_k=0.$$

Now use (3.23) to see

$$\begin{aligned} &[-\alpha_0(q^{-1})b(q^{-1})(1+a(q^{-1}))^{-1}+\alpha_1(q^{-1})]x_k=0 \\ \Rightarrow\quad &[-\alpha_0(q^{-1})b(q^{-1})+\alpha_1(q^{-1})(1+a(q^{-1}))]\tilde{x}_k=0 \end{aligned} \tag{3.25}$$

where

$$\tilde{x}_k=[1+a(q^{-1})]^{-1}x_k.$$

Now the co-primeness of $b(q^{-1})$, $1+a(q^{-1})$ ensures there do not exist filters $\alpha_0(q^{-1})$ of order $p-1$, $\alpha_1(q^{-1})$ of order $m-1$ such that the quantity in square brackets in (3.25) vanishes (why?). Thus there is a non-null FIR filter of order $p+m$ which annihilates $\tilde{x}_k$, and so $\tilde{x}_k$ cannot be PE of order $\geq p+m$. Then according to (3.24), x_k cannot be PE of order $\geq p+m$. If x_k is PE of order $\geq p+m$ we work the above argument in reverse. This completes the proof.

3.2.3 Improvements to Steepest Descent

Let us suppose we could make the iteration index in (3.9) or (3.12) continuous. Denoting this new index by t we would have $\underline{w}=\underline{w}(t)$ and

$$d\underline{w}/dt=\mu(\underline{R}_{xy}-\underline{R}_x\underline{w}) \tag{3.26}$$

or using the weight error $\underline{\tilde{w}}=\underline{\tilde{w}}(t)=\underline{w}(t)-\underline{w}_o$,

$$d\underline{\tilde{w}}/dt=-\mu\underline{R}_x\underline{\tilde{w}}. \tag{3.27}$$

Applying the spectral decomposition (3.3) to this leads to

$$\begin{aligned} d\tilde{v}_u/dt&=-\mu\lambda_u\tilde{v}_u,\quad 1\leq u\leq p \\ \tilde{v}_u&=\tilde{v}_u(t)=\underline{v}_u^T\underline{\tilde{w}}(t). \end{aligned} \tag{3.28}$$

These scalar differential equations have solution

$$\tilde{v}_u(t)=e^{-\mu\lambda_u t}\tilde{v}_u(0),\ 1\leq u\leq p$$

so that, in order that $\tilde{v}_u(t)\to 0$ as $t\to\infty$ we need only $\mu\lambda_u>0$, $1\leq u\leq p$, that is

$$\lambda_{min}>0\quad\Longleftrightarrow\quad \underline{R}_x \text{ has full rank.} \tag{3.29}$$

This can be compared with (3.17) to see that the amplitude bound on μ is not required. If we now view (3.12) as an approximation to (3.27) we see that the amplitude constraint on μ in (3.17) is due to approximating a derivative by a simple difference. It is thus natural to expect that a better approximation would increase the size of the upper bound. This issue arises in the numerical analysis of stiff differential equations and leads to what we shall call differentiation filter methods.

3.2.4 Differentiation Filter Methods

Let us consider then how to approximate the differentiation in (3.26). The simplest method is the forward or explicit scheme (3.9) or (3.12). It is called explicit because the new value is given explicitly in terms of the old. To approximate the differentiation while retaining only (3.29) as a stability requirement, the simplest method is an implicit method as follows

$$\delta \underline{w}_k = \mu(\underline{R}_{xy} - \underline{R}_x \underline{w}_k). \tag{3.30}$$

It is called implicit because $\underline{w}_k$ appears on both sides of (3.30). Substituting (2.5) into (3.30) gives

$$\begin{aligned} \delta \underline{w}_k = \delta \underline{\tilde{w}}_k &= -\mu \underline{R}_x \underline{\tilde{w}}_k \\ \Rightarrow \qquad \underline{\tilde{w}}_k &= \underline{\tilde{w}}_{k-1} - \mu \underline{R}_x \underline{\tilde{w}}_k. \end{aligned}$$

This is easily solved to yield an error system

$$\underline{\tilde{w}}_k = (I + \mu \underline{R}_x)^{-1} \underline{\tilde{w}}_{k-1}. \tag{3.31}$$

Using the eigenvector decomposition (3.3) once more leads to, via (3.14),

$$\tilde{v}_{k,r} = (1 + \mu\lambda_r)^{-1} \tilde{v}_{k-1,r}, \quad 1 \le r \le p$$

and iterating gives

$$\tilde{v}_{k,r} = (1 + \mu\lambda_r)^{-k} \tilde{v}_{0,r}, \quad 1 \le r \le p.$$

We see immediately that convergence follows if only (3.29) holds. So no constraint is placed on the size of μ. Of course in (3.31) we have not achieved much because there is a matrix inversion. But in the adaptive setting we will find that the implicit method does not have this disadvantage.

Unfortunately implicit methods only work in linear settings and so we must look at other ways to expand the stability bound in (3.17).

This is accomplished by means of differentiation filters. In the frequency domain the differentiation operator is $j\omega$ while the difference operator is $1 - e^{j\omega}$. At low frequencies they agree well but not at higher frequencies. The aim then is to find a higher order differentiation filter that matches $j\omega$ over a greater frequency band than $1 - e^{-j\omega}$ does.

In place of (3.30) we are led to try

$$\phi(q^{-1}) \underline{w}_k = \mu(\underline{R}_{xy} - \underline{R}_x \underline{w}_{k-1})$$

where $\phi(e^{-j\omega})$ will be chosen to be a good approximation of $j\omega$. If we take $\phi(q^{-1})$ to be rational (i.e., a ratio of finite order polynomials) then we are led to a differentiation filter method of the form

$$\bar{\alpha}(q^{-1})\underline{w}_k = \mu\beta(q^{-1})(\underline{R}_{xy} - \underline{R}_x\underline{w}_k)$$

$$\bar{\alpha}(q^{-1}) = \sum_{r=0}^{m} \alpha_r q^{-r}$$

$$\beta(q^{-1}) = \sum_{r=1}^{m} \beta_r q^{-r}$$

and to avoid bias we require

$$\bar{\alpha}(1) = \sum_{r=0}^{m} \alpha_r = 0$$

$$\beta(1) = \sum_{r=1}^{m} \beta_r = 1.$$

The first condition tells us that $q = 1$ is a root of $\bar{\alpha}(q^{-1})$ so we can factor $\bar{\alpha}(q^{-1}) = \alpha(q^{-1})(1 - q^{-1})$. This immediately allows us to rewrite the algorithm as

$$\alpha(q^{-1})\delta\underline{w}_k = \mu\beta(q^{-1})(\underline{R}_{xy} - \underline{R}_x\underline{w}_k)$$

and leads to an error system

$$\alpha(q^{-1})\delta\underline{\tilde{w}}_k = -\mu\underline{R}_x\beta(q^{-1})\underline{\tilde{w}}_k.$$

Once more applying the eigenvector decomposition (3.3) leads to a set of scalar difference equations

$$\alpha(q^{-1})\delta\tilde{v}_{k,r} = -\mu\lambda_r\beta(q^{-1})\tilde{v}_{k,r}, \quad 1 \le r \le p$$

which can be rewritten

$$[\alpha(q^{-1})(1 - q^{-1}) + \mu\lambda_r\beta(q^{-1})]\tilde{v}_{k,r} = 0, \quad 1 \le r \le p. \tag{3.32}$$

Stability of these difference equations requires (Appendix C)

$$\begin{aligned} &\text{all roots of } q^p[(1 - q^{-1})\alpha(q^{-1}) + \mu\lambda_r\beta(q^{-1})] \\ &\text{have modulus} < 1, \quad 1 \le r \le p \end{aligned} \tag{3.33}$$

That is, in the complex q plane all these roots must be inside the unit circle.

The position of the roots clearly depends on the values $\mu\lambda_r$, $1 \le r \le p$ and root locus techniques of control theory can be used to plot them out. For the present we just consider a second order example.

A very simple differentiation filter is the so-called momentum filter

$$\alpha(q^{-1}) = (1 - \alpha q^{-1})$$

$$\beta(q^{-1}) = q^{-1}.$$

To investigate stability we consider (3.32) leading to

$$\begin{aligned} & q^2[(1-\alpha q^{-1})(1-q^{-1})+\mu\lambda_r q^{-1}] \\ &= q^2-(\alpha+1)q+\alpha+\mu\lambda_r q \\ &= q^2-[1+\alpha-\mu\lambda_r]q+\alpha. \end{aligned}$$

From here we rapidly find stability ensues when

$$|\alpha|<1, \qquad 0<\mu\lambda_r<2+2\alpha, \quad 1\leq r\leq p.$$

As compared to (3.17) we see the stability boundary is expanded to allow (when $\alpha>0$)

$$\mu\lambda_{max}<2+2\alpha. \tag{3.34}$$

It is worth noting here that, although the stability boundary is expanded, the speed of convergence slows as α nears 1, at least when the roots ξ_1, ξ_2 are complex. Since then

$$|\xi_1|=|\xi_2|=\alpha.$$

In Exercise 3.7 it is pointed out that the momentum algorithm is related to the conjugate gradient iteration.

Example 3.5. Momentum SD

We use the setup of Example 3.1. We take $\alpha=0.5$, $\mu=2/3$ so that for SD

$$\mu\lambda_1=2/3, \quad \mu\lambda_2=2.$$

Thus, (3.17) is just violated. In rotated coordinates (3.15) becomes

$$\begin{aligned} \tilde{v}_{k,1} &= (1-\mu\lambda_1)^k\tilde{v}_{0,1}=(\frac{1}{3})^k\tilde{v}_{0,1} \\ \tilde{v}_{k,2} &= (1-\mu\lambda_2)^k\tilde{v}_{0,2}=(-1)^k\tilde{v}_{0,2}. \end{aligned}$$

So convergence does not occur. This is shown in Fig. 3.4 where SD hops between lines AB, $A'B'$. Note that initial conditions are

$$\begin{aligned} v_{0,1} &= -\underline{v}_1^T\underline{w}_e=-\frac{1}{\sqrt{2}}(1 \quad -1)(2 \quad 3)^T=\frac{1}{\sqrt{2}} \\ v_{0,2} &= -\underline{v}_2^T\underline{w}_e=-\frac{1}{\sqrt{2}}(1 \quad 1)(2 \quad 3)^T=-\frac{5}{\sqrt{2}}. \end{aligned}$$

For the momentum SD algorithm we see that (3.34) is satisfied (since $\mu\lambda_2=2<3$). In rotated coordinates the algorithm is

$$\begin{aligned} & \tilde{v}_{k,r}=(1.5-\mu\lambda_r)\tilde{v}_{k-1,r}+0.5\tilde{v}_{k-2,r}, \quad r=1,2 \\ \Rightarrow \quad & \tilde{v}_{k,1}=0.83\tilde{v}_{k-1,r}+0.5\tilde{v}_{k-2,r}, \quad k\geq 2 \\ & \tilde{v}_{k,2}=-0.5\tilde{v}_{k-1,r}+0.5\tilde{v}_{k-2,r}, \quad k\geq 2. \end{aligned}$$

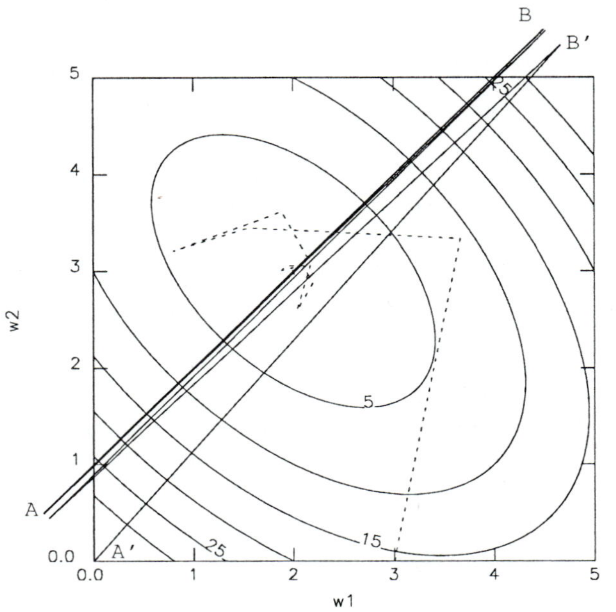

Figure 3.4 Momentum SD (dashed line) and ordinary SD (solid line) which is unstable.

Initial conditions can be taken as $\underline{w}_0 = \underline{w}_1 = \underline{0}$ so that

$$v_{0,1} = v_{1,1} = -\frac{1}{\sqrt{2}}$$

$$v_{0,2} = v_{1,2} = -\frac{5}{\sqrt{2}}.$$

3.2.5 Loop Filter Methods

Aside from being concerned about the size of the stability region we should also be concerned about the speed of convergence. In (3.28) or its solution the speed of convergence is defined by the time constants $\tau_u = (\mu\lambda_u)^{-1}$, $1 \le u \le p$. We can alter these by inserting a filter on the right hand side of (3.26) or (3.27) yielding

$$d\underline{w}/dt = \mu(1 + H(D/\mu))(\underline{R}_{xy} - \underline{R}_x\underline{w}) \tag{3.35}$$

or

$$d\underline{\tilde{w}}/dt = -\mu(1 + H(D/\mu))\underline{R}_x\underline{\tilde{w}}(t) \tag{3.36}$$

where $H(D)$ is a proper transfer function, i.e., one that rolls over at high frequency

$$H(D) = \sum_{u=1}^{p} b_u D^{p-u} / (D^p + \sum_{u=1}^{p} a_u D^{p-u}), \tag{3.37}$$

and $D = d/dt$. Applying the eigenvector decomposition (3.3) once more leads to

$$d\tilde{v}_u/dt = -\mu\lambda_u(1 + H(D/\mu))\tilde{v}_u. \tag{3.38a}$$

The idea now is that $H(D)$ should be chosen to give the system (3.38a) a greater bandwidth (i.e., quicker reaction time) than (3.28). Of course the system (3.38a) still has to be stable.

We can give some insight into this modification by representing it in a feedback control framework.

Fig. 3.5a shows the "open loop" algorithm (3.28). Fig. 3.5b shows the algorithm with a feedback controller of transfer function $\mu\lambda_u H(D/\mu)$. The closed loop transfer relation of Fig. 3.5b is

$$G_u(D) = \frac{(D + \mu\lambda_u)^{-1}}{1 + (D + \mu\lambda_u)^{-1}\mu\lambda_u H(D/\mu)} = [D + \mu\lambda_u + \mu\lambda_u H(D/\mu)]^{-1} \tag{3.38b}$$

which corresponds to (3.38a). So the choice of $H(D/\mu)$ is just the choice of a feedback controller to increase the bandwidth of a first order all pole system. There are many design methods that could be chosen to accomplish this. A simple example is given below.

If $H(D)$ has a state space representation

$$H(D) = \underline{c}^T(DI - \underline{A})^{-1}\underline{b}$$

then (3.35) can be written

$$\dot{\underline{M}} = \mu(\underline{AM} + \underline{b}(\underline{R}_{xy} - \underline{R}_x\underline{w})^T)$$
$$\dot{\underline{w}} = \mu\underline{M}^T\underline{c} + \mu(\underline{R}_{xy} - \underline{R}_x\underline{w}).$$

A discrete version (using a first order differentiation filter) would be

$$\delta\underline{M}_k = \mu(\underline{AM}_{k-1} + \underline{b}(\underline{R}_{xy} - \underline{R}_x\underline{w}_{k-1})^T) \tag{3.39}$$
$$\delta\underline{w}_k = \mu\underline{M}_k^T\underline{c} + \mu(\underline{R}_{xy} - \underline{R}_x\underline{w}_{k-1}). \tag{3.40}$$

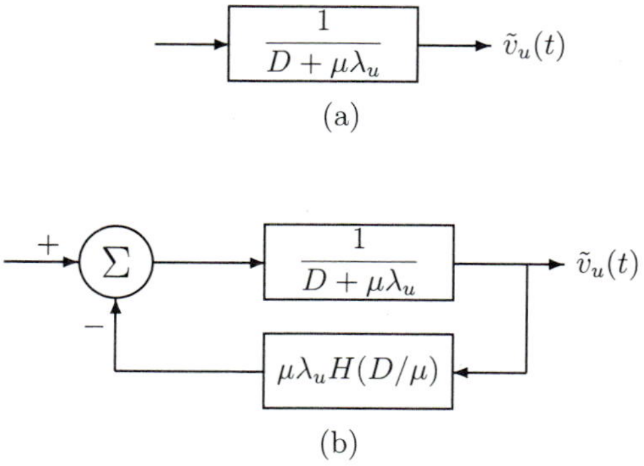

Figure 3.5 (a) SD as an open loop system. (b) Loop Filter as a feedback controller.

This can also be written

$$\delta \underline{w}_k = \mu[H(\delta/\mu) + 1](\underline{R}_{xy} - \underline{R}_x \underline{w}_{k-1}).$$

In rotated coordinates this can be written, for $1 \le u \le p$

$$\delta \underline{V}_{k,u} = \mu(\underline{A}\underline{V}_{k-1,u} - \underline{b}\lambda_u \tilde{v}_{k-1,u}) \tag{3.41}$$

$$\delta \tilde{v}_{k,u} = \mu \underline{c}^T \underline{V}_{k,u} - \mu\lambda_u \tilde{v}_{k-1,u} \tag{3.42}$$

or

$$\delta \tilde{v}_{k,u} = -\mu[H(\delta/\mu) + 1]\lambda_u \tilde{v}_{k-1,u} \qquad 1 \le u \le p. \tag{3.43}$$

To make these calculations more concrete let us consider a first order loop filter algorithm with

$$H(D) = b/(D + a).$$

The closed loop transfer relation is

$$G_u(D) = [D + \mu\lambda_u + \mu\lambda_u b/(D\mu^{-1} + a)]^{-1}$$
$$= \frac{D + \mu a}{D^2 + \mu(\lambda_u + a)D + \mu^2\lambda_u(a + b)}.$$

Before proceeding further it is important to keep an eye on the scaling of μ. Let us rescale $\underline{R}_x$ as

$$\underline{R}_x = \gamma_x \bar{\underline{R}}_x$$

where

$$\gamma_x = \mathrm{tr}(\underline{R}_x)/p = \text{average } x \text{ variance}.$$

Then $\bar{\underline{R}}_x$ has scaled eigenvalues $\bar{\lambda}_u$ with

$$\lambda_u = \gamma_x \bar{\lambda}_u, \quad 1 \le u \le p.$$

It is important to note that this rescaling ensures

$$0 \le \bar{\lambda}_{min} \le 1 \tag{3.44}$$

since

$$\mathrm{tr}(\bar{\underline{R}}_x) = p = \sum_{u=1}^{p} \bar{\lambda}_u.$$

Continuing, we rescale μ as

$$\mu_0 = \gamma_x \mu$$

and rescale a and b as

$$a = \gamma_x \bar{a}$$
$$b = \gamma_x \bar{b}.$$

Then we have

$$G_u(D) = \frac{D + \mu_0 \bar{a}}{D^2 + \mu_0(\bar{\lambda}_u + \bar{a})D + \mu_0^2 \bar{\lambda}_u(\bar{a} + \bar{b})}. \tag{3.45}$$

Now let us insist that

$$0 < \mu_0 \bar{\lambda}_{min} = \mu \lambda_{min} < 1.$$

Then the slowest mode is $1 - \mu_0 \bar{\lambda}_{min}$.

We can now regard the job of the loop filter modification as that of speeding up the slowest mode. Now we insist that the poles of (3.45) be complex, i.e.,

$$\begin{aligned} & [\mu_0(\bar{\lambda}_{min} + \bar{a})]^2 < 4\mu_0^2 \bar{\lambda}_{min}(\bar{a} + \bar{b}) \\ \Rightarrow \quad & (\bar{\lambda}_{min} + \bar{a})^2 < 4\bar{\lambda}_{min}\bar{a} + 4\bar{\lambda}_{min}\bar{b} \\ \Rightarrow \quad & (\bar{\lambda}_{min} - \bar{a})^2 < 4\bar{\lambda}_{min}\bar{b} \end{aligned}$$

which we ensure by insisting that

$$\begin{aligned} & \bar{a} - \bar{\lambda}_{min} < 2(\bar{\lambda}_{min}\bar{b})^{\frac{1}{2}} \\ \Rightarrow \quad & 0 < \bar{a} < \bar{\lambda}_{min} + 2(\bar{\lambda}_{min}\bar{b})^{\frac{1}{2}}. \end{aligned} \tag{3.46}$$

We can thus rewrite (3.45) (for $\bar{\lambda}_u = \bar{\lambda}_{min}$) as

$$G(D) = \frac{D + \mu_0 \bar{a}}{D^2 + 2\zeta\omega_n D + \omega_n^2} \tag{3.47}$$

where

$$\omega_n = \mu_0 \bar{\lambda}_{min}^{\frac{1}{2}} (\bar{a} + \bar{b})^{\frac{1}{2}} \tag{3.48a}$$

$$2\zeta\omega_n = \mu_0(\bar{\lambda}_{min} + \bar{a}). \tag{3.48b}$$

The poles are then

$$-\zeta\omega_n \pm j\omega_n\sqrt{1 - \zeta^2}.$$

If we insist that

$$\bar{\lambda}_{min} \ll \bar{a} \tag{3.49a}$$

then we see from (3.48) that

$$-\mu_0 \bar{a} \ll -\zeta\omega_n \tag{3.49b}$$

so that the zero in (3.47) is far to the left of the poles. Thus the transient characteristics of $G(D)$ are those of a second order system (see section 8.3 of [D1]).

The reaction time of (3.26) or (3.27) is measured by its largest time constant

$$\tau_{SD} = (\mu_0 \bar{\lambda}_{min})^{-1}.$$

The reaction time of (3.47) is measured (in view of (3.49)) by

$$\tau_{LF} = (\zeta\omega_n)^{-1} = \left[\frac{1}{2}\mu_0(\bar{\lambda}_{min} + \bar{a})\right]^{-1}$$

and so it follows from (3.49) that

$$\tau_{LF} \ll \tau_{SD}$$

so that the loop filter method produces a much quicker reaction. The oscillatory nature of (3.47) introduces overshoot and we investigate this by first calculating the damping ratio from (3.48):

$$\zeta_{min} = \frac{\mu_0(\bar{\lambda}_{min} + \bar{a})}{2\mu_0\bar{\lambda}_{min}^{\frac{1}{2}}(\bar{a} + \bar{b})^{\frac{1}{2}}}.$$

Coupling (3.49a) and (3.46) and noting (3.44) we are led to set

$$\bar{a} = 2(\bar{\lambda}_{min}\bar{b})^{\frac{1}{2}} \tag{3.50}$$

so that

$$\begin{aligned}\zeta_{min} &= \frac{\bar{\lambda}_{min} + 2(\bar{\lambda}_{min}\bar{b})^{\frac{1}{2}}}{2\bar{\lambda}_{min}^{\frac{1}{2}}(2(\bar{\lambda}_{min}\bar{b})^{\frac{1}{2}} + \bar{b})^{\frac{1}{2}}} \\ &= \frac{\bar{\lambda}_{min}^{\frac{1}{2}} + 2\bar{b}^{\frac{1}{2}}}{2\bar{b}^{\frac{1}{4}}(2\bar{\lambda}_{min}^{\frac{1}{2}} + \bar{b}^{\frac{1}{2}})^{\frac{1}{2}}}.\end{aligned}$$

Let us set

$$\bar{b} = 1 \tag{3.51}$$

then we find

$$\zeta_{min} = \frac{\bar{\lambda}_{min}^{\frac{1}{2}} + 2}{2(2\bar{\lambda}_{min}^{\frac{1}{2}} + 1)^{\frac{1}{2}}}. \tag{3.52}$$

We need specific values of $\bar{\lambda}_{min}$ to see if this leads to overshoot problems.

We should now look at the effect of these choices on the other modes. However the details are rather tedious. Instead we illustrate with a simple example having just two modes.

Example 3.6. Loop Filter Algorithm

We consider Example 3.3. We have $\gamma_x = 1$, $b = \gamma_x\bar{b} = 1$. With $|\rho| = 0.98$, we have $\kappa = 1.98/0.02 = 99$ and

$$\bar{\lambda}_{min} = \lambda_{min} = 0.02, \quad \bar{\lambda}_{min}^{\frac{1}{2}} = 0.1414$$

$$\lambda_{max} = 1.98$$

$$a = \gamma_x\bar{a} = 2 \times 0.1414 = 0.2828, \quad \text{from (3.50).}$$

From (3.52)

$$\zeta_{min} = \frac{2.1414}{2 \times \sqrt{1.2828}} = \frac{1.0707}{1.1326} = 0.945.$$

The overshoot is thus negligible (see section 3.11 of [D1]). The original time constant was

$$\tau_{SD} = (\mu\lambda_{min})^{-1} = 50\mu^{-1}.$$

The new time constant (of (3.45) and (3.47)) is

$$\tau_{LF} = (0.5 \times \mu(\lambda_{min} + a))^{-1} = 4.71\mu^{-1}.$$

The other mode has complex roots (cf. (3.46) and (3.51)) since

$$\begin{aligned} & |a - \lambda_{max}| < 2(\lambda_{max}\bar{b})^{\frac{1}{2}} = 2\lambda_{max}^{\frac{1}{2}} \\ \Rightarrow \quad & |0.2828 - 1.98| < 2(1.98)^{\frac{1}{2}} \\ \Rightarrow \quad & 1.6972 < 2.814. \end{aligned}$$

The damping ratio of the other mode is

$$\begin{aligned} \zeta_{max} &= \frac{\lambda_{max} + a}{2\lambda_{max}^{\frac{1}{2}}(a + b)^{\frac{1}{2}}} = \frac{1.98 + 0.2828}{2 \times \sqrt{1.98 \times 1.2828}} \\ &= 0.409 \end{aligned}$$

which yields an overshoot of about 20%. The time constant of this other mode is

$$\left(\frac{1}{2}\mu(\lambda_{max} + a)\right)^{-1} = 1.13\mu^{-1} \quad \text{versus} \quad 0.5\mu^{-1}.$$

To implement the discrete version an implicit method was used for simplicity:

$$[\delta^2 + \mu(\lambda_u + a)\delta + \mu^2\lambda_u(a + b)]\tilde{v}_{k,u} = 0, \quad u = 1, 2$$

or

$$\begin{aligned} \tilde{v}_{k,u} &= \frac{B}{A}\tilde{v}_{k-1,u} + \frac{C}{A}\tilde{v}_{k-2,u} \\ A &= 1 + \mu(\lambda_u + a) + \mu^2\lambda_u(a + b) \\ B &= 2 + \mu(\lambda_u + a) \\ C &= -1. \end{aligned}$$

Results are shown in Fig. 3.6 and can be compared with Fig. 3.3.

3.2.6 Relaxation or Leakage Algorithm

Since stability or convergence is such a fundamental concern in adaptive algorithm behavior and since instability is manifested by large values of $\underline{w}_k$, a natural idea is to penalize the size of $\underline{w}_k$ during the iteration. Thus, one is led to minimize not MSE but a modified MSE

$$\mathcal{E}_\alpha(\underline{w}) = \mathcal{E}(\underline{w}) + \alpha\|\underline{w}\|^2$$

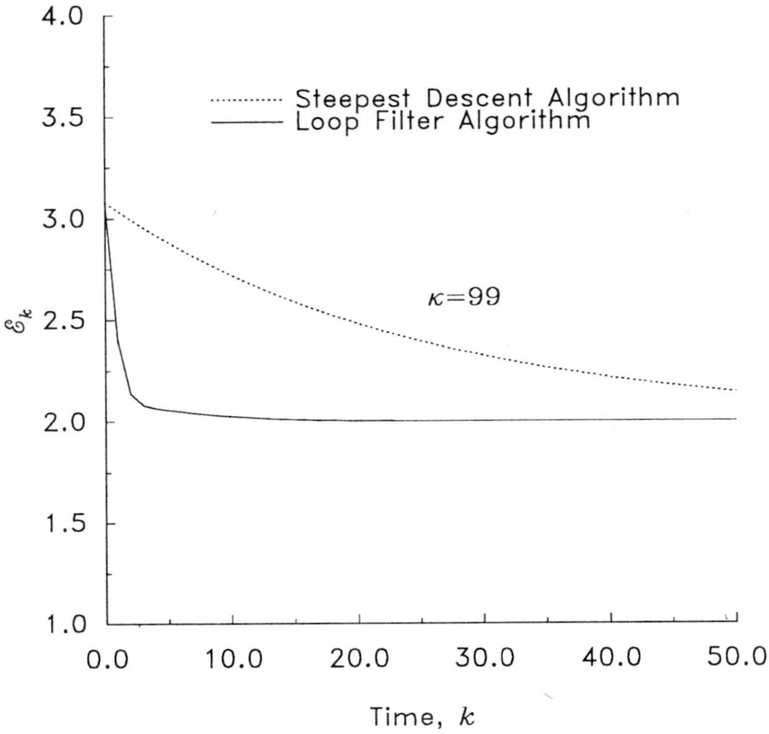

Figure 3.6 Convergence speed comparison of loop filter method and steepest descent method.

where $\alpha > 0$ is a penalty term controlling the weight given to the size of $\underline{w}$. Thus in (3.7) the gradient becomes

$$\begin{aligned} \underline{g}_\alpha(\underline{w}) &= d\mathcal{E}_\alpha/d\underline{w} \\ &= d\mathcal{E}/d\underline{w} + 2\alpha\underline{w} = \underline{g}(\underline{w}) + 2\alpha\underline{w}. \end{aligned}$$

Thus the SD algorithm takes the form

$$\delta\underline{w}_k = -(\frac{1}{2}\mu\underline{g}(\underline{w}_{k-1}) + \alpha\mu\underline{w}_{k-1}).$$

For an FIR filter this becomes

$$\delta\underline{w}_k = \mu(\underline{R}_{xy} - \underline{R}_x\underline{w}_{k-1}) - \alpha\mu\underline{w}_{k-1}.$$

The extra term here is called a relaxation or leakage factor. The error form (3.12) then becomes

$$\delta\underline{\tilde{w}}_k = -\mu\underline{R}_x\underline{\tilde{w}}_{k-1} - \alpha\mu(\underline{\tilde{w}}_{k-1} + \underline{w}_e)$$

or

$$\underline{\tilde{w}}_k = (I(1-\alpha\mu) - \mu\underline{R}_x)\underline{\tilde{w}}_{k-1} - \alpha\mu\underline{w}_e.$$

Applying the eigenvector decomposition (3.3) leads to

$$\tilde{v}_{k,r} = (1-\alpha\mu-\mu\lambda_r)\tilde{v}_{k-1,r} - \alpha\mu\tilde{v}_{e,r}, \quad 1 \le r \le p.$$

Convergence now requires

$$0 < \mu(\lambda_r + \alpha) < 1, \quad 1 \le r \le p$$

and yields a steady state value

$$\tilde{v}_r = \frac{\alpha}{\alpha+\lambda_r}\tilde{v}_{e,r}, \quad 1 \le r \le p.$$

Thus, the relaxation algorithm does not converge to the weight minimizing MSE, i.e., it has a bias. At first sight this makes the algorithm unattractive. But in the adaptive setting where a noisy gradient must be used it can be advantageous to allow a small bias if that makes it possible to reduce the variance substantially and so lead to an overall reduction in MSE.

3.2.7 Newton's Method

In this procedure the iterations are generated by

$$\delta\underline{w}_k = -\mu\underline{H}^{-1}(\underline{w}_{k-1})\underline{g}(\underline{w}_{k-1}) \tag{3.53}$$

where $\underline{g}(\underline{w})$ is the gradient (3.8) and $\underline{H}(\underline{w})$ is the Hessian

$$\underline{H}(\underline{w}) = d^2\mathcal{E}/d\underline{w}d\underline{w}^T \tag{3.54}$$

(see Appendix A for vector differentiation). In the present setting we have from (3.1) that

$$\begin{aligned}\underline{g}(\underline{w}) &= d\mathcal{E}/d\underline{w} = 2\underline{R}_x(\underline{w}-\underline{w}_e)\\ \underline{H}(\underline{w}) &= 2\underline{R}_x.\end{aligned}$$

Thus

$$\underline{H}^{-1}(\underline{w})\underline{g}(\underline{w}) = (\underline{w}-\underline{w}_e).$$

Thus the iteration is

$$\begin{aligned}\delta\underline{w}_k &= -(\underline{w}_{k-1}-\underline{w}_e)\\ \Rightarrow \qquad \underline{w}_k &= \underline{w}_e.\end{aligned}$$

That is, Newton's method (NM) converges in one step! Of course Newton's method requires the inversion of $\underline{R}_x$ so this one step convergence is not so surprising. The one-step convergence does not give us a good sense of the difference between SD and NM. To get such a sense we modify the NM algorithm to

$$\delta\underline{w}_k = -\mu\underline{H}^{-1}(\underline{w}_{k-1})\underline{g}(\underline{w}_{k-1}).$$

In the present case this becomes

$$\begin{aligned} \delta\underline{w}_k &= -\mu(\underline{w}_{k-1} - \underline{w}_e) \\ \Rightarrow \qquad \underline{\tilde{w}}_k &= (1-\mu)\underline{\tilde{w}}_{k-1} \end{aligned}$$

where

$$\underline{\tilde{w}}_k = \underline{w}_k - \underline{w}_e.$$

Iterating yields

$$\underline{\tilde{w}}_k = (1-\mu)^k \underline{\tilde{w}}_0.$$

If we introduce the rotated weight vector (3.14) we get

$$\tilde{v}_{k,r} = (1-\mu)^k \tilde{v}_{0,r}.$$

Comparing this to (3.15) shows that all modes of the modified NM converge at the same rate. Also convergence requires that

$$-1 < 1-\mu < 1 \Longleftrightarrow 0 < \mu < 2.$$

The learning curve becomes

$$\begin{aligned} \mathcal{E}_k &= \mathcal{E}_{min} + \sum_{u=1}^{p} \lambda_u \tilde{v}_{k,u}^2 \\ &= \mathcal{E}_{min} + (1-\mu)^{2k} \sum_{u=1}^{p} \lambda_u \tilde{v}_{0,u}^2. \end{aligned}$$

So we see that the eigenvalue spread has no effect on the rate of convergence. We expect then that the modified NM converges faster than SD. *Although if the eigenvalue spread is mild, there may be only a slight improvement in speed.*

3.3 NONQUADRATIC ERROR SURFACE

There are two ways in which a non-quadratic error surface occurs:

(i) Use of a quadratic error criterion on a problem where the error signal is a nonlinear function of the parameters.

(ii) Use of an error criterion that is not squared error.

As an example of the first sort we look at time delay estimation. For the second type we consider blind equalization.

3.3.1 Steepest Descent

We consider the noise-free time delay estimation problem with no attenuation (see section 2.4). Then in the absence of attenuation the MSE has the form (see (2.43))

$$\mathcal{E}(\theta) = R_y(0) + R_x(0) - 2R_{xy}(\hat{\theta}).$$

The SD algorithm now takes the form

$$\begin{aligned}\delta\hat{\theta}_k &= -\frac{1}{2}\mu d\mathcal{E}/d\hat{\theta}_{k-1}\\ &= \mu R'_{xy}(\hat{\theta}_{k-1}).\end{aligned}$$

To convert this into an error system we appeal to (2.43) and (2.44) which yield that

$$R'_{xy}(\hat{\theta}) = -R'_s(\theta_o - \hat{\theta}) = R'_s(\hat{\theta} - \theta_o) \tag{3.55}$$

so that, introducing the estimation error $\tilde{\theta} = \hat{\theta} - \theta_o$ we find

$$\delta\tilde{\theta}_k = \mu R'_s(\tilde{\theta}_{k-1}). \tag{3.56}$$

In Fig. 3.7 is a sketch of a typical MSE, and a typical normalized autocorrelation function is shown in Fig. 3.9. Let us consider the convergence of the iteration (3.56) using Fig. 3.9 as a guide. In the region $0B$, $R'_s(\tilde{\theta})$ is negative, and so if $\tilde{\theta}_{k-1}$ lies, say, inside $0B$, then if μ is small enough, $\tilde{\theta}_k$ lies closer to the maximum at 0. In a region such as BD, $R'_s(\tilde{\theta})$ is positive, and so $\tilde{\theta}_k$ is driven towards D. From this argument we see that if we start inside $B'0B$ and μ is small enough, then the iteration converges to A as required.

If we write (3.56) as

$$\tilde{\theta}_k = [1 + \mu R'_s(\tilde{\theta}_{k-1})/\tilde{\theta}_{k-1}]\tilde{\theta}_{k-1}$$

then we see that a sufficient condition on μ to guarantee convergence is

$$0 < \mu\lambda < 2$$

$$\lambda = \max_{\tilde{\theta}} \left\{ -\frac{R'_s(\tilde{\theta})}{\tilde{\theta}} \right\}.$$

Let us now turn to the blind equalization problem.

3.3.2 Blind Equalization

We now look at iterative maximization of the non-quadratic criterion $J_2(\underline{w})$ discussed in Section 2.3. There it was found that $J_2(\underline{w})$ (actually $J_2(\underline{t})$) possesses a countably infinite

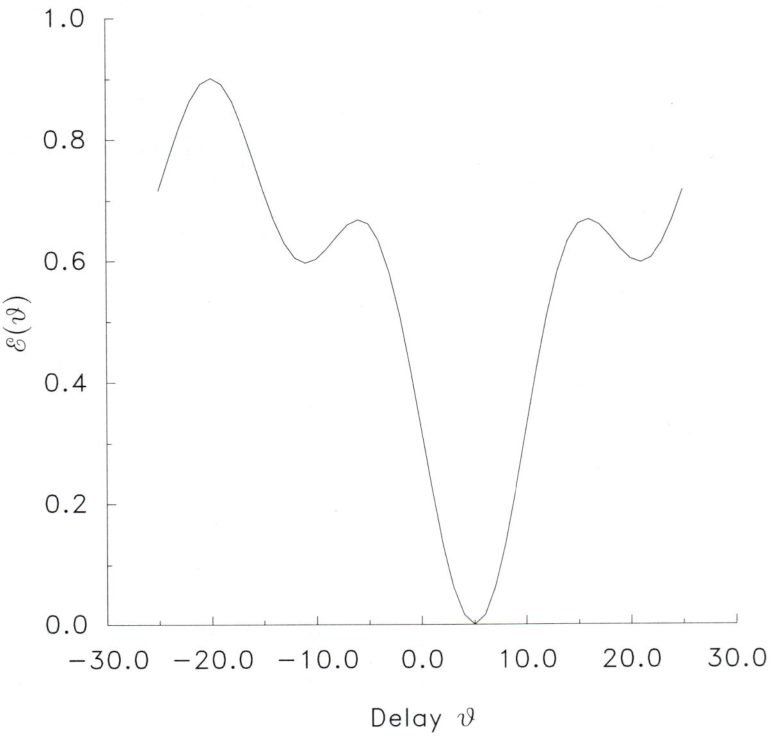

Figure 3.7 Typical $\mathcal{E}(\theta)$ for time delay estimation, optimal delay is $\theta_o = 5.0$.

set of stationary points. It is thus logical to wonder if an iterative algorithm for minimizing $J_2(\underline{w})$ will get stuck at a local minimum.

The iterative algorithm will take the form

$$\delta \underline{w}_k = -\mu \left. \frac{d J_2}{d \underline{w}} \right|_{\underline{w}=\underline{w}_{k-1}} \tag{3.57}$$

and we are interested in the convergence or stability of this nonlinear iteration. Stability is discussed formally in Part II of the book so here we take a heuristic approach. A Taylor series expansion about a stationary point $\underline{w}_*$ tells us that

$$\delta \underline{w}_k \approx \delta \underline{\tilde{w}}_k \approx -\mu J^*_{ww} \underline{\tilde{w}}_{k-1} \tag{3.58}$$

where

$$J^*_{ww} = \left. \frac{d^2 J_2}{d \underline{w} d \underline{w}^T} \right|_{\underline{w}=\underline{w}_*}$$

and we have introduced the weight error

$$\tilde{\underline{w}}_k = \underline{w}_k - \underline{w}_*.$$

From (3.58) it follows that if J^*_{ww} is positive definite then the iteration (3.57) behaves like (3.12) near $\underline{w}_*$ and is thus locally convergent or stable and $\underline{w}_*$ is a local minimum. If J^*_{ww} is not positive definite, then from (3.15) we see that some modes of (3.57) or (3.58) will explode, i.e., the stationary point $\underline{w}_*$ will be a locally unstable point and so the iteration (3.57) will not get stuck there. We now argue that all the stationary points of $J_2(\underline{t})$ (except the global minimum) are locally unstable.

From (2.34a) we find

$$\begin{aligned} J_{rr} &= -12t_r^2(3\mu_2^2 - \mu_4) + 12\mu_2^2 E_2 - 4\mu_4 + 24\mu_2^2 t_r^2 \\ &= 4(3\mu_2^2 E_2 - \mu_4) + 12t_r^2(\mu_4 - \mu_2^2) \end{aligned} \tag{3.59a}$$

$$J_{rs} = 24t_r t_s \mu_2^2, \quad r \neq s \tag{3.59b}$$

where $J_{rs} = d^2 J_2/dw_r dw_s$. Firstly consider the global minimum which occurs when

$$t_r = \delta_{r-r_o} = \begin{cases} 1, & r = r_o \\ 0, & \text{otherwise.} \end{cases}$$

Then

$$\begin{aligned} J_{rr} &= 4(3\mu_2^2 - \mu_4) + 12\delta_{r-r_0}(\mu_4 - \mu_2^2) \\ J_{rs} &= 0, \quad r \neq s. \end{aligned}$$

Clearly the (infinite) matrix $\{J_{rr}\}$ is thus positive definite in view of (2.36).

For any other point we first reorganize (3.59) as follows:

$$\begin{aligned} J_{rs} &= \delta_{r-s} a_r + 24t_r t_s \mu_2^2 \\ a_r &= 4(3\mu_2^2 E_2 - \mu_4) + 12t_r^2(\mu_4 - 3\mu_2^2). \end{aligned}$$

Proceeding informally with our analysis of the infinite matrix $\{J_{rs}\}$, we find that at stationary set S_M

$$\{J_{rs}\} = a_M I + b_M u_M u_M^T$$

where

$$\begin{aligned} a_M &= 4(3\mu_2^2 E_{2,M} - \mu_4) + 12\alpha_M^2(\mu_4 - 3\mu_2^2) \\ b_M &= 24\alpha_M^2 \mu_2^2 \end{aligned}$$

and where I is an infinite identity matrix and u_M an infinite vector of 0's but with 1's in M locations corresponding to the non-zero t_r's. Any vector v_M with 0's in the places where

u_M has 1's is an eigenvector of $\{J_{rs}\}$ with eigenvalue a_M. Also u_M is an eigenvector of $\{J_{rs}\}$ with eigenvalue $a_M + Mb_M$. We now show that

$$a_M < 0, \quad a_M + Mb_M > 0, \quad M \geq 2$$

so that $\{J_{rs}\}$ is not positive definite at any stationary set (with $M \geq 2$). Let us introduce

$$E = \mu_4/3\mu_2^2 \leq 1, \quad \text{by (2.36)}$$

and then from (2.38) and (2.39):

$$\alpha_M^2 = \frac{E}{M-1+E}$$
$$E_{2,M} = \frac{ME}{M-1+E}.$$

Thus

$$\begin{aligned}
a_M &= 12\mu_2^2\Big((E_{2,M} - E) + 3\alpha_M^2(E-1)\Big) \\
&= 12\mu_2^2\left(\left(\frac{ME}{M-1+E} - E\right) + \frac{3E}{M-1+E}(E-1)\right) \\
&= \frac{12\mu_2^2}{M-1+E}(ME - ME + E - E^2 + 3E^2 - 3E) \\
&= \frac{24\mu_2^2}{M-1+E}E(E-2) < 0.
\end{aligned}$$

On the other hand

$$\begin{aligned}
a_M + Mb_M &= \frac{24\mu_2^2 E}{M-1+E}(E-2) + \frac{24ME\mu_2^2}{M-1+E} \\
&= \frac{24\mu_2^2 E}{M-1+E}(E-2+M) > 0, \quad M \geq 2.
\end{aligned}$$

The result is thus established.

Our discussion suggests then that the iteration can be initialized at any value and will converge to the global minimum of $J_2(\underline{t})$. At this stage, however, we have to face up to the fact that our discussion has so far been conceptual. We can not of course carry out an iteration for an infinite dimensional vector such as $\{t_r\}$. We have then to deal with a finite span equalizing filter. While we might have thought our discussion would be useful for a large enough span, it appears nevertheless that stable local minima do occur.

Consider the following example. Suppose a noise free single pole channel

$$x_k = H(q^{-1})\epsilon_k \tag{3.60a}$$

$$H(q^{-1}) = (1 + \alpha q^{-m})^{-1}, \quad |\alpha| < 1 \tag{3.60b}$$

and use an FIR equalizer with m taps

$$\begin{aligned}\hat{s}_k(\underline{w}) &= \sum_{s=0}^{m} w_s x_{k-s} \\ &= W(q^{-1})x_k \\ &= \underline{x}_k^T \underline{w} \\ &= (x_k \cdots x_{k-m})^T (w_0 \cdots w_m).\end{aligned}$$

Clearly ideal equalization occurs with

$$\underline{w} = \underline{w}_e = \pm(1\ 0 \cdots \alpha)^T$$

or

$$W(q^{-1}) = 1 + \alpha q^{-m}.$$

Consider then optimizing the $J_2(\underline{w})$ criterion in terms of the equalizer parameter $\underline{w}$. We have

$$J_2(\underline{w}) = E\left(\hat{s}_k^2(\underline{w}) - R_2\right)^2$$

so that a stationary point must satisfy

$$\begin{aligned}\frac{dJ_2}{dw_r} &= 4E\left((\hat{s}_k^2(\underline{w}) - R_2)x_{k-r}\hat{s}_k(\underline{w})\right) \\ &= 0, \quad 0 \le r \le m.\end{aligned} \tag{3.61}$$

Now we search for solutions to (3.61). Certainly (3.61) is satisfied by $\underline{w}_e$ since we then have

$$\hat{s}_k(\underline{w}_e) = \epsilon_k$$

so that in (3.61) we find (recall that $R_2 = E(\epsilon_k^4)/E(\epsilon_k^2)$)

$$\begin{aligned}&E\left(\hat{s}_k^3(\underline{w}_e)x_{k-r}\right) - R_2 E\left(\hat{s}_k(\underline{w}_e)x_{k-r}\right) \\ &= E(\epsilon_k^3 x_{k-r}) - R_2 E(\epsilon_k x_{k-r}) \\ &= \begin{cases} 0, & 1 \le r \le m \\ E(\epsilon_k^4) - R_2 E(\epsilon_k^2) = 0, & r = 0. \end{cases}\end{aligned}$$

To find other stationary points observe from (3.60) that

$$x_k = \sum_{u=0}^{\infty} (-\alpha)^u \epsilon_{k-mu} \tag{3.62}$$

so that $x_k, x_{k-1}, \cdots, x_{k-(m-1)}$ are statistically independent as are $x_{k-1}, \ldots, x_{k-m}$. Thus only two equations in (3.61) are nontrivial, namely those for $r = 0$, $r = m$. Now we look for a stationary point of the form

$$\underline{w}^* = (0 \cdots 0\, \theta)^T. \tag{3.63}$$

We obtain

$$\begin{aligned} 0 &= E(\theta^2 x_{k-m}^2 - R_2) x_{k-r} x_{k-m}, \qquad r = 0, m \\ \Rightarrow \qquad \theta^2 &= R_2 E(x_{k-m}^2)/E(x_{k-m}^4) \\ &= R_2 E(x_0^2)/E(x_0^4). \end{aligned}$$

Since

$$x_k = \epsilon_k - \alpha x_{k-m} \tag{3.64}$$

the equation for $r = 0$ yields the same expression so we have two stationary points of the form (3.63) corresponding to

$$\theta = \pm \left(\frac{R_2 E(x_0^2)}{E(x_0^4)} \right)^{\frac{1}{2}}. \tag{3.65}$$

The issue now is to consider the local stability of these stationary points. From (3.61) we find

$$\frac{d^2 J_2}{dw_r dw_s} = 4E\left((\hat{s}_k^2(\underline{w}) - R_2) x_{k-r} x_{k-s} \right) + 8E\left(\hat{s}_k^2(\underline{w}) x_{k-r} x_{k-s} \right), \qquad 0 \le r, s \le m.$$

At the stationary points (3.63), (3.65) this becomes

$$12\theta^2 E(x_{k-m}^2 x_{k-r} x_{k-s}) - 4R_2 E(x_{k-r} x_{k-s}), \qquad 0 \le r, s \le m.$$

Because $x_{k-1}, x_{k-2}, \ldots, x_{k-m}$ are independent, this is clearly diagonal for $1 \le r$, $s \le m-1$. For $r = 0$, $1 \le s \le m-1$ or $s = 0$, $1 \le r \le m-1$, use (3.64) to see it is 0. This is also the case for $r = m$, $1 \le s \le m-1$ or $s = m$, $1 \le r \le m-1$. For $r = 0$, $s = m$ or $r = m$, $s = 0$ we get

$$\begin{aligned} & 12\theta^2 E(x_{k-m}^3 x_k) - 4R_2 E(x_k x_{k-m}) \\ &= -\alpha \left(12\theta^2 E(x_{k-m}^4) - 4R_2 E(x_{k-m}^2) \right), \qquad \text{by (3.64)} \\ &= 0, \qquad \text{by (3.64)}. \end{aligned}$$

Thus $d^2 J_2/dw_r dw_s$ is diagonal, with diagonal entries

$$12\theta^2 E(x_{k-m}^2) E(x_{k-m}^2) - 4R_2 E(x_{k-r}^2), \quad 1 \le r \le m-1$$

$$= 4R_2 E(x_0^2) \left(3E(x_0^2)/E(x_0^4) - 1 \right) \tag{3.66a}$$

and also, using (3.64),

$$12\theta^2 E(x_{k-m}^2 x_k^2) - 4R_2 E(x_k^2), \quad r = 0, m$$

$$= 12\theta^2 \left(\alpha^2 E(x_0^4) + \mu_2 E(x_0^2) \right) - 4R_2 E(x_0^2)$$

$$= 4R_2 E(x_0^2) \left(3\alpha^2 + 3\mu_2 E(x_0^2)/E(x_0^4) - 1 \right). \tag{3.66b}$$

The reader is asked to show in Exercise 3.11 that (3.66a), (3.66b) are positive when $\mu_4 < 3\mu_2^2$. Thus we conclude, contrary to the infinite dimensional case, that the stationary points (3.65) are locally stable local minima. Our argument does not depend on m, which can be arbitrarily large.

These calculations may be illustrated with a simple example.

Example 3.7.   Local Minima in Blind Equalization

Consider an AR(1) channel

$$x_k = (1 + 0.6q^{-1})^{-1} \epsilon_k.$$

The message ϵ_k is a binary white noise with probability $P(\epsilon_k = \pm 1) = 0.5$, and a two tap equalizer is used. The ideal equalizer has tap weights $\pm(1, 0.6)$ while local minima are at $\pm(0, 0.5575)$. In Fig. 3.8 is a self-explanatory plot of $J_2(\underline{w})$, clearly showing the local minima.

3.3.3 Improvements to Steepest Descent

The ideas developed in Section 3.2 apply directly here. As an illustration we write out the momentum, loop filter, and relaxation algorithms for time delay estimation.

Momentum SD for time delay estimation

$$\alpha(q^{-1})\delta\hat{\theta}_k = \mu\beta(q^{-1})R'_{xy}(\hat{\theta}_k)$$

Loop filter SD for time delay estimation

$$\delta\underline{m}_k = \mu(\underline{A}\underline{m}_{k-1} - \underline{b}R'_{xy}(\hat{\theta}_{k-1}))$$

$$\delta\hat{\theta}_k = \mu\underline{m}_k^T \underline{c} + \mu R'_{xy}(\hat{\theta}_{k-1})$$

Relaxation SD for time delay estimation

$$\delta\hat{\theta}_k = -\alpha\mu\hat{\theta}_{k-1} + \mu R'_{xy}(\hat{\theta}_{k-1})$$

Using (3.55) each of these may be easily converted into an error system.

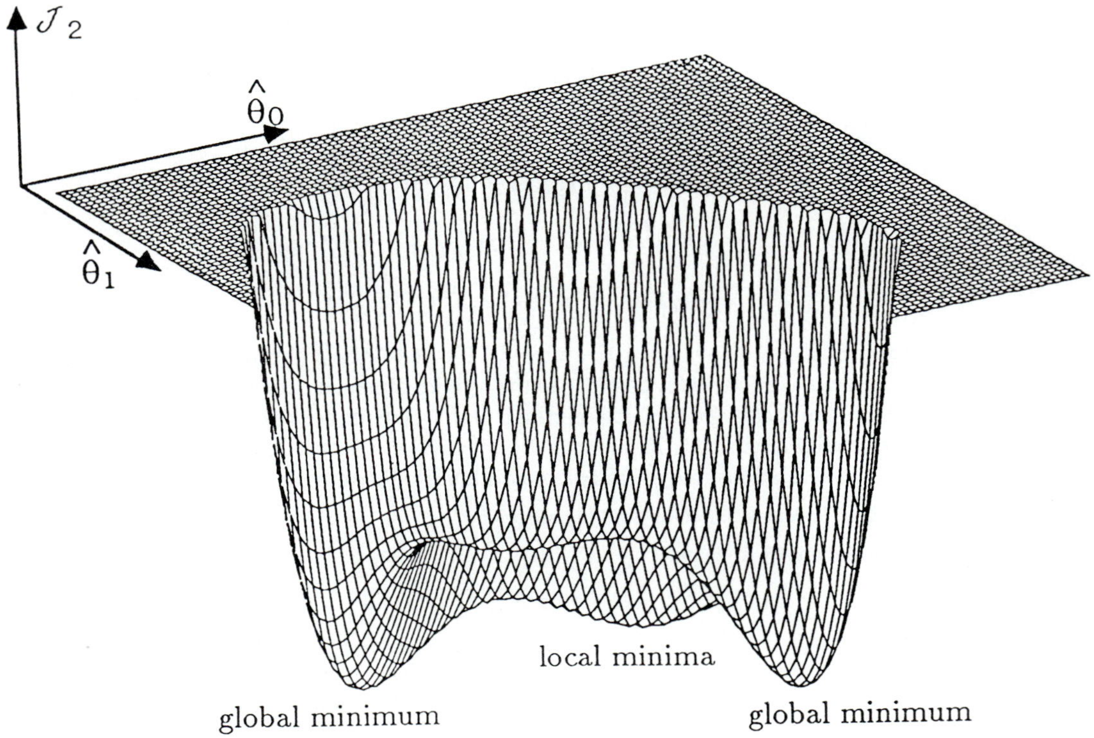

Figure 3.8 $J_2(\underline{w})$ criterion for Example 3.7. Note that $\hat{\theta}_0 = w_0$, $\hat{\theta}_1 = w_1$, after [D4], © September 1991, IEEE.

3.3.4 Newton's Method

Again the idea here is a straightforward extension of the discussion in Section 3.2. In the time delay estimation example (see (2.43), (2.44)) we find that the Hessian matrix is

$$\begin{aligned} H(\hat{\theta}) &= d^2\mathcal{E}/d\hat{\theta}^2 \\ &= -R''_{xy}(\hat{\theta}) \\ &= -R''_s(\hat{\theta} - \theta_o). \end{aligned}$$

The NM algorithm is then

$$\delta\hat{\theta}_k = -\mu(R''_{xy}(\hat{\theta}_{k-1}))^{-1} R'_{xy}(\hat{\theta}_{k-1})$$

with a corresponding error system

$$\delta\tilde{\theta}_k = -\mu(R_s''(\tilde{\theta}_{k-1}))^{-1}R_s'(\tilde{\theta}_{k-1}).$$

Example 3.8. Time Delay Estimation

In this example we consider a noise free, unit attenuation time delay estimation problem. The signal is deterministic and can be expressed as

$$s_k = a(\sin(\omega_s k) + \frac{1}{\sqrt{3}}\sin(2\omega_s k) + \frac{1}{\sqrt{2}}\sin(3\omega_s k)) \tag{3.67}$$

where

$$\omega_s = 2\pi/50, \quad a = 0.5482 \text{ (so } |s_k| \leq 1.0).$$

The normalized autocorrelation function of this signal (namely, $\tilde{E}(s_k s_{k+\tau}) = R_s(\tau)$) is shown in Fig. 3.9. In Fig. 3.10 trajectories of various iteration methods are shown.

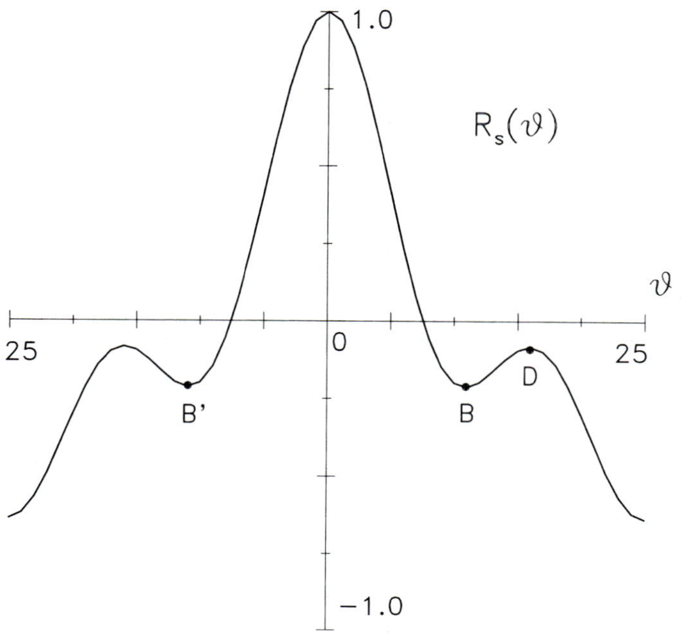

Figure 3.9 Normalized autocorrelation function of s_k given by (3.67).

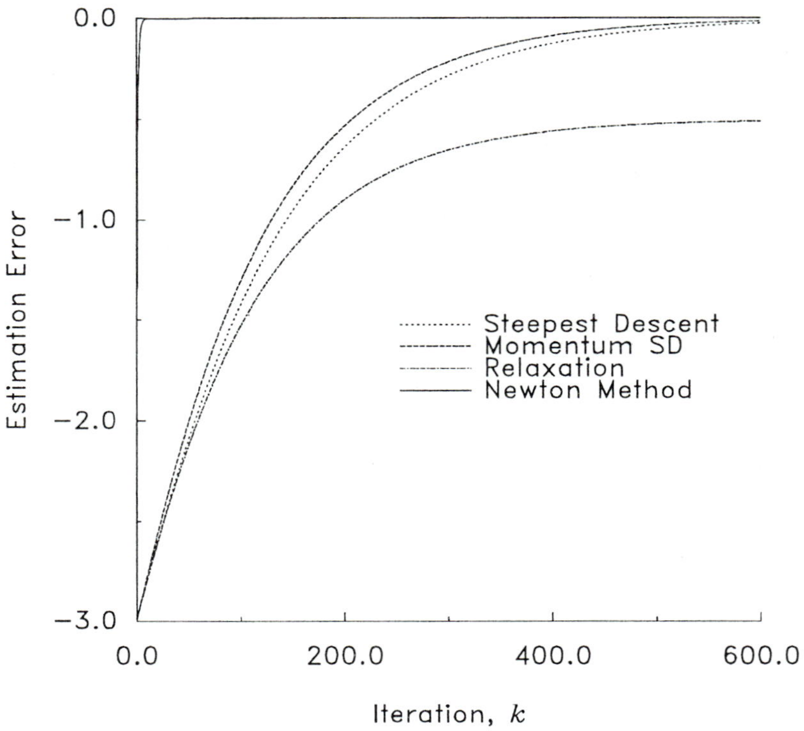

Figure 3.10 Trajectories of various iteration methods.

3.4 NOTES

Our discussion of the MSE surface for LMS is similar to that in [W5]. Differentiation filter methods were introduced into the adaptive area by [P4], [S6], and [S7]. They are discussed generally in [H5]. Loop filter methods were applied to adaptive algorithms by [K3]; however they have long been used in analog phase lock loops (see [L4]). One has to be careful of terminology here. What we have called differentiation filter methods are called multi-step methods in [S6] and [S7]. Also [B7] distinguishes two types of multi-step methods. The first type corresponds to our differentiation filter methods and the second to our loop filter methods. The term multi-step comes from numerical analysis where it denotes methods used to get improved discrete approximations to continuous systems. This corresponds to our differentiation filter methods. Loop filter methods change the underlying system and are totally different from multi-step methods. To avoid confusion we have not used the term multi-step at all. Stiff differential equations are discussed by [G2]. Feedback control design methods such as root locus are discussed in many control engineering books: convenient references are [D1] and [G13]. The MSE surface for the IIR filter is discussed by [S19].

The careful reader may already have noticed that in considering the specifications of a blind equalizer not only is it possible to allow an arbitrary time delay, but also the scaling does not matter. Indeed it seems that the insistence on scaling inherent in minimizing $J_p(\underline{w})$ is what causes locally stable minima to occur. If for example we constrain w_o to be 1, then the filter (3.65) is not possible. Of course the criterion has to be modified. We might for example consider working with $E[(\hat{s}_k(\underline{w}))^4]$. More details on this constrained approach can be found in [S4] and [V2].

EXERCISES

3.1 Consider the adaptive prediction problem of Exercise 2.1 with two weights. Plot contours of MSE. Calculate the stability region for μ. Show the path of SD iterations starting at $\underline{0}$ with two values of μ, one inside the stability region and one outside.

3.2 Consider the minimum phase equalization problem of Exercise 2.6. Plot contours of MSE. Calculate the stability region for μ. Show the path of SD iterations starting at $\underline{0}$, with two values of μ, one inside the stability region and one outside.

3.3 Consider the interference cancelling problem of Exercise 2.11. Plot contours of MSE. Calculate the stability region for μ. Show the path of SD iterations starting at $\underline{0}$, with two values of μ, one inside the stability region and one outside. Take s_k to be a white noise of variance 3. Use three weights and also seven weights.

3.4 For time delay estimation Example 3.8, calculate the MSE analytically and reproduce its plot Fig. 3.9. Calculate a stability region for μ. Then show the path of SD iterations starting at $\tilde{\theta} = 1.0$ for two values of μ, one inside the stability region and one outside.

3.5 Consider the adaptive prediction problem of Exercise 2.5 (with two weights). Plot contours of MSE. Calculate the stability region for μ. Show the path of SD iterations starting at $\underline{0}$ with two values of μ, one inside the stability region, one outside. Use the following parameter values ($a_3 = 0$ in each case):

$$(a_1, a_2, \sigma_\epsilon^2) = (-1.71,\ 0.9,\ 0.361)$$

$$(a_1, a_2, \sigma_\epsilon^2) = (-0.95,\ 0.9,\ 0.143)$$

$$(a_1, a_2, \sigma_\epsilon^2) = (-0.19,\ 0.9,\ 0.188)$$

Compare the results with regard to the effect of eigenvalue spread on speed of convergence.

3.6 **(a)** Repeat Exercise 3.1 using the momentum SD iteration with $\alpha = 0.9$.
(b) Repeat Exercise 3.2 using the momentum SD iteration with $\alpha = 0.9$.
(c) Repeat Exercise 3.3 using the momentum SD iteration with $\alpha = 0.9$.
(d) Repeat Exercise 3.4 using the momentum SD iteration with $\alpha = 0.9$.

3.7 The fixed gains conjugate gradient algorithm for minimizing $\mathcal{E}(\underline{w})$ is

$$\delta \underline{w}_k = \mu \underline{p}_{k-1}$$

$$\underline{p}_{k-1} = \alpha \underline{p}_{k-2} + (\underline{R}_{yx} - \underline{R}_x \underline{w}_{k-1}).$$

Show that this is equivalent to the momentum SD algorithm.

3.8 **(a)** Repeat Exercise 3.1 using a first order loop filter algorithm.
(b) Repeat Exercise 3.2 using a first order loop filter algorithm.

(c) Repeat Exercise 3.3 using a first order loop filter algorithm.
(d) Repeat Exercise 3.4 using a first order loop filter algorithm.

3.9 **(a)** Repeat Exercise 3.1 using an NM algorithm.
(b) Repeat Exercise 3.2 using an NM algorithm.
(c) Repeat Exercise 3.3 using an NM algorithm.
(d) Repeat Exercise 3.4 using an NM algorithm.

3.10 Prove (3.24).

3.11 Show that (3.66a), (3.66b) are positive when $\mu_4 < 3\mu_2^2$.

4

Algorithm Construction

The linear-quadratic iterative procedures in Chapter 3 required that the MSE (or similar criterion function) be known, and this means that the covariance matrix of $(\underline{x}_k, y_k)$ be known; but in a real time setting this statistical information is usually not available. The idea then behind adaptive algorithm construction is to replace MSE with some approximation that can be computed in real time.

The simplest idea is to

(i) replace $E(e_k^2(\underline{w}))$ with an instantaneous approximation, namely, $e_k^2(\underline{w})$ at time k.

This is the basis of most adaptive algorithms. It has the disadvantage that it cannot be used to generate NM algorithms because the Hessian does not have full rank.

Another idea is to use the exponentially weighted squared error criterion (EWSE).

(ii) Replace $E(e_k^2(\underline{w}))$ at time k by

$$J_k(\underline{w}) = \sum_{r=1}^{k} \lambda^{k-r} e_r^2(\underline{w})$$
$$= e_k^2(\underline{w}) + \lambda e_{k-1}^2(\underline{w}) + \lambda^2 e_{k-2}^2(\underline{w}) + \cdots$$

where λ is called the "forgetting factor" and $0 < \lambda < 1$.

The closer λ is to 1, the greater the span of error signal values that "count" towards $J_k(\underline{w})$. The further λ is from 1, the more rapidly are recent errors forgotten. Indeed $\lambda = 0$ makes $J_k(\underline{w})$ collapse to $e_k^2(\underline{w})$.

A third idea is

(iii) model based or signal estimation based adaptation.

Since adaptive algorithms are supposed to track time varying parameters, the idea is to write down a stochastic model for the true (but unknown) time varying parameters. This gives a state equation where the state consists of the unknown time varying parameters. The filter structure is now viewed as an observation equation that relates the auxiliary signal through the true time varying parameters to the primary signal. Thus, we have a real time state estimation problem.

These are perhaps the three main ideas used to generate adaptive algorithms from iterative procedures. They and other methods will be pursued below, but first it is useful to look at the general form of the algorithms obtained.

One other point before we commence. In this chapter we will illustrate many of the algorithms with simulations (see Appendix E for the tools we use). Our consequent discussion of algorithm behavior will be heuristic and guided by the offline computations made in Chapters 2 and 3. Rigorous analysis of algorithm behavior is developed in later chapters.

4.1 ALGORITHM CLASSIFICATION

In a very general sense almost all adaptive algorithms have the same form, namely:

$$\hat{\theta}_{new} = \hat{\theta}_{old} + \text{gain} \times \text{gradient} \times \text{error}. \tag{4.1}$$

The error is a scalar signal; the gradient is a vector signal; the gain is a scalar or a matrix. Note that in order to calculate the gain, gradient, and error at each time, it is required that only a fixed span of previous data be stored. Note also that the size of the gain or step size controls the speed of change of $\hat{\theta}$, i.e., the speed of adaptation. Once the general specification (4.1) is appreciated, it then becomes helpful to classify the various algorithms in several ways based on important behavior and properties. To start the algorithm, $\hat{\theta}$ and the gradient are usually set to zero. Sometimes the gain is a time varying matrix, in which case its initialization will be discussed in specific settings below.

4.1.1 Long or Short Memory Algorithm

A long memory algorithm has gain $\to 0$ as time $\to \infty$. A short memory algorithm has gain $>$ constant > 0 for all time. The significance of the distinction between the two is that only short memory algorithms can track time varying parameters. Since gain $\to 0$ as time $\to \infty$, a long memory algorithm loses its ability to adapt as time $\to \infty$. Since a major reason for using adaptive algorithms is to track time varying parameters, this is an important point. For this reason we only treat short memory algorithms in this book. Long memory algorithms also are called self-tuning algorithms, recursive algorithms, stochastic approximation algorithms, and sometimes, adaptive algorithms; short memory algorithms also are called adaptive algorithms.

4.1.2 Steepest Descent or Newton Algorithm

A steepest descent algorithm uses first order gradient information. A Newton algorithm uses second order derivative information. Guided by our brief look at these methods in the iterative setting in Section 3.2, we expect a Newton algorithm to have better transient behavior (i.e., react more quickly) than a steepest descent algorithm. Newton algorithms often have the descriptor "least squares" but that term is so widely used as to be rather vague.

4.1.3 Single Time Scale or Mixed Time Scale

With a single time scale algorithm the gradient term in (4.1) consists of external signals. For a mixed time scale algorithm the gradient in (4.1) is generated by an auxiliary state equation (driven by an external signal), which may also depend on $\hat{\theta}_{old}$. This state varies much more rapidly than the adaptive estimator (4.1), so the algorithm has a fast time scale and a slow time scale, hence the name. The significance of the distinction between single and mixed time scale algorithms is that the latter are usually much harder to analyze and must often be monitored to ensure stability. This is partly to do with the fact that the error surface has multiple local minima.

4.1.4 Posterior or Prior Algorithm

The distinction here is usually only relevant to steepest descent algorithms. A posterior or implicit algorithm uses an implicit discretisation as in Section 3.2. A prior algorithm uses an explicit discretisation. As was found in Section 3.2, it turns out that posterior algorithms can exhibit convergence without limits on the magnitude of the gain.

4.1.5 Pseudo Gradient

In the FIR case, the error signal is linear in the parameters. In some cases (e.g., IIR), however, even when the error signal is not linear in the parameters, it can be made to appear as if it is. This was illustrated in equation (2.60) where the pseudo linear regression (PLR) idea was introduced. In such a case the gradient in the SD or NM algorithm is replaced by a pseudo gradient.

With the above five classifications in mind we can organize acronyms for the algorithms as follows:

short memory (uppercase, e.g., LMS); long memory (lowercase, e.g., lms);
steepest descent (SD); Newton's method (NM);
single time scale (subscript 1, e.g., SD_1); mixed time scale (subscript 2, e.g., SD_2);
posterior (subscript P, e.g., LMS_{1P}); prior (no subscript);
pseudo gradient (prefix P, e.g., PSD).

We make one exception to this arrangement. Because the LMS algorithm is the most widely known and used adaptive algorithm, we call an SD algorithm whose error signal is linear in parameters or which uses a pseudo gradient an LMS algorithm.

4.2 FILTER STRUCTURE

Prior to algorithm construction is the specification of the filter structure. The most common structures are the FIR filter (also called tapped delay line filter, transversal filter, all zero filter, moving average filter) and the IIR filter (also called rational transfer function, pole zero filter, state space filter, and includes special cases such as the autoregressive or all pole filter). These filter structures have already appeared in Chapter 2.

Filter structure has been most emphasized in system identification where the filter can be regarded as specifying an input/output model for the system being identified. In the adaptive system identification setting the most common models are rational transfer function or state space models. Sometimes the state space model is nonlinear or time varying and so cannot in general be represented by a time invariant IIR filter. Some examples of this will be given in connection with model based algorithms in Section 4.7.

Two popular IIR filter specifications have already appeared. The output error filter or rational transfer function of Section 2.7 and the ARMAX filter of Section 2.7. Another specification is worth mentioning, which seems to incorporate most of the other IIR filter structures. It is the autoregressive transfer function colored noise structure

$$G(q^{-1})y_k = s_k + n_k \tag{4.2}$$

$$s_k = \frac{b(q^{-1})}{A(q^{-1})}x_k \tag{4.3}$$

$$n_k = \frac{C(q^{-1})}{D(q^{-1})}\epsilon_k \tag{4.4}$$

where $C(q^{-1}) = 1 + c(q^{-1})$, where $c(q^{-1}) = \sum_{s=1}^{m} c_s q^{-s}$, etc. The reader is invited to assemble a block diagram for this filter by modifying Fig. 1.13.

4.3 FIR FILTERS AND SINGLE TIME SCALE LMS

We begin with the instantaneous squared error approximation of MSE. So working with the error signal

$$e_k(\underline{w}) = y_k - \underline{x}_k^T \underline{w}$$

we replace $\mathcal{E}(\underline{w}) = E(e_k^2(\underline{w}))$ by $e_k^2(\underline{w})$. Thus to estimate $\underline{w}$ we are led to an adaptive steepest descent algorithm as follows:

$$\hat{\underline{w}}_k - \hat{\underline{w}}_{k-1} = \delta\hat{\underline{w}}_k = -\frac{1}{2}\mu de_k^2(\underline{w})/d\underline{w}\Big|_{\underline{w}=\hat{\underline{w}}_{k-1}} \tag{4.5}$$

$$\Rightarrow \qquad \delta\hat{\underline{w}}_k = \mu \underline{x}_k e_k \tag{4.6}$$

$$e_k = y_k - \underline{x}_k^T \hat{\underline{w}}_{k-1}. \tag{4.7}$$

The pair (4.6), (4.7) is the LMS (least mean square) algorithm. At each iteration or time update, it requires knowledge only of the latest values of $\underline{x}_k$, $\hat{\underline{w}}_{k-1}$, e_k. Often we take $\hat{\underline{w}}_0 = \underline{0}$.

We now look at an example.

Example 4.1. LMS and Its Estimated Learning Curve

In Fig. 4.1 is a plot of the trajectory of the LMS algorithm applied to the channel equalization Example 2.2. Recall for that example the output or primary signal is defined in (2.15) and the reference signal is assembled from (2.16), (2.17). Notice that the presence of noise means that the trajectory does not converge to the minimum of the quadratic bowl (as the iteration did in Fig. 3.2). Rather the trajectory hovers in the vicinity of the minimum. This is a characteristic feature of short memory algorithms. In Fig. 4.2 the equalizer output signal and the error signal are plotted. This shows the equalizer is working well.

In Fig. 4.3 is a plot of the estimated learning curve. The learning curve is a plot of $E(e_k^2)$ versus k. Each plot of an LMS trajectory such as in Fig. 4.1 is called a realization or simulation. To produce Fig. 4.3, 500 independent realizations of the LMS algorithm were generated, and the estimated learning curve was obtained by averaging across these 500 realizations.

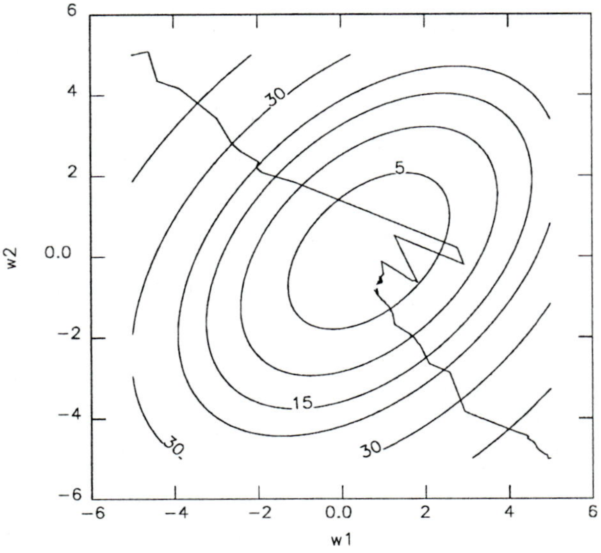

Figure 4.1 LMS trajectories for Example 4.1 corresponding to different initial conditions.

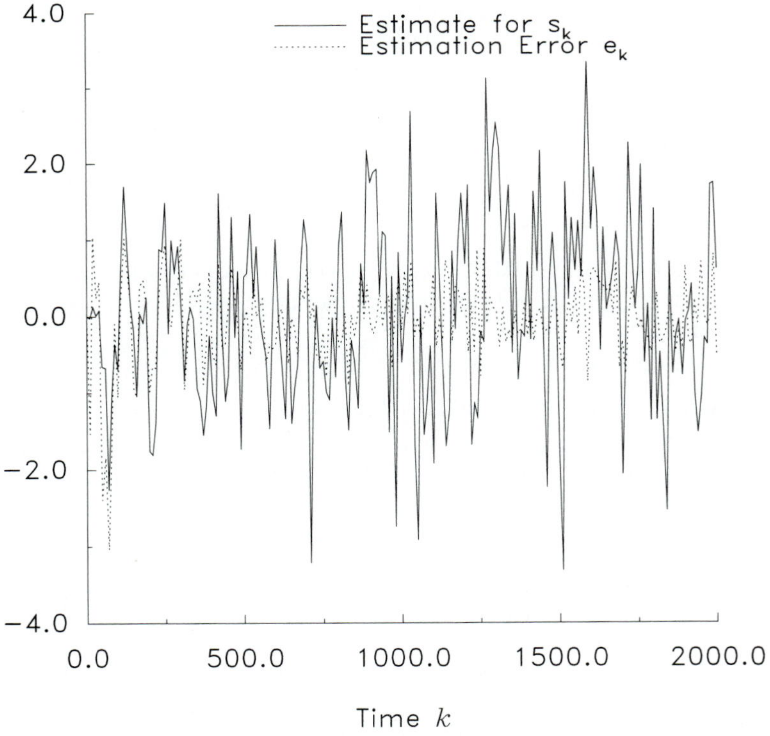

Figure 4.2 Equalizer output signal and error signal.

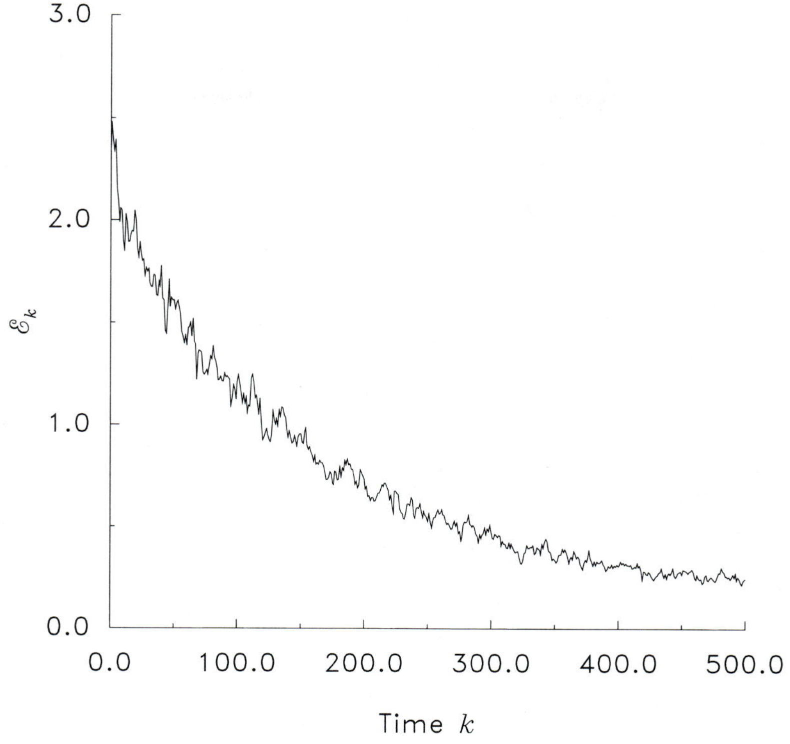

Figure 4.3 Estimated learning curve of LMS algorithm for Example 4.1.

4.4 IIR FILTERS

In this section we discuss basic algorithms for IIR filters as well as some variations of them. Again we work with instantaneous squared error.

4.4.1 Equation Error and Single Time Scale LMS

If we use the equation error form of Section 2.6 to generate the error signal for an IIR filter, then we immediately obtain a single time scale LMS algorithm. From (2.57), (2.64) we obtain

$$\delta\underline{\hat{\theta}}_k = \mu\underline{\psi}_k e_k \tag{4.8}$$

$$e_k = y_k - \underline{\psi}_k^T \underline{\hat{\theta}}_{k-1} \tag{4.9}$$

$$\underline{\psi}_k^T = (-y_{k-1} \cdots - y_{k-p}, \ x_{k-1} \cdots x_{k-m})^T. \tag{4.10}$$

4.4.2 Instrumental Variables

A weakness of the equation error arrangement is that $\underline{\psi}_k$ may be correlated with e_k even when $\hat{\underline{\theta}}_k = \underline{\theta}_o$, the true value of $\underline{\theta}$. The idea of the instrumental variables (IV) method is to replace (y_k in) $\underline{\psi}_k$ with another signal (called the instrumental signal) that is uncorrelated with e_k when $\hat{\underline{\theta}}_k = \underline{\theta}_o$. Thus if $e_k = y_k - \underline{\psi}_k^T \underline{\theta}_o$ is a $MA(r)$ signal, then an instrumental signal is

$$\underline{\zeta}_k = (-y_{k-1-r} \cdots - y_{k-p-r}, x_{k-1-r} \cdots x_{k-m-r})^T$$

and this leads to the PLMS algorithm

$$\delta\hat{\underline{\theta}}_k = \mu \underline{\zeta}_k e_k$$
$$e_k = y_k - \underline{\zeta}_k^T \hat{\underline{\theta}}_{k-1}.$$

4.4.3 Output Error and Mixed Time Scale LMS

If we use the output error, pseudo linear form (2.60), we are led to an LMS_2 algorithm as follows:

$$\delta\hat{\underline{\theta}}_k = \mu \underline{\varphi}_k(\hat{\underline{\theta}}_{k-1}) e_k$$
$$e_k = y_k - \hat{s}_k(\hat{\underline{\theta}}_{k-1})$$
$$\hat{s}_k(\hat{\underline{\theta}}_{k-1}) = -\sum_{r=1}^{p} \hat{s}_{k-r}(\hat{\underline{\theta}}_{k-1})\hat{a}_{r,k-1} + \sum_{r=1}^{m} \hat{b}_{r,k-1} x_{k-r} \tag{4.11}$$
$$= \underline{\varphi}_k^T(\hat{\underline{\theta}}_{k-1})\hat{\underline{\theta}}_{k-1}$$
$$\underline{\varphi}_k(\hat{\underline{\theta}}_{k-1}) = (-\hat{s}_{k-1}(\hat{\underline{\theta}}_{k-1}) \ \ldots \ - \hat{s}_{k-p}(\hat{\underline{\theta}}_{k-1}),\ x_{k-1} \ \ldots \ x_{k-m})^T.$$

This algorithm is unsatisfactory for adaptive use, since to do the filtering to get $\hat{s}_k(\hat{\underline{\theta}}_{k-1})$, it is necessary to store all previous values of x_k. There are two nearly equivalent ways to overcome this problem:

(i) Make the difference equation for $\hat{s}_k(\hat{\underline{\theta}}_{k-1})$ have finite memory.
To do this we replace (4.11) by

$$\hat{s}_k = -\sum_{r=1}^{p} \hat{a}_{r,k-1}\hat{s}_{k-r} + \sum_{r=1}^{m} \hat{b}_{r,k-1} x_{k-r}. \tag{4.12}$$

Note now that to get $\hat{s}_k$ we only need $\hat{\underline{\theta}}_{k-1}$, and then the gradient becomes

$$\hat{\underline{\varphi}}_k = (-\hat{s}_{k-1} \ \ldots \ - \hat{s}_{k-p},\ x_{k-1} \ \ldots \ x_{k-m})^T. \tag{4.13}$$

We are led to the following LMS_2 algorithm:

$$\delta\underline{\hat{\theta}}_k = \mu\underline{\hat{\varphi}}_k e_k \tag{4.14}$$

$$e_k = y_k - \hat{s}_k \tag{4.15}$$

$$\hat{s}_k = \underline{\hat{\varphi}}_k^T \underline{\hat{\theta}}_{k-1}. \tag{4.16}$$

It is important to point out that (4.12), (4.13) can be written in state space form as follows:

$$\underline{\hat{\varphi}}_k = \underline{A}(\underline{\hat{\theta}}_{k-1})\underline{\hat{\varphi}}_{k-1} + \underline{B}(\underline{\hat{\theta}}_{k-1})\underline{x}_{k-1} \tag{4.17}$$

where

$$\underline{A}(\underline{\theta}) = \begin{bmatrix} -a_1 \cdots & -a_p & \underline{0} \\ \mathrm{I} & 0 & \\ & \underline{0}^T & 0 \end{bmatrix}$$

$$\underline{B}(\underline{\theta}) = \begin{bmatrix} b_1 & \cdots & b_m \\ & \underline{\tilde{0}} & \\ & I & \end{bmatrix}$$

$$\underline{x}_{k-1} = [x_{k-1} \ \ldots \ x_{k-m}]^T$$

where $\underline{\tilde{0}}$ is a zero matrix. So we can say that the LMS_2 algorithm consists of the pair (4.14), (4.17) augmented with (4.15) and (4.16). In (4.14) the "slow state" $\underline{\hat{\theta}}_k$ changes slowly (its speed of change is $O(\mu)$). In (4.17) $\underline{\hat{\varphi}}_k$ is a "fast" state that changes at a speed generally much greater than $O(\mu)$. Thus, the two "states" change at different rates, hence the description of the algorithm as being mixed time scale.

(ii) Make the state space form for $\underline{\varphi}_k(\underline{\theta})$ have finite memory.

If we return to (2.60) we can write down a state space model for $\underline{\varphi}_k(\underline{\theta})$ as follows:

$$\underline{\varphi}_k(\underline{\theta}) = \underline{F}(\underline{\theta})\underline{\varphi}_{k-1}(\underline{\theta}) + \underline{b}(\underline{\theta})x_k$$

$$\underline{b}(\underline{\theta}) = [b_1 \ \ldots \ b_m \ \overbrace{0 \cdots 0}^{p-m \text{ times}} \ 1\, 0 \ldots 0]^T$$

$$\underline{F}(\underline{\theta}) = \begin{bmatrix} -a_1 & I & & \\ \vdots & & \underline{\tilde{0}} & \\ -a_p & 0 & & \\ \underline{\tilde{0}} & & \underline{0}^T & 0 \\ & & I & \underline{0} \end{bmatrix}$$

where $\underline{\tilde{0}}$ is a zero matrix. Thus we are led to an algorithm in which (4.17) is replaced by (4.18):

$$\underline{\hat{\varphi}}_k = \underline{F}(\underline{\hat{\theta}}_{k-1})\underline{\hat{\varphi}}_{k-1} + \underline{b}(\underline{\hat{\theta}}_{k-1})x_{k-1}. \tag{4.18}$$

The differences between the two ways of generating $\underline{\hat{\varphi}}_k$ may not be significant.

4.4.4 Output Error and Mixed Time Scale Steepest Descent

Recall from (2.57) that the IIR filter output is

$$e_k(\underline{\theta}) = y_k - \hat{s}_k(\underline{\theta}) \tag{4.19}$$

$$\hat{s}_k(\underline{\theta}) = [1 + a(q^{-1})]^{-1} b(q^{-1}) x_k. \tag{4.20}$$

A steepest descent algorithm will take the form

$$\begin{aligned} \delta\underline{\hat{\theta}}_k &= -\frac{1}{2}\mu d e_k^2(\underline{\hat{\theta}}_{k-1})/d\underline{\hat{\theta}}_{k-1} \\ &= -\mu(de_k/d\underline{\hat{\theta}}_{k-1}) e_k(\underline{\hat{\theta}}_{k-1}). \end{aligned} \tag{4.21}$$

We need then to evaluate the gradient $de_k/d\underline{\theta}$. We differentiate (4.19), (4.20) to find

$$de_k/db_r = -[1 + a(q^{-1})]^{-1} q^{-r} x_k, \qquad 1 \le r \le m$$

$$de_k/da_r = [1 + a(q^{-1})]^{-2} q^{-r} b(q^{-1}) x_k, \quad 1 \le r \le p.$$

If we introduce the auxiliary filtered signals

$$\tilde{x}_k(\underline{\theta}) = [1 + a(q^{-1})]^{-1} x_k \tag{4.22}$$

$$\tilde{s}_k(\underline{\theta}) = -[1 + a(q^{-1})]^{-1} b(q^{-1}) \tilde{x}_k(\underline{\theta}) \tag{4.23}$$

then we can write the gradient as

$$de_k/db_r = -\tilde{x}_{k-r}(\underline{\theta}), \quad 1 \le r \le m \tag{4.24}$$

$$de_k/da_r = -\tilde{s}_{k-r}(\underline{\theta}), \quad 1 \le r \le p. \tag{4.25}$$

So if we introduce the auxiliary signal vector

$$\underline{\tilde{\varphi}}_k(\underline{\theta}) = (-\tilde{s}_{k-1}(\underline{\theta}) \;\ldots\; -\tilde{s}_{k-p}(\underline{\theta})\; \tilde{x}_{k-1}(\underline{\theta}) \;\ldots\; \tilde{x}_{k-m}(\underline{\theta}))^T \tag{4.26}$$

we obtain from (4.21) the algorithm

$$\delta\underline{\hat{\theta}}_k = \mu \underline{\tilde{\varphi}}_k(\underline{\hat{\theta}}_{k-1}) e_k(\underline{\hat{\theta}}_{k-1}) \tag{4.27}$$

$$e_k(\underline{\hat{\theta}}_{k-1}) = y_k - \hat{s}_k(\underline{\hat{\theta}}_{k-1}). \tag{4.28}$$

Some reflection will show that this pair is unsuitable as an adaptive algorithm because, for example, to do the filtering needed to get $\hat{s}_k(\underline{\hat{\theta}}_{k-1})$, it is necessary to store all previous values of x_k. To overcome this problem we can use either of the two approaches used to get the mixed time scale LMS algorithm treated above:

(i) Make the difference equations for the gradient have finite memory.

From (4.20), (4.22), (4.23) the relevant difference equations are

$$\hat{s}_k(\underline{\theta}) = -\sum_{r=1}^{p} a_r \hat{s}_{k-r}(\underline{\theta}) + \sum_{r=1}^{m} b_r x_{k-r}$$

$$\tilde{x}_k(\underline{\theta}) = -\sum_{r=1}^{p} a_r \tilde{x}_{k-r}(\underline{\theta}) + x_k$$

$$\tilde{s}_k(\underline{\theta}) = -\sum_{r=1}^{p} a_r \tilde{s}_{k-r}(\underline{\theta}) + \tilde{x}_k(\underline{\theta}).$$

We make these into finite memory equations as follows:

$$\hat{s}_k = -\sum_{r=1}^{p} \hat{a}_{r,k-1} \hat{s}_{k-r} + \sum_{r=1}^{m} \hat{b}_{r,k-1} x_{k-r} \tag{4.29}$$

$$\tilde{x}_k = -\sum_{r=1}^{p} \hat{a}_{r,k-1} \tilde{x}_{k-r} + x_k \tag{4.30}$$

$$\tilde{s}_k = -\sum_{r=1}^{p} \hat{a}_{r,k-1} \tilde{s}_{k-r} + \tilde{x}_k. \tag{4.31}$$

Initial conditions $\tilde{s}_{-1}, \ldots, \tilde{s}_{-p}, \tilde{x}_{-1}, \ldots, \tilde{x}_{-p}$ must be specified of course—setting them to 0 is usually satisfactory. Continuing, introduce

$$\underline{\tilde{\varphi}}_k = (-\tilde{s}_{k-1} \ \ldots \ -\tilde{s}_{k-p} \ \tilde{x}_{k-1} \ \ldots \ \tilde{x}_{k-m})^T \tag{4.32}$$

$$e_k = y_k - \hat{s}_k \tag{4.33}$$

and finally obtain a finite memory adaptive algorithm

$$\delta \underline{\hat{\theta}}_k = -\mu \underline{\tilde{\varphi}}_k e_k. \tag{4.34}$$

The complete algorithm consists of (4.32), (4.33), (4.34) augmented with (4.29), (4.30), (4.31). Again the last three can be written in state space form thus making clear the mixed time scale nature of the algorithm.

(ii) Make the state space forms for $\hat{s}_k(\underline{\theta})$, $\tilde{x}_k(\underline{\theta})$, $\tilde{s}_k(\underline{\theta})$ have finite memory.

This follows as before and details are left to the reader.

4.4.5 Output Error with Noise Whitening

Now we consider the noise whitening procedure discussed in Sections 1.3 and 2.6. The error signal (see (2.67)–(2.69)) is given by

$$e_k = e_k(\underline{\theta}) = \frac{1 + c(q^{-1})}{1 + d(q^{-1})} n_k \tag{4.35}$$

$$n_k = n_k(\underline{\beta}) = y_k - s_k \tag{4.36}$$

$$s_k = s_k(\underline{\beta}) = \frac{b(q^{-1})}{1 + a(q^{-1})} x_k. \tag{4.37}$$

There are a number of algorithms that can be derived to estimate the parameters $\underline{\theta}$ in (4.35)–(4.37) adaptively. We develop a mixed time scale LMS algorithm and pursue other possibilities in the exercises. Following the development of (4.14)–(4.16), the idea is to write (4.37) as a pseudo linear regression (as in (2.60)). We have immediately, as in (2.60), that

$$s_k = s_k(\underline{\beta}) = \underline{\phi}_k^T(\underline{\beta})\underline{\beta} \tag{4.38}$$

$$\underline{\phi}_k(\underline{\beta}) = (-s_{k-1}(\underline{\beta}) \cdots - s_{k-p}(\underline{\beta})\,,\; x_{k-1} \cdots, x_{k-m})^T \tag{4.39}$$

where $\underline{\beta} = (a_1 \cdots a_p, b_1 \cdots b_m)^T$ as in (2.71). The next step is to reorganize (4.35). We have

$$n_k = -c(q^{-1})n_k + d(q^{-1})e_k + e_k$$

which leads to

$$n_k = \underline{\psi}_k^T(\underline{\theta})\underline{\alpha} + e_k \tag{4.40}$$

$$\underline{\psi}_k(\underline{\theta}) = (-n_{k-1}(\underline{\beta}) \cdots - n_{k-r}(\underline{\beta})\,,\; e_{k-1}(\underline{\theta}) \cdots e_{k-r}(\underline{\theta}))^T \tag{4.41}$$

where $\underline{\alpha} = (c_1 \cdots c_r, d_1 \cdots d_r)^T$ and $\underline{\theta} = (\underline{\beta}^T, \underline{\alpha}^T)^T$ as in (2.70) and (2.72). Thus we obtain

$$\begin{aligned} e_k = e_k(\underline{\theta}) &= n_k(\underline{\beta}) - \underline{\psi}_k^T(\underline{\theta})\underline{\alpha} \\ &= y_k - s_k(\underline{\beta}) - \underline{\psi}_k^T(\underline{\theta})\underline{\alpha} \\ &= y_k - \underline{\phi}_k^T(\underline{\beta})\underline{\beta} - \underline{\psi}_k^T(\underline{\theta})\underline{\alpha}. \end{aligned} \tag{4.42}$$

This equation is the required pseudo linear regression. The rest of the algorithm construction now proceeds much as it did for (4.14)–(4.16), and the details are left to the reader.

4.5 GENERAL ADAPTIVE STEEPEST DESCENT ALGORITHMS

In this section we consider algorithm construction from two general points of view. Firstly we allow criteria other than squared error; secondly we also allow a general parametric dependence for the adjustable filter

$$\hat{s}_k = \hat{s}_k(\underline{\theta}) \tag{4.43}$$

$$e_k = y_k - \hat{s}_k(\underline{\theta}). \tag{4.44}$$

4.5.1 Squared Error Criterion

The adaptive steepest descent algorithm takes the form

$$\delta\underline{\hat{\theta}}_k = -\frac{1}{2}\mu\, de_k^2(\underline{\theta})/d\underline{\theta}\bigg|_{\underline{\theta}=\underline{\hat{\theta}}_{k-1}} \tag{4.45}$$

$$= \mu\underline{\tilde{\varphi}}_k(\underline{\hat{\theta}}_{k-1})e_k(\underline{\hat{\theta}}_{k-1}) \tag{4.46}$$

where we have introduced the gradient

$$\tilde{\underline{\varphi}}_k(\underline{\theta}) = -de_k(\underline{\theta})/d\underline{\theta} = d\hat{s}_k(\underline{\theta})/d\underline{\theta} \tag{4.47}$$

which is sometimes called the sensitivity signal. The algorithm (4.46), (4.47) is sometimes called a stochastic gradient algorithm because the gradient $d\mathcal{E}/d\underline{\theta}$ has been approximated by the noisy gradient $de_k^2(\underline{\theta})/d\underline{\theta}$. We prefer to use the term adaptive steepest descent because we will use (4.46), (4.47) in purely deterministic settings.

As we saw in Section 4.4, the pair (4.46), (4.47) is generally unsuitable as an adaptive algorithm because $e_k(\hat{\underline{\theta}}_{k-1})$, $\tilde{\underline{\varphi}}_k(\hat{\underline{\theta}}_{k-1})$ may not be computable from a fixed window of previous data. Thus some ingenuity is required to make (4.46), (4.47) properly adaptive. In Section 4.4 it was shown how to do this for IIR filters, and here we give two other examples: time delay estimation and a state space model.

We start with time delay estimation. For simplicity we take the noise-free with no attenuation case (see Section 2.4)

$$\hat{s}_k = s_{k-\theta}, \quad y_k = s_k, \quad e_k = (s_{k-\theta_o} - s_{k-\theta}).$$

Then we find

$$\tilde{\varphi}_k(\theta) = -s'_{k-\theta}$$

where we have supposed s_k is sampled from a continuous signal $s(t)$ with unit sample interval so that $s_k = s(k)$ and then $s'_t = ds(t)/dt$. This gives what we will call the idealized algorithm

$$\delta\hat{\theta}_k = -\mu s'_{k-\hat{\theta}_{k-1}}(s_{k-\theta_o} - s_{k-\hat{\theta}_{k-1}}). \tag{4.48}$$

In practice s'_k is not available, so we must approximate it by a differentiation filter applied to s_k. The simplest is a difference filter, and this leads to the SD$_1$ algorithm

$$\delta\hat{\theta}_k = -\mu(s_{k-\hat{\theta}_{k-1}} - s_{k-\hat{\theta}_{k-1}-1})e_k \tag{4.49a}$$

$$e_k = s_{k-\theta_o} - s_{k-\hat{\theta}_{k-1}}. \tag{4.49b}$$

Even now we are not free of problems. There is no reason for $\hat{\theta}_k$, calculated from (4.49), to be a multiple of the sampling interval. So we must quantize $\hat{\theta}_{k-1}$ before using it in (4.49). We then obtain

$$\delta\hat{\theta}_k = -\mu(s_{k-q_{k-1}} - s_{k-q_{k-1}-1})e_k \tag{4.50}$$

$$e_k = s_{k-\theta_o} - s_{k-q_{k-1}} \tag{4.51}$$

$$q_{k-1} = \text{integer nearest to } \hat{\theta}_{k-1}. \tag{4.52}$$

There is still more, however. There is nothing to stop q_{k-1} from being negative. If it were, then in (4.51) we would need to know future values of s_k! To avoid this we have to monitor

(4.50), (4.51). If we set an upper bound b for θ (perhaps from physical considerations) and if we can allow a delay in processing equal to b, then we obtain the algorithm

$$\delta\hat{\theta}_k = -\mu(s_{k-b-q_{k-1}} - s_{k-b-q_{k-1}-1})e_k \tag{4.53}$$

$$e_k = s_{k-b-\theta_o} - s_{k-b-q_{k-1}} \tag{4.54}$$

$$q_{k-1} = \text{integer nearest to } \hat{\theta}_{k-1} \text{ but } \geq -b. \tag{4.55}$$

The triple (4.53)–(4.55) is our final algorithm.

With this extended discussion of the time delay estimation problem, we have tried to indicate how the basic SD_1 algorithm (4.46), (4.47) has to be modified to get a practical algorithm.

Example 4.2. SD—Time Delay Estimation

Here we continue the noise-free unit attenuation Example 3.8. In Fig. 4.4 are plots of the SD_1 algorithm (4.48), (4.49b)

$$\delta\hat{\theta}_k = -\mu s'_{k-\hat{\theta}_{k-1}} e_k$$

and the SD_1 algorithms (4.49) and (4.53)–(4.55). While the first two can be operated in a simulated setting, only the third is a practically feasible algorithm.

Now we consider a state space example

$$y_k = \underline{c}^T(\underline{\theta})\underline{\xi}_k(\underline{\theta}) + n_k \tag{4.56a}$$

$$\underline{\xi}_{k+1}(\underline{\theta}) = \underline{A}(\underline{\theta})\underline{\xi}_k(\underline{\theta}) + \underline{b}(\underline{\theta})x_k + \underline{v}_k \tag{4.56b}$$

where $(\underline{v}_k^T, n_k)$ is a zero mean white noise of covariance

$$\begin{bmatrix} \underline{Q}_v(\underline{\theta}) & \underline{0} \\ \underline{0}^T & Q_n(\underline{\theta}) \end{bmatrix}.$$

The state space model must be put into so-called innovations form (see Appendix B10) to generate an error signal. We have

$$e_k(\underline{\theta}) = y_k - \underline{c}^T(\underline{\theta})\hat{\underline{\xi}}_k(\underline{\theta}) \tag{4.57}$$

$$\begin{aligned} \hat{\underline{\xi}}_{k+1}(\underline{\theta}) &= \underline{A}(\underline{\theta})\hat{\underline{\xi}}_k(\underline{\theta}) + \underline{b}(\underline{\theta})x_k + \underline{K}(\underline{\theta})e_k(\underline{\theta}) \\ &= \left[\underline{A}(\underline{\theta}) - \underline{K}(\underline{\theta})\underline{c}^T(\underline{\theta})\right]\hat{\underline{\xi}}_k(\underline{\theta}) + \underline{b}(\underline{\theta})x_k + \underline{K}(\underline{\theta})y_k \end{aligned} \tag{4.58}$$

where $\underline{K}(\underline{\theta})$, the steady state Kalman gain, is obtained from the gain equations

$$\underline{K} = (\underline{APc})V^{-1} \tag{4.59a}$$

$$\underline{P} = \underline{A}(\underline{P} - \underline{Pcc}^T\underline{P}V^{-1})\underline{A}^T + \underline{Q}_v \tag{4.59b}$$

$$V = Q_n + \underline{c}^T\underline{Pc} \tag{4.59c}$$

in which the argument $\underline{\theta}$ has been dropped for ease.

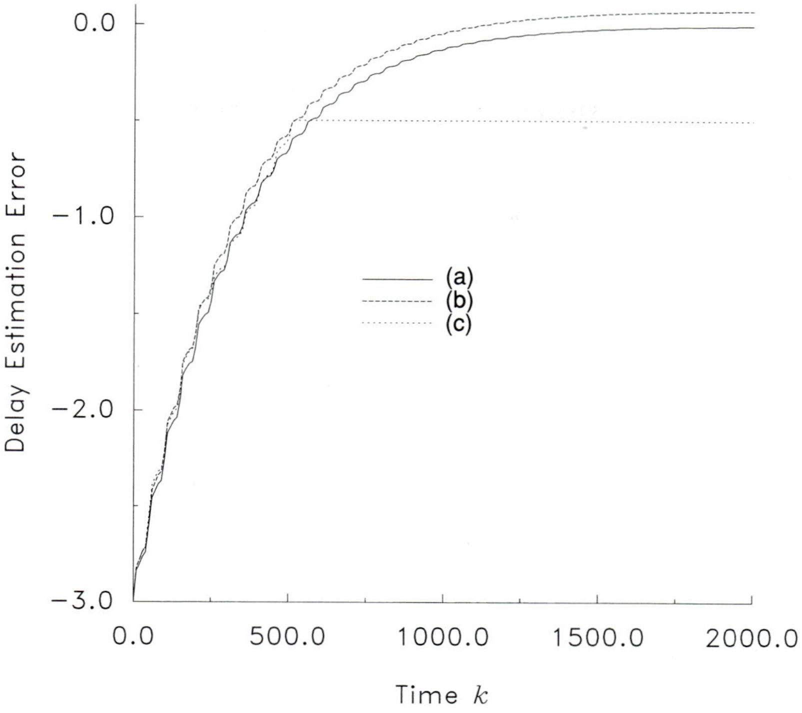

Figure 4.4 Delay estimation error in adaptive time delay estimation: (a) Ideal SD_1 (4.48), (b) SD_1 with approximate gradient (4.49), (c) SD_1 with quantization and monitoring (4.53)–(4.55).

The sensitivity signal is given by

$$\begin{aligned}\underline{\tilde{\varphi}} &= -de_k/d\underline{\theta} = d\hat{s}_k/d\underline{\theta} \\ &= \underline{\xi}^T_{k,\theta}(\underline{\theta})\underline{c}(\underline{\theta}) + \underline{c}^T_{\theta}(\underline{\theta})\underline{\xi}_k(\underline{\theta}) \end{aligned} \tag{4.60}$$

where

$$\underline{\xi}_{k,\theta} = d\underline{\xi}_k/d\underline{\theta}^T, \quad \underline{c}_{\theta} = d\underline{c}/d\underline{\theta}^T \tag{4.61}$$

and the sensitivity state $\underline{\xi}_{k\theta}$ is found by differentiating through the state equation to give

$$\begin{aligned}\underline{\xi}_{k+1,\theta} &= (\underline{A}(\underline{\theta}) - \underline{K}\underline{c}^T)\underline{\xi}_{k,\theta} + \underline{M}(\underline{\theta}, \underline{\xi}_k(\underline{\theta}), x_k) + \underline{K}_{\theta}e_k(\theta) - \underline{K}(\underline{\theta})\underline{\xi}^T_k(\underline{\theta})\underline{c}_{\theta}(\underline{\theta}) \\ &= \underline{A}(\underline{\theta})\underline{\xi}_{k,\theta} + \underline{M}(\underline{\theta}, \underline{\xi}_k(\underline{\theta}), x_k) + \underline{K}_{\theta}e_k(\theta) - \underline{K}(\underline{\theta})\underline{\tilde{\varphi}}^T_k \end{aligned} \tag{4.62}$$

where

$$\underline{M}(\underline{\theta}, \underline{\xi}, x) = \frac{d(\underline{A}(\underline{\theta})\underline{\xi} + \underline{b}(\underline{\theta})x)}{d\underline{\theta}^T}. \tag{4.63}$$

As before, the algorithm is made adaptive by ensuring the state and sensitivity equations have finite memory. This leads to the following algorithm:

$$\delta\hat{\underline{\theta}}_k = \mu\tilde{\underline{\varphi}}_k e_k \tag{4.64a}$$

$$e_k = y_k - \hat{s}_k = y_k - \underline{c}_{k-1}^T \hat{\underline{\xi}}_k \tag{4.64b}$$

augmented by the following auxiliary state equations:

$$\hat{\underline{\xi}}_{k+1} = \underline{A}_k \hat{\underline{\xi}}_k + \underline{b}_k x_k + \underline{K}_k e_k \tag{4.65a}$$

$$\tilde{\underline{\varphi}}_k = \hat{\underline{\xi}}_{k\theta}^T \underline{c}_{k-1} + \underline{c}_{\theta,k-1}^T \hat{\underline{\xi}}_k \tag{4.65b}$$

$$\hat{\underline{\xi}}_{k+1,\theta} = \underline{A}_k \hat{\underline{\xi}}_{k,\theta} + \underline{M}_k + \underline{K}_{\theta,k} e_k - \underline{K}_{k-1}\tilde{\underline{\varphi}}_k^T \tag{4.65c}$$

where

$$(\underline{A}_k, \underline{b}_k, \underline{K}_k, \underline{M}_k) = (\underline{A}(\hat{\underline{\theta}}_k), \underline{b}(\hat{\underline{\theta}}_k), \underline{K}(\hat{\underline{\theta}}_k), \underline{M}(\hat{\underline{\theta}}_k, \hat{\underline{\xi}}_k, x_k)) \tag{4.66a}$$

$$(\underline{c}_k, \underline{c}_{\theta,k}, \underline{K}_{\theta,k}) = (\underline{c}(\hat{\underline{\theta}}_k), \underline{c}_\theta(\hat{\underline{\theta}}_k), \underline{K}_\theta(\hat{\underline{\theta}}_k)). \tag{4.66b}$$

To obtain $\underline{K}_\theta$ it is necessary to differentiate through the gain equations: the tedious details are left to the reader. Of course, if the parameterization is in terms of $\underline{K}$, then there is no problem.

4.5.2 Other Error Criteria

Here for simplicity we consider an FIR filter

$$\hat{s}_k(\underline{w}) = \underline{x}_k^T \underline{w}$$

with error signal

$$e_k(\underline{w}) = y_k - \underline{x}_k^T \underline{w}.$$

But now use a pth order error criterion $|e_k(\underline{w})|^p$, $p > 1$; then the adaptive steepest descent algorithm takes the form

$$\delta \underline{\hat{w}}_k = -\frac{1}{2}\mu \, d|e_k(\underline{w})|^p/d\underline{w}\big|_{\underline{w}=\underline{\hat{w}}_{k-1}}$$

and the gradient is given by

$$\begin{aligned} d/d\underline{w}|e_k(\underline{w})|^p &= d/d\underline{w}[e_k^2(\underline{w})]^{p/2} \\ &= (p/2)[e_k^2(\underline{w})]^{p/2-1}2e_k(\underline{w})de_k/d\underline{w} \\ &= p|e_k(\underline{w})|^{p-1}\,\mathrm{sgn}(e_k(\underline{w}))de_k/d\underline{w} \\ &= -p|e_k(\underline{w})|^{p-1}\,\mathrm{sgn}(e_k(\underline{w}))\underline{x}_k \end{aligned} \tag{4.67}$$

so that the algorithm becomes

$$\delta \underline{\hat{w}}_k = \frac{1}{2}\mu p \underline{x}_k |e_k|^{p-1}\,\mathrm{sgn}(e_k) \tag{4.68}$$

$$e_k = y_k - \underline{x}_k^T \underline{\hat{w}}_{k-1}. \tag{4.69}$$

Note that although the derivation of (4.67) is only valid for $p > 1$, (4.68) makes sense when $p = 1$. In that case this SD_1 algorithm is called the signed-LMS algorithm.

Now we give a blind equalization example. Recall the error criterion discussed in Sections 2.3 and 3.3:

$$J_2(\underline{w}) = E\left(\hat{s}_k^2(\underline{w}) - R_2\right)^2.$$

The natural approximation to this for adaptive use is

$$J_{2,k}(\underline{w}) = \left(\hat{s}_k^2(\underline{w}) - R_2\right)^2$$

and this leads to an adaptive steepest descent algorithm of the form

$$\begin{aligned} \delta \underline{\hat{w}}_k &= -\frac{1}{2}\,\mu d J_{2,k}(\underline{w})/d\underline{w}\big|_{\underline{w}=\underline{\hat{w}}_{k-1}} \\ &= -\mu\left((\hat{s}_k(\underline{\hat{w}}_{k-1}))^2 - R_2\right)\hat{s}_k(\underline{\hat{w}}_{k-1})\underline{x}_k. \end{aligned} \tag{4.70}$$

Fig. 4.5 shows contours of the $J_2(\underline{w})$ criterion and trajectories of the adaptive SD algorithm applied to Example 3.8. Depending on the starting condition, a local or global minimum is reached.

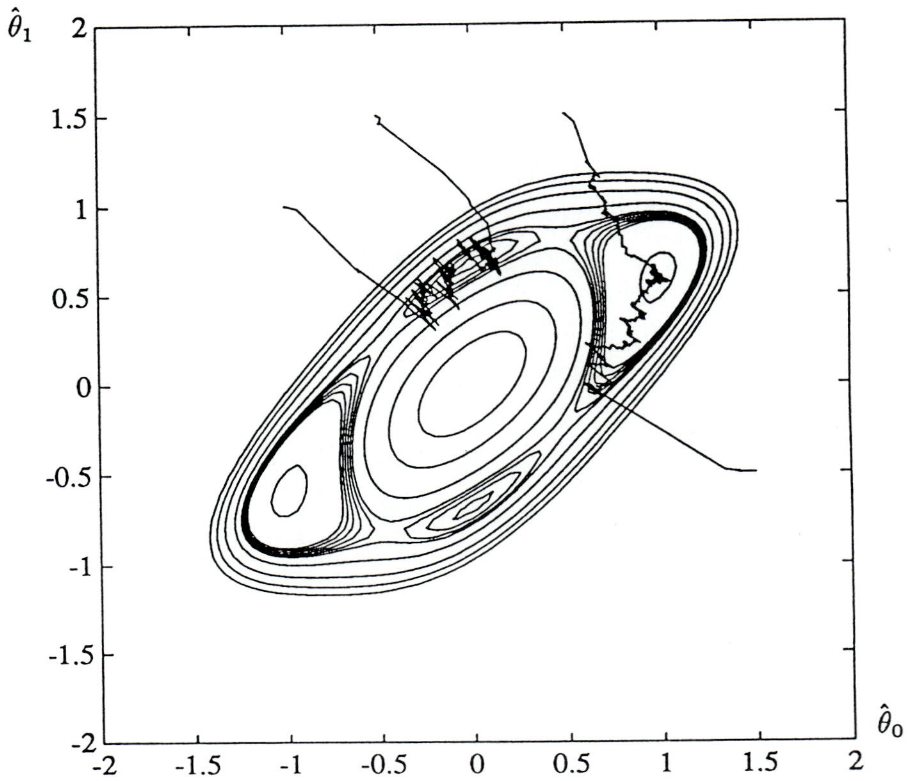

Figure 4.5 Contours of $J_2(\underline{w})$ criterion and adaptive SD trajectories for Example 3.8. Note that $\hat{\theta}_0 = w_0$, $\hat{\theta}_1 = w_1$. From Z. Ding *et al.*, *Ill-Convergence of Goddard Blind Equalizers in Data Communication Systems*, © September 1991, IEEE.

4.6 IMPROVEMENTS TO ADAPTIVE STEEPEST DESCENT

Guided by the discussion in Section 3.2, we look at modifications to the adaptive SD algorithms. To keep things clear we start, as usual, with LMS.

4.6.1 Differentiation Filters

The first idea in Section 3.2 was to implement steepest descent by an implicit procedure. This should increase the allowed range for μ. So we replace (4.6) with

$$\delta \hat{\underline{w}}_k = -\frac{1}{2}\, \mu d e_k^2(\underline{w})/d\underline{w}\Big|_{\underline{w}=\hat{\underline{w}}_k}$$

which leads to

$$\delta \underline{\hat{w}}_k = \mu \underline{x}_k \nu_k \tag{4.71}$$

$$\nu_k = y_k - \underline{x}_k^T \underline{\hat{w}}_k. \tag{4.72}$$

This algorithm cannot be used as it stands, since $\underline{\hat{w}}_k$ appears on both sides of (4.71). Some algebraic rearrangement will sort this out, however. We use (4.71) in (4.72) to find that

$$\begin{aligned} \nu_k &= y_k - \underline{x}_k^T (\underline{\hat{w}}_{k-1} + \mu \underline{x}_k \nu_k) \\ \Rightarrow \qquad \nu_k &= e_k / (1 + \mu \|\underline{x}_k\|^2) \\ e_k &= y_k - \underline{x}_k^T \underline{\hat{w}}_{k-1}. \end{aligned} \tag{4.73}$$

Thus (4.71) can be written

$$\delta \underline{\hat{w}}_k = \mu \underline{x}_k e_k / (1 + \mu \|\underline{x}_k\|^2). \tag{4.74}$$

Now (4.73), (4.74) provides a true adaptive algorithm. It is called the implicit or posterior LMS (LMS_P) algorithm. As expected from the discussion in Section 3.2, it allows large gains. Indeed as $\mu \to \infty$, (4.74) takes the limiting form

$$\delta \underline{\hat{w}}_k = \underline{x}_k e_k / \|\underline{x}_k\|^2. \tag{4.75}$$

The pair (4.73), (4.75) is called the normalized LMS (LMS_N) algorithm.

For nonlinear problems it was noted in Section 3.2 that the implicit method fails. As an example of the explicit methods or differentiation filter methods, we look at the SD time delay estimation algorithm with a momentum differentiation filter. So we replace (4.46) with

$$\delta \underline{\hat{\theta}}_k - \alpha \delta \underline{\hat{\theta}}_{k-1} = \mu \underline{\tilde{\varphi}}(\underline{\hat{\theta}}_{k-1}) e_k (\underline{\hat{\theta}}_{k-1}) \tag{4.76}$$

which becomes in the present case

$$\delta \hat{\theta}_k = \alpha \delta \hat{\theta}_{k-1} - \mu (s_{k-\hat{\theta}_{k-1}} - s_{k-\hat{\theta}_{k-1}-1}) e_k \tag{4.77}$$

$$e_k = s_{k-\theta_o} - s_{k-\hat{\theta}_{k-1}}. \tag{4.78}$$

4.6.2 Loop Filter Methods

The idea here is a straightforward application of the discussion in Section 3.2. We present two loop filter algorithms for illustration.

Loop Filter LMS

$$\delta \underline{M}_k = \mu (\underline{A} \underline{M}_{k-1} + \underline{b} \underline{x}_k^T e_k)$$

$$\delta \underline{\hat{w}}_k = \mu \underline{M}_k^T \underline{c} + \mu \underline{x}_k e_k$$

$$e_k = y_k - \underline{x}_k^T \underline{\hat{w}}_{k-1}$$

Loop Filter Time Delay Estimation (cf. (4.49))

$$\delta \underline{M}_k = \mu(A\underline{M}_{k-1} + \underline{b}\tilde{\varphi}_k e_k)$$

$$\tilde{\varphi}_k = -(s_{k-\hat{\theta}_{k-1}} - s_{k-\hat{\theta}_{k-1}-1})$$

$$e_k = s_{k-\theta_o} - s_{k-\hat{\theta}_{k-1}}$$

$$\delta\hat{\theta}_k = \mu \underline{M}_k^T \underline{c} + \mu\tilde{\varphi}_k e_k$$

4.6.3 Monitoring

It seems that many adaptive algorithms need to be monitored in order to ensure boundedness. For single time scale algorithms the parameter may need to be somehow contained in a bounded region. For mixed time scale algorithms the slow state (i.e., the parameter estimator) may need to be contained, not just in a bounded region, but in a stability region to help ensure that the fast states (i.e., the auxiliary signals) remain bounded.

The simplest way to ensure boundedness of the slow state is by an orthogonal projection. Thus, the algorithm takes the form

$$\underline{\theta}'_k = \hat{\underline{\theta}}_{k-1} + \mu\underline{\phi}_k e_k \tag{4.79}$$

$$\hat{\underline{\theta}}_k = P_M(\underline{\theta}'_k)$$

where $\underline{\phi}_k$ is a gradient, e_k an error signal, and $P_M(\underline{\theta}')$ denotes orthogonal projection onto a "convenient" compact convex set M. Note that

$$\|P_M(\underline{\theta}') - \underline{\theta}\| \leq \|\underline{\theta}' - \underline{\theta}\|, \quad \underline{\theta} \in M$$

$$\Rightarrow \quad \|\hat{\underline{\theta}}_k - \hat{\underline{\theta}}_{k-1}\| \leq \|\underline{\theta}'_k - \hat{\underline{\theta}}_{k-1}\|.$$

The most common case is that M is a ball of radius R; then

$$P_M(\underline{\theta}') = \begin{cases} \underline{\theta}', & \text{for } \|\underline{\theta}'\| \leq R \\ \underline{\theta}'R/\|\underline{\theta}'\|, & \text{otherwise.} \end{cases}$$

We consider this case in more detail. Suppose at time k, $\hat{\underline{\theta}}_{k-1}$ is inside M while $\underline{\theta}'_k$ is outside M: then we have $\|\hat{\underline{\theta}}_k\| = R$. Now consider the next stage:

$$\hat{\underline{\theta}}'_{k+1} = \hat{\underline{\theta}}_k + \mu\underline{\phi}_{k+1}e_{k+1}$$

$$\Rightarrow \quad \|\hat{\underline{\theta}}'_{k+1}\|^2 = \|\hat{\underline{\theta}}_k\|^2 + \mu^2\|\underline{\phi}_{k+1}\|^2 e_{k+1}^2 + 2\mu e_{k+1}\hat{\underline{\theta}}_k^T\underline{\phi}_{k+1}.$$

Thus in order for the estimator to get off the boundary of M it is necessary that

$$2e_{k+1}\hat{\underline{\theta}}_k^T\underline{\phi}_{k+1} < -\mu\|\underline{\phi}_{k+1}\|^2 e_{k+1}^2 \tag{4.80}$$

which may or may not occur. So the algorithm may stick on the boundary for many time steps. If we use an oblique projection, it is possible to get the algorithm off the boundary in one step. Consider then the following scheme:

$$\hat{\underline{\theta}}_k = P_p(\hat{\underline{\theta}}_{k-1}, \mu\underline{\phi}_k e_k) \tag{4.81}$$

where

$$\underline{\hat{\theta}} = P_p(\underline{\theta}, \underline{d})$$
$$= \begin{cases} \underline{\theta} + \underline{d}, & \text{if } \|\underline{\theta} + \underline{d}\| < R \\ (\underline{\theta} + \underline{d})(1 - \underline{d}^T(\underline{\theta} + \underline{d})/\|\underline{\theta} + \underline{d}\|^2, & \text{otherwise.} \end{cases}$$

Then consider that if $\|\underline{\theta} + \underline{d}\| \geq R$

$$\begin{aligned} \|\underline{\hat{\theta}}\|^2 &= \|\underline{\theta} + \underline{d}\|^2 \left[1 - \frac{2\underline{d}^T(\underline{\theta} + \underline{d})}{\|\underline{\theta} + \underline{d}\|^2} + \frac{(\underline{d}^T(\underline{\theta} + \underline{d}))^2}{\|\underline{\theta} + \underline{d}\|^4} \right] \\ &= \|\underline{\theta} + \underline{d}\|^2 - 2\underline{d}^T(\underline{\theta} + \underline{d}) + (\underline{d}^T(\underline{\theta} + \underline{d}))^2/\|\underline{\theta} + \underline{d}\|^2 \\ &= \|\underline{\theta}\|^2 - \|\underline{d}\|^2 \left[1 - \frac{(\underline{d}^T(\underline{\theta} + \underline{d}))^2}{\|\underline{\theta} + \underline{d}\|^2 \|\underline{d}\|^2} \right] \\ &\leq \|\underline{\theta}\|^2. \end{aligned}$$

Thus in (4.81)

$$\|\underline{\hat{\theta}}_k\|^2 \leq \|\underline{\hat{\theta}}_{k-1}\|^2. \tag{4.82}$$

Thus if $\|\underline{\hat{\theta}}_{k-1}\| < R$ and $\|\underline{\theta}'_k\| \geq R$, then $\|\underline{\hat{\theta}}_k\| < R$. So if we start with $\|\underline{\hat{\theta}}_0\| < R$, then we always have $\|\underline{\hat{\theta}}_k\| < R$. Furthermore, the algorithm moves off the boundary in one step in view of (4.82). Although the two projection methods are similar in other ways (see Exercise 4.10), we thus prefer the oblique projection (4.81).

Now we turn to boundedness of mixed time scale algorithms. Because the fast state is time varying (say, as in (4.65a)), it is not enough to only ensure that its characteristic polynomial is stable at each time k. The time variation must also be slow (see Appendix C6(v)). This means that the gain μ in (4.79) must be small.

With regard, then, to keeping a parameter trajectory inside a stability region, several methods have been suggested:

(i) The stability region can be partitioned into a finite collection of nonconnected convex compact sets. Project the parameters orthogonally at each time k into the nearest such set. The method is hard to use in dimensions > 2 because the convex compact sets are hard to describe.

(ii) At each time k find the roots of the characteristic polynomial of the fast state and reflect (into the unit circle) those roots whose modulus is > 1. Root finding is not a numerically well-conditioned computation and must be done iteratively. Thus this procedure is not so straightforward in real time.

(iii) Reparameterize the polynomial in terms of reflection coefficients (also called partial autocorrelation coefficients) and estimate these new parameters adaptively. Stability is attained when all reflection coefficients have modulus < 1. As the time variation is slow, at each time k just reflect those coefficients which have a modulus > 1. The difficulty here is that the reparameterization is very messy and leads to a complicated and highly nonlinear algorithm.

(iv) We now present a method that can be easily used in real time and does ensure stability. It is based on Wilson's algorithm [W7] for spectral factorization (see Appendix B11). Wilson's algorithm is an iterative (Newton) algorithm for spectral factorization of a positive semi-definite MA polynomial. It has the important feature that if initialized with a stable spectral factor, then all subsequent iterates are stable.

In Appendix B10 it is pointed out that the Kalman Filter performs spectral factorization of an ARMA spectrum. When applied to MA spectral factorization it reduces to Bauer's algorithm [B3], but successive iterates are only guaranteed to be stable near convergence. The procedure, then, is as follows.

(a) Suppose we calculate the p-vector $\underline{\beta}'_k$ at time k from an adaptive algorithm and that the corresponding polynomial (which is desired to be stable) is

$$\beta'_k(q) = \sum_{u=0}^{p} \beta_{ku} q^u.$$

(b) Choose a shrinkage factor ρ, $0 < \rho < 1$, ρ near 1.

(c) Perform a stability test on $\beta'_k(q\rho^{-1})$ (e.g., using the Schur-Cohn test of Appendix C4(v)). If $\beta'_k(q\rho^{-1})$ is stable, set

$$\beta_k(q) = \beta'_k(q).$$

Otherwise continue to the next step.

(d) Form the MA polynomial

$$\gamma_k(q) = \beta'_k(q^{-1}\rho)\beta'_k(q\rho^{-1}) = \sum_{u=-p}^{p} \gamma_{ku} q^{-u}.$$

By construction $\gamma_k(q)$ is positive semi-definite on the unit circle, i.e., $\gamma_k(e^{-j\omega}) \geq 0$.

(e) Suppose $\beta_{k-1}(q)$ is the stable polynomial obtained at time $k-1$ (with corresponding parameter $\hat{\underline{\beta}}_{k-1}$). Then do one step (or even a finite number of steps) of Wilson's algorithm on $\beta_{k-1}(q)$, $\gamma_k(q)$ to get $\beta''_k(q)$ which has all roots of modulus > 1. That is

$$\underline{\beta}^*_k = \underline{T}^{-1}(\hat{\underline{\beta}}_{k-1})\underline{\gamma}_k$$

$$\underline{\gamma}_k = (\gamma_{k0} \cdots \gamma_{kp})^T$$

and then

$$\underline{\beta}''_k = \frac{1}{2}\hat{\underline{\beta}}_{k-1} + \underline{\beta}^*_k$$

and $\underline{T}(\underline{\beta})$ is given in (B11.7), Appendix B. Now set

$$\beta_k(q) = \beta''_k(\rho^{-1}q).$$

Then $\beta_k(q)$ has all roots of modulus $> \rho > 1$. Return to (a).

In view of (c) and (e), all iterates have roots of modulus $> \rho > 1$.

Example 4.3. Improved LMS Algorithms

We continue with the channel equalization Example 4.1. In Fig. 4.6 three trajectories are shown: (a) LMS, (b) LMS_P, (c) LMS_N. The LMS trajectory diverges because the gain μ is too large. Guided by our analysis of differentiation filter methods in Section 3.2, we expect that LMS_P (and LMS_N), being an implicit differentiation filter modification, will be well behaved whatever the value of μ.

Example 4.4. Improved SD—Time Delay Estimation

Here we continue with Example 4.2. In Fig. 4.7 are plots of the SD (4.48) and momentum SD algorithms (4.77), (4.78). Again, the discussion of Section 3.2 would lead us to expect that momentum-SD would remain well behaved for a wider range of μ values than would the SD algorithm.

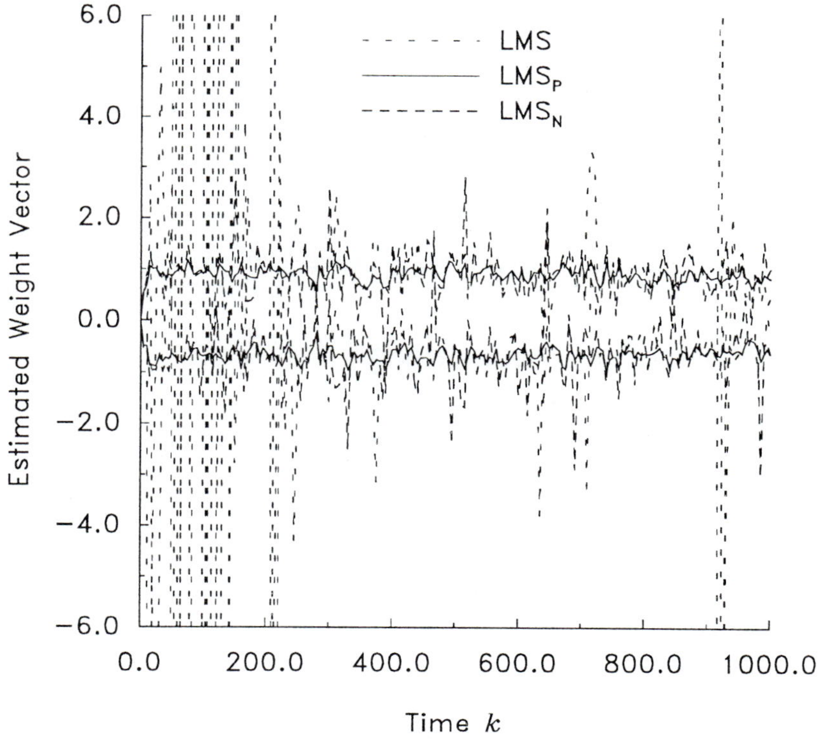

Figure 4.6 Improved LMS algorithms (a) LMS, (b) LMS_P, (c) LMS_N.

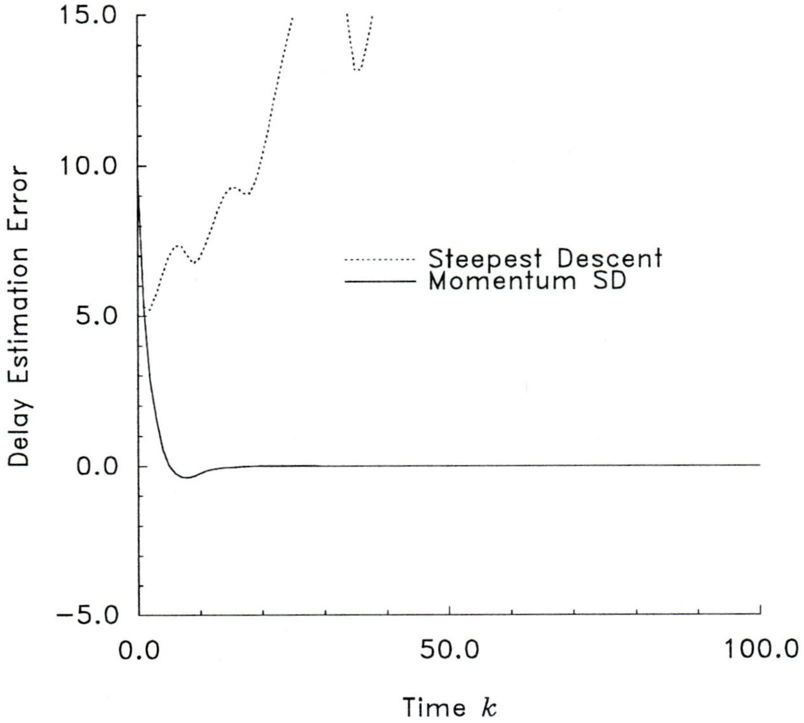

Figure 4.7 Delay estimation error with $\mu = 10.0$. (a) SD (4.48), (b) momentum-SD (4.77).

4.7 NEWTON'S METHOD ALGORITHMS

As remarked earlier (Section 3.2), the obvious weakness of steepest descent as compared to Newton's method is the slow rate of convergence. If we try to apply Newton's method directly to $e_k^2(\underline{\theta})$, it fails because the second derivative is a rank one matrix. It seems necessary then to supply a modified criterion, and this is the starting point for the use of EWSE, which leads to exponentially weighted least squares (EWLS).

An alternative procedure is to acknowledge explicitly the idea that $\underline{\theta}$ is really time varying, by trying to model its variation in a simple way. Such model based methods are discussed below.

4.7.1 Exponentially Weighted Least Squares

To keep the discussion straightforward, we work with an FIR filter and its corresponding error signal

$$e_k(\underline{w}) = y_k - \underline{x}_k^T \underline{w}$$

and the EWSE criterion

$$J_k(\underline{w}) = \frac{1}{2}\sum_{r=1}^{k} \lambda^{k-r} e_r^2(\underline{w})$$
$$= \frac{1}{2}(e_k^2(\underline{w}) + \lambda e_{k-1}^2(\underline{w}) + \lambda^2 e_{k-2}^2(\underline{w}) + \cdots).$$

The idea now is to minimize $J_k(\underline{w})$. For the Newton method we need the gradient

$$\underline{g}_k(\underline{w}) = dJ_k(\underline{w})/d\underline{w} = -\sum_{r=1}^{k} \lambda^{k-r} \underline{x}_r e_r(\underline{w}) = -\sum_{r=1}^{k} \lambda^{k-r} \underline{x}_r (y_r - \underline{x}_r^T \underline{w})$$

and the Hessian

$$\underline{H}(\underline{w}) = \underline{H}_k = d^2 J_k / d\underline{w} d\underline{w}^T = \sum_{r=1}^{k} \lambda^{k-r} \underline{x}_r \underline{x}_r^T.$$

Note that we can thus write

$$\underline{g}_k(\underline{w}) = -(\underline{s}_k - \underline{H}_k \underline{w})$$
$$\underline{s}_k = \sum_{r=1}^{k} \lambda^{k-r} \underline{x}_r y_r.$$

The Newton algorithm (cf. Section 3.2) is then

$$\delta \underline{\hat{w}}_k = -\underline{H}_k^{-1} \underline{g}_k(\underline{\hat{w}}_{k-1}) = \underline{H}_k^{-1}(\underline{s}_k - \underline{H}_k \underline{\hat{w}}_{k-1})$$
$$\Rightarrow \qquad \underline{\hat{w}}_k = \underline{H}_k^{-1} \underline{s}_k.$$

As it stands this is not suitable as an adaptive algorithm, because apparently all previous data must be stored to compute $\underline{H}_k$ and $\underline{s}_k$. However, we can solve this problem with a little algebra. Observe that

$$\underline{H}_k = \lambda \underline{H}_{k-1} + \underline{x}_k \underline{x}_k^T \tag{4.83}$$

while

$$\underline{s}_k = \lambda \underline{s}_{k-1} + \underline{x}_k y_k. \tag{4.84}$$

Thus we find using the above relations that

$$\underline{\hat{w}}_k = \underline{H}_k^{-1}(\lambda \underline{s}_{k-1} + \underline{x}_k y_k)$$
$$= \lambda \underline{H}_k^{-1}(\underline{H}_{k-1} \underline{\hat{w}}_{k-1}) + \underline{H}_k^{-1} \underline{x}_k y_k$$
$$= \underline{H}_k^{-1}(\underline{H}_k - \underline{x}_k \underline{x}_k^T)\underline{\hat{w}}_{k-1} + \underline{H}_k^{-1} \underline{x}_k y_k$$

$$\Rightarrow \qquad \underline{\hat{w}}_k = \underline{\hat{w}}_{k-1} + \underline{H}_k^{-1} \underline{x}_k e_k \tag{4.85}$$

$$e_k = y_k - \underline{x}_k^T \underline{\hat{w}}_{k-1}. \tag{4.86}$$

We have achieved a true (single time scale) adaptive algorithm with the set (4.83), (4.85), (4.86), but the inversion of $\underline{H}_k$ at each step is computationally expensive. However, since the update in (4.83) is only a rank one change, we can convert it into an update for $\underline{H}_k^{-1}$. Using the matrix inversion lemma of Appendix A we find from (4.83) that

$$\underline{H}_k^{-1} = \lambda^{-1}\underline{H}_{k-1}^{-1} - \lambda^{-1}\underline{H}_{k-1}^{-1}\underline{x}_k\underline{x}_k^T\underline{H}_{k-1}^{-1}\lambda^{-1}/(1 + \underline{x}_k^T\underline{H}_{k-1}^{-1}\underline{x}_k\lambda^{-1}).$$

If we denote $\underline{R}_k = \underline{H}_k^{-1}$ we can rewrite this as

$$\underline{R}_k = \lambda^{-1}\underline{R}_{k-1} - \lambda^{-1}\underline{R}_{k-1}\underline{x}_k\underline{x}_k^T\underline{R}_{k-1}/(\lambda + \underline{x}_k^T\underline{R}_{k-1}\underline{x}_k) \tag{4.87}$$

so that (4.85) becomes

$$\delta\underline{\hat{w}}_k = \underline{R}_k\underline{x}_k e_k. \tag{4.88}$$

From (4.87) we find

$$\underline{R}_k\underline{x}_k = \underline{R}_{k-1}\underline{x}_k/(\lambda + \underline{x}_k^T\underline{R}_{k-1}\underline{x}_k)$$

so that (4.88) can be written

$$\delta\underline{\hat{w}}_k = \underline{R}_{k-1}\underline{x}_k e_k/(\lambda + \underline{x}_k^T\underline{R}_{k-1}\underline{x}_k). \tag{4.89}$$

This saves explicitly calculating $\underline{R}_k\underline{x}_k$.

Now the algorithm consists of the set (4.86), (4.87), (4.89). This is not the most computationally efficient set of equations but it is nearly so and is fully adaptive. One important point to mention is the initial condition $\underline{R}_0$ for $\underline{R}_k$. This is often taken as a $\underline{R}_0 = cI$ where c is large. This ensures that $\underline{R}_k$ has full rank for all $k \geq 0$ and that the effect of the initial condition dies out quickly. Another important point to note is the possibility that the update formula (4.87) will cause $\underline{R}_k$ to lose its positive definiteness. To avoid this, a square root algorithm is usually used (see Exercise 4.5).

It is straightforward to develop the EWLS algorithm for a general error signal specification such as (4.43) and (4.44). Clearly $\underline{x}_k$ is replaced by $d\hat{s}_k/d\underline{\hat{\theta}}_{k-1}$ in (4.87) and (4.89) and e_k in (4.86) is replaced by $y_k - \hat{s}_k(\underline{\hat{\theta}}_{k-1})$. Then, as discussed in Sections 4.4 and 4.5, we replace $d\hat{s}_k/d\underline{\hat{\theta}}_{k-1}$, $\hat{s}_k(\underline{\hat{\theta}}_{k-1})$ with finite memory approximations.

4.7.2 Model Based Algorithms—Kalman Filter Algorithm

The idea of model based procedures of algorithm construction is to recognize explicitly that parameters may be time varying by positing a model for them. Parameter adaptation then becomes state estimation, and stochastic filtering methods can be used to solve these problems. Since the time varying parameters are not directly observed, it is preferred to use models that capture the idea of parameter variation without being too elaborate. As usual we work with the FIR case for simplicity.

The simplest idea is to model the true time varying weight vector as a first order autoregression

$$\underline{w}_{o,k} = a\underline{w}_{o,k-1} + \underline{\delta}_k, \quad |a| < 1$$

where $\underline{\delta}_k$ is a white noise of variance $\sigma_\delta^2 I$. This is a simple stationary model described by two so-called hyper-parameters a, σ_δ^2. The time varying weight is simultaneously regarded as a state and a parameter. To complete the model we add in the FIR specification

$$y_k = \underline{x}_k^T \underline{w}_{o,k} + \epsilon_k \tag{4.90}$$

where we suppose ϵ_k is a white noise of variance σ_ϵ^2. Also suppose $\{\underline{x}_k\}$, $\{\epsilon_k\}$, $\{\underline{\delta}_k\}$ are jointly independent.

The problem now is to estimate the stochastic signal $\underline{w}_{o,k}$ from the measured signals $\underline{x}_k, y_k$. The solution is obtained from the Kalman Filter (see Appendix B9) as follows:

$$\underline{\hat{w}}_k = a\underline{\hat{w}}_{k-1} + \underline{g}_k e_k \tag{4.91}$$

$$e_k = y_k - \underline{x}_k^T \underline{\hat{w}}_{k-1} \tag{4.92}$$

$$\underline{g}_k = a\underline{P}_k \underline{x}_k / V_k \tag{4.93}$$

$$V_k = \sigma_\epsilon^2 + \underline{x}_k^T \underline{P}_k \underline{x}_k \tag{4.94}$$

$$\underline{P}_{k+1} = a^2(\underline{P}_k - \underline{P}_k \underline{x}_k \underline{x}_k^T \underline{P}_k V_k^{-1}) + \sigma_\delta^2 I. \tag{4.95}$$

Notice that (4.91) and (4.92) have the standard form (4.1), but the gradient signal is rather more complicated than it is for LMS. It is useful to rescale these equations as follows. Introduce

$$\underline{M}_k = \underline{P}_k / \sigma_\epsilon^2, \qquad V_{ok} = V_k / \sigma_\epsilon^2. \tag{4.96}$$

Then (4.93)–(4.95) can be replaced by

$$\underline{g}_k = a\underline{M}_k \underline{x}_k / V_{ok} \tag{4.97}$$

$$V_{ok} = 1 + \underline{x}_k^T \underline{M}_k \underline{x}_k \tag{4.98}$$

$$\underline{M}_{k+1} = a^2(\underline{M}_k - \underline{M}_k \underline{x}_k \underline{x}_k^T \underline{M}_k V_{ok}^{-1}) + \sigma_\delta^2 \sigma_\epsilon^{-2} I. \tag{4.99}$$

It is also most important to note that the use of (4.97)–(4.99) requires knowledge of the so-called hyper-parameters

$$\underline{\gamma} = (\sigma_\delta^2 / \sigma_\epsilon^2,\ a)^T.$$

Even if we believe the autoregressive model (it is really intended only to capture the time varying behavior in a general way), $\underline{w}_{o,k}$ is not observed so $\underline{\gamma}$ is not usually known. Still we need to provide a value for $\underline{\gamma}$ and, noting that e_k depends on $\underline{\gamma}$, the natural thing to do is to estimate $\underline{\gamma}$ online (by a long memory algorithm) so as to minimize $E(e_k^2(\underline{\gamma}))$. Such a procedure is rather complicated and at the time of writing there is not much reliable experience or any analysis to speak of. Much more research needs to be done in this area.

By way of comparison it is helpful to note that other methods can often be reinterpreted from a model based point of view. We give two examples:

(i) EWLS

Take $a = 1$ (which corresponds to a random walk)

$\underline{P}_k = \underline{R}_{k-1}$

$\sigma_\epsilon^2 = \lambda$

$\underline{\Sigma}_{\delta k} = (1-\lambda)\lambda^{-1}[\underline{R}_{k-1} - \underline{R}_{k-1}\underline{x}_k\underline{x}_k^T\underline{R}_{k-1}/(\lambda + \underline{x}_k^T\underline{R}_{k-1}\underline{x}_k)] = (1-\lambda)\underline{R}_k$

(by (4.87))

where $\underline{\Sigma}_{\delta k}$ is a time varying covariance matrix that replaces $\sigma_\delta^2 I$ in (4.95).

(ii) LMS_P

Take $a = 1$ (Random walk model again)

$\underline{P}_k = \mu I$

$\sigma_\epsilon^2 = 1$

$\underline{\Sigma}_{\delta k} = \mu^2\underline{x}_k\underline{x}_k^T/(1 + \mu\|\underline{x}_k\|^2)$

Example 4.5. An NM Algorithm and Its Estimated Learning Curve

Once again we pursue Examples 2.2, 4.1. A plot of the trajectory of the NM_1 algorithm (4.86)–(4.88) is shown in Fig. 4.8 ($\lambda = 0.9$). Note the characteristic hovering around the optimal weight.

In Fig. 4.9 is an estimated learning curve together with plots repeated from Fig. 4.2. We see that the NM_1 algorithm settles down more quickly. This is what we could expect from the discussion at the end of Section 3.2.

There are two approaches possible to develop a model based algorithm for a general error signal specification such as (4.43), (4.44). The first idea is to replace $\underline{x}_k$ in (4.93)–(4.95) by $d\hat{s}_k/d\hat{\underline{\theta}}_{k-1}$ and replace e_k in (4.91), (4.92) by $y_k - \hat{s}_k(\hat{\underline{\theta}}_{k-1})$. Then, as discussed earlier, replace $d\hat{s}_k/d\hat{\underline{\theta}}_{k-1}$, $\hat{s}_k(\hat{\underline{\theta}}_{k-1})$ with finite memory approximations. The second approach is based on the so-called extended Kalman filter (EKF) (see Appendix B12).

The idea of the EKF is illustrated with a state space model, namely (4.56). To apply the EKF we append the parameter model

$$\underline{\theta}_k = a\underline{\theta}_{k-1} + \underline{\delta}_k \tag{4.100}$$

(where $\underline{\delta}_k$ is a white noise of variance $\sigma_\delta^2 I$ and is independent of $\underline{v}_k$, n_k in (4.56)) to the state equation thus obtaining a nonlinear system

$$\underline{\zeta}_{k+1} = \underline{f}(\underline{\zeta}_k) + \underline{\bar{v}}_k \tag{4.101a}$$

$$y_k = h(\underline{\zeta}_k) + n_k \tag{4.101b}$$

where

$$\underline{\zeta}_k = (\underline{\xi}_k^T, \underline{\theta}_k^T)^T$$

$$\underline{\bar{v}}_k = (\underline{v}_k^T, \underline{\delta}_k^T)^T$$

$$h(\underline{\zeta}_k) = \underline{c}^T(\underline{\theta}_k)\underline{\xi}_k$$

$$\underline{f}(\underline{\zeta}_k) = \begin{bmatrix} \underline{A}(\underline{\theta}_k)\underline{\xi}_k + \underline{b}(\underline{\theta}_k)x_k \\ a\underline{\theta}_k \end{bmatrix}.$$

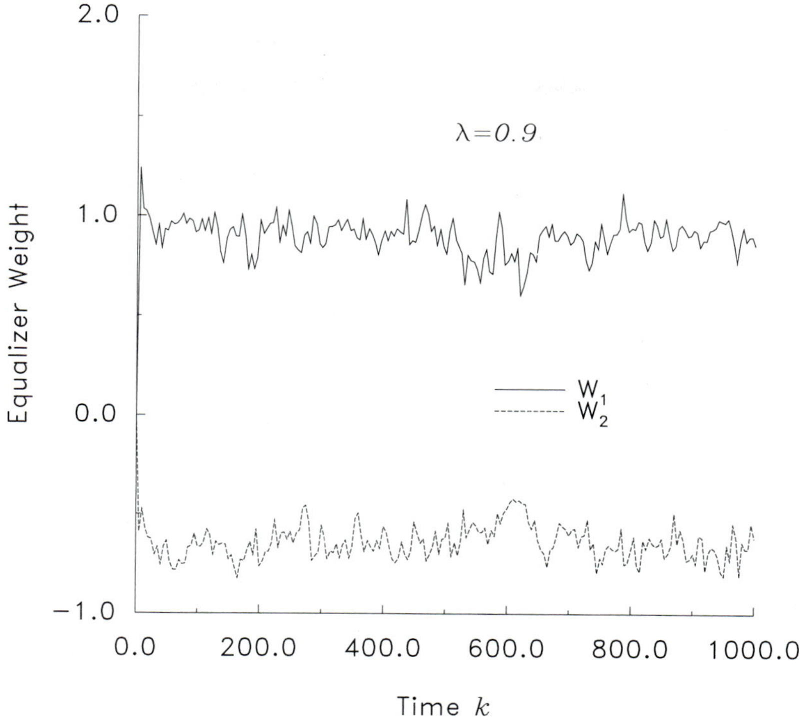

Figure 4.8 NM_1 trajectory for Example 4.5.

We next calculate the gradients

$$\frac{\partial \underline{f}}{\partial \underline{\xi}^T} = \begin{bmatrix} \underline{A}(\theta) & \underline{M}(\underline{\theta}, \underline{\xi}, x) \\ 0 & a\underline{I} \end{bmatrix}$$

$$\frac{\partial h}{\partial \underline{\xi}^T} = [\, \underline{c}^T(\underline{\theta}) \quad \underline{\xi}^T \underline{c}_\theta(\underline{\theta}) \,]$$

where $\underline{M}(\underline{\theta}, \underline{\xi}, x)$ is given in (4.63). Now we can apply the EKF and are led to the following adaptive algorithm

$$\begin{pmatrix} \hat{\underline{\xi}}_{k+1} \\ \hat{\underline{\theta}}_{k+1} \end{pmatrix} = \begin{pmatrix} \underline{A}_k \hat{\underline{\xi}}_k \\ a\hat{\underline{\theta}}_k \end{pmatrix} + \begin{pmatrix} \underline{G}_k \\ \underline{L}_k \end{pmatrix} e_{k+1} + \begin{pmatrix} \underline{b}_k x_k \\ \underline{0} \end{pmatrix}$$

$$e_{k+1} = y_{k+1} - \underline{c}_k^T \hat{\underline{\xi}}_k$$

$$\begin{pmatrix} \underline{G}_k \\ \underline{L}_k \end{pmatrix} = \begin{pmatrix} \underline{A}_k & \underline{M}_k \\ 0 & a\underline{I} \end{pmatrix} \begin{pmatrix} \underline{P}_{k,1} & \underline{P}_{k,c} \\ \underline{P}_{k,c}^T & \underline{P}_{k,2} \end{pmatrix} \begin{pmatrix} \underline{c}_k \\ \underline{c}_{\theta,k}^T \hat{\underline{\xi}}_k \end{pmatrix} V_k^{-1} \qquad (4.102)$$

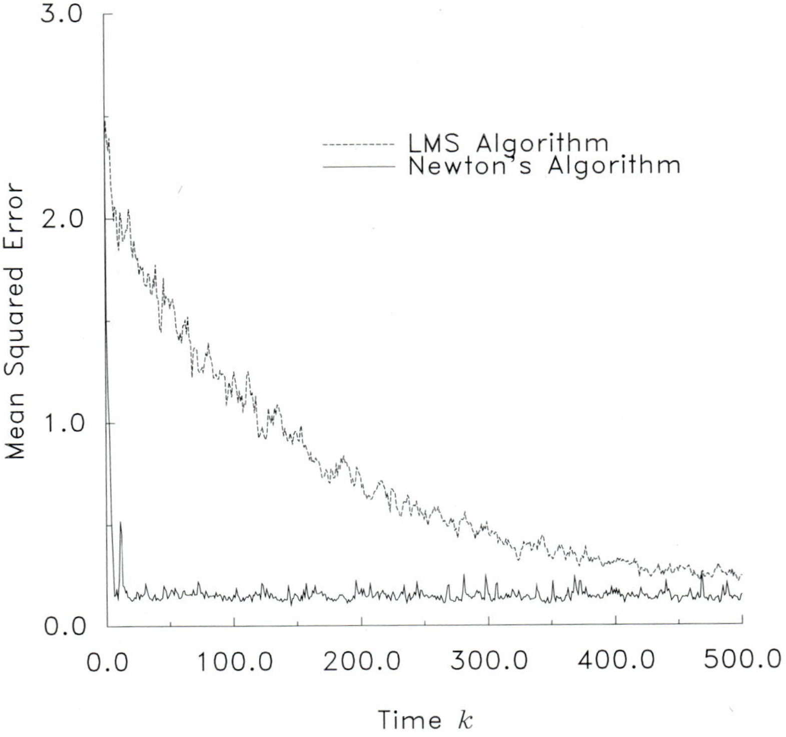

Figure 4.9 Estimated learning curves for Example 4.5. (a) LMS, $\mu = 0.002$; (b) NM_1, (4.86)–(4.88), $\lambda = 0.9$

where $\underline{A}_k$, $\underline{M}_k$, $\underline{c}_k$, and $\underline{c}_{\theta,k}$ are given in (4.66). Also

$$\begin{pmatrix} \underline{P}_{k+1,1} & \underline{P}_{k+1,c} \\ \underline{P}^T_{k+1,c} & \underline{P}_{k+1,2} \end{pmatrix} = \begin{pmatrix} \underline{A}_k & \underline{M}_k \\ \underline{0}^T & a\underline{I} \end{pmatrix} \begin{pmatrix} \underline{P}_{k,1} & \underline{P}_{k,c} \\ \underline{P}^T_{k,c} & \underline{P}_{k,2} \end{pmatrix} \begin{pmatrix} \underline{A}^T_k & \underline{0} \\ \underline{M}^T_k & aI \end{pmatrix} - \begin{pmatrix} \underline{G}_k \\ \underline{L}_k \end{pmatrix} V_k \,(\underline{G}^T_k \quad \underline{L}^T_k) + \begin{pmatrix} \underline{Q}_k & \underline{0} \\ \underline{0}^T & \sigma^2_\delta I \end{pmatrix}$$

$$V_k = \sigma^2_k + (\underline{c}_k \quad \underline{c}^T_{\theta,k}\hat{\underline{\xi}}_k) \begin{pmatrix} \underline{P}_{k,1} & \underline{P}_{k,c} \\ \underline{P}^T_{k,c} & \underline{P}_{k,2} \end{pmatrix} \begin{pmatrix} \underline{c}^T_k \\ \hat{\underline{\xi}}^T_k \underline{c}_{\theta,k} \end{pmatrix}$$

where $(\underline{Q}_k \ \sigma^2_k) = (\underline{Q}_v(\hat{\underline{\theta}}_k) \ Q_n(\hat{\underline{\theta}}_k))$ and $\hat{\underline{\xi}}_k$ is generated in (4.64b) and (4.65a).

It is interesting to compare this algorithm with the NM algorithm obtained by replacing $\underline{x}_k$ by $d\hat{\underline{s}}_k/\underline{\theta}$ in (4.93)–(4.95). This alternative algorithm takes the following form:

$$\underline{\hat{\theta}}_k = a\underline{\hat{\theta}}_{k-1} + \underline{g}_k e_k$$
$$e_k = y_k - \hat{s}_k = y_k - \underline{c}_{k-1}^T \underline{\hat{\xi}}_k$$
$$\underline{g}_k = a\underline{P}_k \underline{\tilde{\varphi}}_k V_k^{-1}$$
$$V_k = \sigma_k^2 + \underline{\tilde{\varphi}}_k^T \underline{P}_k \underline{\tilde{\varphi}}_k$$
$$\underline{P}_{k+1} = a^2(\underline{P}_k - \underline{P}_k \underline{\tilde{\varphi}}_k \underline{\tilde{\varphi}}_k^T \underline{P}_k V_k^{-1}) + \sigma_\delta^2 I$$

and $\underline{\hat{\xi}}_k$, $\underline{\tilde{\varphi}}_k$ are given in (4.64)–(4.66).

Returning to the EKF, notice that the equation for $\underline{P}_{k,c}$ is

$$\underline{P}_{k+1,c} = a(\underline{A}_k \underline{P}_{k,c} + \underline{M}_k \underline{P}_{k,2}) - \underline{G}_k V_k \underline{L}_k^T. \tag{4.103}$$

This can be compared to the state sensitivity equation for $\underline{\hat{\xi}}_{k,\theta}$, (4.65c). Notice that $\underline{P}_{k,2}$ post-multiplies $\underline{M}_k$ here, but more significantly, the term $\underline{K}_{\theta,k} e_k$ is missing here. Continuing, we see that the parameter gain is

$$\underline{L}_k = a(\underline{P}_{k,c}^T \underline{c}_k + \underline{P}_{k,2} \underline{c}_{\theta,k}^T \underline{\hat{\xi}}_k) V_k^{-1} \tag{4.104}$$

and this can be compared to the sensitivity signal $\underline{\tilde{\varphi}}_k$ generated in (4.65a), (4.65b). Again a $\underline{P}_{k,2}$ multiplier appears here.

4.8 GENERALIZATIONS

In this section we briefly discuss algorithm construction in two settings not so far included.

4.8.1 Continuous Discrete Algorithms

It is often the case, particularly in system identification, that the relation between primary and reference signals is best expressed in an analog (i.e., continuous time) way and the filter parameters are thus best specified as analog filter parameters. We illustrate the ideas with a system identification example. The idea in what follows is not to discretise the system equations (among other things this will cause linearly appearing parameters to appear nonlinearly) and then construct an algorithm. Rather the idea is to save the discretisation until the last possible stage; that is to discretise the sensitivity equations.

Suppose, then, we have an analog relation between two signals $s(t)$, $x(t)$

$$s(t) = B(D)(D^m + A(D))^{-1} x(t) \tag{4.105}$$

where $D = d/dt$ and

$$B(D) = \sum_{s=0}^{m-1} b_s D^s$$
$$A(D) = \sum_{s=0}^{m-1} a_s D^s.$$

Suppose also that we collect discrete noisy data y_k at a sampling rate Δ^{-1} samples per unit time so that

$$y_k = s_k + n_k$$
$$x_k = x(k\Delta), \qquad s_k = s(k\Delta)$$

where n_k is a discrete time noise. An adaptive steepest descent algorithm will have the preliminary form

$$\delta\underline{\hat{\theta}}_k = -\frac{1}{2}\mu \left. \frac{de_k^2(\underline{\theta})}{d\underline{\theta}} \right|_{\underline{\theta}=\underline{\hat{\theta}}_{k-1}}$$
$$e_k(\underline{\theta}) = y_k - s_k(\underline{\theta})$$
$$\underline{\theta} = (a_0 \cdots a_{m-1}, b_0 \cdots b_{m-1})^T.$$

This leads to

$$\delta\underline{\hat{\theta}}_k = \mu\underline{\phi}(\underline{\hat{\theta}}_{k-1})e_k(\underline{\hat{\theta}}_{k-1})$$

where $\underline{\phi}_k(\underline{\theta})$ is the sensitivity signal

$$\underline{\phi}_k(\underline{\theta}) = ds_k/d\underline{\theta}.$$

We now generate a differential equation for the sensitivity signal as follows. Differentiating (4.105) gives

$$\frac{ds(t)}{db_r} = D^r x_F(t), \qquad 1 \le r \le m-1$$
$$x_F(t) = (D^m + A(D))^{-1}x(t) \tag{4.106}$$
$$\frac{ds(t)}{da_r} = -D^r s_F(t), \qquad 1 \le r \le m-1$$
$$s_F(t) = (D^m + A(D))^{-1}s(t) \tag{4.107}$$

where the subscript F stands for filtered.

Although we have not supposed $x(t)$ is noisy, the unwritten rule, that in practice it always would be, makes it unwise to differentiate $x(t)$ unless it has first been filtered. This requirement is satisfied in (4.106), (4.107). If a delay is allowed in processing, the problem can otherwise be overcome with the use of differentiation filters. Continuing, it is convenient to put (4.106), (4.107) into state space form as

$$\underline{\dot{q}}_x = \underline{A}(\underline{\theta})\underline{q}_x + \underline{c}x(t)$$
$$\underline{\dot{q}}_s = \underline{A}(\underline{\theta})\underline{q}_s + \underline{c}s(t)$$

where

$$\underline{q}_x = (ds(t)/db_0 \cdots ds(t)/db_{m-1})^T$$

$$\underline{q}_s = (ds(t)/da_0 \cdots ds(t)/da_{m-1})^T$$

$$\underline{A}(\theta) = \begin{bmatrix} 0 & 1 & 0 & \cdots & 0 \\ 0 & 0 & 1 & \cdots & 0 \\ \vdots & \vdots & \vdots & \ddots & \vdots \\ 0 & 0 & 0 & \cdots & 1 \\ -a_0 & -a_1 & -a_2 & \cdots & -a_{m-1} \end{bmatrix}, \quad \underline{c} = \begin{bmatrix} 0 \\ 0 \\ \vdots \\ 0 \\ 1 \end{bmatrix}.$$

Also we add a state equation for $s(t)$:

$$\dot{\underline{\xi}} = \underline{A}(\underline{\theta})\underline{\xi} + \underline{b}(\underline{\theta})x(t)$$

$$s(t) = \underline{c}^T\underline{\xi}$$

$$\underline{b}(\underline{\theta}) = [b_0 \cdots b_{m-1}]^T.$$

We are thus led to the following algorithm:

$$\delta\hat{\underline{\theta}}_k = \mu\underline{\phi}_k e_k$$

$$e_k = y_k - \hat{s}_k$$

$$\underline{\phi}_k = \left[\underline{q}_s^T(k\Delta), \underline{q}_x^T(k\Delta)\right]^T$$

$$\hat{s}_k = \underline{c}^T\underline{\xi}_k$$

$$\underline{\xi}_k = \underline{\xi}(k\Delta)$$

and for $(k-1)\Delta \le t \le k\Delta$

$$\begin{pmatrix} \dot{\underline{\xi}} \\ \dot{\underline{q}}_s \end{pmatrix} = \begin{pmatrix} \underline{A}(\hat{\underline{\theta}}_{k-1}) & \underline{0} \\ \underline{c}\underline{c}^T & \underline{A}(\hat{\underline{\theta}}_{k-1}) \end{pmatrix} \begin{pmatrix} \underline{\xi} \\ \underline{q}_s \end{pmatrix} + \begin{pmatrix} \underline{b}(\hat{\underline{\theta}}_{k-1}) \\ \underline{0} \end{pmatrix} x(t) \tag{4.108}$$

$$\dot{\underline{q}}_x = \underline{A}(\hat{\underline{\theta}}_{k-1})\underline{q}_x + \underline{c}x(t).$$

Finally in (4.108) if we can sample $x(t)$ more rapidly than the rate Δ^{-1}, we can integrate (4.108) accordingly. Otherwise in (4.108) we can interpolate $x(t)$ in various ways. The simplest of course is with a step signal:

$$x(t) = x_{k-1}, \quad (k-1)\Delta \le t \le k\Delta.$$

Then (4.108) integrates to

$$\begin{pmatrix} \underline{\xi}(k\Delta) \\ \underline{q}_s(k\Delta) \end{pmatrix} = \underline{F}_k \begin{pmatrix} \underline{\xi}((k-1)\Delta) \\ \underline{q}_s((k-1)\Delta) \end{pmatrix} + (\underline{F}_k - I) \begin{pmatrix} \underline{b}(\hat{\underline{\theta}}_{k-1}) \\ \underline{0} \end{pmatrix} x_{k-1}$$

$$\underline{F}_k = e^{\underline{A}_k\Delta}$$

$$\underline{A}_k = \begin{bmatrix} \underline{A}(\hat{\underline{\theta}}_{k-1}) & \underline{0} \\ \underline{c}\underline{c}^T & \underline{A}(\hat{\underline{\theta}}_{k-1}) \end{bmatrix}.$$

4.8.2 Algorithms with Vector Signals

Throughout the book we have considered scalar signals; however, the methods can all be extended, usually quite easily, to cover vector signals. The basic idea is to reorganize parameters into vector form (using "vec" calculus—see Appendix A, Section A9) and then follow the scalar ideas. We illustrate with an example.

Consider a matrix of FIR filters for vector signals. The error signal is

$$\underline{e}_k(\underline{w}) = \underline{y}_k - \sum_{s=1}^{p} \underline{W}_s \underline{x}_{k-s}$$

where now $\underline{y}_k$ is a p-vector primary signal, $\underline{x}_k$ an r-vector of auxiliary signals, $\underline{W}_s$ a set of matrix filter weights. If we stack the columns of $\underline{W}_s$ as vec($\underline{W}_s$) (see Appendix A, Section A9) we can write

$$\underline{e}_k(\underline{w}) = \underline{y}_k - \sum_{s=1}^{p} \underline{x}_{k-s}^T \otimes I \operatorname{vec}(\underline{W}_s)$$

$$\underline{w} = [\operatorname{vec}^T(\underline{W}_1) \cdots \operatorname{vec}^T(\underline{W}_p)]^T$$

or

$$\underline{e}_k(\underline{w}) = \underline{y}_k - \mathcal{X}_k \underline{w}$$

where we have introduced the matrix of reference signals

$$\mathcal{X}_k = [\underline{x}_{k-1}^T \otimes I \cdots \underline{x}_{k-p}^T \otimes I].$$

The adaptive SD algorithm has the form

$$\delta \underline{\hat{w}}_k = -\frac{1}{2}\mu \, d\|\underline{e}_k\|^2/d\underline{w}\Big|_{\underline{w}=\underline{\hat{w}}_{k-1}}$$

which gives

$$\delta \underline{\hat{w}}_k = -\mu \mathcal{X}_k \underline{e}_k$$

$$\underline{e}_k = \underline{y}_k - \mathcal{X}_k \underline{\hat{w}}_{k-1}.$$

This is, then, the vector LMS algorithm.

4.9 NOTES

The earliest discrete-time (short memory) adaptive algorithm seems to be LMS developed by [W2]. Many long memory algorithms were developed in the late 1960s and early 1970s. For example [T2] discussed stochastic approximation in the context of parameter estimation. A comprehensive discussion of many long memory algorithms as well as a large set of references is available in Chapters 2 and 3 of [L8] where long memory NM algorithms are called prediction error recursions. The output error mixed time scale SD algorithm comes from the work of [W1]. Instrumental variable algorithms are discussed in some detail (in a long memory setting) in [S10] and [Y2]. Further details on vector signals and continuous discrete algorithms can be found in [S13]. There is an interesting variation on the output error mixed time scale LMS algorithm in which e_k of (4.15) is replaced by a filtered error

$$e_{fk} = e_k + \sum_{s=1}^{m} c_s e_{k-s}.$$

It is sometimes possible to judiciously choose the filter $1 + c(q^{-1})$ to improve the stability properties of the algorithm. The idea is due to [L2] and has been pursued by [L3]. The SD adaptive time delay estimation algorithm (4.48), (4.49) was developed by [E2]. Differentiation filter based algorithms were introduced by [P4], [S6], and [S7]. A discussion of momentum LMS can be found in [R2]; LMS_N comes from the work of [N1] and [A1]; the material of Section 4.8 draws on [S13]; and signed LMS was introduced by [S1]. A survey of IIR algorithms can be found in [S8]. Throughout our discussion of SD algorithms, we have taken the gain in (4.1) to be a scalar. But as mentioned below (4.1), it can be a matrix. In particular a popular form is a diagonal matrix with different step sizes μ for each parameter component. However deciding on the relative sizes of these steps is not an easy matter. In any case the optimal choice of gain matrix is discussed in Chapter 5. The extended Kalman filter based algorithms are discussed in [L8]. An analysis of the EKF in a long memory setting in [L7] and [L8] shows that the absence of $\underline{K}_{\theta,k} e_k$ in (4.104) causes ill effects for the convergence of the EKF. The idea of estimating the hyperparameters $\underline{\gamma}$ by online minimization of $E(e_k^2(\underline{\gamma}))$ appears in Section 4.42 of [B7]. The oblique projection method is a discrete version of the continuous time method from the works of [P3]. Orthogonal projection onto ellipses is discussed in [G9]. The Wilson based algorithm appears in [S14]. Adaptive estimation of roots of a polynomial is discussed in [N3] and [S18]. One issue in algorithm development we have not discussed is fast implementations such as adaptive lattice algorithms; these are well covered in [A2].

EXERCISES

4.1 For the ARMAX model (2.75) find a PLR representation and hence write down a PSD algorithm (see [P1], [Y1], [Y2]). Also develop an NM algorithm (see [S9]).

4.2 For the TFCN model (2.67)–(2.69) or (4.35)–(4.37) develop an NM algorithm (see [S13]).

4.3 For the ARARX (autoregressive, autoregressive exogenous) model specified as follows

$$A(q^{-1}) y_k = b(q^{-1}) x_k + D^{-1}(q^{-1}) \epsilon_k$$

$$A(q^{-1}) = 1 + \sum_{s=1}^{p} a_s q^{-s}, \quad \text{etc.}$$

derive an NM algorithm (see [H6], [G4]).

4.4 For the AR-TFCN structure (4.2)–(4.4) develop a PLR representation and hence construct (i) a PSD algorithm and (ii) an NM algorithm (see [L8]).

4.5 Develop a PLR based algorithm for the continuous discrete example discussed in Section 4.8.

4.6 Square root algorithm for EWLS:
The aim is to replace the update (4.87) with an update for a square root. Suppose $\underline{Q}_k$ is a square root of $\underline{R}_k$ so that

$$\underline{Q}_k\underline{Q}_k^T = \underline{R}_k.$$

Then look for an update of the form

$$\underline{Q}_k = \lambda^{-\frac{1}{2}}\underline{Q}_{k-1}(I - \underline{b}_k\underline{b}_k^T/c_k).$$

Show that, to make (4.87) consistent with this, we need to choose

$$\underline{b}_k = \lambda^{-\frac{1}{2}}\underline{Q}_{k-1}^T\underline{x}_k$$

$$c_k = 1 + \xi_k + \sqrt{1+\xi_k}$$

$$\xi_k = \|\underline{b}_k\|^2.$$

4.7 Loop filter algorithm from a model based point of view: Show how the loop filter LMS algorithm can be viewed from a model based perspective by positing a state space model for the time varying weight vector.

4.8 **(a)** For the adaptive prediction example of Exercises 2.1 and 3.1, plot trajectories of the LMS algorithm (both weights) and also e_k, y_k and the estimated learning curve $\mathcal{E}_k$ (averaged over 10 realizations). Use two values of μ, one leading to stable behavior and one leading to instability.
(b) Repeat (a) for the minimum phase equalization example of Exercise 3.2.
(c) Repeat (a) for the interference cancelling example of Exercise 3.3.
(d) Repeat (a) for the time delay estimation example of Exercise 3.4 (for LMS read SD_1).

4.9 **(a)** Repeat Exercise 4.8(a) with
 (i) momentum LMS, $\alpha = 0.9$.
 (ii) EWLS, $\lambda = 0.9$.
 (iii) a loop filter algorithm.
(b) Repeat Exercise 4.8(b) as per Exercise 4.9(a).
(c) Repeat Exercise 4.8(c) as per Exercise 4.9(a).
(d) Repeat Exercise 4.8(d) as per Exercise 4.9(a).

4.10 Consider the projection schemes discussed in Section 4.6. Introduce the indicator variable

$$J(\underline{\theta}) = \begin{cases} 0, & \|\underline{\theta}\| < R \\ 1, & \text{otherwise.} \end{cases}$$

Show that for orthogonal or oblique projection, the algorithm can be written in the form

$$\begin{aligned}\hat{\underline{\theta}}_k &= \underline{\theta}_k'(1 - \mu\alpha_k J_k)\\ &= (\hat{\underline{\theta}}_{k-1} + \mu\underline{\phi}_k e_k)(1 - \mu\alpha_k J_k)\end{aligned}$$

where

$$|\alpha_k| \le \|\underline{\phi}_k\||e_k|/\|\underline{\theta}_k'\|.$$

This allows a view of these monitoring schemes as a time variant relaxation algorithm where the relaxation factor depends on the size of the error signal.

5

Algorithm Analysis: Gaussian White Noise Setting

5.1 GAUSSIAN WHITE NOISE ASSUMPTIONS

In this chapter we analyze the behavior of the single time scale LMS algorithm under Gaussian white noise assumptions. The advantage of this special setting is threefold. Firstly, the calculations are very straightforward. Secondly, under these simple assumptions we are able to get detailed information about the algorithm. For example, we can analyze algorithm behavior in the presence of time varying parameters. Thirdly, although the analysis method we use with these Gaussian white noise assumptions does not extend to more general situations, the insight we get about adaptive algorithm behavior and the conclusions we draw do extend. In fact much of the research literature on the LMS algorithm works under the Gaussian white noise assumptions used in this chapter. The reason for this is at least partly a "folklore" to the effect that the results hold more generally than the assumptions would suggest: in particular that the results are still valid when $\underline{x}_k$ below is a colored signal, but ϵ_k below is still a white noise and μ is small. In Parts II and III of the book we will see to what extent averaging methods put some substance behind the folklore.

5.2 LMS WITH TIME INVARIANT PARAMETERS

In this section we assume there is a true time invariant parameter or weight. Consider then the LMS algorithm (4.6), (4.7)

$$\delta \underline{\hat{w}}_k = \mu \underline{x}_k e_k \tag{5.1}$$

$$e_k = y_k - \underline{x}_k^T \underline{\hat{w}}_{k-1}. \tag{5.2}$$

We introduce the following signal assumptions:

(5.2A1) $\{\underline{x}_k\}$ is a vector, zero mean Gaussian white noise with covariance matrix

$$\underline{R}_x = E(\underline{x}_k \underline{x}_k^T).$$

(5.2A2) There is a true weight $\underline{w}_o$ such that the error signal

$$\epsilon_k = y_k - \underline{x}_k^T \underline{w}_o$$

is a zero mean Gaussian white noise of variance σ_ϵ^2.

(5.2A3) $\{\underline{x}_k\}$, $\{\epsilon_k\}$ are statistically independent.

It follows from (5.2A2), (5.2A3) that

$$\begin{aligned}\underline{R}_{xy} = E(\underline{x}_k y_k) &= E(\underline{x}_k \epsilon_k) + E(\underline{x}_k \underline{x}_k^T)\underline{w}_o \\ &= \underline{R}_x \underline{w}_o.\end{aligned}$$

So that indeed $\underline{w}_o$ solves the discrete Wiener-Hopf equation (2.5).

We are now in a position to study the properties of the LMS algorithm and we do this by calculating its mean and variance. Also of interest is the learning curve, which is the actual MSE as a function of k. To analyze any adaptive algorithm it is usually convenient to convert it to a so-called error form; indeed, from (5.2A2) and (5.2), we can write

$$\begin{aligned}e_k &= \epsilon_k + \underline{x}_k^T \underline{w}_o - \underline{x}_k^T \underline{\hat{w}}_{k-1} \\ &= \epsilon_k - \underline{x}_k^T \underline{\tilde{w}}_{k-1}\end{aligned} \tag{5.3}$$

where we have introduced the weight error vector

$$\underline{\tilde{w}}_k = \underline{\hat{w}}_k - \underline{w}_o.$$

Now since $\delta \underline{\tilde{w}}_k = \delta \underline{\hat{w}}_k$, we can rewrite (5.1) as

$$\delta \underline{\tilde{w}}_k = -\mu \underline{x}_k \underline{x}_k^T \underline{\tilde{w}}_{k-1} + \mu \underline{x}_k \epsilon_k. \tag{5.4}$$

The error system is then the pair (5.3), (5.4). Continuing, it helps to rewrite (5.4) as

$$\underline{\tilde{w}}_k = (I - \mu \underline{x}_k \underline{x}_k^T)\underline{\tilde{w}}_{k-1} + \mu \underline{x}_k \epsilon_k. \tag{5.5}$$

This is a time varying forced or nonhomogeneous stochastic difference equation. We now consider how to address the issue of stability and performance. If there is no noise (ϵ_k) we have a homogeneous stochastic difference equation and we would be interested in convergence. With noise there can be no convergence (instead we expect that the weight error "hovers" near zero), so our concern shifts to boundedness or input-output stability. About the minimal stochastic boundedness property we can have is boundedness in probability. That is

$$\lim_{B\to\infty} \sup_k P(\|\underline{\tilde{w}}_k\| \geq B) = 0.$$

This says that, uniformly in k, no probability mass escapes to infinity. This property is also called tightness. By Chebyshev's inequality (Appendix B5)

$$P(\|\underline{\tilde{w}}_k\| \geq B) \leq E\|\underline{\tilde{w}}_k\|^2/B^2$$
$$= (\|E(\underline{\tilde{w}}_k)\|^2 + \text{var}(\underline{\tilde{w}}_k))/B^2.$$

So if we can show

$$\|E(\underline{\tilde{w}}_k)\| \to 0 \text{ as } k \to \infty, \text{ or is uniformly bounded in } k \text{ and}$$
$$\text{var}(\underline{\tilde{w}}_k) \text{ uniformly bounded in } k$$

then boundedness in probability follows (actually so does mean square boundedness). We are thus led to a first order (mean) and second order (variance) analysis of (5.5).

With regard to performance, the main concern is to calculate the MSE as a function of k (the so-called "learning curve"). Of secondary interest will be the variance of (5.5). In any case, to compute the learning curve it will be necessary to calculate the variance.

5.2.1 First Order Analysis—Time Invariant Parameter

To calculate $E(\underline{\tilde{w}}_k)$ we use an iterated conditional expectation procedure. By induction, we see from (5.5) that $\underline{\tilde{w}}_k$ depends on the "history" X_k, E_k, where $X_k = (\underline{x}_k, \underline{x}_{k-1}, \ldots, \underline{x}_1)$, $E_k = (\epsilon_k, \epsilon_{k-1}, \ldots, \epsilon_1)$. So let us take conditional expectations in (5.5) to find via (5.2A3) that

$$E(\underline{\tilde{w}}_k|X_{k-1}, E_{k-1}) = \underline{\tilde{w}}_{k-1} - \mu E(\underline{x}_k\underline{x}_k^T\underline{\tilde{w}}_{k-1}|X_{k-1}, E_{k-1}).$$

However from (5.2A1), $\underline{x}_k$ is independent of X_{k-1}, E_{k-1} and so

$$E(\underline{\tilde{w}}_k|X_{k-1}, E_{k-1}) = \underline{\tilde{w}}_{k-1} - \mu E(\underline{x}_k\underline{x}_k^T|X_{k-1}, E_{k-1})\underline{\tilde{w}}_{k-1}$$
$$= \underline{\tilde{w}}_{k-1} - \mu\underline{R}_x\underline{\tilde{w}}_{k-1} \qquad \text{(by (5.2A1))}.$$

Taking expectations over X_{k-1}, E_{k-1} (see Appendix B3) gives

$$\underline{\tilde{m}}_k = E(\underline{\tilde{w}}_k) = (I - \mu\underline{R}_x)\underline{\tilde{m}}_{k-1}. \tag{5.6}$$

This is exactly the same difference equation that we analyzed in Section 3.2. We conclude that the LMS algorithm exhibits first order convergence (convergence in the mean) if

$$\text{(i) } \underline{R}_x \text{ is positive definite, i.e., } \mu\lambda_{min} > 0;$$
$$\text{(ii) } \mu\lambda_{max} < 2 \tag{5.7}$$

where λ_{min}, λ_{max} are, respectively, the smallest and largest eigenvalues of $\underline{R}_x$. If either of this pair of conditions fails then some modes of (5.6) will not converge or may diverge.

While the first order analysis does not completely describe the stability of the stochastic algorithm (5.5), it does show that some constraints must be put on allowed μ values. Note also that the first order analysis links up with the earlier offline analysis. Now we turn to the variance calculation.

5.2.2 Second Order Analysis—Time Invariant Parameter

We begin by noting that

$$var(\underline{\tilde{w}}_k) = E(\underline{\tilde{w}}_k - \underline{\tilde{m}}_k)(\underline{\tilde{w}}_k - \underline{\tilde{m}}_k)^T$$
$$= E(\underline{\tilde{w}}_k \underline{\tilde{w}}_k^T) - \underline{\tilde{m}}_k \underline{\tilde{m}}_k^T.$$

So provided (5.7) holds, $\underline{\tilde{m}}_k \to 0$ as $k \to \infty$, and so we can directly work with the second moment

$$\Gamma_k = E(\underline{\tilde{w}}_k \underline{\tilde{w}}_k^T).$$

Returning to (5.5) we find

$$\underline{\tilde{w}}_k \underline{\tilde{w}}_k^T = (I - \mu \underline{x}_k \underline{x}_k^T) \underline{\tilde{w}}_{k-1} \underline{\tilde{w}}_{k-1}^T (I - \mu \underline{x}_k \underline{x}_k^T) + \underline{p}_k + \underline{p}_k^T + \mu^2 \epsilon_k^2 \underline{x}_k \underline{x}_k^T$$

where

$$\underline{p}_k = \mu \epsilon_k \underline{x}_k \underline{\tilde{w}}_{k-1}^T (I - \mu \underline{x}_k \underline{x}_k^T).$$

Recalling that X_k denotes the $\underline{x}_k$ "history" $(\underline{x}_k, \underline{x}_{k-1} \cdots \underline{x}_0)$ and E_k denotes the ϵ_k "history" $(\epsilon_k, \epsilon_{k-1} \cdots \epsilon_1)$, take conditional expectations to see that $E(\underline{p}_k | X_{k-1}, E_{k-1}) = 0$ so that

$$E(\underline{\tilde{w}}_k \underline{\tilde{w}}_k^T | X_{k-1}, E_{k-1}) = \underline{\tilde{w}}_{k-1} \underline{\tilde{w}}_{k-1}^T - \mu \underline{\tilde{w}}_{k-1} \underline{\tilde{w}}_{k-1}^T \underline{R}_x - \mu \underline{R}_x \underline{\tilde{w}}_{k-1} \underline{\tilde{w}}_{k-1}^T$$
$$+ \mu^2 E(\underline{x}_k \underline{x}_k^T (\underline{x}_k^T \underline{\tilde{w}}_{k-1})^2) + \mu^2 \sigma_\epsilon^2 \underline{R}_x.$$

Taking unconditional expectations and using the result in Appendix B4(vi) gives

$$\Gamma_k = \Gamma_{k-1} - \mu(\Gamma_{k-1} \underline{R}_x + \underline{R}_x \Gamma_{k-1}) + 2\mu^2 \underline{R}_x \Gamma_{k-1} \underline{R}_x$$
$$+ \mu^2 \operatorname{tr}(\Gamma_{k-1} \underline{R}_x) \underline{R}_x + \mu^2 \sigma_\epsilon^2 \underline{R}_x. \tag{5.8}$$

To analyze the stability of this difference equation we again apply the eigenvector decomposition. Introduce the transformed variance matrix

$$\underline{C}_k = \underline{Q} \Gamma_k \underline{Q}^T$$
$$\underline{Q} = (\underline{v}_1 \ \ldots \ \underline{v}_p)$$

where $\underline{v}_u$, $1 \le u \le p$, are the eigenvectors of $\underline{R}_x$ so that

$$\underline{Q} \underline{R}_x \underline{Q}^T = \Lambda = \operatorname{diag}(\lambda_1 \ \ldots \ \lambda_p) \tag{5.9}$$
$$\underline{Q}^T \underline{Q} = I$$

and λ_u, $1 \le u \le p$, the eigenvalues of $\underline{R}_x$. Then (5.8) becomes

$$\underline{C}_k = \underline{C}_{k-1} - \mu(\underline{C}_{k-1} \Lambda + \Lambda \underline{C}_{k-1}) + 2\mu^2 \Lambda \underline{C}_{k-1} \Lambda + \mu^2 \alpha_{k-1} \Lambda + \mu^2 \sigma_\epsilon^2 \Lambda \tag{5.10}$$

where

$$\alpha_k = \operatorname{tr}(\underline{R}_x \Gamma_k) = \operatorname{tr}(\Lambda \underline{C}_k). \tag{5.11}$$

Thus the off-diagonal elements of $\underline{C}_k$ obey

$$
\begin{aligned}
C_{k,ij} &= C_{k-1,ij} - \mu(\lambda_i + \lambda_j)C_{k-1,ij} + 2\mu^2\lambda_i\lambda_j C_{k-1,ij} \quad , \quad i \neq j \\
&= \rho_{ij}C_{k-1,ij} \quad , \quad i \neq j
\end{aligned}
$$

where

$$\rho_{ij} = 1 - \mu(\lambda_i + \lambda_j) + 2\mu^2\lambda_i\lambda_j \quad , \quad i \neq j.$$

Clearly

$$C_{k,ij} \to 0, \quad \text{as } k \to \infty, \quad \text{if and only if } |\rho_{ij}| < 1.$$

This certainly occurs if

$$0 < \mu\lambda_i < 1, \quad 1 \le i \le p. \tag{5.12}$$

To see this note that

$$\rho_{ij} = 1 - \mu\lambda_i(1 - \mu\lambda_j) - \mu\lambda_j(1 - \mu\lambda_i).$$

From (5.10), the diagonal terms of $\underline{C}_k$, say, $\gamma_{k,u}$, $1 \le u \le p$, obey

$$
\begin{aligned}
\gamma_{k,u} &= \gamma_{k-1,u} - 2\mu\lambda_u\gamma_{k-1,u} + 2\mu^2\lambda_u^2\gamma_{k-1,u} + \mu^2\alpha_{k-1}\lambda_u + \mu^2\sigma_\epsilon^2\lambda_u \\
&= (1 - 2\mu\lambda_u + 2\mu^2\lambda_u^2)\gamma_{k-1,u} + \mu^2\lambda_u(\alpha_{k-1} + \sigma_\epsilon^2), \quad 1 \le u \le p. \qquad (5.13)
\end{aligned}
$$

A proper stability analysis of this set of coupled difference equations is somewhat messy, so we will proceed somewhat heuristically (reference to a rigorous analysis is given in the notes at the end of this chapter). If (5.13) has a steady state, i.e., if $\gamma_{k,u}$ has a limit value γ_u, it will obey, from (5.13)

$$\gamma_u = (1 - 2\mu\lambda_u + 2\mu^2\lambda_u^2)\gamma_u + \mu^2\lambda_u(\alpha + \sigma_\epsilon^2), \quad 1 \le u \le p$$

where α is the steady state value of α_k,

$$\alpha = \sum_{u=1}^{p} \lambda_u\gamma_u.$$

Solving, we find

$$\gamma_u = \mu(\alpha + \sigma_\epsilon^2)/[2(1 - \mu\lambda_u)], \qquad 1 \le u \le p \tag{5.14}$$

$$\Rightarrow \qquad \alpha = \sum_{u=1}^{p} \lambda_u\gamma_u = c(\alpha + \sigma_\epsilon^2)$$

where

$$c = \frac{1}{2}\sum_{u=1}^{p} \mu\lambda_u/(1 - \mu\lambda_u) \tag{5.15}$$

$$\Rightarrow \qquad \alpha = c\sigma_\epsilon^2/(1 - c). \tag{5.16}$$

Thus a necessary condition for the convergence of $\gamma_{k,u}$ as $k \to \infty$ is that

$$0 < c < 1. \tag{5.17}$$

It turns out that this is also a sufficient condition. Since (5.17) is not easy to interpret, we look for something simpler. If $\mu\lambda_u$ is small (for $1 \le u \le p$), then (5.15) indicates that (5.17) becomes $\mu \sum_{u=1}^{p} \lambda_u < 2$. So let us look for a sufficient condition of the form

$$\mu \operatorname{tr}(\underline{R}_x) = \mu \sum_{u=1}^{p} \lambda_u < \beta < 1.$$

This of course implies

$$\mu\lambda_u < \beta < 1, \quad 1 \le u \le p$$

and thus (5.12) holds. Next

$$\begin{aligned} & 1 - \beta < 1 - \mu\lambda_u \\ \Rightarrow \quad & (1 - \mu\lambda_u)^{-1} < (1 - \beta)^{-1} \\ \Rightarrow \quad & \sum_{u=1}^{p} \mu\lambda_u (1 - \mu\lambda_u)^{-1} < \sum_{u=1}^{p} \mu\lambda_u (1 - \beta)^{-1} < \beta(1 - \beta)^{-1}. \end{aligned}$$

Thus, the best value of β occurs, from (5.15), (5.17), when

$$\beta(1 - \beta)^{-1} = 2 \quad \Rightarrow \quad \beta = 2/3.$$

So a sufficient condition to replace (5.17) is

$$\mu \operatorname{tr}(\underline{R}_x) < 2/3. \tag{5.18}$$

This condition should be compared with the first order condition (5.7).

Returning to (5.14) we find

$$\gamma_u = \frac{\mu\sigma_\epsilon^2}{2(1 - \mu\lambda_u)(1 - c)}, \qquad 1 \le u \le p. \tag{5.19a}$$

It is convenient to use an overall measure of parameter error variance such as

$$\begin{aligned} \gamma &= \lim_{k \to \infty} \operatorname{tr}(\Gamma_k)/p = \lim_{k \to \infty} \operatorname{tr}(\underline{C}_k)/p = \sum_{u=1}^{p} \gamma_u/p \\ &= \frac{\mu\sigma_\epsilon^2 b/p}{2(1 - c)} \end{aligned} \tag{5.19b}$$

where we used the γ_u formula above and introduced

$$b = \sum_{u=1}^{p} (1 - \mu\lambda_u)^{-1}. \tag{5.20}$$

The "steady state" weight error covariance matrix is asked for in Exercise 5.5.

5.2.3 Learning Curve—Time Invariant Parameter

Now we turn to the MSE calculation. We have from (5.3) and (5.2A1)–(5.2A3)

$$\begin{aligned}\mathcal{E}_k &= \sigma_\epsilon^2 + E(\underline{\tilde{w}}_{k-1}^T \underline{x}_k \underline{x}_k^T \underline{\tilde{w}}_{k-1})\\ &= \sigma_\epsilon^2 + E(\underline{\tilde{w}}_{k-1}^T \underline{R}_x \underline{\tilde{w}}_{k-1})\\ &= \sigma_\epsilon^2 + \operatorname{tr}(\underline{R}_x \Gamma_{k-1})\\ &= \sigma_\epsilon^2 + \alpha_{k-1}. \end{aligned} \tag{5.21}$$

From (5.16) we see there is a limiting value for $\mathcal{E}_k$, namely

$$\begin{aligned}\mathcal{E}_\infty &= \sigma_\epsilon^2 + \alpha\\ &= \sigma_\epsilon^2 + c\sigma_\epsilon^2/(1-c)\\ &= \sigma_\epsilon^2/(1-c). \end{aligned} \tag{5.22}$$

Clearly $\mathcal{E}_\infty > \sigma_\epsilon^2$ so that the noisy adaptation has led to an inflation of MSE over the ideal value σ_ϵ^2. This inflation is called misadjustment and is measured on a nondimensional scale as

$$\begin{aligned}M &= (\mathcal{E}_\infty - \mathcal{E}_{min})/\mathcal{E}_{min}\\ &= (\mathcal{E}_\infty - \sigma_\epsilon^2)/\sigma_\epsilon^2\\ &= \mathcal{E}_\infty/\sigma_\epsilon^2 - 1\\ &= \alpha/\sigma_\epsilon^2\\ &= c/(1-c). \end{aligned} \tag{5.23}$$

We would of course like to keep M small.

5.2.4 Learning Speed and Misadjustment Tradeoff

To get some insight into the effect of μ on algorithm behavior let us suppose

$$\mu \operatorname{tr}(\underline{R}_x) \ll 1. \tag{5.24}$$

Then $0 < \mu\lambda_u \ll 1$, $1 \le u \le p$, so from (5.15), (5.20)

$$\begin{aligned}c &\simeq \frac{1}{2}\mu \operatorname{tr}(\underline{R}_x) \ll 1 \qquad (5.25)\\ b &\simeq p.\end{aligned}$$

Thus from (5.19), (5.23)

$$\begin{aligned}M &\simeq c \simeq \frac{1}{2}\mu \operatorname{tr}(\underline{R}_x) \qquad (5.26)\\ \gamma &\simeq \frac{1}{2}\mu\sigma_\epsilon^2.\end{aligned}$$

On the other hand the speed of convergence is controlled by

$$1 - \mu\lambda_{min}, \quad 1 - \mu\lambda_{max}.$$

We see that

fast speed requires μ large,
small misadjustment or variance requires μ small.

Thus there is a tradeoff between speed of adjustment and size of misadjustment.

Example 5.1. Speed-Misadjustment Tradeoff

We will pursue the Channel Equalization Example 2.2. However it is necessary to change the specifications we have used previously to satisfy some of our current assumptions.

Referring to Fig. 2.4, suppose the delay $\Delta = 1$ and the primary noise m_k is a zero mean white noise of variance σ_m^2. Then the primary signal is

$$y_k = s_{k-1} + m_k. \tag{5.27}$$

We suppose as before that the message signal s_k is an AR(1) process

$$s_k = (1 + \beta q^{-1})^{-1}\epsilon_k \tag{5.28}$$

where ϵ_k is a zero mean white noise of variance σ_ϵ^2. The variance of s_k is then

$$\sigma_s^2 = \sigma_\epsilon^2/(1 - \beta^2).$$

With the choices $\sigma_m^2 = 0.5$, $\sigma_\epsilon^2 = 0.3825$, $\beta = 0.915$ we get a signal to noise ratio of

$$\sigma_s^2/\sigma_m^2 = 4.7.$$

Next we suppose no receiver noise and a first order channel ($\theta = -0.765$) so that the reference signal is

$$x_k = \xi_k = g(1 + \theta q^{-1})^{-1} s_k \tag{5.29}$$

where g is the channel gain ($g = 1$). Thus

$$x_k + \theta x_{k-1} = g s_k. \tag{5.30}$$

Thus we find from (5.27)

$$y_k = \underline{w}_o^T \underline{x}_k + m_k \tag{5.31}$$

$$\underline{w}_o = (g^{-1} \quad \theta g^{-1})^T, \quad \underline{x}_k = (x_{k-1} \quad x_{k-2})^T.$$

It also follows from (5.28), (5.29) that x_k is an AR(2) process with

$$x_k = (1 + a_1 q^{-1} + a_2 q^{-2})^{-1}\epsilon'_k \tag{5.32}$$

$$(a_1 \quad a_2) = (\theta + \beta \quad \theta\beta) = (0.15 \quad -0.70)$$

$$\epsilon'_k = g\epsilon_k \quad \Rightarrow \quad \sigma_{\epsilon'}^2 = g^2\sigma_\epsilon^2 = 0.3825.$$

From (5.31), (5.32), Appendix B

$$\mathcal{E}_{min} = \sigma_m^2 = 0.5$$

$$\underline{R}_x = R_{x0}\begin{pmatrix} 1 & \rho_1 \\ \rho_1 & 1 \end{pmatrix} \tag{5.33}$$

$$\rho_1 = -\frac{a_1}{1+a_2} = -0.5$$

$$R_{x0} = \left(\frac{1+a_2}{1-a_2}\right)\frac{\sigma_{\epsilon'}^2}{(1+a_2)^2 - a_1^2} = 1.0. \tag{5.34}$$

It is readily checked (cf. Example 3.3) that the two eigenvalues of $\underline{R}_x$ are

$$\lambda_{min}, \lambda_{max} = (1 - |\rho_1|)R_{x0}, (1 + |\rho_1|)R_{x0}$$

so that the eigenvalue spread condition number is

$$\kappa = \frac{1+|\rho_1|}{1-|\rho_1|} = 3.$$

Now while (5.31) satisfies assumptions (5.2A2), (5.2A3), it does not satisfy (5.2A1). Nevertheless we will pursue the example and thereby illustrate the folklore mentioned in Section 5.1 that the fact that x_k is colored does not affect some of our results.

An LMS trajectory is shown in Fig. 5.1. Note the characteristic hovering of the trajectory around the optimal weight.

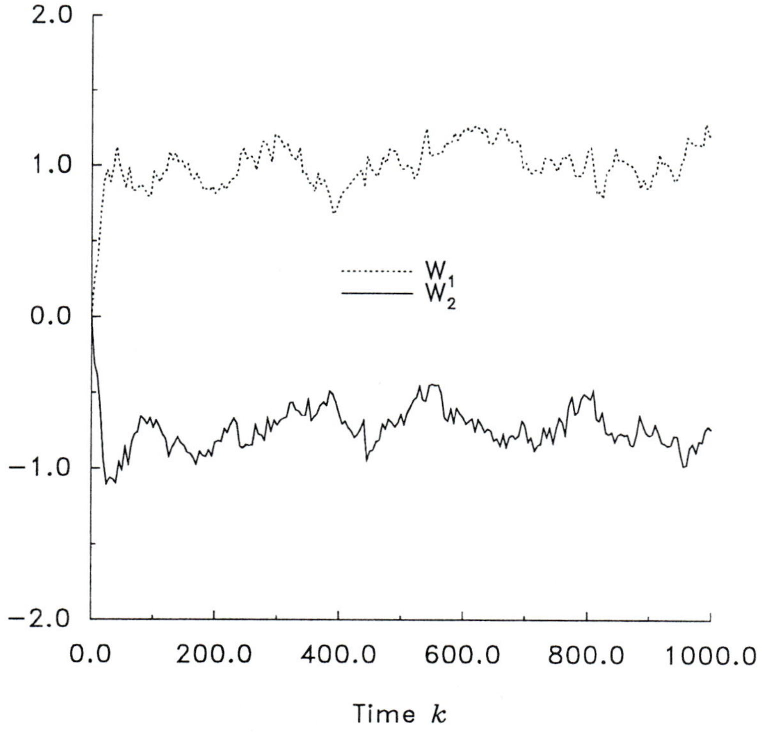

Figure 5.1 LMS trajectory for Example 5.1.

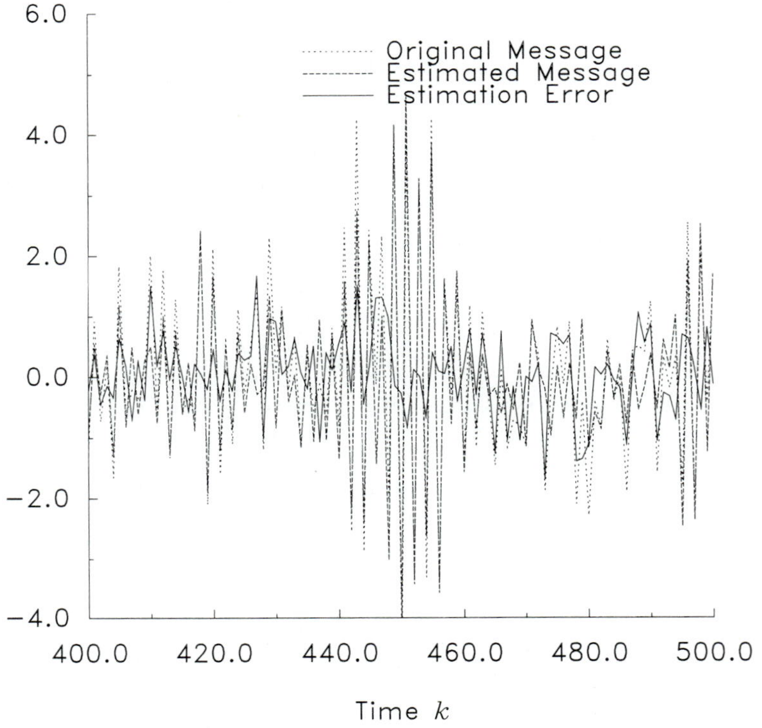

Figure 5.2 Message signal, estimated message signal, error signal for Example 5.1.

Fig. 5.2 shows message signal, estimated message signal, and error signal trajectories. It is clear that the adaptive equalizer is performing well.

Fig. 5.3 shows estimated learning curves for two different values of μ. The more rapid convergence of the larger μ trajectory is apparent as is its larger steady state misadjustment. This is illustrated more quantitatively in the following table:

μ	Estimated misadjustment	Theoretical misadjustment
0.01	0.016	0.01
0.05	0.063	0.05

From (5.26) the theoretical misadjustment is

$$\begin{aligned} M &= \mu \mathrm{tr}(\underline{R}_x)/2 \\ &= \mu R_{x0}, \quad \text{(from (5.33))} \\ &= 1.0\mu, \quad \text{(from (5.34))}. \end{aligned}$$

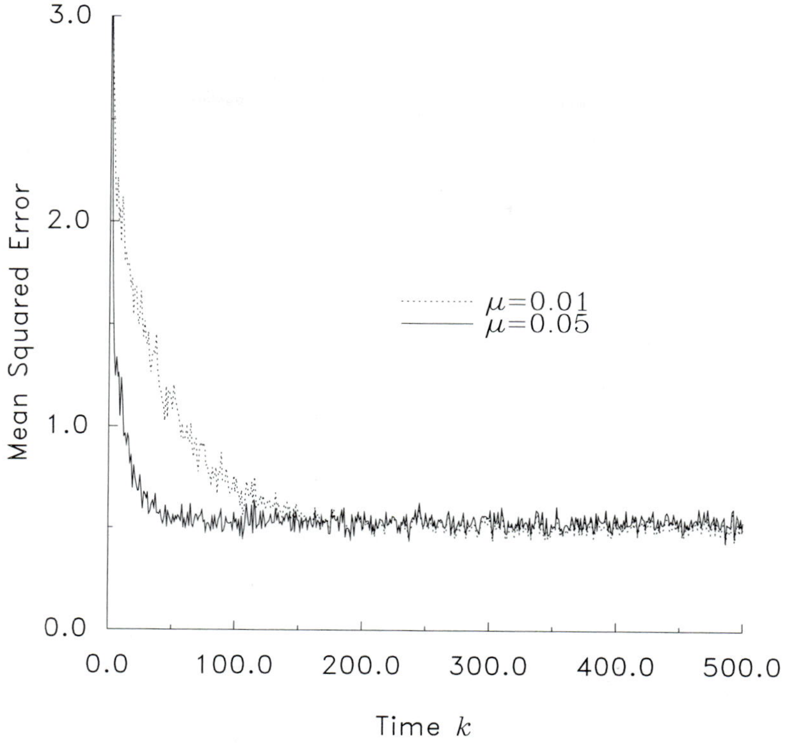

Figure 5.3 Estimated learning curves for Example 5.1 (averaged over 500 realizations).

The empirical values are obtained by averaging over the steady state part of 500 realizations. Even though $\underline{x}_k$ is colored we see that the theory is still reliable.

Finally in Fig. 5.4 the effect of eigenvalue spread on speed is illustrated for $\mu = 0.05$. The parameter settings for the $\kappa = 100$ case are

$$(\theta, \quad \beta) = (-0.900, \quad 0.999)$$
$$(a_1, \quad a_2) = (0.099, \quad -0.899)$$
$$(\sigma_m^2, \quad \sigma_\epsilon^2) = (0.5, \quad 0.0075).$$

This produces a signal to noise ratio

$$\sigma_s^2/\sigma_m^2 = 7.5.$$

Also, then,

$$\rho_1 = -0.9802, \qquad R_{x0} = 0.9972, \qquad \kappa = 100.$$

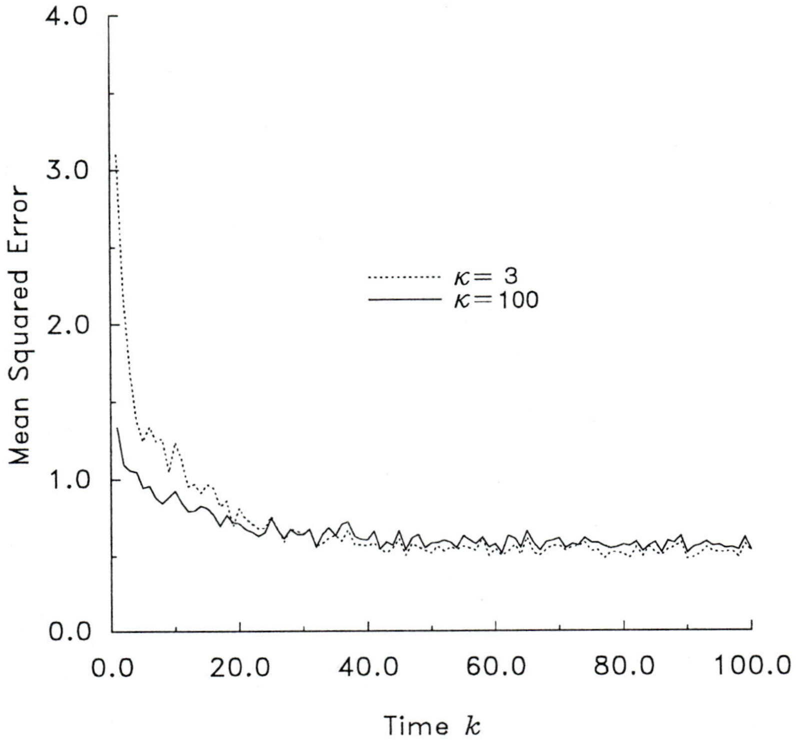

Figure 5.4 Effect of eigenvalue spread on speed for Example 5.1 as shown in estimated learning curves.

5.3 LMS WITH TIME VARYING PARAMETERS

Now we consider the behavior of the LMS algorithm when the weight vector is time varying. We modify our assumption as follows:

(5.3A1) $\{\underline{x}_k\}$ is a vector, zero mean white Gaussian noise of variance $\underline{R}_x$.

(5.3A2)

$$y_k = \epsilon_k + \underline{x}_k^T \underline{w}_{o,k-1}$$

where $\underline{w}_{o,k-1}$ is the true time varying weight; while ϵ_k is a zero mean white Gaussian noise of variance σ_ϵ^2.

(5.3A3) $\{\underline{x}_k\}$, $\{\epsilon_k\}$, $\{\underline{w}_{o,k}\}$ are statistically independent.

We need to be more explicit about the statistical behavior of $\underline{w}_{o,k}$. We will work with two different parameter models:

(5.3A4)(i) $\underline{w}_{o,k}$ is a random walk

$$\delta \underline{w}_{o,k} = \underline{\delta}_k, \quad \underline{w}_{o,0} \text{ deterministic}$$

where $\underline{\delta}_k$ is a Gaussian white noise of variance $\sigma_\delta^2 I$.

(5.3A4)(ii) $\underline{w}_{o,k}$ is a stationary first order autoregressive process

$$\underline{w}_{o,k} = \underline{w}_o + \underline{w}_{f,k}$$
$$\underline{w}_{f,k} = \theta \underline{w}_{f,k-1} + \underline{\eta}_k, \quad |\theta| < 1$$

where $\underline{\eta}_k$ is a zero mean Gaussian white noise of variance $\sigma_\eta^2 I$. Note that we have split $\underline{w}_{o,k}$ into a 'dc' component $\underline{w}_o$ and an 'ac' or fluctuation component $\underline{w}_{f,k}$.

The random walk parameter model can be thought of as a stochastic ramp (it has a variance $\to \infty$ as $k \to \infty$) and models a failing system. It can only be valid on a finite time interval since it violates the physical constraint that $var(y_k)$ be uniformly bounded in k. The second parameter model has a bounded variance and is a simple model for a continually changing parameter. Later we will see that the speed at which $\underline{w}_{o,k}$ varies is an important parameter.

Our aim now is to analyze the behavior of LMS under the regime (5.3A1)–(5.3A4). As usual, we first construct the error system. Returning to (5.1), (5.2) we find that the error system (5.3) becomes

$$e_k = \epsilon_k + \underline{x}_k^T \underline{w}_{o,k-1} - \underline{x}_k^T \hat{\underline{w}}_{k-1}$$
$$= \epsilon_k - \underline{x}_k^T \tilde{\underline{w}}_{k-1}$$

where $\tilde{\underline{w}}_k$ is the estimation error

$$\tilde{\underline{w}}_k = \hat{\underline{w}}_k - \underline{w}_{o,k}.$$

Thus the error system (5.4) becomes

$$\delta \tilde{\underline{w}}_k = -\mu \underline{x}_k \underline{x}_k^T \tilde{\underline{w}}_{k-1} + \mu \underline{x}_k \epsilon_k - \delta \underline{w}_{o,k} \tag{5.35}$$

where we have introduced the parameter velocity

$$\delta \underline{w}_{o,k} = \underline{w}_{o,k} - \underline{w}_{o,k-1}. \tag{5.36}$$

The error system (5.35) has a new perturbation term as compared to (5.4). We turn now to the first and second order analyses. We first give an analysis under the Random Walk model (5.3A4(i)) and then the stationary model (5.3A4(ii)). Analysis under the stationary model is rather long but entirely straightforward.

5.3.1 First Order Analysis—Random Walk Model (5.3A4(i))

By induction, we see from (5.35) that $\tilde{\underline{w}}_k$ depends on the history (X_k, E_k, Δ_k), $\Delta_k = (\underline{\delta}_k, \underline{\delta}_{k-1}, \ldots \underline{\delta}_1)$, etc. Repeating the type of conditional expectation calculation made in Section 5.2 leads us again to (5.6) and so the first order analysis is unchanged.

5.3.2 Second Order Analysis—Random Walk Model (5.3A4(i))

Proceeding as in Section 5.2 we find that since by (5.3A4)(i), $\underline{\delta}_k$ is independent of the other terms in (5.35), we get the same equation as (5.8) for Γ_k, but with an extra term added, namely $\sigma_\delta^2 I$. Continuing with the spectral decomposition calculation, we find the off diagonal variance equation unchanged and so (5.12) is again required.

The equations for diagonal elements are then the same as (5.13) but with σ_δ^2 added, so that we now find

$$\gamma_{k,u} = (1 - 2\mu\lambda_u + 2\mu^2\lambda_u^2)\gamma_{k-1,u} + \mu^2\lambda_u(\alpha_{k-1} + \sigma_\epsilon^2) + \sigma_\delta^2 \tag{5.37}$$

where

$$\alpha_k = \sum_{u=1}^{p} \lambda_u \gamma_{k,u}.$$

Much as before, the steady state value will be

$$\begin{aligned}\gamma_u &= [\mu^2\lambda_u(\alpha + \sigma_\epsilon^2) + \sigma_\delta^2]/(2\mu\lambda_u(1 - \mu\lambda_u)) \\ &= \frac{1}{2}\frac{\mu(\alpha + \sigma_\epsilon^2)}{(1 - \mu\lambda_u)} + \frac{\sigma_\delta^2}{2\mu\lambda_u}\frac{1}{(1 - \mu\lambda_u)}.\end{aligned} \tag{5.38}$$

From this we can get the steady state value α:

$$\begin{aligned}\alpha &= \sum_{u=1}^{p} \lambda_u \gamma_u \\ &= c(\alpha + \sigma_\epsilon^2) + \sigma_\delta^2 b/2\mu\end{aligned}$$

where c is given by (5.15) and b as in (5.20),

$$b = \sum_{u=1}^{p} (1 - \mu\lambda_u)^{-1}. \tag{5.39}$$

Solving gives

$$\alpha = \frac{c\sigma_\epsilon^2}{(1 - c)} + \frac{\sigma_\delta^2}{2\mu}\frac{b}{(1 - c)}. \tag{5.40}$$

Again, stability ensues when (5.17) or (5.18) hold.

5.3.3 Learning Curve—Random Walk Model (5.3A4(i))

The MSE calculation is similar to before. We find (5.19) still holds and that

$$\begin{aligned}\mathcal{E}_k \to \mathcal{E}_\infty &= \sigma_\epsilon^2 + \alpha \\ &= \sigma_\epsilon^2 + \frac{c}{(1 - c)}\sigma_\epsilon^2 + \frac{\sigma_\delta^2}{2\mu}\frac{b}{(1 - c)} \\ &= \frac{\sigma_\epsilon^2}{(1 - c)} + \frac{\sigma_\delta^2}{2\mu}\frac{b}{(1 - c)}.\end{aligned} \tag{5.41}$$

Note here that we have taken $k \to \infty$ even though this violates our physical constraint $var(y_k) < \infty$. The steady state values will be approached closely for large finite k however. Note now, that in (5.41) there is an additional inflation in MSE due to the parameter variation; it is called excess lag. The total misadjustment is then

$$M = (\mathcal{E}_\infty - \sigma_\epsilon^2)/\sigma_\epsilon^2 = \alpha/\sigma_\epsilon^2$$
$$= \frac{c}{1-c} + \frac{\sigma_\delta^2}{\sigma_\epsilon^2}\frac{1}{2\mu}\frac{b}{1-c}. \tag{5.42}$$

To get some insight into this formula, let us suppose as before that $\mu\lambda_u \ll 1$, $1 \leq u \leq p$, and then using Taylor series expansions (see Exercise 5.5) we can show

$$M \simeq \frac{1}{2}\mu t_1 + \frac{1}{2}(\sigma_\delta^2/\sigma_\epsilon^2)\left(\frac{p}{\mu} + t_1(\frac{p}{2} + 1) + \mu(p\alpha_2 + \frac{t_1^2}{2} + t_2)\right) + O(\mu^2) \tag{5.43}$$

where

$$t_s = \sum_{u=1}^{p} \lambda_u^s, \quad s = 1, 2 \tag{5.44}$$

$$\alpha_2 = \frac{1}{2}(t_2 + t_1^2/2). \tag{5.45}$$

From (5.43) misadjustment exhibits a very different behavior with respect to μ than it did (see (5.26)) in the absence of parameter variation. There is indeed a minimizing value of μ. Recalling our interpretation of the Random Walk model as a model of failure, we see that the adaptive algorithm will track the Random Walk with a bounded error variance; although if μ is too small this error will be large and so the tracking will not be very good.

Let us finally look at the case of a slowly changing Random Walk. Note that σ_δ^2 measures the speed of parameter change (since $var(\underline{w}_{o,k} - \underline{w}_{o,k-1}) = \sigma_\delta^2 I$). So let us put

$$\sigma_\delta^2 = \rho\sigma_\nu^2$$

where $\sigma_\nu^2 = O(1)$ and $\rho = o(1)$, i.e., $\rho \ll 1$. A small ρ thus entails a slowly changing parameter. If we know ρ then (5.43) is approximately minimized when the first two terms are equal, i.e.,

$$\mu t_1 = \rho\sigma_\nu^2 p/(\mu\sigma_\epsilon^2)$$
$$\Rightarrow \quad \mu = a\rho^{\frac{1}{2}} \tag{5.46}$$
$$a = \sigma_\nu(p/t_1)^{\frac{1}{2}}/\sigma_\epsilon. \tag{5.47}$$

Then (5.43) becomes

$$M = \mu t_1 + O(\mu^2).$$

So a slow drift can be successfully tracked.

5.3.4 First Order Analysis—Stationary Model (5.3A4(ii))

In this case $\delta \underline{w}_{o,k}$ of (5.36) is no longer a white noise, so it is necessary to consider jointly (5.35), (5.3A4)(ii). We find

$$\underline{\tilde{w}}_k = (I - \mu \underline{x}_k \underline{x}_k^T)\underline{\tilde{w}}_{k-1} + \mu \underline{x}_k \epsilon_k - (\underline{w}_{f,k} - \underline{w}_{f,k-1}) \tag{5.48}$$

$$\underline{w}_{f,k} = \theta \underline{w}_{f,k-1} + \underline{\eta}_k \tag{5.49}$$

where we have used the fact that $\delta \underline{w}_{o,k} = \delta \underline{w}_{f,k}$. Taking iterated conditional expectations as before, we find (5.6) holds again and so the first order analysis is unchanged.

5.3.5 Second Order Analysis—Stationary Model (5.3A4(ii))

As mentioned earlier, the computation in this subsection is somewhat tedious but entirely straightforward. Substitute (5.49) into (5.48) to get the following pair of equations:

$$\begin{pmatrix} \underline{\tilde{w}}_k \\ \underline{w}_{f,k} \end{pmatrix} = \begin{pmatrix} I - \mu \underline{x}_k \underline{x}_k^T & (1-\theta)I \\ \underline{0} & \theta I \end{pmatrix} \begin{pmatrix} \underline{\tilde{w}}_{k-1} \\ \underline{w}_{f,k-1} \end{pmatrix} + \begin{pmatrix} \mu \underline{x}_k \epsilon_k - \underline{\eta}_k \\ \underline{\eta}_k \end{pmatrix}. \tag{5.50}$$

This pair can be written in the compact form

$$\underline{W}_k = (\underline{B} - \mu \underline{\xi}_k \underline{\xi}_k^T)\underline{W}_{k-1} + \underline{\zeta}_k \tag{5.51}$$

where

$$\underline{W}_k = (\underline{\tilde{w}}_k^T,\ \underline{w}_{f,k}^T)^T, \quad \underline{\xi}_k = (\underline{x}_k^T,\ \underline{0}^T)^T$$

$$\underline{B} = \begin{bmatrix} I & (1-\theta)I \\ \underline{0} & \theta I \end{bmatrix}, \quad \underline{\zeta}_k = \begin{bmatrix} \mu \underline{x}_k \epsilon_k - \underline{\eta}_k \\ \underline{\eta}_k \end{bmatrix}.$$

Since $\underline{\xi}_k$ is a white noise, we can treat (5.51) much as we treated (5.5) to derive (5.8). The only real difference is that $\underline{\zeta}_k$ does not have a diagonal covariance matrix. We are led then to find that $\underline{\Sigma}_k = E(\underline{\zeta}_k \underline{\zeta}_k^T)$ obeys the equation

$$\begin{aligned} \underline{\Sigma}_{k+1} = {} & \underline{B}\underline{\Sigma}_k\underline{B}^T - \mu(\underline{B}\underline{\Sigma}_k\underline{R}_\xi + \underline{R}_\xi\underline{\Sigma}_k\underline{B}^T) + 2\mu^2\underline{R}_\xi\underline{\Sigma}_k\underline{R}_\xi \\ & + \mu^2 \operatorname{tr}(\underline{\Sigma}_k\underline{R}_\xi)\underline{R}_\xi + \mu^2\sigma_\epsilon^2\underline{R}_\xi + \sigma_\eta^2 U \end{aligned}$$

where

$$\underline{R}_\xi = \begin{pmatrix} R_x & \underline{0} \\ \underline{0} & \underline{0} \end{pmatrix}, \quad U = \begin{pmatrix} I & -I \\ -I & I \end{pmatrix}.$$

Now as before we use an eigenvector decomposition procedure. Recalling (5.9) we introduce

$$\underline{Q}_\Sigma = \begin{pmatrix} \underline{Q} & \underline{0} \\ \underline{0} & \underline{Q} \end{pmatrix}$$

and note that

$$\underline{Q}_\Sigma^T \underline{B} \underline{Q}_\Sigma = \underline{B}, \quad \underline{Q}_\Sigma^T U \underline{Q}_\Sigma = U.$$

Then setting

$$\underline{D}_k = \underline{Q}_\Sigma^T \Sigma_k \underline{Q}_\Sigma$$

we find

$$\underline{D}_k = \underline{B}\,\underline{D}_{k-1}\underline{B}^T - \mu(\underline{B}\,\underline{D}_{k-1}\Lambda_\xi + \Lambda_\xi \underline{D}_{k-1}\underline{B}^T) + 2\mu^2\Lambda_\xi \underline{D}_{k-1}\Lambda_\xi$$
$$+ \mu^2\Lambda_\xi(\alpha_{k-1} + \sigma_\epsilon^2) + \sigma_\eta^2 U \tag{5.52}$$

where (with Λ defined in (5.9))

$$\alpha_k = \operatorname{tr}(\underline{D}_k\Lambda_\xi)$$
$$\Lambda_\xi = \begin{pmatrix} \Lambda & \underline{0} \\ \underline{0} & \underline{0} \end{pmatrix}.$$

Let us write $\underline{D}_k$ in block form as

$$\underline{D}_k = \begin{pmatrix} \underline{D}_{a,k} & \underline{D}_{b,k}^T \\ \underline{D}_{b,k} & \underline{D}_{c,k} \end{pmatrix}.$$

Then from (5.52) we find on equating blocks that

$$\underline{D}_{a,k} = [\underline{D}_{a,k-1} + (1-\theta)(\underline{D}_{b,k-1} + \underline{D}_{b,k-1}^T) + (1-\theta)^2\underline{D}_{c,k-1}]$$
$$- \mu(\underline{D}_{a,k-1} + (1-\theta)\underline{D}_{b,k-1})\Lambda - \mu\Lambda(\underline{D}_{a,k-1} + (1-\theta)\underline{D}_{b,k-1}^T)$$
$$+ \mu^2(2\Lambda\underline{D}_{a,k-1}\Lambda + \Lambda(\alpha_{k-1} + \sigma_\epsilon^2)) + \sigma_\eta^2 I \tag{5.53}$$
$$\underline{D}_{b,k} = -\mu\theta\underline{D}_{b,k-1}\Lambda + \theta\underline{D}_{b,k-1} + \theta(1-\theta)\underline{D}_{c,k-1} - \sigma_\eta^2 I$$
$$\underline{D}_{c,k} = \theta^2\underline{D}_{c,k-1} + \sigma_\eta^2 I.$$

As before we look for a steady state solution. We find firstly

$$\underline{D}_c = D_c I \tag{5.54}$$
$$D_c = \sigma_\eta^2(1-\theta^2)^{-1} \tag{5.55}$$
$$\underline{D}_b = [\theta(1-\theta)\underline{D}_c - \sigma_\eta^2 I]((1-\theta)I + \mu\theta\Lambda)^{-1}.$$

Thus $\underline{D}_b$ is in fact diagonal with u^{th} entry

$$D_b^{(u)} = \left[\frac{\theta(1-\theta)\sigma_\eta^2}{1-\theta^2} - \sigma_\eta^2\right]\frac{1}{1-\theta+\mu\theta\lambda_u}, \quad 1 \le u \le p$$

$$\Rightarrow \quad D_b^{(u)} = -\frac{\sigma_\eta^2}{(1+\theta)(1-\theta+\mu\theta\lambda_u)}, \quad 1 \le u \le p. \tag{5.56}$$

Now we see by inspection from (5.53), (5.55), and (5.56) that the steady state variance $\underline{D}_a$ will be diagonal with its uth diagonal entry $D_a^{(u)}$ obeying

$$\begin{aligned} &D_a^{(u)}(2\mu\lambda_u - 2\mu^2\lambda_u^2) \\ &= 2(1-\theta)D_b^{(u)} + (1-\theta)^2 D_c - 2\mu(1-\theta)D_b^{(u)}\lambda_u + \mu^2(\alpha+\sigma_\epsilon^2)\lambda_u + \sigma_\eta^2. \end{aligned}$$

Substituting (5.55), (5.56) into this gives

$$\begin{aligned} &(2\mu\lambda_u - 2\mu^2\lambda_u^2)D_a^{(u)} \\ &= -\frac{2(1-\theta)(1-\mu\lambda_u)\sigma_\eta^2}{(1+\theta)(1-\theta+\mu\theta\lambda_u)} + \frac{(1-\theta)^2\sigma_\eta^2}{1-\theta^2} + \mu^2(\alpha+\sigma_\epsilon^2)\lambda_u + \sigma_\eta^2 \\ &= \frac{\sigma_\eta^2}{(1+\theta)(1-\theta+\mu\theta\lambda_u)}\left[\begin{array}{c} -2(1-\theta)(1-\mu\lambda_u) + (1-\theta)(1-\theta+\mu\theta\lambda_u) \\ +(1+\theta)(1-\theta+\mu\theta\lambda_u) \end{array}\right] \\ &\quad + \mu^2(\alpha+\sigma_\epsilon^2)\lambda_u \\ &= \frac{2\mu\lambda_u\sigma_\eta^2}{(1+\theta)(1-\theta+\mu\theta\lambda_u)} + \mu^2(\alpha+\sigma_\epsilon^2)\lambda_u. \end{aligned}$$

Thus we get for $1 \le u \le p$,

$$D_a^{(u)} = \frac{1}{1-\mu\lambda_u}\left[\mu(\alpha+\sigma_\epsilon^2)/2 + \frac{\sigma_\eta^2}{(1+\theta)(1-\theta+\mu\theta\lambda_u)}\right]. \tag{5.57}$$

Now we can calculate the steady state value of α:

$$\begin{aligned} \alpha &= \sum_{u=1}^{p} \lambda_u D_a^{(u)} \\ &= c(\alpha+\sigma_\epsilon^2) + \sigma_\eta^2 d \end{aligned} \tag{5.58}$$

where c is given as before by (5.15) and

$$d = \frac{1}{1+\theta}\sum_{u=1}^{p}\frac{\lambda_u}{(1-\theta+\mu\theta\lambda_u)(1-\mu\lambda_u)}. \tag{5.59}$$

Thus we find that

$$\alpha = \frac{c}{1-c}\sigma_\epsilon^2 + \frac{d}{1-c}\sigma_\eta^2.$$

We are now able to calculate the learning curve.

5.3.6 Learning Curve—Stationary Model (5.3A4(ii))

We have, as previously, and on using (5.58) found that

$$\begin{aligned} \mathcal{E}_k &= E(e_k^2) \\ &= \sigma_\epsilon^2 + \alpha_{k-1} \\ \Rightarrow \qquad \mathcal{E}_\infty &= \sigma_\epsilon^2 + \alpha \\ &= \frac{\sigma_\epsilon^2}{1-c} + \frac{\sigma_\eta^2 d}{1-c}. \end{aligned}$$

Thus the misadjustment is

$$\begin{aligned} M &= (\mathcal{E}_\infty - \sigma_\epsilon^2)/\sigma_\epsilon^2 = \alpha/\sigma_\epsilon^2 \\ &= \frac{c}{1-c} + \frac{\sigma_\eta^2}{\sigma_\epsilon^2}\frac{d}{1-c}. \end{aligned} \tag{5.60}$$

The stability condition (5.17) is then unchanged. As we have done before we expand M in powers of μ by assuming $\mu \ \mathrm{tr}(\underline{R}_x) \ll 1$ so that $\mu\lambda_u \ll 1$, $1 \le u \le p$. Then if

$$1 - \theta \text{ is not small} \tag{5.61}$$

we find the following expansion (see Exercise 5.6):

$$M \simeq \frac{1}{2}\mu t_1 + \frac{\sigma_\eta^2}{\sigma_\epsilon^2}\frac{1}{1-\theta}(t_1 + \mu(t_2 + t_1^2/2)) + O(\mu^2) \tag{5.62}$$

where t_1, t_2 are defined in (5.44). Note from (5.3A4)(ii) that

$$var(\underline{w}_{f,k}) = \sigma_\eta^2 I/(1-\theta^2) \tag{5.63}$$

so that

$$var(\underline{w}_{f,k} - \underline{w}_{f,k-1}) = 2\sigma_\eta^2 I/(1+\theta). \tag{5.64}$$

That is, the speed of parameter change, as measured by the square root of the norm of (5.64), is $O(1)$. With $\mu \ \mathrm{tr}(\underline{R}_x)$ small, we have slow adaptation speed relative to the true speed of parameter change.

We see that when (5.61) holds, the misadjustment has a dominating term that is unaffected by μ, namely, $\sigma_\eta^2 t_1/(\sigma_\epsilon^2(1-\theta))$, and consequently the misadjustment cannot be made small. To put it more clearly, the conclusion is that slow adaptation fails. We are led to suspect that successful adaptation requires the adaptation speed and true speed be matched. To investigate this issue we have to turn to slowly varying parameter models.

5.3.7 Slowly Varying Parameter—Stationary Model (5.3A4(ii))

We consider a large-amplitude, slow-speed model as follows. We reexpress the model in (5.3A4) (ii) in terms of small parameter ρ as follows:

$$\begin{aligned} 1-\theta &= \rho, \quad \rho \ll 1 \\ \sigma_\eta^2 &= \rho\sigma_\nu^2, \quad \sigma_\nu^2 = O(1). \end{aligned} \tag{5.65}$$

We see from (5.63), (5.65) that

$$\begin{aligned} var(\underline{w}_{f,k}) &\simeq \sigma_\nu^2 I/2 = O(1) \\ var(\underline{w}_{f,k} - \underline{w}_{f,k-1}) &\simeq \rho\sigma_\nu^2 I. \end{aligned}$$

So indeed this model has large amplitude but slow speed ($= \rho^{\frac{1}{2}}\sigma_\nu \ll 1$). Now the condition (5.61) no longer holds so we must redo the Taylor series expansion of M in (5.60). From (5.59) we see that

$$\begin{aligned} d &= \sum_{u=1}^{p} \frac{\lambda_u}{(2-\rho)(\rho(1-\mu\lambda_u)+\mu\lambda_u)(1-\mu\lambda_u)} \\ &\simeq \frac{1}{2\mu}\sum_{u=1}^{p}\frac{1}{1-\mu\lambda_u} + O(1), \quad \rho \ll \mu \ll 1. \end{aligned} \tag{5.66}$$

Thus anticipating that $\mu = O(\rho^{\frac{1}{2}})$ we find (see Exercise 5.7)

$$\begin{aligned} M &\simeq \frac{1}{2}\mu t_1 + \frac{\rho\sigma_\nu^2}{2\sigma_\epsilon^2}\frac{1}{\mu}\left[\sum_{u=1}^{p}(1+\mu\lambda_u)\right][1+\mu t_1/2] + O(\mu^2) \\ &\simeq \frac{1}{2}\mu t_1 + \frac{\rho\sigma_\nu^2}{2\sigma_\epsilon^2}\frac{1}{\mu}[p + \mu t_1(p/2+1)] + O(\mu^2). \end{aligned} \tag{5.67}$$

From this expression there is an optimal value for μ which occurs approximately when the first two terms are equal, i.e.,

$$\begin{aligned} \mu t_1 &= \rho p\sigma_\nu^2/(\sigma_\epsilon^2\mu) \\ \Rightarrow \quad \mu &= a\rho^{\frac{1}{2}} \quad \text{(speed match condition)} \end{aligned} \tag{5.68}$$

where (cf. (5.47))

$$a = (p\sigma_\nu^2/(\sigma_\epsilon^2 t_1))^{\frac{1}{2}}.$$

Note that (5.68) is consistent with (5.66), i.e., $\rho \ll \mu$. With this optimal setting we find

$$\begin{aligned} M &\simeq \mu t_1 + O(\mu^2) \\ &= \rho^{\frac{1}{2}} a t_1 + O(\mu^2). \end{aligned}$$

That is, the misadjustment is small. Note how the results obtained with the present model are the same as those obtained with the Random Walk model. In Exercise 5.10 the reader is asked to show that when the speed match condition (5.68) holds,

$$\text{var}(\underline{\tilde{w}}_k) = O(\rho^{\frac{1}{2}}).$$

This should be compared to

$$\text{var}(\underline{w}_{fk}) = O(1)$$

and it tells us that LMS can successfully track a large amplitude slowly varying parameter if the adaptation speed is matched to the parameter speed (via (5.68)).

Example 5.2. Tracking a Large Amplitude Slowly Time Varying Parameter with LMS

Again we use the Channel Equalization Example 2.2. As in Example 5.1 we suppose

$$y_k = s_{k-1} + m_k \tag{5.69}$$

$$s_k = (1 + \beta q^{-1})^{-1} \epsilon_k. \tag{5.70}$$

But now we assume a time varying channel

$$x_k = \xi_k = -\theta_{k-1}\xi_{k-1} + g s_k \tag{5.71}$$

where θ_k is the time varying parameter (we can think of $1/\ln(1/|\theta_k|)$ as a time constant). From (5.69), (5.71)

$$y_k = \underline{w}_k^T \underline{x}_k + m_k \tag{5.72}$$

$$\underline{w}_k = (\, g^{-1} \quad \theta_{k-2}/g\,)^T; \quad \underline{x}_k = (\, x_{k-1} \quad x_{k-2}\,). \tag{5.73}$$

We suppose θ_k consists of two parts: a "dc" component θ_{dc} and an "ac" or fluctuation component $\theta_{f,k}$, which is a zero mean AR(1) process

$$\theta_k = \theta_{dc} + \theta_{f,k} \tag{5.74}$$

$$\theta_{f,k} = (1 - \rho)\theta_{f,k-1} + \rho^{1/2}\nu_k, \quad \rho \ll 1 \tag{5.75}$$

where ν_k is a white noise of zero mean and variance σ_ν^2. Also $\{\nu_k\}$ is independent of $\{m_k\}$, $\{\epsilon_k\}$. Note that

$$\text{var}(\theta_{f,k}) \simeq \sigma_\nu^2/2; \quad \text{var}(\delta\theta_{f,k}) \simeq \rho\sigma_\nu^2.$$

From (5.73)–(5.75) we find 5.3A4(ii) is satisfied. Note that in (5.71) we need a condition on $\{\theta_k\}$ to ensure x_k has bounded sample paths. This issue is not straightforward so we take a pragmatic view. Clearly the smaller the variance of $\theta_{f,k}$, the less likely that x_k will blow up. So in simulations we try successively smaller σ_ν^2 values until we get a stable x_k simulation. The parameter settings chosen for the simulations are

$$(\,\theta_{dc}, \quad \beta\,) = (\,1.5, \quad -0.7\,)$$

$$(\,\sigma_\nu^2, \quad \sigma_m^2, \quad \sigma_\epsilon^2, \quad g\,) = (\,0.02, \quad 0.5, \quad 1.0, \quad 1.0\,)$$

$$\rho = 0.01.$$

In Figs. 5.5–5.10 we show slow adaptation ($\mu = o(\rho^{1/2})$), matched adaptation ($\mu = O(\rho^{1/2})$), and fast adaptation ($\rho^{1/2} = o(\mu)$).

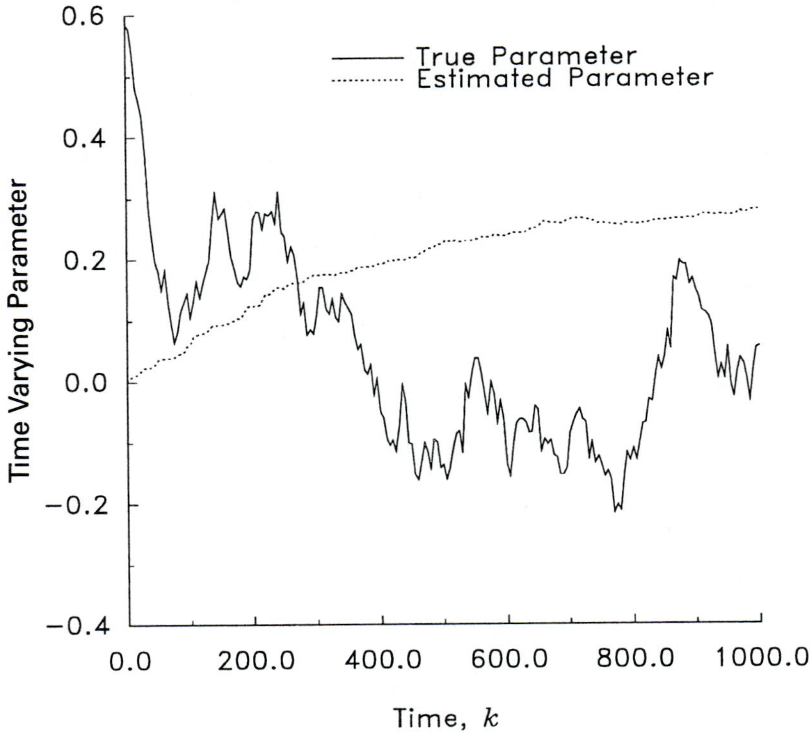

Figure 5.5 Tracking a slowly time varying parameter: slow adaptation $(\mu, \rho) = (0.0005, 0.01)$.

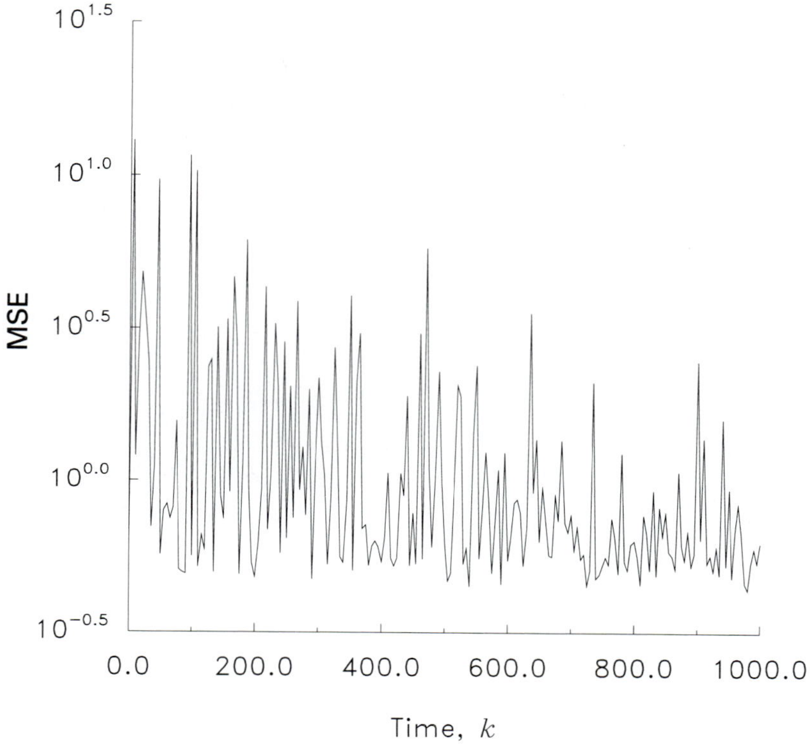

Figure 5.6 Estimated learning curve: slow adaptation $(\mu, \rho) = (0.0005, 0.01)$.

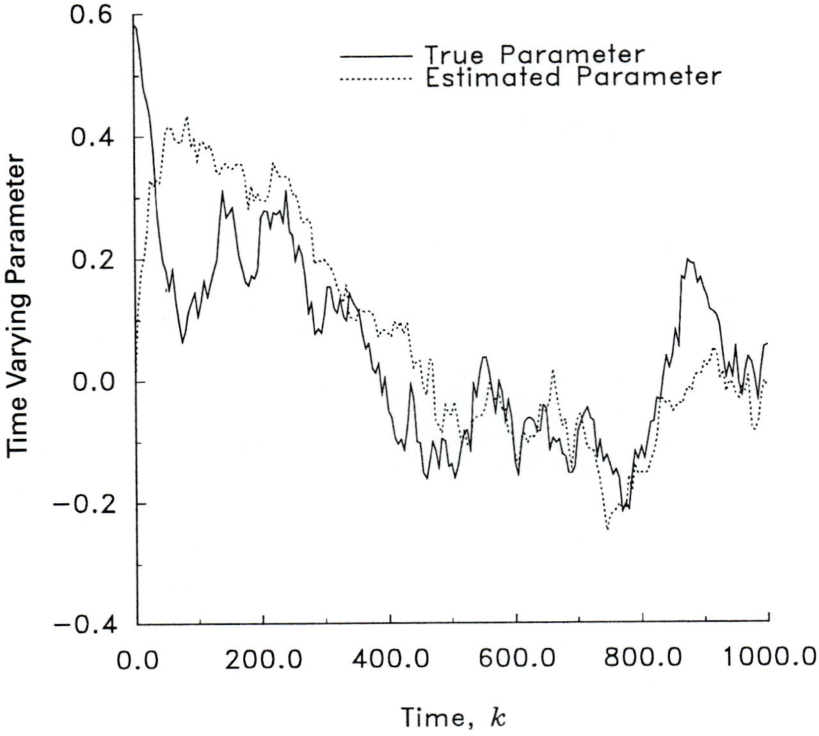

Figure 5.7 Tracking a slowly time varying parameter: matched adaptation $(\mu, \rho) = (0.01, 0.01)$.

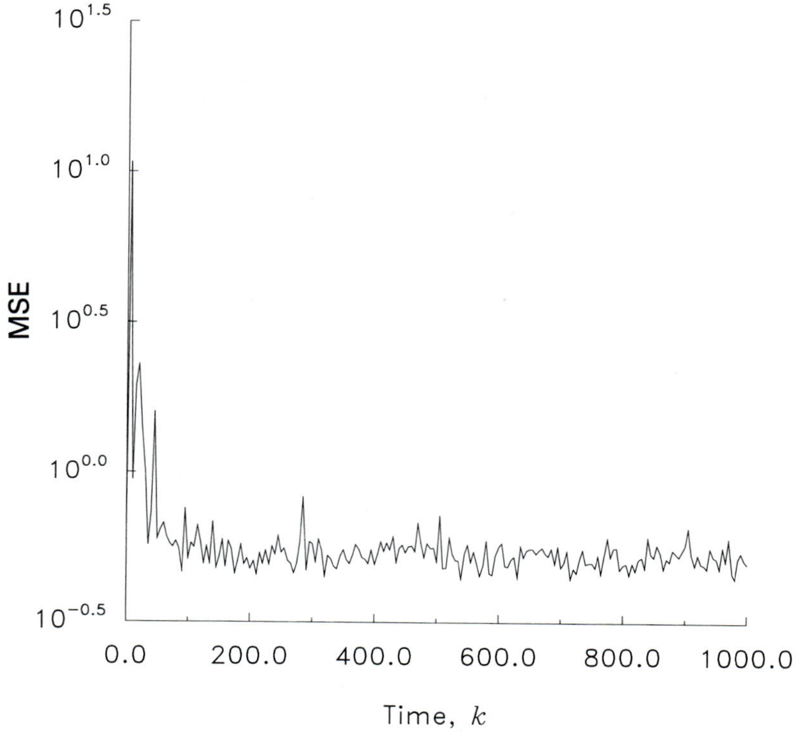

Figure 5.8 Estimated learning curve: matched adaptation $(\mu, \rho) = (0.01, 0.01)$.

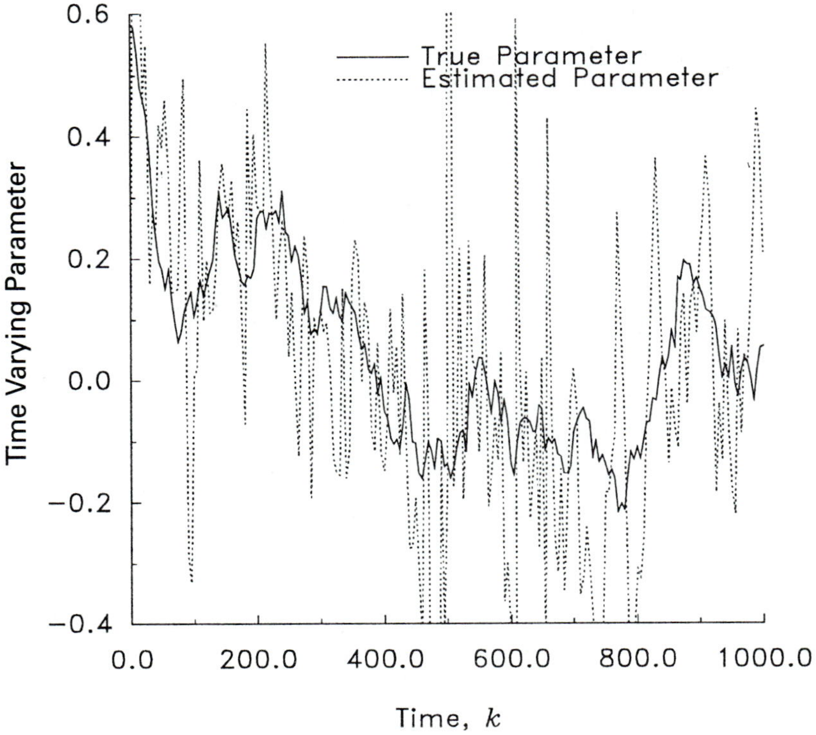

Figure 5.9 Tracking a slowly time varying parameter: fast adaptation $(\mu, \rho) = (0.1, 0.01)$.

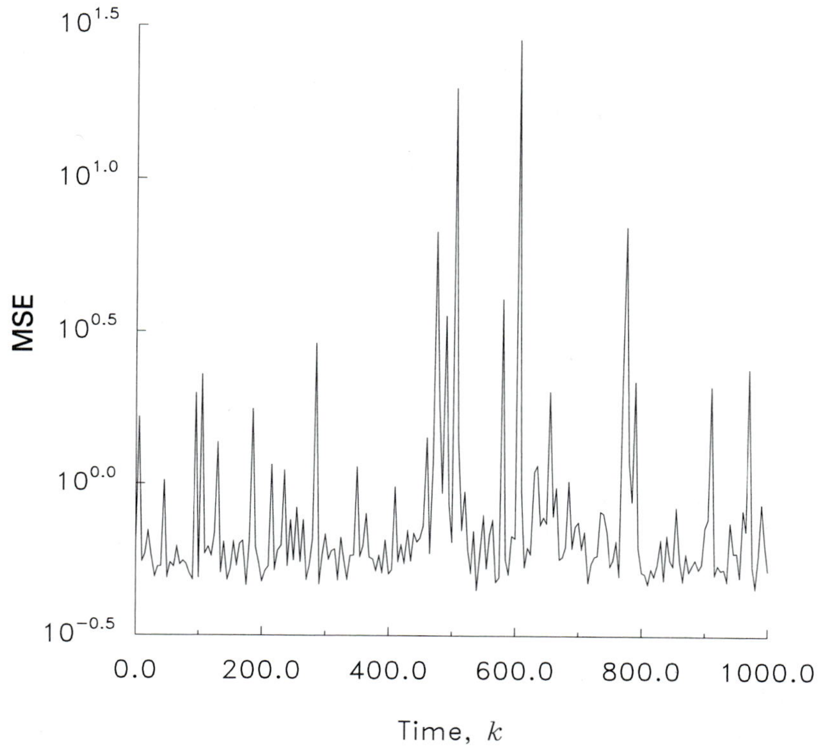

Figure 5.10 Estimated learning curve: fast adaptation $(\mu, \rho) = (0.1, 0.01)$.

5.3.8 Optimal Choice of Gain

In Exercise 5.8 the reader is asked to show that for the large amplitude slowly time varying model (5.65) with the speed match condition $\mu = \rho^{\frac{1}{2}}$, the steady state weight error variance

$$\underline{P} = \lim_{k\to\infty} \text{var}(\underline{\tilde{w}}_k)$$

obeys

$$\underline{P}\mu^{-1} \to \underline{\Pi} \quad \text{as } \mu \to 0$$

where

$$-\underline{R}_x\underline{\Pi} - \underline{\Pi}\underline{R}_x + \underline{R}_x\sigma_\epsilon^2 + \sigma_v^2 I = \underline{0}. \tag{5.76}$$

Now consider the LMS algorithm with a matrix gain,

$$\delta\underline{\hat{w}}_k = \mu\underline{G}\underline{x}_k e_k$$
$$e_k = y_k - \underline{x}_k^T\underline{\hat{w}}_{k-1}.$$

Under the same conditions as above, Exercise 5.9 asks to show that the steady state weight error variance

$$\underline{P} = \lim_{k\to\infty} \text{var}(\underline{\tilde{w}}_k)$$

obeys

$$\underline{P}\mu^{-1} \to \underline{\Pi} \quad \text{as } \mu \to 0$$

where

$$-\underline{G}\underline{R}_x\underline{\Pi} - \underline{\Pi}(\underline{G}\underline{R}_x)^T + \underline{G}\underline{R}_x\underline{G}\sigma_\epsilon^2 + \sigma_\nu^2 I = \underline{0}. \tag{5.77}$$

From this expression Exercise 5.9 asks to show that the choice of $\underline{G}$ that minimizes $\underline{\Pi}$ is

$$\underline{G}_o = (\sigma_\nu/\sigma_\epsilon)\underline{R}_x^{-\frac{1}{2}} \tag{5.78}$$

where $\underline{R}_x^{\frac{1}{2}}$ is the symmetric square root of $\underline{R}_x$. The corresponding optimized variance is

$$\underline{\Pi}_o = (\sigma_\nu\sigma_\epsilon)\underline{R}_x^{-\frac{1}{2}}. \tag{5.79}$$

Although we have concentrated on LMS in this chapter, it is worth making some comments here on NM algorithms. It is implicit in the construction of the EWLS algorithm (Section 4.7) that the optimal vector gain should be $\underline{R}_x^{-1}$ (see Exercise 7.1). We see from the current analysis that this is untrue. This casts the EWLS algorithm in a negative light, although it does have excellent transient properties (i.e., it reacts quickly to parameter change). In Exercise 7.7, it is shown that, with the large amplitude slowly time varying model (5.65), the Kalman filter algorithm of Section 4.7 yields the optimal gain as $\mu = \rho^{\frac{1}{2}} \to 0$. Further discussion of NM algorithms must wait until Chapter 9.

5.4 NOTES

The results of Section 5.2 are due to [F2] in which a rigorous stability analysis is provided of (5.13) which shows that (5.17) is indeed necessary and sufficient for stability. An analysis of LMS_N under white Gaussian noise assumptions is given by [B8]. The results of Section 5.3 were developed in class notes of VS but calculations under (5.3A1)–(5.3A3), 5.3A4(ii) (but with $\underline{R}_x = \sigma_x^2 I$) have been made by [H2].

The first attempt to analyze LMS in the presence of time varying parameters is in [W4]. They found an $O(\mu^{-1})$ term in the misadjustment but their analysis is rather heuristic. It was [M1] where this term was shown to be due to a Random Walk model. The large amplitude slowly varying parameter model was introduced in [S15] and [S16]. In [B6] and [B7] a small amplitude slowly varying parameter model is used (see Exercise 5.8). This model fails, however, to show the fundamental feature that an adaptive algorithm is capable of tracking a large amplitude slowly varying parameter.

Learning curve and weight error variance calculation (for the LMS type algorithm) under more general signal assumptions are discussed in [L11] for a Random Walk model and in [S15] and [S16] for more general parameter variation models. From [B6], [B7], [S15], and [S16] it follows that the results of this chapter are unaffected to order μ if $\underline{x}_k$ is a colored stationary signal (independent of ϵ_k which is a white noise). The results are changed if ϵ_k is a colored signal. Further discussion of these points is made in Chapter 9.

The optimal matrix gain for tracking a slowly varying parameter is also given by (5.78) for the Random Walk model: this fact is from the work of [B6] (see also [B7]).

He also observes that the Kalman filter provides the optimal gain (5.79) for the Random Walk model. So the results on the Kalman filter for the Random Walk model and the large amplitude slowly varying AR model are the same, and they cast EWLS in a negative light. The lesson here is that, at the second order level, there is an interaction between learning and signal processing, so that heuristics based on a separation of them is not completely reliable. There is no such interaction at the first order level. Other useful observations about EWLS are made in [C6].

In this chapter we have concentrated on ensemble average performance by calculating first and second moments. From a practical point of view, the user has only one realization of the algorithm trajectory to work with and would prefer a realization wise analysis. Such calculations have been made for LMS under general signal assumptions in [S15]. Averaging analysis produces realization wise results also. Other analysis under non-stationary signal assumptions include [C4], [G1], and [L12] on LMS; and [C4], [G14], and [L12] on the Kalman filter.

In implementing the LMS algorithm an important practical issue is the effect of roundoff: this has been analyzed by [C2] where it is shown that it produces a μ^{-1} term in the misadjustment.

EXERCISES

5.1 **(a)** Repeat Exercise 4.7(a), this time estimating the misadjustment by averaging over the steady state part of the estimated learning curve for the choice of μ leading to stability. Compare this with the theoretical result.

(b) Repeat Exercise 4.7(b) as per 5.1(a).

(c) Repeat Exercise 4.7(c) as per 5.1(a).

5.2 Consider the momentum LMS algorithm

$$\delta\underline{\hat{w}}_k = \alpha\delta\underline{\hat{w}}_{k-1} + \mu\underline{x}_k e_k, \quad e_k = y_k - \underline{x}_k^T\underline{\hat{w}}_{k-1}$$

under assumptions (5.2A1)–(5.2A3):

(a) Find first order stability conditions.

(b) Find second order stability conditions.

(c) Calculate the misadjustment.

5.3 Consider the relaxation LMS algorithm

$$\delta\underline{\hat{w}}_k = -\alpha\mu\underline{\hat{w}}_{k-1} + \mu\underline{x}_k e_k$$

under assumptions (5.2A1)–(5.2A3):

(a) Find first order stability conditions and show that $E(\underline{\hat{w}}_k) \to \underline{w}_e \neq \underline{w}_o$. Give an expression for $\underline{w}_e$.

(b) Consider now, for tractability, the scalar case $p = 1$. Calculate the MSE and show there is a range of small μ values where relaxation LMS, although biased, has a lower MSE than LMS.

5.4 Consider the time varying channel equalization setup (sketched in Fig. 5.11) with the following signal specifications:

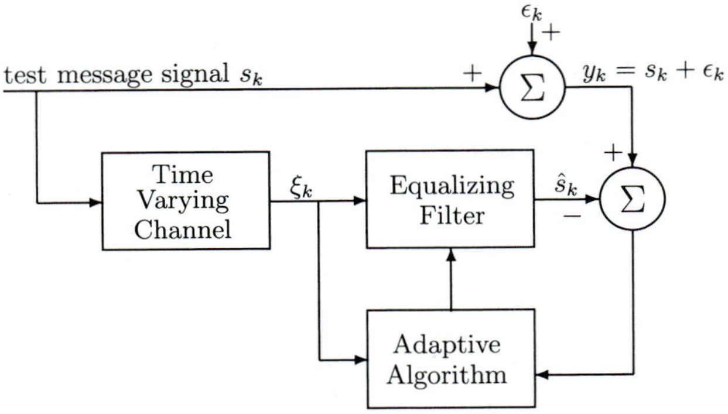

Figure 5.11 Adaptive channel equalizer with time varying channel

$$\xi_k = a_{1k}\xi_{k-1} + g_k s_k$$

$$g_k = g_{dc} + g_{fk}, \qquad a_{1k} = a_{dc} + a_{fk}$$

$$(g_{dc}, a_{dc}) = (1, -1.5)$$

g_{fk} is zero mean white noise of variance 0.1

a_{fk} is an AR(1) process

$$a_{fk} = (1-\rho)a_{f,k-1} + \rho^{\frac{1}{2}}\nu_k, \qquad \rho = 0.1$$

ν_k is a zero mean white noise of variance 0.32

and use a single tap equalizer.

(a) Simulate the system using three values of μ that show (i) slow adaptation; (ii) matched adaptation; (iii) fast adaptation. From the simulation estimate the misadjustment in each case.

(b) Calculate the theoretical misadjustment (using the white noise formulae even though "$\underline{x}_k$" is not a white noise).

5.5 From (5.19) show that the steady state weight error covariance matrix is

$$\Gamma_\infty = \frac{\mu\sigma_\epsilon^2}{2(1-c)}(I - \mu\underline{R}_x)^{-1}.$$

Find the first two terms in a Taylor series expansion.

5.6 Derive the expansion (5.43) from (5.42), (5.39), and (5.15).

5.7 Derive the expansion (5.62) from (5.60) under (5.61).

5.8 Derive the expansion (5.67) from (5.60) under (5.64). Show that $M = M(\mu)$ has a minimum if

$$\mu\lambda_{max} < (1-2\rho)/(1-\rho).$$

5.9 Assume the large amplitude slowly time varying model (5.65) and the speed match condition $\mu = \rho^{\frac{1}{2}}$. Using the expressions for D_a, D_b, D_c in Section 5.3 show that the steady state weight error variance

$$\underline{P} = \underline{P}(\mu) = \lim_{k\to\infty} \text{var}(\underline{\tilde{w}}_k)$$

obeys

$$\underline{P}(\mu)/\mu \to \underline{\Pi}, \qquad \text{as } \mu \to 0$$

where $\underline{\Pi}$ obeys the Lyapunov equation (see Appendix C6)

$$-\underline{R}_x\underline{\Pi} - \underline{\Pi}\underline{R}_x + \underline{R}_x\sigma_\epsilon^2 + \sigma_v^2 I = \underline{0}.$$

5.10 Repeat the analysis of Exercise 5.8 for the Random Walk model with $\sigma_\delta^2 = \mu^2\sigma_v^2$.

5.11 Assume the large amplitude slowly time varying model (5.65) and the speed match condition (5.68). Consider the LMS algorithm with matrix gain

$$\delta\underline{\hat{w}}_k = \mu\underline{G}\underline{x}_k e_k$$

$$e_k = y_k - \underline{x}_k^T\underline{\hat{w}}_{k-1}.$$

Repeat the analysis of Section 5.3 to show that the steady weight error variance

$$\underline{P} = \underline{P}(\mu) = \lim_{k\to\infty} \operatorname{var}(\underline{\tilde{w}}_k)$$

obeys

$$\underline{P}(\mu)/\mu \to \underline{\Pi} \quad \text{as } \mu \to 0$$

where

$$-\underline{G}\underline{R}_x\underline{\Pi} - \underline{\Pi}(\underline{G}\underline{R}_x)^T + \underline{G}\underline{R}_x\underline{G}^T\sigma_\epsilon^2 + \sigma_v^2 I = \underline{0}.$$

Show that the choice of $\underline{G}$ minimizing $\underline{\Pi}$ is

$$\underline{G}_o = (\sigma_v/\sigma_\epsilon)\underline{R}_x^{-\frac{1}{2}}$$

where $\underline{R}_x^{-\frac{1}{2}}$ is the symmetric square root of $\underline{R}_x$. Show that the minimized value of $\underline{\Pi}$ is

$$\underline{\Pi}_o = (\sigma_v\sigma_\epsilon)\underline{R}_x^{-\frac{1}{2}}.$$

5.12 Repeat Exercise 5.11 for the Random Walk model with $\sigma_\delta^2 = \mu^2\sigma_\mu^2$.

5.13 Consider the following small amplitude slowly time varying parameter model:

$$\delta\underline{w}_{fk} = -\rho(\underline{w}_{f,k-1} + \underline{v}_k), \qquad \rho \ll 1$$

where $\underline{v}_k$ is a white noise of zero mean, variance $\sigma_v^2 I$.

(a) Show that

$$\operatorname{var}(\underline{w}_{fk}) \sim (\rho\sigma_v^2/2)I$$

$$\operatorname{var}(\delta\underline{w}_{fk}) \sim (\rho^2\sigma_v^2/2)I$$

so that both amplitude and speed are small.

(b) Calculate a Taylor series expansion for the misadjustment of LMS (assuming (5.3A1)–(5.3A3)) and compare your results with (5.43), (5.62), and (5.67).

(c) Show the approximately optimal value for μ is $\mu = E\rho$ and give an expression for E.

(d) With the approximate optimal setting for μ show that

$$\operatorname{var}(\underline{\tilde{w}}_k) = O(\rho)$$

which, when compared with $\operatorname{var}(\underline{w}_{fk}) = O(\rho)$, shows that an adaptive algorithm is not capable of tracking a small amplitude slowly varying parameter.

6

Algorithm Analysis: Deterministic Global Theory

6.1 SINGLE TIME SCALE LMS AND PERSISTENT EXCITATION

In this chapter we look at the behavior of several adaptive algorithms in a deterministic setting. The stability problem reduces to the problem of convergence and, aside from the speed of convergence issue, the performance question does not really arise. As mentioned in the introduction to Part I, the convergence issue is not as narrow as it may seem because it throws considerable light on boundedness or input-output stability in a noisy setting. This point will become clear in Part III.

In this chapter we deal with a limited group of algorithms for which a global convergence theory can be developed under mild signal assumptions. While the methods used do not generalize, the analysis given here has proved important in the development and understanding of adaptive algorithms.

We start with the LMS algorithm

$$\delta \underline{\hat{w}}_k = \mu \underline{x}_k e_k$$

$$e_k = y_k - \underline{x}_k^T \underline{\hat{w}}_{k-1}$$

and introduce the following assumptions:

(6.1A1) Amplitude constraint

$$\sup_k \|\underline{x}_k\|^2 \leq B_x < \infty$$

(6.1A2) Modeling assumption. There is no output noise and there is a true weight $\underline{w}_o$ with

$$y_k = \underline{x}_k^T \underline{w}_o.$$

As usual we proceed to construct an error system. Using (6.1A2) we can rewrite the error signal as

$$e_k = -\underline{x}_k^T \tilde{\underline{w}}_{k-1} \tag{6.1}$$

where $\tilde{\underline{w}}_k$ is the weight estimation error

$$\tilde{\underline{w}}_k = \hat{\underline{w}}_k - \underline{w}_o.$$

Thus the LMS error system is

$$\tilde{\underline{w}}_k = (I - \mu \underline{x}_k \underline{x}_k^T)\tilde{\underline{w}}_{k-1}. \tag{6.2}$$

To analyze the convergence of this deterministic time varying difference equation, we will use a Lyapunov function (see Appendix C). A Lyapunov function is typically a system "energy" and is often a quadratic form in the system state. The idea is that the energy should $\rightarrow 0$ as time $\rightarrow \infty$ if the system is stable. The simplest candidate Lyapunov function here is

$$V_k = \|\tilde{\underline{w}}_k\|^2.$$

If we can show $V_k \rightarrow 0$ then clearly we can deduce $\|\tilde{\underline{w}}_k\|^2 \rightarrow 0$. We have from (6.2) that

$$\begin{aligned} V_k &= \tilde{\underline{w}}_{k-1}^T (I - \mu \underline{x}_k \underline{x}_k^T)(I - \mu \underline{x}_k \underline{x}_k^T)\tilde{\underline{w}}_{k-1} \\ &= V_{k-1} - \mu(\underline{x}_k^T \tilde{\underline{w}}_{k-1})^2 (2 - \mu \underline{x}_k^T \underline{x}_k). \end{aligned}$$

We would like to deduce that $V_k \leq V_{k-1}$; to this end we assume, in view of (6.1A1) that μ is chosen so that

(6.1A3) $\mu B_x < 2$. (B_x is defined in (6.1A1))

Then we find

$$V_k \leq V_{k-1} - \rho\mu(\underline{x}_k^T \tilde{\underline{w}}_{k-1})^2 \tag{6.3}$$

where

$$\rho = 2 - \mu B_x > 0.$$

We now deduce

$$V_k \leq V_{k-1}.$$

Thus V_k is a bounded, positive, non-increasing sequence of numbers and so must have a finite limit (see Appendix D); call it V_∞. Next using (6.1) and summing (6.3) we find

$$V_k + \rho\mu \sum_{t=1}^{k} e_t^2 \leq V_0.$$

Since we already know $V_k \to V_\infty < \infty$, we deduce

$$\sum_{t=1}^{\infty} e_t^2 < \infty \tag{6.4}$$

so that (see Appendix D)

$$e_t \to 0 \quad \text{as } t \to \infty.$$

That is, we deduce the following remarkable result.

Theorem 6.1. Under (6.1A1)–(6.1A3), LMS has the "error nulling" property, namely, $e_t \to 0$ as time $t \to \infty$.

To deduce parameter convergence, i.e., $V_k \to 0$, we need an additional assumption. If we reflect on the Gaussian white noise analysis of Section 5.2 this should not be surprising. There we saw (in condition 5.7) that $\underline{R}_x$ must be positive definite, and so clearly what is needed here is a deterministic version of that condition. The required condition is called persistence of excitation.

(6.1A4) $\{\underline{x}_k\}$ is strongly persistently exciting (PE) if there are positive constants α_1, α_2 and a window span N, such that

$$\alpha_1 I \le \frac{1}{N} \sum_{k=k_0+1}^{k_0+N} \underline{x}_k \underline{x}_k^T \le \alpha_2 I, \qquad \text{for all } k_0 \ge 0.$$

(If U, V are two matrices, $U \le V$ is defined in Appendix A1(vi)). The condition says that an average of span N is positive definite no matter what time point the average is started from. See Exercise 6.3 for some examples.

To use (6.1A4) in our analysis we return to (6.3) and sum to get, for any given k_0,

$$V_{k_0+N} \le V_{k_0} - \rho\mu \sum_{k=k_0+1}^{k_0+N} \underline{\tilde{w}}_{k-1}^T \underline{x}_k \underline{x}_k^T \underline{\tilde{w}}_{k-1}.$$

Next iterate (6.2) to find

$$\underline{\tilde{w}}_{k-1} = \Pi_{k_0+1,k-1} \underline{\tilde{w}}_{k_0}$$

where we have introduced the system transition matrix of (6.2):

$$\Pi_{k_0+1,k-1} = (I - \mu \underline{x}_{k-1} \underline{x}_{k-1}^T) \cdots (I - \mu \underline{x}_{k_0+1} \underline{x}_{k_0+1}^T). \tag{6.5}$$

Thus we can write

$$V_{k_0+N} \le V_{k_0} - \rho\mu \underline{\tilde{w}}_{k_0}^T \underline{G}_{k_0} \underline{\tilde{w}}_{k_0} \tag{6.6}$$

where

$$\underline{G}_{k_0} = \sum_{k=k_0+1}^{k_0+N} \Pi_{k_0+1,k-1}^T \underline{x}_k \underline{x}_k^T \Pi_{k_0,k-1}.$$

Next we prove the following lemma.

Lemma 6.1. If (6.1A1) and (6.1A4) hold, then there is an $\eta > 0$ such that $\underline{G}_{k_0} \geq \eta I$, for any k_0.

Applying this lemma to (6.6) gives

$$V_{k_0+N} \leq V_{k_0} - \rho\mu\eta V_{k_0} = (1 - \rho\mu\eta)V_{k_0}.$$

We can take η so small that $\rho\mu\eta < 1$ and then iterate to find that, for any $m \geq 0$,

$$V_{mN+r} \leq (1 - \rho\mu\eta)^m V_r, \quad 0 \leq r \leq N - 1.$$

From this we see

$$V_k \to 0, \quad \text{exponentially or geometrically fast, as } k \to \infty.$$

More precisely, we can show that (see Exercise 6.5) for some M

$$V_k \leq M\beta^k V_0, \quad k \geq 1 \tag{6.7}$$

where

$$0 < \beta = (1 - \rho\mu\eta)^{1/N} < 1. \tag{6.8}$$

This is of course the (exponential) convergence of $\underline{\tilde{w}}_k$ that we sought. We have then

Theorem 6.2. Under (6.1A1)–(6.1A4), $\underline{\tilde{w}}_k \to 0$ exponentially fast. To put it another way (see Appendix C), $\underline{0}$ is an exponentially stable equilibrium point of (6.2).

Proof of Lemma 6.1. The proof is by contradiction. If the result is false, then for some k_0 there exists an $\underline{\alpha}$ with $\|\underline{\alpha}\| = 1$, such that

$$\underline{\alpha}^T \underline{G}_{k_0}\underline{\alpha} \leq c^2 = \alpha_1/2NP^2,$$

where P is a quantity to be chosen, α_1 is defined in (6.1A4). Thus we have

$$\sum_{k=k_0+1}^{k_0+N} (\underline{x}_k^T \Pi_{k_0+1,k-1}\underline{\alpha})^2 \leq c^2$$

$$\Rightarrow \quad |\underline{x}_k^T \Pi_{k_0+1,k-1}\underline{\alpha}| \leq c, \quad k_0 + 1 \leq k \leq k_0 + N.$$

Setting $k = k_0 + 1$ we find

$$|\underline{x}_{k_0+1}^T\underline{\alpha}| \leq c.$$

Setting $k = k_0 + 2$ gives

$$
\begin{aligned}
&|\underline{x}_{k_0+2}^T(I - \mu\underline{x}_{k_0+1}\underline{x}_{k_0+1}^T)\underline{\alpha}| \leq c \\
\Rightarrow \quad &|\underline{x}_{k_0+2}^T\underline{\alpha}| \leq |\underline{x}_{k_0+2}^T(I - \mu\underline{x}_{k_0+1}\underline{x}_{k_0+1}^T)\underline{\alpha}| \\
&\qquad + \mu|\underline{x}_{k_0+2}^T\underline{x}_{k_0+1}||\underline{x}_{k_0+1}^T\underline{\alpha}| \\
&\qquad \leq c(1 + \mu B_x), \quad \text{by (6.1A1)} \\
&\qquad \leq 3c \quad \text{(by (6.1A3))}.
\end{aligned}
$$

Continuing we find for some quantity P

$$|\underline{x}_k^T \underline{\alpha}| \le Pc, \quad k_0 + 1 \le k \le k_0 + N.$$

Thus we deduce

$$\begin{aligned}\underline{\alpha}^T \sum_{k=k_0+1}^{k_0+N} \underline{x}_k \underline{x}_k^T \underline{\alpha} &= \sum_{k=k_0+1}^{k_0+N} (\underline{x}_k^T \underline{\alpha})^2 \\ &\le N P^2 c^2 \le \alpha_1/2\end{aligned}$$

which contradicts (6.1A4). The result is thus established.

6.1.1 Persistent Excitation (PE)

In (6.1A4) we introduced strong PE for deterministic signals, while in Section 3.2, PE was discussed for periodic deterministic signals. To make the connection a little clearer, we extend the deterministic theory to allow almost periodic signals. For this purpose we will treat a discrete signal x_k as a sampled continuous time signal $x_k = x(k\Delta)$ where Δ is the sampling interval.

Definition 6.1 (see Appendix C) A deterministic signal $x(t)$ is called uniformly stationary with autocovariance $R_x(u)$ if

$$\lim_{T\to\infty} T^{-1} \int_t^{t+T} x(s)x(s+u)ds = R_x(u)$$

exists uniformly in t. In this case there is also a spectral representation (see Appendix C)

$$R_x(u) = \int_{-\infty}^{\infty} e^{j\omega u} dG_x(\omega)/2\pi$$

where $G_x(\omega)$ is a nonnegative spectral distribution function.

A general class of uniformly stationary signals are the almost periodic signals. If $x(t)$ is almost periodic (see Appendix C) then it is uniformly stationary and $R_x(u)$ is almost periodic. Furthermore, $G_x(\omega)$ consists solely of jumps so that the spectral representation becomes

$$R_x(u) = \sum_{n=-\infty}^{\infty} G_n e^{j\omega_n u}$$

where $\{\omega_n\}$ is a sequence of discrete frequencies and $G_n > 0$ are the spectral line values. Since $R_x(u)$ is real, $G_n = G_{-n}$.

Consider now the case that

$$\underline{x}_k = (x_{k-1} \cdots x_{k-p})^T$$

and x_k is sampled from

$$x_k = x(k\Delta).$$

Then if $x(t)$ is almost periodic and so uniformly stationary, we have, uniformly in k

$$\lim_{N\to\infty} N^{-1}\sum_{t=k}^{k+N} \underline{x}_t\underline{x}_t^T = \underline{R}_x$$

$$\underline{R}_x = \sum_{n=-\infty}^{\infty} G_n \underline{\xi}_n \underline{\xi}_n^H$$

$$\underline{\xi}_n = (1\ e^{j\omega_n}\ \cdots\ e^{j\omega_n(p-1)})^T$$

where superscript H denotes complex conjugate transpose. Thus we introduce:

Definition 6.2.

$\underline{x}_k$ is weakly PE if $\underline{R}_x$ is positive definite.

We relate weak PE back to x_k by the following:

Definition 6.3. We say x_k is PE of order $2r$ (or $2r+1$) if it has at least r distinct spectral lines (and a "dc" component).

Now we claim that

$\underline{x}_k$ is weakly PE if and only if

x_k is PE of order $2[(p+1)/2](=2\times$ integer part of $(p+1)/2)$.

Proof. We take p even for simplicity. Suppose $\underline{x}_k$ is weakly PE, yet x_k has only m spectral lines, $2m < p$. Then

$$\underline{R}_x = \sum_{n=-m}^{m} G_n \underline{\xi}_n \underline{\xi}_n^H, \quad G_0 = 0.$$

But being the sum of only $2m$ dyads, $\underline{R}_x$ has rank $\leq 2m$.

Suppose on the other hand $\underline{x}_k$ is not weakly PE, yet x_k has $p/2$ distinct spectral lines; then there is a vector $\underline{\alpha}$ such that $\underline{\alpha}^T\underline{R}_x\underline{\alpha} = 0$. Thus

$$\sum_{n=-\infty}^{\infty} G_n |\underline{\xi}_n^H \underline{\alpha}|^2 = 0$$

$$\Rightarrow \qquad G_n[\underline{\xi}_n^H \underline{\alpha}] = 0, \quad \text{for all } n.$$

But there are at least p nonzero G_n's so

$$\underline{\xi}_n^H \underline{\alpha} = 0, \quad \text{for } p \text{ values of } n.$$

So the $p/2$ non-null $\underline{\xi}_n$'s cannot be linearly independent, and this contradicts x_k being PE of order $\geq p$.

6.2 MIXED TIME SCALE LMS AND THE POSITIVE REAL CONDITION

In this section we study the stability of the posterior version of the LMS_2 algorithm presented in Section 4.4, equations (4.13)–(4.16). The algorithm has the form

$$\delta\underline{\hat{\theta}}_k = \mu\underline{\hat{\varphi}}_k\nu_k \tag{6.9a}$$

$$\nu_k = y_k - \hat{s}_k \tag{6.9b}$$

$$\hat{s}_k = \underline{\hat{\varphi}}_k^T\underline{\hat{\theta}}_k \tag{6.9c}$$

$$\underline{\hat{\varphi}}_k = (-\hat{s}_{k-1} \cdots - \hat{s}_{k-p},\ x_{k-1} \cdots x_{k-m})^T. \tag{6.9d}$$

To form an error system suited to stability analysis we start as usual with (6.9c). We suppose

(6.2A1) There is a true parameter $\underline{\theta}_o$ such that

$$y_k = s_k = -\sum_{t=1}^{p} a_{to}s_{k-t} + \sum_{t=1}^{m} b_{to}x_{k-t}.$$

Again there is no output noise.

Using (6.2A1) along with (6.9) gives

$$\begin{aligned}\nu_k &= y_k - \hat{s}_k = s_k - \hat{s}_k\\ &= -\sum_{r=1}^{p} a_{ro}s_{k-r} + \sum_{r=1}^{m} b_{ro}x_{k-r} + \sum_{r=1}^{p} \hat{a}_{rk}\hat{s}_{k-r} - \sum_{r=1}^{m} \hat{b}_{rk}x_{k-r}\\ &= \sum_{r=1}^{p}(\hat{a}_{rk} - a_{ro})\hat{s}_{k-r} - \sum_{r=1}^{m}(\hat{b}_{rk} - b_{ro})x_{k-r} - \sum_{r=1}^{p} a_{ro}(s_{k-r} - \hat{s}_{k-r})\end{aligned}$$

which we can write compactly as

$$[1 + a_o(q^{-1})]\nu_k = -\underline{\hat{\varphi}}_k^T\underline{\tilde{\theta}}_k \tag{6.10}$$

where

$$a_o(q^{-1}) = \sum_{r=1}^{p} a_{ro}q^{-r}$$

and we have introduced the parameter estimation error

$$\underline{\tilde{\theta}}_k = \underline{\hat{\theta}}_k - \underline{\theta}_o.$$

Then we also rewrite (6.9a) as

$$\delta\underline{\tilde{\theta}}_k = \mu\underline{\hat{\varphi}}_k\nu_k. \tag{6.11}$$

So the error system consists of the pair (6.10) and (6.11) augmented by (6.9c) and (6.9d). The mixed time scale structure of the algorithm is clearly seen in this pair. If we represent the algorithm in a block diagram as in Fig. 6.1, we can view it as a feedback

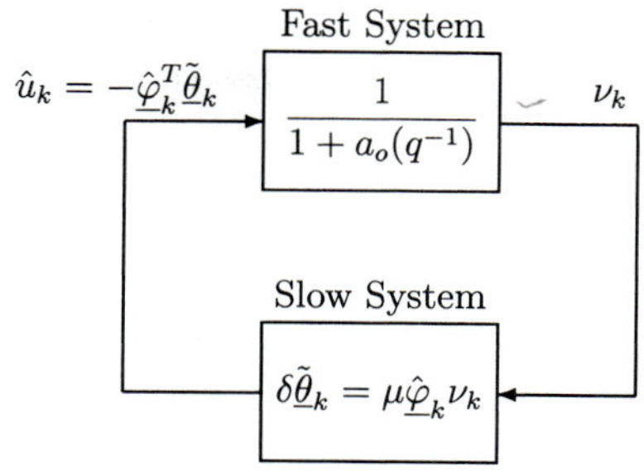

Figure 6.1 A block diagram of the mixed time scale LMS algorithm.

interconnection of two blocks: a time invariant fast system and a time varying slow system. This view of the mixed time scale algorithm is fairly general although the fast system is not usually time invariant. The idea now is to develop a Lyapunov-based stability analysis of this error system. Three such analyses are possible: using a Lyapunov function suitable to the slow system; using a Lyapunov function suitable to the fast system; and using a joint Lyapunov function for both systems. We pursue the first approach; the other two are discussed further in the Notes.

We rewrite (6.11) as

$$\tilde{\underline{\theta}}_k = \tilde{\underline{\theta}}_{k-1} + \mu\hat{\underline{\varphi}}_k\nu_k$$
$$\Rightarrow \qquad \tilde{\underline{\theta}}_{k-1} = \tilde{\underline{\theta}}_k - \mu\hat{\underline{\varphi}}_k\nu_k.$$

Now introducing the Lyapunov function $V_k = \|\tilde{\underline{\theta}}_k\|^2$ we find

$$V_{k-1} = V_k - 2\mu\tilde{\underline{\theta}}_k^T\hat{\underline{\varphi}}_k\nu_k + \mu^2\|\hat{\underline{\varphi}}_k\|^2\nu_k^2$$
$$\Rightarrow \qquad V_k = V_{k-1} + 2\mu\tilde{\underline{\theta}}_k^T\hat{\underline{\varphi}}_k\nu_k - \mu^2\|\hat{\underline{\varphi}}_k\|^2\nu_k^2$$
$$= V_{k-1} - 2\mu\tilde{u}_k\nu_k - \mu^2\|\hat{\underline{\varphi}}_k\|^2\nu_k^2 \qquad (6.12)$$

where with (6.10) in mind we have introduced the pseudo input

$$\tilde{u}_k = -\tilde{\underline{\theta}}_k^T\hat{\underline{\varphi}}_k$$

so that (6.10) can be written as

$$\nu_k = H_0(q^{-1})\tilde{u}_k \qquad (6.13)$$
$$H_0(q^{-1}) = \left[1 + a_0(q^{-1})\right]^{-1}.$$

Now introduce the input/output cross-correlation

$$J_k = \sum_{t=1}^{k}\tilde{u}_t\nu_t + J_0 \qquad (6.14)$$

(where J_0 is an initial condition to be determined) so that (6.12) can be rewritten

$$V_k + 2\mu J_k = V_{k-1} + 2\mu J_{k-1} - \mu^2 \|\hat{\underline{\varphi}}_k\|^2 \nu_k^2 \tag{6.15a}$$

$$\Rightarrow \quad V_k + 2\mu J_k \le V_{k-1} + 2\mu J_{k-1}. \tag{6.15b}$$

We would like J_k to be positive. If $\tilde{u}_k$ was a current and ν_k a voltage, then from (6.14) J_k would be a power and so could be positive if we put some appropriate constraint on $H_0(q^{-1})$ in (6.13). To see what this should be, we use Fourier analysis to express J_k in frequency domain terms. To do this we use the following trick. Fix N and introduce the truncated or "shut-down" pseudo input

$$\bar{u}_{kN} = \begin{cases} \tilde{u}_k, & 1 \le k \le N \\ 0, & k \le 0 \text{ or } k > N \end{cases}$$

and its corresponding output

$$\bar{\nu}_{kN} = H_0(q^{-1})\bar{u}_{kN}.$$

After time N, $\bar{\nu}_{kN}$ becomes a decaying transient (if $H_0(q^{-1})$ is stable). We then have, aside from some initial condition transients

$$\bar{\nu}_{kN} = \nu_k, \quad 1 \le k \le N.$$

Thus we find

$$J_N = \sum_{t=1}^{N} \tilde{u}_t \nu_t = \sum_{t=1}^{N} \bar{u}_{tN} \bar{\nu}_{tN}$$

$$= \sum_{t=1}^{\infty} \bar{u}_{tN} \bar{\nu}_{tN}.$$

However, by Parseval's Theorem, this is

$$J_N = \int_{-\pi}^{\pi} \bar{u}_N(e^{-j\omega}) \bar{\nu}_N(e^{j\omega}) d\omega / 2\pi$$

$$= \int_{-\pi}^{\pi} |\bar{u}_N(e^{-j\omega})|^2 H_0(e^{j\omega}) d\omega / 2\pi, \quad \text{by (6.13)}$$

$$= \int_{-\pi}^{\pi} |\bar{u}_N(e^{-j\omega})|^2 Re(H_0(e^{j\omega})) d\omega / 2\pi$$

where for instance

$$\bar{u}_N(e^{-j\omega}) = \sum_{t=1}^{\infty} \bar{u}_{tN} e^{-j\omega t}.$$

We now introduce the following assumption:

(6.2A2) $H_0(q^{-1})$ is strictly positive real (SPR), i.e., $H_0(q^{-1})$ is stable and there is a small $\eta > 0$ so that

$$Re(H_0(e^{-j\omega})) \ge \eta.$$

Note that $H_0^{-1}(q^{-1})$ is also SPR; the positivity is easily seen; the stability follows by the Nyquist criterion. The SPR condition places a constraint on the class of stable transfer functions. In Fig. 6.2 the SPR region is shown as a subset of the triangular stability region for a second order all pole model.

Continuing, use (6.2A2) to find

$$\begin{aligned} J_N &\geq \eta \int_{-\pi}^{\pi} |\bar{u}_N(e^{-j\omega})|^2 d\omega/2\pi \\ &= \eta \sum_{k=1}^{\infty} \bar{u}_{kN}^2 \\ &= \eta \sum_{k=1}^{N} \bar{u}_{kN}^2 \\ &= \eta \sum_{k=1}^{N} \tilde{u}_k^2. \end{aligned} \tag{6.16}$$

Now return to (6.15) to find that $V_k + 2\mu J_k$ is a positive non-increasing sequence and so must have a finite limit. Thus, we deduce $\|\underline{\tilde{\theta}}_k\|$ is bounded and via (6.16) that

$$\sum_{k=1}^{\infty} \tilde{u}_k^2 < \infty.$$

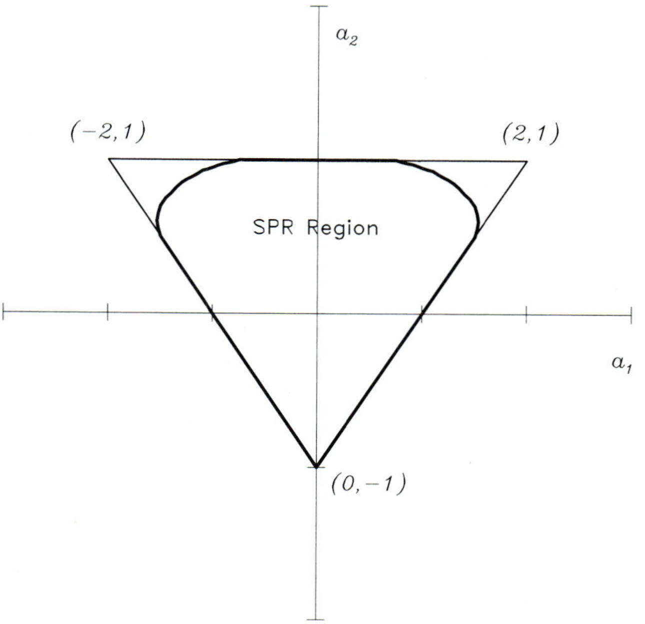

Figure 6.2 SPR region (bounded by thicker lines) as a subset of the triangular stability region for a second order all pole filter.

From this we deduce via (6.13) (see Appendix C) that

$$\sum_{k=1}^{\infty} \nu_k^2 < \infty$$

which yields (see Appendix D)

$$\nu_k \to 0 \quad \text{as } k \to \infty. \tag{6.17}$$

We would like however to have $e_k \to 0$. From (6.10) we have

$$\nu_k = e_k/(1 + \mu\|\underline{\hat{\varphi}}_k\|^2).$$

So we need to show $\|\underline{\hat{\varphi}}_k\|^2$ is bounded. We have

$$\underline{\hat{\varphi}}_k = (-y_{k-1} + \nu_{k-1}, -y_{k-2} + \nu_{k-2}, \ldots, -y_{k-p} + \nu_{k-p}, x_{k-1}, \ldots, x_{k-m})^T$$

and the boundedness of $\{x_k, y_k\}$ together with (6.17) ensures that $\|\underline{\hat{\varphi}}_k\|$ is bounded. So finally we conclude

$$e_k \to 0. \tag{6.18}$$

We thus have that under (6.2A1), (6.2A2) the tracking error $\to 0$ as $k \to \infty$.

As with the single time scale LMS algorithm, we expect some type of persistent excitation condition is needed to show convergence of $\underline{\tilde{\theta}}_k$. This issue is pursued in the notes.

6.3 NOTES

The first analysis of the LMS algorithm was given (in continuous time) by [A5] and [M6]. The (continuous time version of the) mixed time scale LMS algorithm of Section 6.2 is due to [L1] and was analyzed by him using the second Lyapunov approach mentioned earlier in Section 6.2: this is the so-called hyper-stability method. The first Lyapunov approach given here was used by [S12] for a long memory algorithm. To deal with parameter convergence, the third Lyapunov approach mentioned earlier is needed; it is based on a state space construction and naturally requires a PE condition: details are in [A4]. It is important to understand that if a posterior algorithm is not used in Section 6.2 then stability cannot be demonstrated (see Exercise 6.4) using the present argument. In the setting of long memory algorithms the SPR condition was derived by [L6]. An important point not so far made is that PE depends on the algorithm. This point has been emphasized by [J3, Chapter 5].

EXERCISES

6.1 Consider the LMS_P algorithm of Section 4.6.

$$\delta\underline{\hat{w}}_k = \mu\underline{x}_k e_k/(1 + \mu\|\underline{x}_k\|^2).$$

Give a Lyapunov stability analysis to show that under (6.1A1), (6.1A2), $e_k \to 0$ as $k \to \infty$. So that no bound on μ such as (6.1A3) is needed. This illustrates the strength of the implicit method of algorithm construction.

6.2 Adaptive Interference Canceller as a notch filter. Consider a two tap LMS algorithm with a "mains" reference signal $\underline{x}_k = (x_{1k}\ x_{2k})^T$

$$x_{1k} = a\cos(\omega_o k + \phi)$$
$$x_{2k} = a\sin(\omega_o k + \phi)$$

and an arbitrary noise free output $y_k = s_k$

$$\delta\underline{\hat{w}}_k = \mu \underline{x}_k e_k$$
$$e_k = s_k - \underline{x}_k^T \underline{\hat{w}}_{k-1}.$$

Show that although the LMS algorithm is time varying, it acts as a time invariant notch filter in that

$$e_k = H(q^{-1}) s_k$$
$$H(q^{-1}) = \frac{1 - 2q^{-1}\cos\omega_o + q^{-2}}{1 - q^{-1}(2 - \mu a^2)\cos\omega_o + q^{-2}(1 - \mu a^2)}.$$

(*Hint:* Begin by applying the operator $1 - 2q^{-1}\cos\omega_o + q^{-2}$ to e_k.)

6.3 Test the following signals for PE:

(a) $x_k = c$

(b) $x_k = c/(k+1)$

(c) $x_k = \sin(\omega k)$: Is this signal PE of order 2? Is it PE of order 3?

(d) $x_k = \sin(\omega_0 k) + \cos(\omega_1 k), \quad \omega_0 \neq \omega_1$. What order of PE is this signal?

6.4 Set up an error system for the "prior" LMS_2 algorithm of Section 4.4 ((4.13)–(4.16)) and attempt to repeat the stability argument given in Section 6.2. Explain what problem arises.

6.5 Prove expressions (6.7) and (6.8).

6.6 Consider the smoothed error version of algorithm (6.9) where ν_k is replaced by

$$\nu_{sk} = \nu_k + \sum_{s=1}^{m} c_s \nu_{k-s}.$$

Show that the error system is

$$\delta\underline{\tilde{\theta}}_k = \mu \underline{\hat{\varphi}}_k \nu_{sk}$$
$$\nu_{sk} = -H(q^{-1}) \underline{\hat{\varphi}}_k^T \underline{\tilde{\theta}}_k$$
$$H(q^{-1}) = c(q^{-1})/(1 + a_o(q^{-1}))$$
$$c(q^{-1}) = 1 + \sum_{s=1}^{m} c_s q^{-s}.$$

Hence show that tracking error $\to 0$ as $k \to \infty$ when $H_0(q^{-1})$ in (6.2A2) is replaced by $H(q^{-1})$.

PART II

DETERMINISTIC AVERAGING

Our stability and performance analyses started with the Gaussian white noise case. In this setting we could treat both stability and performance in the presence of time varying parameters, but the techniques used are essentially limited to that setting.

In Chapter 6, in a noise-free setting, we were able to develop a convergence (stability) analysis (with time invariant parameters) that allows fairly general deterministic signal characteristics. We analyzed a single time scale LMS algorithm and a mixed time scale LMS algorithm. These two cases essentially exhaust the algorithms to which the techniques of Chapter 6 apply.

Yet of course the list of adaptive algorithms requiring analysis is much longer. What is needed then is a general tool for algorithm analysis, and averaging methods provide just such a tool. The averaging technique relies on the presence of a small parameter (the gain) in the difference equation to be studied. It is then possible to associate an averaged system with the algorithm of interest. The averaged system is, however, easier to analyze, and its behavior can often be connected sufficiently strongly with that of the original algorithm to allow stability to be proved and to allow performance computations to be made.

7

Deterministic Averaging: Single Time Scale

7.1 INTRODUCTION

Averaging methods for the approximation of the trajectory of a differential or difference equation dependent upon a small parameter have a long history in astronomy and dynamical systems (see notes at the end of this chapter). But their application to the analysis of adaptive algorithms is relatively recent.

There are three types of averaging theorem:

Finite Time "trajectory locking" results

Infinite Time "trajectory locking" results

Infinite Time Stability results

The first type of result shows how to approximate, uniformly in time, on a finite stretched time interval, the trajectory of the algorithm of interest by the trajectory of a simpler averaged system. The second type of result extends this approximation to an infinite time interval. It gives only indirect information about stability but can nevertheless be quite useful.

The third type of result is concerned with the question of how stability (or convergence or boundedness) of the averaged system is related to stability of the original system. One has to be careful here because the averaged system may have equilibrium points not possessed by the primary system and vice versa. In any case, stability is often analyzed by means of a Lyapunov function, and in many cases if a Lyapunov function can be found for the averaged system, then a perturbation of it can yield a Lyapunov function for the original system.

Averaging analysis is a very general technique quite capable of handling error systems derived from algorithms with time varying parameters as well as noise. In this chapter, however, only a noise-free setting is considered.

7.2 FINITE TIME AVERAGING—TIME INVARIANT PARAMETERS

Let us motivate the setup to be used by claiming that the single time scale algorithms of Chapter 4 all have error systems of the general form

$$\delta \underline{z}_k = \mu \underline{f}(k, \underline{z}_{k-1}, \mu). \tag{7.1}$$

We call (7.1) a primary system. For example we have the following cases:
LMS:

$$\underline{f}(k, \underline{\tilde{\theta}}, \mu) = -\underline{x}_k \underline{x}_k^T \underline{\tilde{\theta}} \tag{7.2}$$

Idealized time delay estimation via steepest descent:

$$f(k, \tilde{\theta}, \mu) = (s_{k-\tilde{\theta}+1} - s_{k-\tilde{\theta}-1})(s_k - s_{k-\tilde{\theta}}) \tag{7.3}$$

LMS$_P$:

$$\underline{f}(k, \underline{\tilde{\theta}}, \mu) = -\underline{x}_k \underline{x}_k^T \underline{\tilde{\theta}} / (1 + \mu \|\underline{x}_k\|^2) \tag{7.4}$$

In the first two cases, $\underline{f}(k, \underline{\tilde{\theta}}, \mu)$ does not depend on μ. Indeed the first two have the simpler form

$$\delta \underline{z}_k = \mu \underline{f}(k, \underline{z}_{k-1}). \tag{7.5}$$

To keep the development as clear and simple as possible we will initially work with (7.5). The way in which an averaged system can be associated with the primary system (7.5) can be seen from the following heuristic discussion.

Sum (7.5) over a fixed time span N to find

$$\underline{z}_{k+N} - \underline{z}_k = \mu \sum_{s=k+1}^{k+N} \underline{f}(s, \underline{z}_{s-1}). \tag{7.6}$$

Now if μ is small, then $\underline{z}_s$ changes slowly and so it should be close to $\underline{z}_k$ over the interval $k+1 \le s \le k+N$, if N is not too large. Thus we could approximate the right hand side of (7.6) by replacing $\underline{z}_s$ by $\underline{z}_k$ to find

$$\underline{z}_{k+N} - \underline{z}_k \approx \mu \sum_{s=k+1}^{k+N} \underline{f}(s, \underline{z}_k). \tag{7.7}$$

Now suppose the following average exists:

$$\underline{f}_{av}(\underline{z}) = \lim_{N\to\infty} N^{-1} \sum_{s=k+1}^{k+N} \underline{f}(s, \underline{z}), \quad \text{uniformly in } k. \tag{7.8}$$

Then if N is large enough (but not too large), (7.7) becomes approximately

$$\underline{z}_{k+N} - \underline{z}_k \approx \mu N f_{av}(\underline{z}_k). \tag{7.9}$$

Now arguing in reverse that $\underline{z}_s$ is close to $\underline{z}_k$ we can approximate (7.9) by

$$\underline{z}_{k+N} - \underline{z}_k \approx \mu \sum_{s=k+1}^{k+N} \underline{f}_{av}(\underline{z}_{s-1})$$

and this finally suggests that the trajectory of $\underline{z}_k$ is close to that of the following averaged system:

$$\delta\bar{\underline{z}}_k = \mu \underline{f}_{av}(\bar{\underline{z}}_{k-1}). \tag{7.10}$$

Note that (7.10) is an autonomous system (i.e., $\underline{f}_{av}(\underline{z})$ is time invariant) and so is simpler than (7.5).

Example 7.1. Averaging the LMS Algorithm

To see what the averaged system looks like consider the LMS algorithm. From (7.2), (7.8) we find

$$\underline{f}_{av}(\tilde{\underline{\theta}}) = \underline{R}_x \tilde{\underline{\theta}}$$

where we have supposed the following limit exists

$$\underline{R}_x = \lim_{N\to\infty} N^{-1} \sum_{s=k+1}^{k+N} \underline{x}_s \underline{x}_s^T, \quad \text{uniformly in } k.$$

For example, if $\underline{x}_k$ is composed of periodic signals then $\underline{R}_x$ exists. More generally, a signal $\underline{x}_k$ for which $\underline{R}_x$ exists is called uniformly stationary. From (7.10) we see then that the averaged system is

$$\delta\bar{\underline{\theta}}_k = -\mu \underline{R}_x \bar{\underline{\theta}}_{k-1}. \tag{7.11}$$

This difference equation is identical to the one we obtained in our first order analysis under Gaussian white noise assumptions in Section 5.2.

Now we pursue Example 3.2. The averaged system is identical to the iteration (3.12) plotted in Fig. 3.2! In Fig. 7.1 is a plot of the primary LMS algorithm and the averaged system for several values of μ. We see that as μ gets smaller the primary trajectory "locks" on to the averaged trajectory. The signal used is given as follows:

$$\begin{pmatrix} x_{1k} \\ x_{2k} \end{pmatrix} = \begin{pmatrix} \sqrt{2}\sin\omega_1 k + \sqrt{2}\cos(\omega_2 k + \phi) \\ 2\cos\omega_2 k \end{pmatrix}$$

where $\omega_1 = 2\pi/25.0$, $\omega_2 = 2\pi/30.0$, and $\phi = \pi/4$.

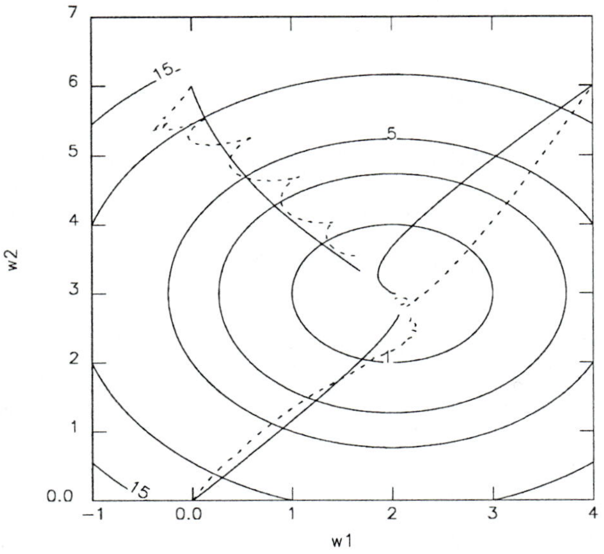

Figure 7.1 LMS and averaged LMS: Solid lines are the trajectories of LMS and dashed lines are the trajectories of the corresponding averaged LMS. For the trajectories starting from (0, 0), $\mu = 0.01$; for the trajectories starting from (0, 6), $\mu = 0.02$; and for the trajectories starting from (4, 6), $\mu = 0.1$.

7.2.1 Analysis in a Simplified Setting

We will give further examples below, but let us now turn to a rigorous discussion of the connection between the primary system (7.5) and the averaged system (7.10). We sum (7.5), (7.10) to obtain

$$\underline{z}_k = \underline{z}_0 + \mu \sum_{s=1}^{k} \underline{f}(s, \underline{z}_{s-1})$$

$$\underline{\bar{z}}_k = \underline{\bar{z}}_0 + \mu \sum_{s=1}^{k} \underline{f}_{av}(\underline{\bar{z}}_{s-1}).$$

Now subtract these to find

$$\begin{aligned} \underline{\Delta}_k &= \underline{z}_k - \underline{\bar{z}}_k \\ &= \underline{\Delta}_0 + \mu \sum_{s=1}^{k} \underline{\tilde{f}}_s \end{aligned} \tag{7.12}$$

where

$$\underline{\tilde{f}}_s = \underline{f}(s, \underline{z}_{s-1}) - \underline{f}_{av}(\underline{\bar{z}}_{s-1}).$$

Now we split (7.12) into two pieces and obtain

$$\underline{\Delta}_k = \underline{\Delta}_0 + \mu \sum_{s=1}^{k} \left(\underline{f}_{av}(\underline{z}_{s-1}) - \underline{f}_{av}(\bar{\underline{z}}_{s-1}) \right) + \mu \underline{J}_k \tag{7.13}$$

$$\underline{J}_k = \sum_{s=1}^{k} \left(\underline{f}(s, \underline{z}_{s-1}) - \underline{f}_{av}(\underline{z}_{s-1}) \right). \tag{7.14}$$

To proceed further, some regularity conditions must be introduced. Fix a time interval $[0, T]$ and consider the following conditions to hold in a ball of radius h:

(7.2A1) $\|\underline{f}(k, \underline{z})\| \leq B_f$ for $\|\underline{z}\| \leq h$

(7.2A2) Introduce the total deviation or perturbation

$$\underline{p}(k, \underline{z}) = \sum_{s=1}^{k} \left(\underline{f}(s, \underline{z}) - \underline{f}_{av}(\underline{z}) \right).$$

We suppose that for $\|\underline{z}\| \leq h$, $\|\underline{z}'\| \leq h$

$$\|\underline{p}(k, \underline{z}) - \underline{p}(k, \underline{z}')\| \leq L_p \|\underline{z} - \underline{z}'\|$$
$$\|\underline{p}(k, 0)\| \leq B_p$$

(7.2A3) $\underline{f}_{av}(\underline{z})$ obeys a Lipschitz condition

$$\|\underline{f}_{av}(\underline{z}) - \underline{f}_{av}(\underline{z}')\| \leq L_f \|\underline{z} - \underline{z}'\|, \quad \text{for } \|\underline{z}\|, \|\underline{z}'\| \leq h.$$

(7.2A4) $\|\bar{\underline{z}}_0\|$ is so small that on $[0, T/\mu]$

$$\|\bar{\underline{z}}_k\| \leq h/2.$$

Assumptions (7.2A1), (7.2A3), (7.2A4) are self explanatory. To see what (7.2A2) entails consider the scalar LMS example so that from (7.2), (7.11)

$$p(k, \theta) = \sum_{s=1}^{k} (x_s^2 - R_x)\theta.$$

Now suppose

$$x_s = \cos(\omega_o s), \quad \omega_o \neq 0.$$

Then we have $R_x = \frac{1}{2}$ and

$$\begin{aligned} p(k,\theta) - p(k,\theta') &= \sum_{s=1}^{k}(\cos^2(\omega_o s) - \frac{1}{2})(\theta - \theta') \\ &= \sum_{s=1}^{k}(\frac{1}{2} + \frac{1}{2}\cos(2\omega_o s) - \frac{1}{2})(\theta - \theta') \\ &= \frac{1}{2}\sum_{s=1}^{k}\cos(2\omega_o s)(\theta - \theta') \\ &= \cos(\omega_o k)\frac{\sin(k+1)\omega_o}{\sin\omega_o}(\theta - \theta') \end{aligned}$$

$$\Rightarrow |p(k,\theta) - p(k,\theta')| \le (2|\sin\omega_o|)^{-1}|\theta - \theta'|$$

as required. Clearly (7.2A2) holds in general if $\underline{x}_k$ is composed of periodic signals. Actually below we will weaken (7.2A2), but we have chosen the present strengthened form to make the proof as direct and transparent as possible.

To apply these regularity conditions we will have to know that $\|\underline{z}_k\| \le h$ for $1 \le k \le T/\mu$. But it may be that $\|\underline{z}_k\|$ gets as big as h long before $k = T/\mu$. So introduce

$$T'_\mu = \text{first time } k \text{ that } \|\underline{z}_k\| \le h,\ \|\underline{z}_{k+1}\| > h$$

and let

$$T_\mu = \min\{T'_\mu, T/\mu\}. \tag{7.15}$$

We will ultimately be able to use an argument by contradiction to show $T_\mu = T/\mu$.

With conditions (7.2A1)–(7.2A4), we will show below that

$$\|\underline{J}_k\| \le a + bT, \quad 1 \le k \le T_\mu. \tag{7.16}$$

Using (7.16) in (7.13) we find via (7.2A3) that

$$\|\underline{\Delta}_k\| \le \|\underline{\Delta}_0\| + \mu L_f \sum_{s=1}^{k}\|\underline{\Delta}_{s-1}\| + \mu(a + bT), \quad 1 \le k \le T_\mu. \tag{7.17}$$

Now apply the discrete Bellman-Gronwall lemma (Appendix C2) to find that

$$\begin{aligned} \|\underline{\Delta}_k\| &\le (\|\underline{\Delta}_0\| + \mu(a+bT))(1+\mu L_f)^{k-1}, \quad 1 \le k \le T_\mu \\ &\le (\|\underline{\Delta}_0\| + c_T(\mu))e^{\mu L_f k}, \quad 1 \le k \le T_\mu \end{aligned}$$

where

$$c_T(\mu) = \mu(a + bT) \downarrow 0 \text{ as } \mu \downarrow 0. \tag{7.18}$$

Thus, we find

$$\begin{aligned} \max_{1\le k\le T_\mu} \|\underline{\Delta}_k\| &\le (\|\underline{\Delta}_0\| + c_T(\mu))e^{\mu L_f T_\mu} \\ &\le (\|\underline{\Delta}_0\| + c_T(\mu))e^{T L_f}. \end{aligned} \tag{7.19}$$

For simplicity we suppose $\underline{\Delta}_0 = 0$, i.e., the primary and averaged systems start at the same point.

With this inequality we can prove the following theorem.

Theorem 7.1. For the primary system (7.5) and averaged system (7.10) with identical initial conditions, suppose (7.2A1)–(7.2A4) hold. Then given $T > 0$ there exist $b_T > 0$, $\mu_T > 0$ such that

$$\max_{1 \le k \le T/\mu} \|\underline{z}_k - \bar{\underline{z}}_k\| \le c_T(\mu) b_T, \quad 0 \le \mu \le \mu_T$$

where

$$c_T(\mu) \to 0 \quad \text{as } \mu \to 0.$$

Proof. Take μ_T so small that

$$c_T(\mu_T) e^{TL} = h/4, \quad \mu B_f < h/4.$$

Then from (7.2A4), (7.18), (7.19)

$$\begin{aligned} \|\underline{z}_k\| &\le \|\underline{\Delta}_k\| + \|\bar{\underline{z}}_k\|, \quad 1 \le k \le T_\mu, 0 \le \mu \le \mu_T \\ &\le h/4 + h/2 = 3h/4 \end{aligned}$$

while from (7.5)

$$\|\underline{z}_{T_\mu + 1}\| \le \|\underline{z}_{T_\mu}\| + \mu B_f \le h.$$

But from the definition of T_μ this means $T_\mu = T/\mu$. Thus the result follows from (7.19) with $b_T = e^{TL}$.

This is our trajectory locking result because it ensures

$$\max_{1 \le k \le T/\mu} \|\underline{z}_k - \bar{\underline{z}}_k\| \to 0, \quad \text{as} \quad \mu \to 0.$$

To complete the discussion we must supply a proof of (7.16). This in fact is the subtle part of the averaging argument. The idea is to approximate $\underline{J}_k$ by

$$\underline{P}_k = \underline{p}(k, \underline{z}_{k-1}).$$

Since $\bar{\underline{z}}_k$ is slowly changing when μ is small this approximation should do well. We note that

$$\begin{aligned} \underline{P}_k - \underline{P}_{k-1} &= \underline{p}(k, \underline{z}_{k-1}) - \underline{p}(k-1, \underline{z}_{k-2}) \\ &= \underline{p}(k, \underline{z}_{k-1}) - \underline{p}(k-1, \underline{z}_{k-1}) + \underline{\eta}_k \\ &= \underline{f}(k, \underline{z}_{k-1}) - \underline{f}_{av}(\underline{z}_{k-1}) + \underline{\eta}_k \\ \underline{\eta}_k &= \underline{p}(k-1, \underline{z}_{k-1}) - \underline{p}(k-1, \underline{z}_{k-2}). \end{aligned}$$

Referring to (7.14) and (7.2A2) we see on the other hand that

$$\underline{P}_k - \underline{P}_{k-1} = \underline{J}_k - \underline{J}_{k-1} + \underline{\eta}_k.$$

Setting $\underline{P}_0 = \underline{0} = \underline{J}_0$ and summing this gives

$$\underline{P}_k = \underline{J}_k + \sum_{s=1}^{k} \underline{\eta}_s. \tag{7.20}$$

From (7.2A1), (7.2A2), for $1 \le k \le T_\mu$

$$\begin{aligned}\|\underline{\eta}_k\| &\le L_p \|\underline{z}_{k-1} - \underline{z}_{k-2}\| \\ &\le L_p \mu \|\underline{f}(k-1, \underline{z}_{k-2})\| \\ &\le L_p \mu B_f, \quad \text{by (7.2A1)} \\ &= \mu b\end{aligned}$$

where $b = L_p B_f$. Also from (7.2A2), for $1 \le k \le T_\mu$

$$\begin{aligned}\|\underline{P}_k\| &\le \|\underline{p}(k, \underline{z}_{k-1}) - \underline{p}(k, 0)\| + \|\underline{p}(k, 0)\| \\ &\le L_p \|\underline{z}_{k-1}\| + B_p \\ &\le L_p h/2 + B_p.\end{aligned}$$

So from (7.20)

$$\begin{aligned}\|\underline{J}_k\| &\le (L_p h/2 + B_p) + \mu k b \\ &\le a + bT, \quad 1 \le k \le T_\mu\end{aligned}$$

which is (7.16).

Example 7.2. Time Delay Estimation

Consider the idealized time delay estimation algorithm of Section 4.5 (with Δ the sampling interval)

$$\delta\hat{\theta}_k = -\mu s'_{k\Delta - \hat{\theta}_{k-1}} (s_{k\Delta - \theta_o} - s_{k\Delta - \hat{\theta}_{k-1}}) \tag{7.21}$$

and introduce two regularity conditions:

(7.2B1) s_k is sampled from a continuous time signal $s(t)$, so $s_k = s(k\Delta)$

(7.2B2) $s(t)$ has a uniform autocorrelation function

$$\lim_{N\to\infty} N^{-1} \sum_{t=k}^{k+N} s(t\Delta + u) s(t\Delta + v) = R(u - v), \quad \text{uniformly in } k.$$

Then the averaged function is (with $\theta_o = 0$)

$$\begin{aligned}f_{av}(\theta) &= \lim_{N\to\infty} -\frac{1}{N} \sum_{t=k}^{k+N} s'_{t\Delta-\theta}(s_{t\Delta} - s_{t\Delta-\theta}) \\ &= -R'(-\theta) + R'(0) \\ &= R'(\theta).\end{aligned}$$

The averaged system is then

$$\delta\bar{\theta}_k = \mu R'(\bar{\theta}_{k-1}). \tag{7.22}$$

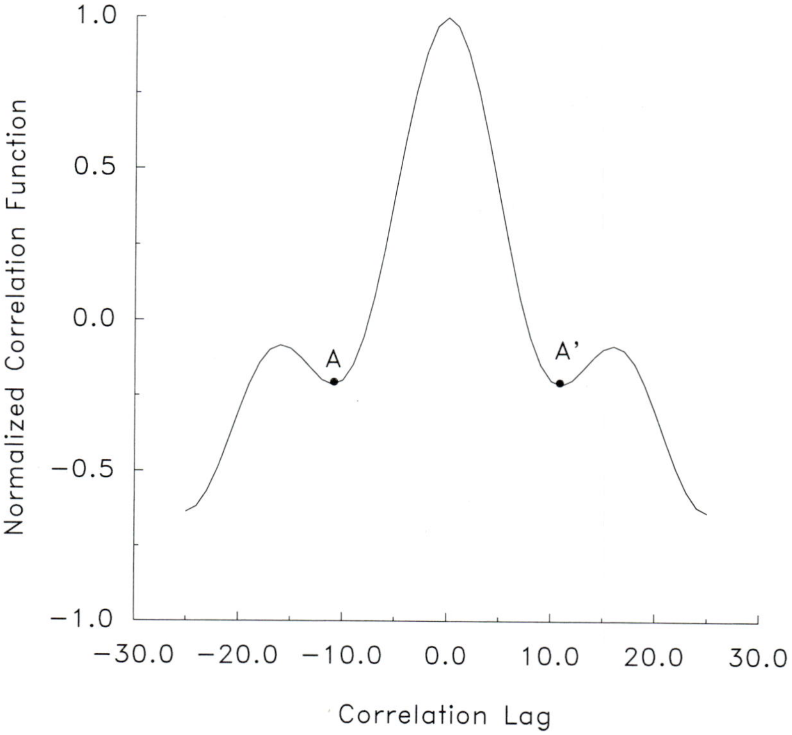

Figure 7.2 Typical correlation function with first local minima at AA'.

It is left as an exercise for the reader to show that provided the initial point lies in the interval AA' in Fig. 7.2 then the averaged system converges to the maximum of the correlation function. The reader is also invited to check conditions (7.2A1)–(7.2A4).

We pursue the noise-free unit attenuation case of Example 3.8. The correlation function Fig. 3.9 is repeated here for convenience in Fig. 7.2. In Fig. 7.3 is plotted the delay error trajectory (for two values of μ) for both the primary system and the averaged system.

7.2.2 Analysis in a More General Setting

Now we return to the more general primary system (7.1) and we introduce an associated averaged function

$$f_{av}(\underline{z}) = \lim_{N\to\infty} N^{-1} \sum_{s=k+1}^{k+N} \underline{f}(s, \underline{z}, 0) \tag{7.23}$$

and a corresponding averaged system (7.10). We also introduce the corresponding assumptions:

(7.2C1) $\|\underline{f}(k, \underline{z}, \mu)\| \le B_f, \ \|\underline{z}\| \le h$

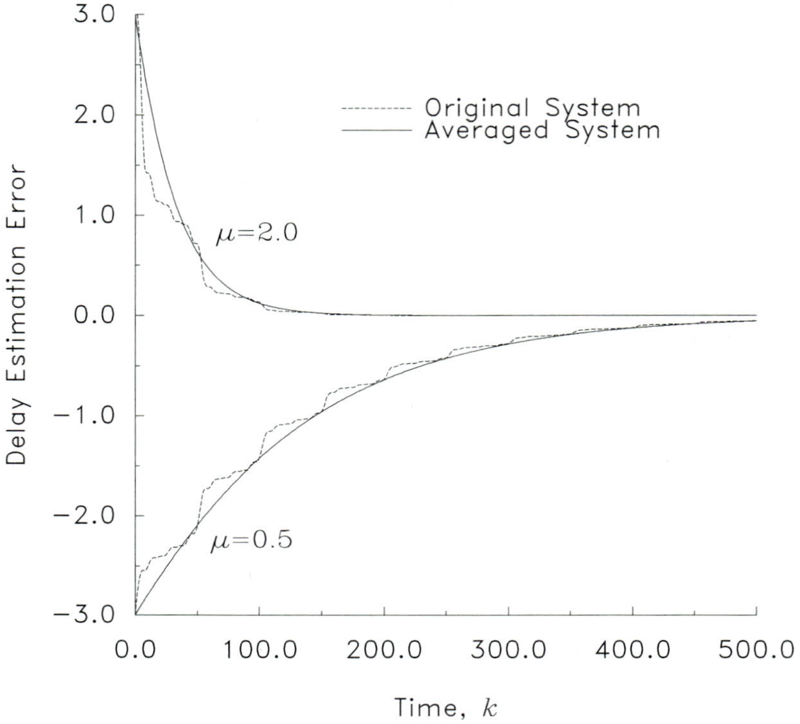

Figure 7.3 Estimation error trajectories of (a) SD algorithm and (b) averaged algorithm for time delay estimation.

(7.2C2) Introduce the total perturbation

$$\underline{p}(k, \underline{z}) = \sum_{s=1}^{k} \left(\underline{f}(s, \underline{z}, \underline{0}) - f_{av}(\underline{z}) \right)$$

and suppose that

$$\|\underline{p}(k, \underline{0})\| \leq B_p$$
$$\|\underline{p}(k, \underline{z}) - \underline{p}(k, \underline{z}')\| \leq L_p \|\underline{z} - \underline{z}'\|, \quad \|\underline{z}\|, \|\underline{z}'\| \leq h.$$

(7.2C3) $\underline{f}_{av}(\underline{z})$ obeys a Lipschitz condition

$$\|\underline{f}_{av}(\underline{z}) - \underline{f}_{av}(\underline{z}')\| \leq L_f \|\underline{z} - \underline{z}'\|, \quad \|\underline{z}\|, \|\underline{z}'\| \leq h.$$

(7.2C4) $\|\bar{\underline{z}}_0\|$ is so small that $\|\bar{\underline{z}}_k\| \leq h/2$, $1 \leq k \leq T/\mu$.

Now proceed exactly as before to find (7.13) where now

$$\underline{J}_k = \sum_{s=1}^{k} \Big(\underline{f}(s, \underline{z}_{s-1}, \mu) - \underline{f}_{av}(z_{s-1}) \Big). \tag{7.24}$$

Now define T_μ as in (7.15). If we can show (7.16) holds in the present setting then the extension of Theorem 7.1 to cover (7.1) will be complete. Now split $\underline{J}_k$ into two pieces:

$$\underline{J}_k = \underline{J}_{k1} + \underline{J}_{k2}$$

$$\underline{J}_{k1} = \sum_{s=1}^{k} \Big(\underline{f}(s, \underline{z}_{s-1}, \mu) - \underline{f}(s, \underline{z}_{s-1}, 0) \Big)$$

$$\underline{J}_{k2} = \sum_{s=1}^{k} \Big(\underline{f}(s, \underline{z}_{s-1}, 0) - \underline{f}_{av}(\underline{z}_{s-1}) \Big).$$

Now $\underline{J}_{k2}$ can be treated exactly as before. To deal with $\underline{J}_{k1}$, we need another assumption:

(7.2C5) $\underline{f}(k, \underline{z}, \mu)$ obeys a Lipschitz condition in μ, linearly in $\underline{z}$:

$$\|\underline{f}(k, \underline{z}, \mu) - \underline{f}(k, \underline{z}, \mu')\| \le L|\mu - \mu'|\|\underline{z}\|, \qquad \|\underline{z}\| \le h.$$

Then we find immediately

$$\begin{aligned} \|\underline{J}_{k1}\| &\le hkL\mu, \qquad 1 \le k \le T_\mu \\ &\le hLT, \qquad 1 \le k \le T_\mu. \end{aligned}$$

Adding this bound to the bound $\|\underline{J}_{k2}\|$ gives (7.16) as required. We thus have:

Theorem 7.2. For the primary system (7.1) and averaged system (7.10) with identical initial conditions, suppose (7.2C1)–(7.2C5) hold. Then given $T > 0$, there exist b_T, $\mu_T > 0$ so that

$$\max_{1 \le k \le T/\mu} \|\underline{z}_k - \bar{\underline{z}}_k\| \le c_T(\mu) b_T, \qquad 0 \le \mu \le \mu_T$$

where $c_T(\mu) \to 0$ as $\mu \to 0$.

7.2.3 Analysis under Weaker Regularity Conditions

Finally we consider a weakening of (7.2A2), (7.2C2). We suppose

(7.2D2)

$$\|\underline{p}(k, \underline{z}) - \underline{p}(k, \underline{z}')\|/k \le L_p \|\underline{z} - \underline{z}'\| \gamma(k), \qquad \|\underline{z}\|, \|\underline{z}'\| \le h$$

$$\|\underline{p}(k, 0)\|/k \le B_p \gamma(k)$$

where $\gamma(k) \ge 0$ is a so-called convergence function such that $\gamma(k)$ is non-increasing and

$$\gamma(k) \to 0, \quad \text{as} \quad k \to \infty.$$

Then repeating the proof of (7.16) we find

$$\|\underline{P}_k\| \le (L_p h/2 + B_p)\gamma(k)k.$$

A similar calculation with $\underline{\eta}_k$ leads to

$$\|\underline{J}_k\| \le ka\gamma(k) + \mu b \sum_{t=1}^{k} t\gamma(t)$$

$$\Rightarrow \qquad \mu\|\underline{J}_k\| \le c_T(\mu)$$

$$c_T(\mu) = \mu \max_{1\le k\le T\mu^{-1}} k\gamma(k)(a+Tb)$$

$$= \mu k_\mu \gamma(k_\mu)(a+Tb)$$

where k_μ is the index at which the maximum occurs. There are now two cases: (i) k_μ is bounded, $k_\mu \le B$, say; (ii) $k_\mu \to \infty$ as $\mu \to 0$. In case (i)

$$c_T(\mu) \le \mu B\gamma(0)(a+Tb) \to 0, \quad \text{as } \mu \to 0.$$

In case (ii)

$$c_T(\mu) \le \gamma(k_\mu)T(a+Tb) \to 0, \quad \text{as } \mu \to 0$$

since $\gamma(k) \to 0$ as $k \to \infty$. The rest of the argument now proceeds as before. So we have the following result:

Theorem 7.3. Theorems 7.1, 7.2 remain true when (7.2A2), (7.2C2), respectively, are replaced by (7.2D2).

7.3 FINITE TIME AVERAGING—TIME VARYING PARAMETERS

To motivate the type of problem to be faced here, let us consider the LMS algorithm with time varying parameters. The algorithm is

$$\delta\underline{\hat{w}}_k = \mu\underline{x}_k e_k$$

$$e_k = y_k - \underline{x}_k^T \underline{\hat{w}}_{k-1}$$

and we suppose

$$y_k = \underline{x}_k^T \underline{w}_{ok}$$

where $\underline{w}_{o,k}$ are the time varying weights. We then obtain an error system

$$\delta\underline{\tilde{w}}_k = -\mu\underline{x}_k\underline{x}_k^T\underline{\tilde{w}}_{k-1} - \delta\underline{w}_{ok}$$

where, as usual, $\underline{\tilde{w}}_k = \underline{\hat{w}}_k - \underline{w}_{ok}$ is the parameter estimation error.

We see that the presence of time varying parameters results in a forced (linear) system. The issue then will be with input-output stability and then tracking stability. However to get a general system form we need to look at a nonlinear example. So we pursue the time delay estimation problem of Example 7.2.

In the presence of a time varying delay θ_{ok} the idealized algorithm (7.21) becomes

$$\delta\hat{\theta}_k = -\mu s'_{k\Delta-\hat{\theta}_{k-1}}\left(s_{k\Delta-\theta_{ok}} - s_{k\Delta-\tilde{\theta}_{k-1}-\theta_{o,k-1}}\right).$$

Introducing the delay estimation error $\tilde{\theta}_k = \hat{\theta}_k - \theta_{ok}$ yields the following error system

$$\delta\tilde{\theta}_k = -\mu s'_{k\Delta-\tilde{\theta}_{k-1}-\theta_{o,k-1}}(s_{k\Delta-\theta_{ok}} - s_{k\Delta-\tilde{\theta}_{k-1}-\theta_{o,k-1}}) - \delta\theta_{ok}.$$

To get a general system form here we need to be more specific about the parameter variation. We use the following very general model:

$$\theta_{ok} \text{ is a large amplitude slowly varying signal}$$
$$\text{i.e., } \theta_{ok} = \theta_o(\epsilon k),\ \epsilon \ll 1,\ \theta_o(\bullet) \text{ differentiable.}$$

This purely deterministic specification of parameter variation differs somewhat from the specification in Section 5.3 where the parameter variation is put completely into the stochastic part of the parameter model. A simple example is

$$\theta_{ok} = 2 + \sin(\epsilon k).$$

Note that

$$\delta\theta_{ok} \simeq \epsilon\dot{\theta}_o(\epsilon k), \quad \dot{\theta}_o(\tau) = d\theta_o(\tau)/d\tau$$

so that ϵ measures the speed of parameter change. We are led then to the following general system

$$\delta\underline{z}_k = \mu\underline{f}(k, \underline{z}_{k-1}, \epsilon k, \mu) - \delta\underline{z}_o(\epsilon k). \tag{7.25}$$

The new features as compared to (7.1) are: (i) the presence of a "slow time" ϵk and (ii) the presence of a forcing term. By singling out the forcing term we have arranged that $\underline{f}(k, \underline{0}, \epsilon k, \mu) = \underline{0}$. We could have subsumed the forcing term under the ϵk argument (whereupon we would have $\underline{f}(k, \underline{0}, \epsilon k, \mu) \neq \underline{0}$) but it proves convenient to separate it out.

To do an averaging analysis it is necessary to specify a relation between the two time scales μ^{-1} and ϵ^{-1}. There are three cases:

(i) Slow Adaptation

$$\mu \ll \epsilon, \quad \text{e.g., } \mu = \epsilon^r,\ r > 1$$

Here the speed of adaptation is slower than the true speed of parameter change.

(ii) Matched Adaptation

$$\mu = O(\epsilon), \quad \text{e.g., } \mu = \epsilon$$

The speed of adaptation is matched to the true speed of parameter change.

(iii) Fast Adaptation

$$\mu \gg \epsilon, \quad \text{e.g., } \mu = \epsilon^r,\ r < 1$$

The speed of adaptation is faster than the true speed of parameter change.

In case (i) averaging is not possible on a time scale μ^{-1}. It is, however, possible on a time scale ϵ^{-1} and leads to an averaged system

$$\delta\bar{\underline{z}}_k = -\delta\underline{z}_o(\epsilon k).$$

This shows that adaptation is useless because the estimation error is as big as the parameter being estimated.

In case (iii), (7.25) collapses to the form (7.1) and the averaged system is as in the time invariant case. We might think this means adaptation is super-effective, but the presence of noise will upset this feature, so we are left to concentrate on (ii).

The system is now

$$\delta \underline{z}_k = \mu \underline{f}(k, \underline{z}_{k-1}, \mu k, \mu) - \delta \underline{z}_o(\mu k) \tag{7.26}$$

and to associate an averaged system we proceed in several stages. Firstly consider the homogeneous or unforced system (dropping the μ argument for further simplicity)

$$\delta \underline{z}'_k = \mu \underline{f}(k, \underline{z}'_{k-1}, \mu k) \tag{7.27}$$

and now follow the earlier heuristic argument, first summing up to find

$$\underline{z}'_{k+N} - \underline{z}'_k = \mu \sum_{s=k+1}^{k+N} \underline{f}(s, \underline{z}'_{s-1}, \mu s).$$

If μ is small, N not too large, then $\underline{z}_s$, μs change slowly and should be close to, respectively, $\underline{z}_k$, μk over the interval $k+1 \leq s \leq k+N$. Thus we find

$$\underline{z}'_{k+N} - \underline{z}'_k \simeq \mu \sum_{s=k+1}^{k+N} \underline{f}(s, \underline{z}'_k, \mu k)$$

and we are led to introduce the averaged function

$$\underline{f}_{av}(\underline{z}, \tau) = \lim_{N\to\infty} \frac{1}{N} \sum_{s=k+1}^{k+N} \underline{f}(s, \underline{z}, \tau). \tag{7.28}$$

Continuing the heuristic discussion as before, we find the homogeneous or unforced averaged system is

$$\delta \bar{\underline{z}}'_k = \mu \underline{f}_{av}(\bar{\underline{z}}'_{k-1}, \mu k). \tag{7.29}$$

Note that this is in general a time varying system. Similarly, for the system

$$\delta \underline{z}'_k = \mu \underline{f}(k, \underline{z}'_{k-1}, \mu k, \mu) \tag{7.30}$$

the averaged system will be (7.29), where now

$$\underline{f}_{av}(\underline{z}, \tau) = \lim_{N\to\infty} \frac{1}{N} \sum_{s=k+1}^{k+N} \underline{f}(s, \underline{z}, \tau, 0). \tag{7.31}$$

As before, a proof begins by subtracting (7.29) from (7.30) and summing to find

$$\underline{\Delta}'_k = \underline{z}'_k - \underline{\bar{z}}'_k = \underline{\Delta}'_0 + \mu \sum_{s=1}^{k} \left(\underline{f}_{av}(\underline{z}'_{s-1}, \mu s) - \underline{f}_{av}(\underline{\bar{z}}'_{s-1}, \mu s) \right) + \mu \underline{J}_k \tag{7.32a}$$

$$\underline{J}_k = \sum_{s=1}^{k} \left(\underline{f}(s, \underline{z}'_{s-1}, \mu s, \mu) - \underline{f}_{av}(\underline{z}'_{s-1}, \mu s) \right). \tag{7.32b}$$

The argument now proceeds much as before and requires the following regularity conditions:

(7.3A1) $\|\underline{f}(k, \underline{z}, \tau, \mu)\| \leq B_f$, $\|\underline{z}\| \leq h$

(7.3A2) Introduce the total perturbation

$$\underline{p}(k, \underline{z}, \tau) = \sum_{s=1}^{k} \left(\underline{f}(s, \underline{z}, \tau, 0) - \underline{f}_{av}(\underline{z}, \tau) \right)$$

and suppose that

$$\|\underline{p}(k, \underline{0}, \tau)\| \leq B_p k \gamma(k)$$

$$\|\underline{p}(k, \underline{z}, \tau) - \underline{p}(k, \underline{z}', \tau')\| \leq \left(L_{pa}\|\underline{z} - \underline{z}'\| + L_{pb}|\tau - \tau'| \right) k\gamma(k), \quad \|\underline{z}\|, \|\underline{z}'\| \leq h$$

where $\gamma(k)$ is a non-increasing convergence function $\to 0$ as $k \to \infty$.

(7.3A3) $\underline{f}_{av}(\underline{z}, \tau)$ obeys a Lipschitz condition

$$\|\underline{f}_{av}(\underline{z}, \tau) - \underline{f}_{av}(\underline{z}', \tau)\| \leq L_{fa}\|\underline{z} - \underline{z}'\|, \quad \|\underline{z}\|, \|\underline{z}'\| \leq h.$$

(7.3A4) $\|\underline{\bar{z}}'_0\|$ is so small that $\|\underline{\bar{z}}'_k\| \leq h/2$, $1 \leq k \leq T/\mu$.

(7.3A5) $\underline{f}(k, \underline{z}, \tau, \mu)$ obeys a Lipschitz condition in μ, linearly in $\|\underline{z}\|$

$$\|\underline{f}(k, \underline{z}, \tau, \mu) - \underline{f}(k, \underline{z}, \tau, \mu')\| \leq L_a|\mu - \mu'|\|\underline{z}\|, \quad \|\underline{z}\| \leq h.$$

(7.3A6) $\dot{\underline{z}}_o(\tau) = d\underline{z}_o(\tau)/d\tau$ is bounded for all τ. $\|\dot{\underline{z}}_o(\tau)\| \leq \dot{B}$.

Before we pursue the proof let us return to the forced system (7.26) and claim that the associated averaged system is

$$\delta\underline{\bar{z}}_k = \mu \underline{f}_{av}(\underline{\bar{z}}_{k-1}, \mu k) - \delta\underline{z}_o(\mu k). \tag{7.33}$$

This follows simply by subtracting (7.33) from (7.26) and summing which leads to (7.32) again with $\underline{z}'_k$, $\underline{\bar{z}}'_k$ replaced by $\underline{z}_k$, $\underline{\bar{z}}_k$. We call (7.33) a partially averaged system because the forcing term is not averaged. We are led to the following result:

Theorem 7.4. For the primary forced system (7.26) and associated averaged system (7.33) with identical initial conditions, suppose (7.3A1)–(7.3A5) hold. Then given $T > 0$, there are $b_T > 0$, $\mu_T > 0$ such that

$$\max_{1 \leq k \leq T/\mu} \|\underline{z}_k - \underline{\bar{z}}_k\| \leq c_T(\mu) b_T, \quad 0 \leq \mu \leq \mu_T$$

and $c_T(\mu) \to 0$ as $\mu \to 0$.

Proof. Follows nearly the same lines as the proof of Theorem 7.3. We need only look at $\underline{J}_k$ of (7.32) in detail and we split it into two pieces:

$$\underline{J}_k = \underline{J}_{k,1} + \underline{J}_{k,2}$$

$$\underline{J}_{k,1} = \sum_{s=1}^{k} \left(\underline{f}(s, \underline{z}_{s-1}, \mu s, \mu) - \underline{f}(s, \underline{z}_{s-1}, \mu s, 0) \right)$$

$$\underline{J}_{k,2} = \sum_{s=1}^{k} \left(\underline{f}(s, \underline{z}_{s-1}, \mu s, 0) - \underline{f}_{av}(\underline{z}_{s-1}, \mu s) \right).$$

Now $\underline{J}_{k,1}$ is treated as before, while (much as before) we approximate $\underline{J}_{k,2}$ by

$$\underline{P}_k = \underline{p}(k, \underline{z}_{k-1}, \mu k)$$

and find that

$$\begin{aligned} \underline{P}_k - \underline{P}_{k-1} &= \underline{p}(k, \underline{z}_{k-1}, \mu k) - \underline{p}(k-1, \underline{z}_{k-2}, \mu k - \mu) \\ &= \underline{p}(k, \underline{z}_{k-1}, \mu k) - \underline{p}(k-1, \underline{z}_{k-1}, \mu k) + \underline{\eta}_k \\ &= \underline{f}(k, \underline{z}_{k-1}, \mu k, 0) - \underline{f}_{av}(\underline{z}_{k-1}, \mu k) + \underline{\eta}_k \end{aligned}$$

where

$$\underline{\eta}_k = \underline{p}(k-1, \underline{z}_{k-1}, \mu k) - \underline{p}(k-1, \underline{z}_{k-2}, \mu k - \mu).$$

Thus summing up gives

$$\underline{P}_k = \underline{J}_{k,2} + \sum_{s=1}^{k} \underline{\eta}_k.$$

Next, by (7.3A2), (7.3A6) and the mean value theorem

$$\begin{aligned} (k\gamma(k))^{-1} \|\underline{\eta}_k\| &\le L_{pa} \|\underline{z}_{k-1} - \underline{z}_{k-2}\| + L_{pb}\mu \\ &\le L_{pa}\mu \left(\|\underline{f}(k-1, \underline{z}_{k-2}, \mu k - \mu, \mu)\| + \dot{B} \right) + \mu L_{pb} \\ &\le \mu(L_{pa}B_f + \dot{B} + L_{pb}) \end{aligned}$$

which is the type of bound required. The rest of the argument proceeds as before.

To get some insight into these results we consider the LMS algorithm in more detail. The averaged system (7.33) for LMS is

$$\delta\underline{\bar{w}}_k = -\mu \underline{R}_x \underline{\bar{w}}_{k-1} - \delta \underline{w}_{ok}. \tag{7.34}$$

This is a forced linear system and we now consider its input-output stability and tracking behavior. To gain some insight here we orthogonalize as usual; so suppose $\underline{R}_x$ has eigenvalue decomposition (3.3) with eigenvectors $\underline{v}_r$, $1 \le r \le p$, then set

$$\bar{v}_{k,r} = \underline{v}_r^T \bar{\underline{w}}_k, \quad 1 \le r \le p$$
$$v_{ok,r} = \underline{v}_r^T \underline{w}_{ok}$$

so that (7.34) becomes

$$\delta \bar{v}_{k,r} = -\mu\lambda_r \bar{v}_{k-1,r} - \delta v_{ok,r} \quad 1 \le r \le p. \tag{7.35}$$

This is a linear first order forced difference equation, and to see what behavior to expect, it is natural to do a Fourier analysis of (7.35) and suppose $v_{ok,r}$ is a sine/cosine of frequency ω, say,

$$v_{ok,r} = A_r e^{j\omega k}$$
$$\Rightarrow \quad \delta v_{ok,r} = A_r e^{j\omega k}(1 - e^{-j\omega}).$$

From this it rapidly follows that (providing $0 < \mu\lambda_{min} < \mu\lambda_{max} < 2$, i.e., (3.17) holds)

$$\bar{v}_{k,r} = \frac{(1 - e^{-j\omega})e^{j\omega k} A_r}{1 - e^{-j\omega}(1 - \mu\lambda_r)}$$
$$= \frac{(1 - e^{-j\omega})e^{j\omega k} A_r}{\mu\lambda_r + (1 - e^{-j\omega})(1 - \mu\lambda_r)}.$$

To gain some insight into this expression, consider a low frequency parameter variation; i.e., suppose $v_{ok,r}$ is large amplitude but slowly varying and put $\omega = \epsilon$ where ϵ is small. Then (with $A_r = O(1)$) the speed of parameter variation is

$$\delta v_{ok,r} \simeq j\epsilon A_r e^{j\omega k} = O(\epsilon)$$

so that

$$\bar{v}_{k,r} \simeq \frac{j\epsilon A_r e^{j\omega k}}{\mu\lambda_r + j\epsilon}$$
$$= \frac{j(\epsilon/\mu\lambda_r) A_r e^{j\epsilon k}}{1 + j(\epsilon/\mu\lambda_r)}.$$

Recalling that $\bar{v}_{k,r}$ is an averaged error signal, one would like it to be small relative to $v_{ok,r}$. This leads to the requirement

$$\epsilon \ll \mu\lambda_r, \quad 1 \le r \le p.$$

For a first order system, $1/(\mu\lambda_r)$ is the system time constant, or, in the frequency domain, $\mu\lambda_r$ measures the system bandwidth. So for successful tracking, the bandwidth of parameter variation must be less than the bandwidth that the adaptive system can master to react to such variation. For given ϵ this means that larger μ's are desirable. These conclusions are fundamental to understanding the tracking behavior of adaptive systems and they show that successful tracking requires a careful choice of μ and that the bandwidth of true parameter change, ϵ, must be known.

These results will now be illustrated with two examples.

Example 7.3. LMS Algorithm with Time Varying Parameters

We pursue Examples 3.2, 7.1. We suppose the weights vary according to

$$\underline{w}_{ok} = (\sin \epsilon k, \cos \epsilon k)^T.$$

From Example 3.2 we have $(\lambda_1, \lambda_2) = (1, 3)$. Fig. 7.4 shows plots of $\hat{w}_{k1}$ for several values of μ. Fig. 7.5 gives similar plots for $\hat{w}_{k2}$. In each case $\epsilon = 0.01$ while $\mu(\lambda_1, \lambda_2) = (0.001, 0.003), (0.01, 0.03), (0.1, 0.3)$. Clearly the relatively larger μ values do best.

Although our discussion was based on the averaged system, the results apply well to the actual system.

Example 7.4. Time Varying Delay Estimation

We pursue Examples 3.8 and 7.2, but now allow time varying delay. The averaged system is

$$\delta\bar{\theta}_k = -\mu R'(\bar{\theta}_{k-1}) - \delta\theta_{ok}. \tag{7.36}$$

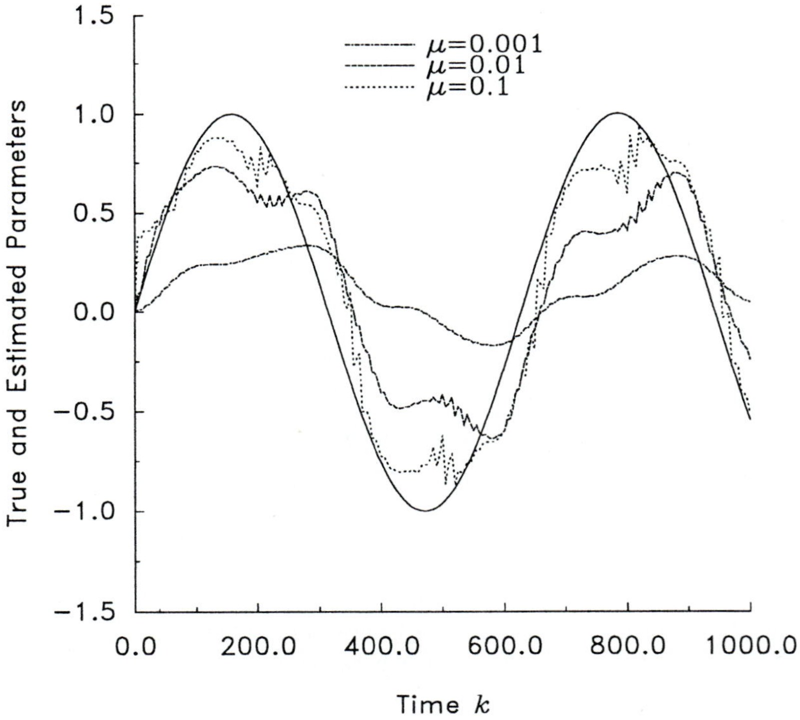

Figure 7.4 $\hat{w}_{k1}$, tracking behavior of LMS ($\epsilon = 0.01$), solid line is the true parameter.

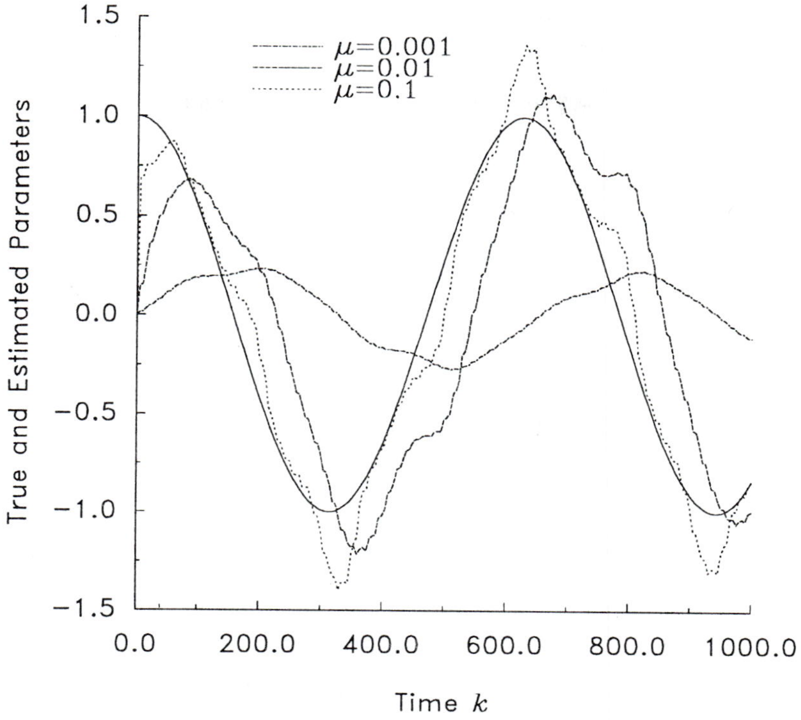

Figure 7.5 $\hat{w}_{k2}$, tracking behavior of LMS ($\epsilon = 0.01$), solid line is the true parameter.

This is a forced nonlinear time varying system. We cannot draw as precise a set of conclusions regarding its behavior as we could with LMS, but we should expect that good tracking is at least possible even if θ_{ok} is large amplitude as long as it is slowly varying. So suppose

$$\theta_{ok} = \theta_{dc} + \theta_{fk}$$
$$\theta_{fk} = \theta(\epsilon k)$$

where $\theta(\bullet)$ is a continuously differentiable function (e.g., $\theta(\epsilon k) = \sin(\epsilon k)$). Then if ϵ is small

$$\delta\theta_{ok} = \delta\theta_{fk} \simeq \epsilon\dot{\theta}(\epsilon k)$$

where $\dot{\theta}(t) = d\theta(t)/dt$. Thus (7.36) becomes

$$\delta\bar{\theta}_k = -\mu R'(\bar{\theta}_{k-1}) - \epsilon\dot{\theta}(\epsilon k).$$

We can now mimic the calculation we did with LMS by a linearization (For more on this see Section 7.4) as follows. Near $\bar{\theta} = 0$ a Taylor series gives

$$R'(\bar{\theta}) \simeq \bar{\theta}R''(0) + O(\bar{\theta}^3)$$

since $R'(0) = R'''(0) = 0$. So near $\bar{\theta} = 0$

$$\delta\bar{\theta}_k = -\mu R''(0)\bar{\theta}_{k-1} - \epsilon\dot{\theta}(\epsilon k).$$

We conclude that successful tracking requires

$$\mu R''(0) \gg \epsilon.$$

For Example 3.8 we find from (3.67) that

$$R''(0) = \frac{41}{12}\omega_o^2.$$

In the simulation below we use

$$\omega_o = \frac{2\pi}{50} \quad \Rightarrow \quad R''(0) = 5.4 \times 10^{-2}.$$

Also we take $\epsilon = (2\pi/5) \times 10^{-2} = 1.26 \times 10^{-2}$. Fig. 7.6 shows tracking behavior for several values of μ: $\mu R''(0) = (0.02, 0.2, 2.0)$.

In Fig. 7.7 we look at the error between the true delay and the averaged system. The magnitude of the errors is smaller for larger μ as can be seen by comparing Figs. 7.6 and 7.7.

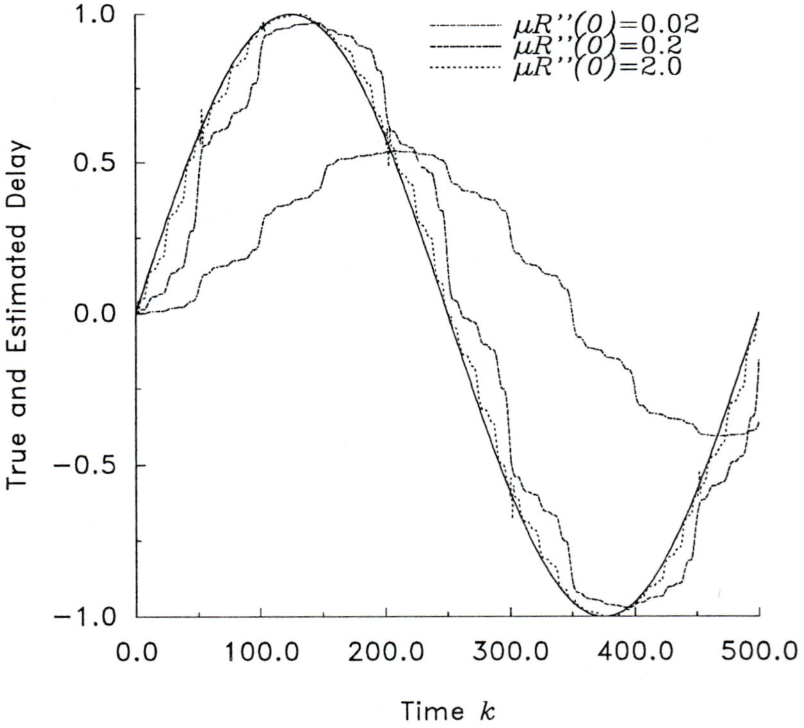

Figure 7.6 Tracking time varying delay with the SD algorithm, solid line is the true delay.

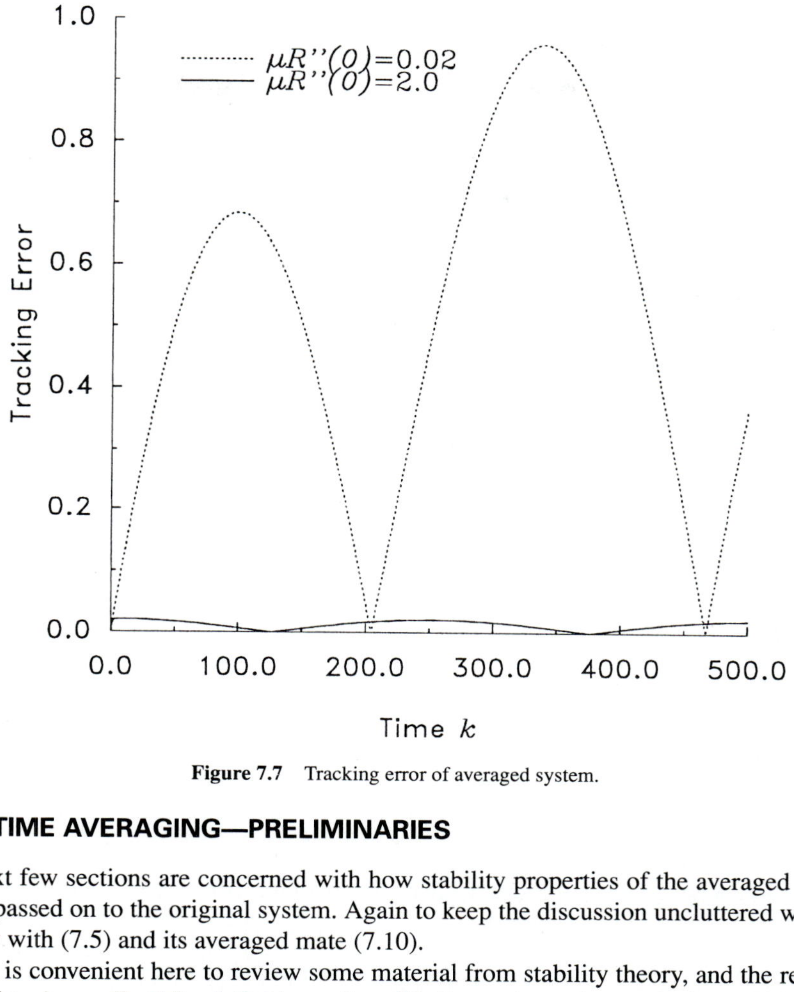

Figure 7.7 Tracking error of averaged system.

7.4 INFINITE TIME AVERAGING—PRELIMINARIES

The next few sections are concerned with how stability properties of the averaged system can be passed on to the original system. Again to keep the discussion uncluttered we work initially with (7.5) and its averaged mate (7.10).

It is convenient here to review some material from stability theory, and the reader is directed to Appendix C for definitions of equilibrium point and exponential stability.

Proving convergence or stability of a nonlinear difference equation such as (7.5) or (7.10) is generally quite hard. It is often achieved, when it can be, by means of a Lyapunov function which is typically a system energy that decreases with time. With this in mind the following result is useful in the sequel.

Theorem 7.5. Consider the system

$$\delta \underline{z}_k = \mu \underline{f}(k, \underline{z}_{k-1}) \tag{7.37}$$

and suppose $\underline{f}(k, \underline{z})$ obeys a uniform Lipschitz condition

$$\|\underline{f}(k, \underline{z}) - \underline{f}(k, \underline{z}')\| \leq L\|\underline{z} - \underline{z}'\|, \quad \|\underline{z}\|, \|\underline{z}'\| \leq h. \tag{7.38}$$

Then the following two statements are equivalent:

(i) $\underline{z} = \underline{0}$ is a locally exponentially stable equilibrium point of (7.37) of stability radius h and decay rate $O(\mu)$.

(ii) There is a Lyapunov function $V(k, \underline{z})$ such that for $\|\underline{z}\| \le h$

(7.4A1) $\alpha_1 \|\underline{z}\|^2 \le V(k, \underline{z}) \le \alpha_2 \|\underline{z}\|^2$

(7.4A2) $V(k, \underline{z} + \mu \underline{f}(k, \underline{z})) - V(k-1, \underline{z}) \le -\alpha_3 \mu \|\underline{z}\|^2$

(7.4A3) $\|V_z(k, \underline{z})\| \le \alpha_4 \|\underline{z}\|$, $V_z(k, \underline{z}) = \partial V(k, \underline{z})/\partial \underline{z}$

(7.4A4) If in addition,

$$\sup_k \|\underline{f}_{zi}(k, \underline{z})\| \le d, \quad \|\underline{z}\| \le h, \quad 1 \le i \le p$$

where $\underline{f}_{zi}(k, \underline{z}) = \partial^2 \underline{f}(k, \underline{z})/\partial \underline{z}^T \partial z_i$, $1 \le i \le p$,

$$\text{then} \|V_{zz}(k, \underline{z})\| \le \alpha_5 + \alpha_6 \|\underline{z}\|, \quad V_{zz}(k, \underline{z}) = \partial^2 V(k, \underline{z})/\partial \underline{z} \partial \underline{z}^T$$

where α_1–α_6 are positive.

Proof. This is Theorem C7.1 of Appendix C7. Note that the constants α_1–α_6 are functions of μ and converge to positive limits as $\mu \to 0$. The reason for properties (7.4A3) and (7.4A4) will become apparent.

If in (i) the stability is global then the conditions in (ii) hold for any finite h. The same result holds for systems of the form (7.1): The proof is the same with only notational changes; even under (7.37), $V(k, \underline{z})$ depends on μ although we have not explicitly indicated that. The result that will be of most use is the version of this theorem when (7.37) is autonomous:

Theorem 7.6. Consider the system

$$\delta \underline{z}_k = \mu \underline{f}_{av}(\underline{z}_{k-1}) \tag{7.39}$$

and suppose $\underline{f}_{av}(\underline{z})$ obeys a Lipschitz condition (for $\|\underline{z}\| \le h$). Then the following two statements are equivalent:

(i) $\underline{z} = \underline{0}$ is a locally exponentially stable equilibrium point of (7.39) with stability radius h and decay rate $O(\mu)$.

(ii) There is a Lyapunov function $V(\underline{z})$ such that for $\|\underline{z}\| \le h$

(7.4a1) $\alpha_1 \|\underline{z}\|^2 \le V(\underline{z}) \le \alpha_2 \|\underline{z}\|^2$

(7.4a2) $V(\underline{z} + \mu \underline{f}_{av}(\underline{z})) - V(\underline{z}) \le -\alpha_3 \mu \|\underline{z}\|^2$

(7.4a3) $\|V_z(\underline{z})\| \le \alpha_4 \|\underline{z}\|$, $V_z = dV/d\underline{z}$

(7.4a4) If also $\|\underline{f}_{av,zi}(\underline{z})\| \le d$, $\|\underline{z}\| \le h$ where $\underline{f}_{av,zi}(\underline{z}) = d^2 \underline{f}_{av}(\underline{z})/d\underline{z}^T dz_i$, $1 \le i \le p$, then

$$\|V_{zz}(\underline{z})\| \le \alpha_5 + \alpha_6 \|\underline{z}\|, \quad V_{zz} = d^2 V/d\underline{z} d\underline{z}^T$$

where α_1–α_6 are positive constants.

Proof. This is Theorem C7.2 of Appendix C7. Note that the constants α_1–α_6 are functions of μ and converge to positive limits as $\mu \to 0$.

Again global stability in (i) entails any finite h in (ii).

Theorem 7.6 is used in the following way. Suppose (7.39) describes an averaged system corresponding to (7.37) and that somehow it can be shown to be exponentially stable (i.e., $\underline{0}$ is an exponentially stable equilibrium point of (7.39)). Then we have a Lyapunov function with the prescribed properties and we try to use this Lyapunov function as a Lyapunov function for (7.37). This fails, but then a modification is found for the Lyapunov function to get a perturbed Lyapunov function that does work on (7.37).

One could of course try to show exponential stability of (7.37) directly, but generally this has been found possible only in certain specific cases (basically those covered in Part I of this book). How then can one show exponential stability of a system such as (7.39) (or perhaps (7.37))? A common way is by linearization.

Theorem 7.7. Consider the system (7.37), and suppose $\underline{0}$ is an equilibrium point and that there is a sequence of matrices $\underline{A}_k$ such that

$$\lim_{\|\underline{z}\|\to 0} \sup_{k\geq 0} \|\underline{f}(k,\underline{z}) - \underline{A}_k\underline{z}\|/\|\underline{z}\| = 0. \tag{7.40}$$

Then if $\underline{0}$ is an exponentially stable equilibrium point of the linearized system

$$\delta\underline{z}_k = \mu\underline{A}_k\underline{z}_{k-1} \tag{7.41}$$

then $\underline{0}$ is an exponentially stable equilibrium point of (7.37).

Proof. This is Theorem C7.3 of Appendix C7.

Sometimes $\underline{A}_k$ can be guessed, or else it can be calculated as

$$\underline{A}_k = \partial\underline{f}(k,\underline{z})/\partial\underline{z}^T\Big|_{\underline{z}=\underline{0}}.$$

Of course if we are dealing with (7.39) then

$$\underline{A}_k = \underline{A} = d\underline{f}_{av}(\underline{z})/d\underline{z}^T\Big|_{\underline{z}=\underline{0}}$$

and there is the following corollary:

Theorem 7.8. Consider the system (7.39), and suppose $\underline{0}$ is an equilibrium point and there is a matrix $\underline{A}$ such that

$$\lim_{\|\underline{z}\|\to 0} \|\underline{f}_{av}(\underline{z}) - \underline{A}\underline{z}\|/\|\underline{z}\| = 0. \tag{7.42}$$

Then if $\underline{0}$ is an exponentially stable equilibrium point of the linearized system

$$\delta\underline{z}_k = \mu\underline{A}\underline{z}_{k-1} \tag{7.43}$$

then $\underline{0}$ is an exponentially stable equilibrium point of (7.39). A Taylor series expansion (see (D12)) shows that condition (7.42) simply says that $d\underline{f}_{av}/d\underline{z}$ is continuous at $\underline{0}$.

7.5 INFINITE TIME AVERAGING—HOVERING THEOREM

As was indicated in Section 7.1, the finite time averaging results, while useful and interesting, do not answer the question of algorithm stability. They are concerned with finite time behavior whereas stability is an infinite time phenomenon.

The simplest infinite time result one could ask for is to extend the trajectory approximation results of Theorems 7.1, 7.2, 7.3 to hold for all time. To do this the idea is to piece together finite time results over an infinite sequence of intervals, but for this to be possible regularity conditions must hold uniformly in time and local ES is needed for the averaged system. The resulting "Hovering Theorem" is not however a stability result, since it does not for example satisfy the stability definitions in Appendix C7. Nevertheless, it proves to be quite useful and provides indirect information about stability.

Consider then the system (7.1), the associated averaged function (7.23), and averaged system (7.10). Introduce also the following regularity condition which is a sort of time invariant version of (7.2C2).

(7.5A1) Consider the total perturbation

$$\underline{p}(k, \underline{z}) = \sum_{s=1}^{k} [\underline{f}(s, \underline{z}, \underline{0}) - \underline{f}_{av}(\underline{z})]$$

$$\underline{p}_r(k, \underline{z}) = \underline{p}(k, \underline{z}) - \underline{p}(r-1, \underline{z}), \quad k \geq r$$

and suppose that there is a convergence function $\gamma(k) \geq 0$ so that for any r

$$k^{-1}\|\underline{p}_r(k+r, \underline{z}) - \underline{p}_r(k+r, \underline{z}')\| \leq \gamma(k) L_p \|\underline{z} - \underline{z}'\|, \quad \|\underline{z}\|, \|\underline{z}'\| \leq h.$$

Then we have the following result:

Theorem 7.9. (Hovering Theorem). Consider the primary system (7.1) and averaged system (7.10) with identical initial conditions and suppose (7.2C1), (7.5A1), (7.2C3)–(7.2C5) hold. Suppose $\underline{z}_e$ is an ES equilibrium point of the averaged system with decay rate $O(\mu)$ (but we do not assume $\underline{z}_e$ is an equilibrium point of the primary system) and the initial condition is in the stability region for $\underline{z}_e$. Then

$$\sup_{k\geq 1} \|\underline{z}_k - \bar{\underline{z}}_k\| \leq c(\mu)$$

where $c(\mu) \to 0$ as $\mu \to 0$.

Proof. Let us partition the time axis as $\bigcup_{n=0}^{\infty} I_{n,n+1}$ where the intervals $I_{n,n+1}$ are

$$I_{n,n+1} = \{k : nT\mu^{-1} \leq k \leq (n+1)T\mu^{-1}\}$$

and $T > 0$ is a value to be chosen. On each interval $I_{n,n+1}$ define

$$\bar{\underline{z}}_{n,k} \text{ is the solution to (7.10) with initial value } \bar{\underline{z}}_{n,nT\mu^{-1}} = \underline{z}_{nT\mu^{-1}}.$$

Then by the Averaging theorem 7.2 and the uniformity in (7.5A1) we have

$$\|\underline{z}_k - \bar{\underline{z}}_{n,k}\| \leq c_T(\mu), \quad k \in I_{n,n+1}$$

where $c_T(\mu) \to 0$ as $\mu \to 0$.

Next, the continuity property of ES (see Appendix C7) for the averaged system ensures, for some α, $0 < \alpha < 1$, and k in $I_{n,n+1}$

$$\begin{aligned}\|\bar{\underline{z}}_k - \bar{\underline{z}}_{n,k}\| &\leq m\alpha^{(k-nT\mu^{-1})}\|\bar{\underline{z}}_{nT\mu^{-1}} - \bar{\underline{z}}_{n,nT\mu^{-1}}\| \\ &\leq m\alpha^{(k-nT\mu^{-1})}\{\|\bar{\underline{z}}_{nT\mu^{-1}} - \bar{\underline{z}}_{n-1,nT\mu^{-1}}\| + \|\bar{\underline{z}}_{n,nT\mu^{-1}} - \bar{\underline{z}}_{n-1,nT\mu^{-1}}\|\}.\end{aligned}$$

However

$$\|\bar{\underline{z}}_{n,nT\mu^{-1}} - \bar{\underline{z}}_{n-1,nT\mu^{-1}}\| = \|\underline{z}_{nT\mu^{-1}} - \bar{\underline{z}}_{n-1,nT\mu^{-1}}\| \leq c_T(\mu).$$

Thus, introducing

$$\Delta_{n-1} = \|\bar{\underline{z}}_{nT\mu^{-1}} - \bar{\underline{z}}_{n-1,nT\mu^{-1}}\|$$

we have, for $k \in I_{n,n+1}$

$$\|\bar{\underline{z}}_k - \bar{\underline{z}}_{n,k}\| \leq m\alpha^{(k-nT\mu^{-1})}(\Delta_{n-1} + c_T(\mu)).$$

Thus

$$\max_{k\in I_{n,n+1}} \|\bar{\underline{z}}_k - \bar{\underline{z}}_{n,k}\| \leq m(\Delta_{n-1} + c_T(\mu)). \tag{7.44}$$

Also, setting $k = (n+1)T\mu^{-1}$ gives

$$\Delta_n \leq m\alpha^{T\mu^{-1}}(\Delta_{n-1} + c_T(\mu)). \tag{7.45}$$

Now choose T so large that

$$m\alpha^{T\mu^{-1}} = k_T < 1. \tag{7.46}$$

Then iterating gives for any n (noting that $\Delta_0 = 0$)

$$\Delta_n \leq k_T c_T(\mu)/(1 - k_T).$$

Placing this in (7.44) gives

$$\max_{k\in I_{n,n+1}} \|\bar{\underline{z}}_k - \bar{\underline{z}}_{n,k}\| \leq c_T(\mu)m(1 - k_T)^{-1}. \tag{7.47}$$

However for $k \in I_{n,n+1}$

$$\begin{aligned}\|\underline{z}_k - \bar{\underline{z}}_k\| &\leq \|\underline{z}_k - \bar{\underline{z}}_{n,k}\| + \|\bar{\underline{z}}_{n,k} - \bar{\underline{z}}_k\| \\ &\leq c_T(\mu) + c_T(\mu)m(1 - k_T)^{-1} \\ &= c_T(\mu)[1 + m/(1 - k_T)] = c(\mu)\end{aligned} \tag{7.48}$$

and since n is arbitrary this is the required result. Note that with decay rate $O(\mu)$ so that $\alpha \approx 1 - a\mu$ as $\mu \to 0$, then given $\mu_0 > 0$ choose T so that

$$\sup_{0\leq\mu\leq\mu_0} m(\mu)e^{-aT} < \frac{1}{2}.$$

This ensures that (7.46) holds and that $c(\mu) \to 0$ as $\mu \to 0$.

We call this result the Hovering theorem because of the following scenario. Suppose the primary system (7.1) has only one equilibrium point, say $\underline{z} = \underline{0}$, while the averaged system has other equilibrium points as well. Let $\underline{z}_e$ be an ES equilibrium point of (7.10) but not an equilibrium point of (7.1). If we start close to $\underline{z}_e$, then if μ is small enough the Hovering theorem tells us that $\underline{z}_k$ always remains close to $\bar{\underline{z}}_k$. But $\bar{\underline{z}}_k$ will be converging to $\underline{z}_e$. Since $\underline{z}_e$ is not an equilibrium point of (7.1), then $\underline{z}_k$ must "hover" around $\underline{z}_e$. We now illustrate this.

Example 7.5. Time Delay Estimation

We pursue further the noise-free unit attenuation case of Examples 3.8, 7.2. In Fig. 7.2 we see a local maximum of the correlation function at $\tilde{\theta} = \pm 16$, and this is an equilibrium point of the averaged system (7.22). But clearly (see (3.67)) the only equilibrium points of the primary system (7.21) are $\theta = \theta_o + 50m$, where m is an integer. In Fig. 7.8 is plotted the delay error trajectory (for three values of μ) for both the primary and averaged systems. The initial delay is just to the right of A' in Fig. 7.2 at $\tilde{\theta} = -12$. It is left to the reader to check that (7.5A1) holds if (7.2B1), (7.2B2) hold.

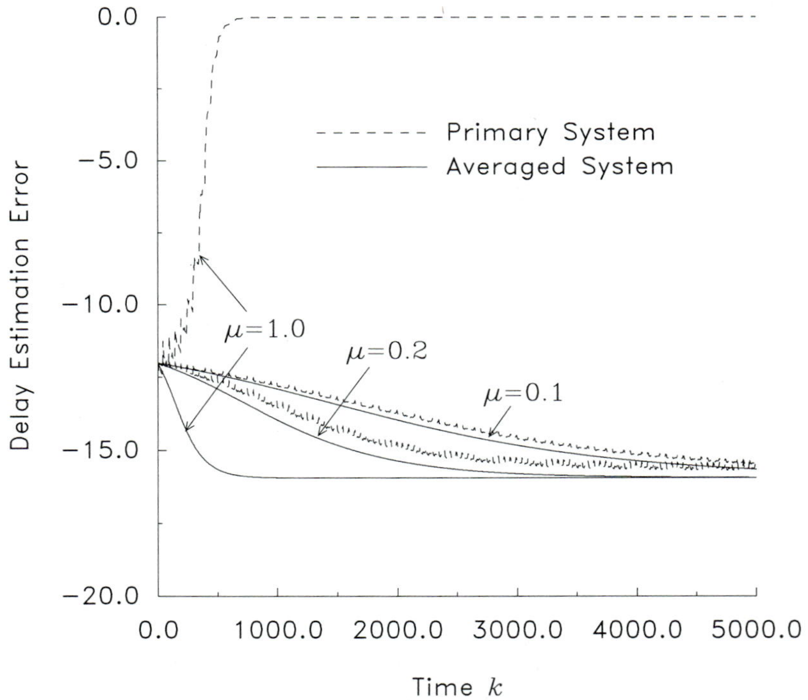

Figure 7.8 An example showing the original system hovering around the equilibrium point of the averaged system when μ is small ($\mu = 0.1, 0.2$). When μ is large, the original system escapes the stability region of this equilibrium point for the averaged system and converges to an equilibrium point possessed by both primary and averaged systems.

Notice that if μ is small the primary system hovers near $\tilde{\theta} = -16$, while if μ is large the primary system escapes the stability region of this equilibrium point of the averaged system and converges to $\tilde{\theta} = 0$. So the region of attraction of the equilibrium point $\tilde{\theta}_o$ depends on μ.

7.6 INFINITE TIME AVERAGING—STABILITY VIA PERTURBED LYAPUNOV FUNCTION

Now it is time to consider how to use stability properties of the averaged system to deduce stability of the primary system. For adaptive algorithms it is rarely possible to use Theorem 7.7 directly on (7.37); rather we use Theorem 7.8 on the averaged system (7.39). Thus, as indicated earlier, we first show, if we can, that $\underline{0}$ is a locally exponentially stable equilibrium point of (7.39). Then Theorem 7.6 tells us a Lyapunov function exists and we find a way to modify this Lyapunov function so that it works on (7.37). To begin the discussion we introduce some regularity conditions.

(7.6A1) $\underline{0}$ is an equilibrium point of (7.37), i.e., $\underline{f}(k, \underline{0}) = \underline{0}$.

(7.6A2) $\underline{0}$ is a locally exponentially stable equilibrium point of (7.39) of stability radius h and decay rate $O(\mu)$.

(7.6A3) $\underline{f}(k, \underline{z})$, $\underline{f}_{av}(\underline{z})$ obey a Lipschitz condition (without loss of generality, the same Lipschitz constant is used for both) for $\|\underline{z}\|$, $\|\underline{z}'\| \le h$:

$$\|\underline{f}(k, \underline{z}) - \underline{f}(k, \underline{z}')\| \le L\|\underline{z} - \underline{z}'\|$$

$$\|\underline{f}_{av}(\underline{z}) - \underline{f}_{av}(\underline{z}')\| \le L\|\underline{z} - \underline{z}'\|$$

(7.6A4)

$$\|\underline{f}_{av,zi}(\underline{z})\| \le d, \quad \|\underline{z}\| \le h, \quad 1 \le i \le p$$

where $\underline{f}_{av,zi}(\underline{z}) = d^2 \underline{f}_{av}(\underline{z})/d\underline{z}^T dz_i$

Another regularity condition will be added later. Then by Theorem 7.6 there is a Lyapunov function $V(\underline{z})$ obeying (7.4a1)–(7.4a4).

Now we try to use the Lyapunov function $V(\underline{z})$ on (7.37). We insist

$$\|\underline{z}_0\| \le H, \quad \text{where } H \text{ is to be chosen} \tag{7.49}$$

and now employ an inductive argument and suppose

$$\|\underline{z}_{t-1}\| \le h/(1 + 4\mu L), \quad 1 \le t \le k. \tag{7.50}$$

Provided $H \le h/(1 + 4\mu L)$ this holds for $t = 1$. It now follows from (7.37), (7.6A3), (7.39) that

$$\|\underline{z}_k\| \le (1 + \mu L)\|\underline{z}_{k-1}\| \le h \tag{7.51a}$$

$$\|\tilde{\underline{z}}_{k-1}\| = \|\underline{z}_{k-1} + \mu \underline{f}_{av}(\underline{z}_{k-1})\| \le h\frac{1 + \mu L}{1 + 4\mu L} \le h. \tag{7.51b}$$

We will establish later that $\underline{z}_k$ also satisfies (7.50). Now turning to the Lyapunov computation we have

$$\begin{aligned} V(\underline{z}_k) &= V(\underline{z}_{k-1} + \mu \underline{f}(k, \underline{z}_{k-1})) \\ &= V(\underline{z}_{k-1} + \mu \underline{f}_{av}(\underline{z}_{k-1}) + \mu \tilde{\underline{f}}_k) \\ &= V(\tilde{\underline{z}}_{k-1} + \mu \tilde{\underline{f}}_k) \end{aligned}$$

where $\tilde{\underline{z}}_k$ is defined above in (7.51) and

$$\tilde{\underline{f}}_k = \underline{f}(k, \underline{z}_{k-1}) - \underline{f}_{av}(\underline{z}_{k-1}). \tag{7.52a}$$

Note that, from (7.6A1), (7.6A2), and (7.6A3)

$$\|\tilde{\underline{f}}_k\| \leq 2L\|\underline{z}_{k-1}\|. \tag{7.52b}$$

Now apply a Taylor series (Appendix D13) to find

$$V(\underline{z}_k) = V(\underline{z}_{k-1} + \mu \underline{f}_{av}(\underline{z}_{k-1})) + \mu \tilde{\underline{f}}_k^T V_z(\underline{z}_k^*)$$

where (with $0 < \lambda_k < 1$)

$$\underline{z}_k^* = \tilde{\underline{z}}_{k-1} + \mu \lambda_k \tilde{\underline{f}}_k. \tag{7.52c}$$

So from (7.51), (7.52)

$$\|\underline{z}_k^*\| \leq \|\tilde{\underline{z}}_{k-1}\| + \mu\|\tilde{\underline{f}}_k\| \leq (1 + 3\mu L)\|\underline{z}_{k-1}\| \leq h. \tag{7.53}$$

Now apply (7.4a2), (7.6a4) to find

$$V(\underline{z}_k) \leq V(\underline{z}_{k-1}) - \alpha_3 \mu \|\underline{z}_{k-1}\|^2 + \mu \tilde{\underline{f}}_k^T V_z(\underline{z}_k^*). \tag{7.54}$$

So far the argument is proceeding nicely but now observe that the third term is too big; it is $O(\mu)$ and of indeterminate sign. The trick now is to modify the Lyapunov function so as to cancel it out. To do this we introduce a perturbation related to the one we used earlier (see (7.2A2)) but a little more carefully constructed.

Consider the discounted or exponentially weighted perturbation ($\tilde{\underline{f}}(s, \underline{z}) = \underline{f}(s, \underline{z}) - \underline{f}_{av}(\underline{z})$)

$$\underline{P}(k, \underline{z}) = \sum_{s=k+1}^{\infty} (1 - \mu)^{s-k} \tilde{\underline{f}}(s, \underline{z}). \tag{7.55}$$

We will first show a basic perturbation approximation identity, namely

$$\delta \underline{P}(k, \underline{z}) = -\tilde{\underline{f}}(k, \underline{z}) + \mu \left(\underline{P}(k, \underline{z}) + \tilde{\underline{f}}(k, \underline{z}) \right). \tag{7.56}$$

In fact we have

$$\underline{P}(k, \underline{z}) - \underline{P}(k-1, \underline{z}) = \sum_{s=k+1}^{\infty} (1-\mu)^{s-k} \underline{\tilde{f}}(s, \underline{z}) - \sum_{s=k}^{\infty} (1-\mu)^{s-(k-1)} \underline{\tilde{f}}(s, \underline{z})$$

$$= \mu \sum_{s=k+1}^{\infty} (1-\mu)^{s-k} \underline{\tilde{f}}(s, \underline{z}) - (1-\mu) \underline{\tilde{f}}(k, \underline{z})$$

which is (7.56). The idea is to use (7.56) to ultimately cancel out the third term in (7.54). This is useful because the second term in (7.56) is $O(\mu)$ while as we will show $\mu \underline{P}(k, \underline{z})$ itself is small. First, however, some other conditions and computations are needed. We suppose a uniform version of the averaging assumption (7.2A2) (i.e., (7.5A1)):

(7.6A5) With $\underline{\tilde{f}}(s, \underline{z}) = \underline{f}(s, \underline{z}) - \underline{f}_{av}(\underline{z})$,

$$\| \sum_{s=k+1}^{k+N} \underline{\tilde{f}}(s, \underline{z}) - \sum_{s=k+1}^{k+N} \underline{\tilde{f}}(s, \underline{z}') \| \le L_P \| \underline{z} - \underline{z}' \| \gamma(N) N, \quad \|\underline{z}\|, \|\underline{z}'\| \le h$$

uniformly in k. Here $\gamma(N)$ is a non-increasing convergence function, which tends to 0 as $N \to \infty$.

Using (7.6A5) it is shown below that

$$\mu \| \underline{P}(k, \underline{z}) - \underline{P}(k, \underline{z}') \| \le \xi(\mu) \| \underline{z} - \underline{z}' \|, \quad \|\underline{z}\|, \|\underline{z}'\| \le h \tag{7.57}$$

where $\xi(\mu) \to 0$ as $\mu \to 0$. Note that from (7.6A1), (7.6A2)

$$\underline{P}(k, \underline{0}) = \underline{0}$$

so that (7.57) also yields

$$\mu \| \underline{P}(k, \underline{z}) \| \le \xi(\mu) \| \underline{z} \|, \quad \|\underline{z}\| \le h. \tag{7.58}$$

This result shows that the second term in (7.56) is much smaller than the first. Now we can introduce the perturbed Lyapunov function

$$V(k, \underline{z}) = V(\underline{z}) + \mu V_1(k, \underline{z})$$

$$V_1(k, \underline{z}) = \underline{P}^T(k, \underline{z}) V_z(\underline{z}).$$

Then we have from (7.4a3), (7.58) that

$$\mu |V_1(k, \underline{z})| \le \alpha_4 \xi(\mu) \|\underline{z}\|^2.$$

Thus if we take μ small enough we have

$$0 < [\alpha_1 - \alpha_4 \xi(\mu)] \|\underline{z}\|^2 \le V(k, \underline{z}) \le [\alpha_2 + \alpha_4 \xi(\mu)] \|\underline{z}\|^2. \tag{7.59}$$

Consider then using $V(k, \underline{z})$ in a Lyapunov calculation on (7.37). We have

$$V(k, \underline{z}_k) - V(k-1, \underline{z}_{k-1})$$

$$= V(\underline{z}_k) - V(\underline{z}_{k-1}) + \mu (V_1(k, \underline{z}_k) - V_1(k-1, \underline{z}_{k-1}))$$

$$= A_k + B_k + C_k + D_k$$

where

$$A_k = V(\underline{z}_k) - V(\underline{z}_{k-1})$$

$$B_k = \mu\left(\underline{P}(k, \underline{z}_{k-1}) - \underline{P}(k-1, \underline{z}_{k-1})\right)^T V_z(\underline{z}_{k-1}) \tag{7.60a}$$

$$C_k = \mu(\underline{P}(k, \underline{z}_k) - \underline{P}(k, \underline{z}_{k-1}))^T V_z(\underline{z}_{k-1}) \tag{7.60b}$$

$$D_k = \mu \underline{P}^T(k, \underline{z}_k)(V_z(\underline{z}_k) - V_z(\underline{z}_{k-1})). \tag{7.60c}$$

Now from (7.54)

$$A_k \leq -\mu\alpha_3\|\underline{z}_{k-1}\|^2 + \mu\underline{\tilde{f}}_k^T V_z(\underline{z}_k^*) \tag{7.61}$$

while the perturbation approximation identity (7.56) yields

$$B_k = -\mu\underline{\tilde{f}}_k^T V_z(\underline{z}_{k-1}) + \mu^2\left(\underline{P}(k, \underline{z}_{k-1}) + \underline{\tilde{f}}_k\right)^T V_z(\underline{z}_{k-1}). \tag{7.62}$$

We see that the first term in B_k will nearly cancel the second term in A_k, which is what we required. Thus, using the regularity conditions and properties (7.56)–(7.58), we are able to show below that for some positive function $\psi(\mu) \to 0$ as $\mu \to 0$

$$V(k, \underline{z}_k) - V(k-1, \underline{z}_{k-1}) \leq -\mu(\alpha_3 - \psi(\mu))\|\underline{z}_{k-1}\|^2. \tag{7.63}$$

So if we take μ small enough we get

$$\begin{aligned} V(k, \underline{z}_k) - V(k-1, \underline{z}_{k-1}) &\leq -\frac{1}{3}\mu\alpha_3\|\underline{z}_{k-1}\|^2 \\ &\leq -\frac{1}{3}\mu\alpha_3(\alpha_2 + \alpha_4\xi(\mu))V(k-1, \underline{z}_{k-1}) \quad \text{(by (7.59))} \\ \Rightarrow \qquad V(k, \underline{z}_k) &\leq (1 - b\mu)V(k-1, \underline{z}_{k-1}) \end{aligned}$$

where

$$b = \frac{1}{3}\alpha_3(\alpha_2 + \alpha_4\xi(\mu)).$$

By our inductive assumption we can iterate this to get

$$V(k, \underline{z}_k) \leq (1 - b\mu)^k V(0, \underline{z}_0).$$

Now apply (7.59) to find

$$\begin{aligned} \|\underline{z}_k\| &\leq (1 - b\mu)^{k/2}\|\underline{z}_0\|[(\alpha_2 + \alpha_4\xi(\mu))/(\alpha_1 - \alpha_4\xi(\mu))]^{1/2} \\ &= (1 - b\mu)^{k/2} M\|\underline{z}_0\|, \quad \text{say.} \end{aligned} \tag{7.64}$$

So in (7.49), choosing

$$H = \frac{h}{M(1 + 4\mu L)}$$

ensures the inductive step is completed, since then

$$\|\underline{z}_k\| \leq M\|\underline{z}_0\| \leq \frac{h}{1 + 4\mu L}.$$

Also then (7.64) establishes local exponential stability. The following result has thus been established.

Theorem 7.10. Consider the primary system (7.37) and its averaged mate (7.39). Suppose (7.6A1)–(7.6A5) hold: then (7.37) is locally exponentially stable provided μ is sufficiently small.

Proof of (7.57). Using the definition of $\underline{p}(k, \underline{z})$ in (7.2A2)

$$\begin{aligned}
\underline{P}(k, \underline{z}) &= \sum_{s=k+1}^{\infty} (1-\mu)^s [\underline{p}(s, \underline{z}) - \underline{p}(s-1, \underline{z})](1-\mu)^{-k} \\
&= \sum_{s=k+1}^{\infty} \left\{ \begin{matrix} (1-\mu)^s \underline{p}(s, \underline{z}) - (1-\mu)^{s-1} \underline{p}(s-1, \underline{z}) \\ +(1-\mu)^{s-1} \mu \underline{p}(s-1, \underline{z}) \end{matrix} \right\} (1-\mu)^{-k} \\
&= -\underline{p}(k, \underline{z}) + \mu \sum_{s=k+1}^{\infty} (1-\mu)^{s-k-1} \underline{p}(s-1, \underline{z}) \\
&= \mu \sum_{s=k+1}^{\infty} (1-\mu)^{s-k-1} \left(\underline{p}(s-1, \underline{z}) - \underline{p}(k, \underline{z}) \right).
\end{aligned}$$

Thus

$$\underline{P}(k, \underline{z}) - \underline{P}(k, \underline{z}') = \mu \sum_{s=k+1}^{\infty} (1-\mu)^{s-k-1} \left[\underline{p}_k(s-1, \underline{z}) - \underline{p}_k(s-1, \underline{z}') \right]$$

where

$$\underline{p}_k(s-1, \underline{z}) = \underline{p}(s-1, \underline{z}) - \underline{p}(k, \underline{z}) = \sum_{r=k+1}^{s-1} \tilde{\underline{f}}(r, \underline{z}).$$

Now apply (7.6A5) to find

$$\begin{aligned}
\mu \|\underline{P}(k, \underline{z}) - \underline{P}(k, \underline{z}')\| &\leq \mu^2 \sum_{s=k+1}^{\infty} (1-\mu)^{s-k-1} (s-k-1) \gamma(s-k-1) L_p \|\underline{z} - \underline{z}'\| \\
&= \xi(\mu) \|\underline{z} - \underline{z}'\|
\end{aligned}$$

where

$$\xi(\mu) = \mu^2 L_p \sum_{r=1}^{\infty} (1-\mu)^r r \gamma(r).$$

We have only to show $\xi(\mu) \to 0$ as $\mu \to 0$. We use the Toeplitz lemma of Appendix D7. We have

$$\mu^2 r(1-\mu)^r \to 0 \quad \text{as } \mu \to 0, \quad r \text{ fixed}$$

$$\mu^2 \sum_{r=1}^{\infty} r(1-\mu)^r = 1 - \mu \le 1$$

$$\mu^2 \sum_{r=1}^{\infty} r(1-\mu)^r = 1 - \mu \to 1 \quad \text{as } \mu \to 0.$$

Thus, since $\gamma(k) \to 0$ as $k \to \infty$, we find that

$$\xi(\mu) \to 0 \quad \text{as } \mu \to 0.$$

So the proof of (7.57) is complete.

Proof of (7.63). We begin with A_k, B_k. From (7.61), (7.62)

$$A_k + B_k \le -\mu\alpha_3\|\underline{z}_{k-1}\|^2 + \mu^2[\underline{P}(k, \underline{z}_{k-1}) + \underline{\tilde{f}}_k]^T V_z(\underline{z}_{k-1})$$
$$+\mu\underline{\tilde{f}}_k^T (V_z(\underline{z}_k^*) - V_z(\underline{z}_{k-1})). \tag{7.65a}$$

Now apply (7.58), (7.51), (7.52), (7.4a3) to find

$$A_k + B_k \le -\mu\alpha_3\|\underline{z}_{k-1}\|^2 + \mu\alpha_4\xi(\mu)\|\underline{z}_{k-1}\|^2 + 2\mu^2 L\alpha_4\|\underline{z}_{k-1}\|^2$$
$$+2\mu L\|\underline{z}_{k-1}\|\|V_z(\underline{z}_k^*) - V_z(\underline{z}_{k-1})\|.$$

A Taylor series (Appendix D13) now gives

$$\|V_z(\underline{z}_k^*) - V_z(\underline{z}_{k-1})\| \le \|\underline{z}_k^* - \underline{z}_{k-1}\|\|V_{zz}(\hat{\underline{z}}_k)\|$$

where from (7.52) $\hat{\underline{z}}_k = \underline{z}_k^* + \lambda_{k-1}\mu\underline{f}_{av}(\underline{z}_{k-1})$, $0 < \lambda_{k-1} < 1$. So from (7.53)

$$\|\hat{\underline{z}}_k\| \le (1 + 4\mu L)\|\underline{z}_{k-1}\| \le h.$$

Thus, by (7.4a4), (7.51), (7.52)

$$\|V_z(\underline{z}_k^*) - V_z(\underline{z}_{k-1})\| \le 3\mu L\|\underline{z}_{k-1}\|(\alpha_5 + \alpha_6 h).$$

So by (7.51)

$$A_k + B_k \le -\mu\alpha_3\|\underline{z}_{k-1}\|^2 + \mu\xi(\mu)\alpha_4\|\underline{z}_{k-1}\|^2$$
$$+ 2\mu^2 L\alpha_4\|\underline{z}_{k-1}\|^2 + 6\mu^2 L^2(\alpha_5 + \alpha_6 h)\|\underline{z}_{k-1}\|^2. \tag{7.65b}$$

Turning to C_k and D_k we find from (7.57), (7.50), (7.51)

$$C_k \le \xi(\mu)\|\underline{z}_k - \underline{z}_{k-1}\|\|V_z(\underline{z}_{k-1})\|$$
$$\le \xi(\mu)\mu L\|\underline{z}_{k-1}\|^2\alpha_4, \quad \text{by (7.6A3), (7.51), (7.4a3)} \tag{7.66}$$
$$D_k \le \xi(\mu)\|\underline{z}_k\|\|\underline{z}_k - \underline{z}_{k-1}\|\|V_{zz}(\hat{\underline{z}}_k)\|$$

where $\hat{\underline{z}}_k$ is an intermediate point with

$$\|\hat{\underline{z}}_k - \underline{z}_{k-1}\| \le \|\underline{z}_k - \underline{z}_{k-1}\| \le \mu L\|\underline{z}_{k-1}\|$$
$$\Rightarrow \quad \|\hat{\underline{z}}_k\| \le \|\underline{z}_{k-1}\|(1 + \mu L) \le h.$$

Thus by (7.4a4), (7.6A1), (7.6A3)

$$D_k \le \xi(\mu)\|\underline{z}_{k-1}\|^2 \mu L(1+\mu L)(\alpha_5 + \alpha_6 h). \tag{7.67}$$

Collecting these bounds (7.65)–(7.67) together yields (7.63).

So far in this section we have developed the perturbed Lyapunov function technique for a system of the form (7.5). For a more general system of the form (7.1) the results are much the same.

Theorem 7.11. Consider the primary system (7.1) with its associated averaged system (7.10) where now $\underline{f}_{av}(\underline{z})$ is given by (7.23). Suppose (7.6a1)–(7.6a5), (7.2C5) hold. Then (7.1) is locally ES for small μ.

(7.6a1) $\underline{0}$ is an equilibrium point of (7.1).

(7.6a2) $\underline{0}$ is a locally ES equilibrium point of (7.10).

(7.6a3) $\underline{f}(k, \underline{z}, 0)$, $\underline{f}_{av}(\underline{z})$ obey Lipschitz conditions

$$\|\underline{f}(k, \underline{z}, 0) - \underline{f}(k, \underline{z}'0)\| \le L\|\underline{z} - \underline{z}'\|$$
$$\|\underline{f}_{av}(\underline{z}) - \underline{f}_{av}(\underline{z}')\| \le L\|\underline{z} - \underline{z}'\|$$

for $\|\underline{z}\|, \|\underline{z}'\| \le h$.

(7.6a4)

$$\|\underline{f}_{av,zi}(\underline{z})\| \le d, \quad \|\underline{z}\| \le h \text{ and } 1 \le i \le p$$

where $\underline{f}_{av,zi}(\underline{z}) = d^2\underline{f}_{av}(\underline{z})/d\underline{z}^T dz_i$.

(7.6a5) With $\tilde{\underline{f}}(s, \underline{z}) = \underline{f}(s, \underline{z}, 0) - \underline{f}_{av}(\underline{z})$

$$\|\sum_{s=k+1}^{k+N} \tilde{\underline{f}}(s, \underline{z}) - \sum_{s=k+1}^{k+N} \tilde{\underline{f}}(s, \underline{z}')\| \le L_p\|\underline{z} - \underline{z}'\| N\gamma(N), \quad \|\underline{z}\|, \|\underline{z}'\| \le h$$

uniformly in k. Here $\gamma(N)$ is a non-increasing convergence function $\to 0$ as $N \to \infty$.

Proof. Much the same as the proof of Theorem 7.10 with the following changes. Now we have

$$\tilde{\underline{f}}_k = \underline{f}(k, \underline{z}_{k-1}, \mu) - \underline{f}_{av}(\underline{z}_{k-1})$$
$$= \tilde{\underline{f}}_k^a + \tilde{\underline{f}}_k^b$$
$$\tilde{\underline{f}}_k^a = \underline{f}(k, \underline{z}_{k-1}, \mu) - \underline{f}(k, \underline{z}_{k-1}, 0)$$
$$\tilde{\underline{f}}_k^b = \underline{f}(k, \underline{z}_{k-1}, 0) - \underline{f}_{av}(\underline{z}_{k-1})$$
$$\tilde{\underline{f}}(k, \underline{z}) = \underline{f}(k, \underline{z}, 0) - \underline{f}_{av}(\underline{z}).$$

In (7.52) the $\underline{\tilde{f}}_k$ term is now $\underline{\tilde{f}}_k^b$. Thus the $\underline{\tilde{f}}_k$ term in the third term of (7.65a) is $\underline{\tilde{f}}_k^b$. Also (7.65a) then has another term $\mu \underline{\tilde{f}}_k^{aT} V_z(\underline{z}_k^*)$. But this term has norm bounded by $\mu^2 L\alpha_4 \|\underline{z}_{k-1}\| \|\underline{z}_k^*\|$ and so does not cause any problems. The rest of the argument proceeds as before with $\underline{\tilde{f}}_k$ everywhere replaced by $\underline{\tilde{f}}_k^b$.

7.7 INFINITE TIME AVERAGING—STABILITY WITH TIME INVARIANT PARAMETERS

The result of the previous section will now be applied to some examples.

Example 7.6. Convergence of LMS

Consider the LMS error system (7.2)

$$\delta \underline{\tilde{w}}_k = -\mu \underline{x}_k \underline{x}_k^T \underline{\tilde{w}}_{k-1} \tag{7.68}$$

and its averaged mate (7.11)

$$\delta \underline{\bar{w}}_k = -\mu \underline{R}_x \underline{\bar{w}}_{k-1}. \tag{7.69}$$

First check that $\underline{0}$ is an equilibrium point of (7.68). We have

$$\begin{aligned} & \underline{f}(k, \underline{\tilde{w}}) = -\mu \underline{x}_k \underline{x}_k^T \underline{\tilde{w}} \\ \Rightarrow \quad & \underline{f}(k, \underline{0}) = \underline{0}. \end{aligned}$$

Next check that (7.69) is locally ES. We have

$$\begin{aligned} & \underline{f}_{av}(\underline{\bar{w}}) = -\underline{R}_x \underline{\bar{w}} \\ \Rightarrow \quad & \underline{f}_{av}(\underline{0}) = \underline{0} \end{aligned}$$

so that $\underline{0}$ is an equilibrium point. The ES of (7.69) can be demonstrated in two ways:

(i) We already analyzed the convergence of (7.69) in Section 3.2 (see (3.12)) and found ES holds if (3.17) holds, i.e.,

(7.7A1) $\lambda_{min} > 0$, $\mu \lambda_{max} < 2$.

(ii) Construct a Lyapunov function for (7.69). Let

$$V(\underline{\bar{w}}) = \|\underline{\bar{w}}\|^2.$$

Then

$$\begin{aligned} V_k = V(\underline{\bar{w}}_k) &= \underline{\bar{w}}_k^T \underline{\bar{w}}_k \\ &= \underline{\bar{w}}_{k-1}^T (I - \mu \underline{R}_x)^2 \underline{\bar{w}}_{k-1} \\ &= V_{k-1} - 2\mu \underline{\bar{w}}_{k-1}^T \underline{R}_x \underline{\bar{w}}_{k-1} + \mu^2 \underline{\bar{w}}_{k-1}^T \underline{R}_x^2 \underline{\bar{w}}_{k-1} \\ &\leq V_{k-1} - \mu(2 - \lambda_{max}\mu) \underline{\bar{w}}_{k-1}^T \underline{R}_x \underline{\bar{w}}_{k-1} \\ &\leq V_{k-1} - \mu \rho \lambda_{min} V_{k-1}, \quad \rho = 2 - \mu \lambda_{max} \\ &= (1 - \mu \rho \lambda_{min})^{k-k_0} V_{k_0}, \quad k \geq k_0. \end{aligned}$$

So ES follows provided (7.7A1) holds (because $0 < \mu\rho\lambda_{min} \leq \mu\lambda_{min}(2 - \mu\lambda_{min}) < 1$).

Next observe that $\underline{f}(k, \underline{z})$, $\underline{f}_{av}(\underline{z})$ obey Lipschitz conditions if (7.7A2) holds.

(7.7A2) $\sup_k \|\underline{x}_k\| \le B_x < \infty$.

Finally consider (7.6A4) which reads

$$\| \sum_{s=k+1}^{k+N} (\underline{x}_s \underline{x}_s^T - \underline{R}_x)(\bar{\underline{w}} - \bar{\underline{w}}') \| \le L_p \|\bar{\underline{w}} - \bar{\underline{w}}'\| \gamma(N) N$$

and so follows from

(7.7A3) $\| \sum_{s=k+1}^{k+N} (\underline{x}_s \underline{x}_s^T - \underline{R}_x) \| \le L_p \gamma(N) N$.

This condition may be compared with the PE condition (6.1A4) which clearly is somewhat weaker. Certainly (7.7A3) is satisfied if $\underline{x}_s$ is composed of periodic or almost periodic signals.

In any case the conclusion is that for μ small enough the LMS algorithm (7.68) is ES if (7.7A1)–(7.7A3) hold. This may be compared to Theorem 6.2. That theorem is a better result, but the present theorem is nearly as good, and the proof technique generalizes whereas that used for Theorem 6.2 does not.

Example 7.7. Convergence of Adaptive Time Delay Estimation

Consider the idealized time delay estimation algorithm of Section 4.5 (4.48)

$$\delta\tilde{\theta}_k = -\mu s'_{k-\theta_o-\tilde{\theta}_{k-1}} (s_{k-\theta_o} - s_{k-\theta_o-\tilde{\theta}_{k-1}}). \tag{7.70}$$

Clearly $\tilde{\theta} = 0$ is an equilibrium point. The averaged system is (see Example 7.2)

$$\delta\bar{\theta}_k = \mu R'_s(\bar{\theta}_{k-1}). \tag{7.71}$$

In Section 3.3 we already saw conditions under which (7.71) (which is the same as (3.56)) is locally ES so that (7.6A2) holds. Now we look at (7.6A3). We have

$$\tilde{f}(k, \tilde{\theta}) - \tilde{f}(k, \tilde{\theta}') = f(k, \tilde{\theta}) - f(k, \tilde{\theta}') = \int_{\tilde{\theta}'}^{\tilde{\theta}} d/d\theta (f(t, \theta)) d\theta$$

with

$$d/d\theta(f(t, \theta)) = s''_{k-\theta_o-\theta}(s_{k-\theta_o} - s_{k-\theta_o-\theta}) - (s'_{k-\theta_o-\theta})^2. \tag{7.72}$$

Let us now suppose

(7.7C1) $|s(t)| \le a$, $|s'(t)| \le a'$, $|s''(t)| \le a''$.

Then

$$|d/d\theta(f(t, \theta))| \le 2aa'' + (a')^2$$

from which it follows that

$$\begin{aligned} |f(k, \tilde{\theta}) - f(k, \tilde{\theta}')| &\le 2aa''|\tilde{\theta} - \tilde{\theta}'| + (a')^2 |\tilde{\theta} - \tilde{\theta}'| \\ &= L_f |\tilde{\theta} - \tilde{\theta}'|, \quad \text{say.} \end{aligned}$$

For the Lipschitz condition on the averaged system we need

(7.7C2) $|R_s''(\tilde{\theta})| \le R'' < \infty$.

Now consider (7.6A5). We have

$$\left|\sum_{t=k+1}^{k+N} (f(t,\tilde{\theta}) - f(t,\tilde{\theta}'))\right| = \int_{\tilde{\theta}'}^{\tilde{\theta}} \sum_{t=k+1}^{k+N} d/d\theta(f(t,\theta))d\theta.$$

In view of (7.72) let us suppose

(7.7C3)

$$\begin{cases} \text{(i)} & \left|\sum_{t=k+1}^{k+N} (s_{t+u}'' s_t - R_s''(u))du\right| \le L\gamma(N)N \\ \text{(ii)} & \left|\sum_{t=k+1}^{k+N} \left((s_{t+u}')^2 - R_s''(0)\right)\right| \le L\gamma(N)N. \end{cases}$$

Then we find

$$\left|\sum_{t=k+1}^{k+N} \left(f(t,\tilde{\theta}) - f(t,\tilde{\theta}')\right)\right| \le 3LN\gamma(N)|\tilde{\theta} - \tilde{\theta}'|$$

as required. So we conclude that, for μ small enough, the ideal algorithm (7.70) is ES if (7.7C1)–(7.7C3) hold.

Some comments are in order on this example. Firstly we know of no other way to prove local ES for this example. Secondly the averaged system has equilibrium points wherever the acf $R_s(\tilde{\theta})$ has a local maximum; but the original system has equilibrium points only where $R_s(\tilde{\theta})$ has a global maximum.

7.8 INFINITE TIME AVERAGING—TIME VARIANT PARAMETERS

We have already introduced the averaged system (7.33) associated with the forced time varying system (7.26). We gained some insight into the input-output stability (IOS) of (7.33) by means of an example. Now we develop a general analysis.

The idea is simply to develop IOS properties of (7.26) from stability properties of the homogeneous or unforced primary system (7.30). Suppose we somehow establish that the primary homogeneous system (7.30) is ES; then the associated Lyapunov function is easily used to deliver the required IOS results for (7.26).

To begin, then, we need the following "slow time" versions of earlier results on Lyapunov functions.

Theorem 7.12. The results of Lyapunov Theorem 7.5 hold for the system (7.30).

Proof. The proof is the same with only notational changes. The Lipschitz condition is

$$\|\underline{f}(k, \underline{z}, \mu k, \mu) - \underline{f}(k, \underline{z}', \mu k, \mu)\| \le L\|\underline{z} - \underline{z}'\|.$$

The Lyapunov function can be denoted $V(k, \underline{z}, \mu k)$.

Theorem 7.13. The results of Lyapunov Theorem 7.5 hold for the averaged system (7.29).

Proof. The proof is the same with only notational changes. The Lipschitz condition is

$$\|\underline{f}_{av}(\underline{z}, \tau) - \underline{f}_{av}(\underline{z}', \tau)\| \leq L\|\underline{z} - \underline{z}'\|.$$

The Lyapunov function can be denoted $V(\underline{z}, \mu k)$.

Theorem 7.14. The results of perturbed Lyapunov Theorem 7.11 hold for the primary system (7.30) and the associated averaged system (7.29) when (7.8A1)–(7.8A5) hold.

(7.8A1) $\underline{0}$ is an equilibrium point of (7.30).

(7.8A2) $\underline{0}$ is a locally ES equilibrium point of (7.29) of stability radius h.

(7.8A3) $\underline{f}(k, \underline{z}, \tau, \mu)$, $\underline{f}_{av}(\underline{z}, \tau)$ obey Lipschitz conditions

$$\|\underline{f}(k, \underline{z}, \tau, \mu) - \underline{f}(k, \underline{z}', \tau, \mu)\| \leq L\|\underline{z} - \underline{z}'\|$$
$$\|\underline{f}_{av}(\underline{z}, \tau) - \underline{f}_{av}(\underline{z}', \tau)\| \leq L\|\underline{z} - \underline{z}'\|$$

for $\|\underline{z}\|, \|\underline{z}'\| \leq h$.

(7.8A4)

$$\|\underline{f}_{av,z_i}(\underline{z}, \tau)\| \leq d, \quad \|\underline{z}\| \leq h,\ 1 \leq i \leq p$$
$$\underline{f}_{av,z_i}(\underline{z}) = d^2\underline{f}_{av}(\underline{z})/d\underline{z}^T dz_i$$

(7.8A5) With $\underline{\tilde{f}}(s, \underline{z}, \tau) = \underline{f}(s, \underline{z}, \tau, 0) - \underline{f}_{av}(\underline{z}, \tau)$,

$$\|\sum_{s=k+1}^{k+N} \underline{\tilde{f}}(s, \underline{z}, \tau) - \sum_{s=k+1}^{k+N} \underline{\tilde{f}}(s, \underline{z}', \tau)\| \leq L_{pa}\|\underline{z} - \underline{z}'\| N\gamma(N)$$
$$\|\sum_{s=k+1}^{k+N} \underline{\tilde{f}}(s, \underline{z}, \tau) - \sum_{s=k+1}^{k+N} \underline{\tilde{f}}(s, \underline{z}, \tau')\| \leq L_{pb}\|\underline{z}\||\tau - \tau'| N\gamma(N)$$

for $\|\underline{z}\|, \|\underline{z}'\| \leq h$, and where $\gamma(N)$ is a non-increasing convergence function $\to 0$ as $N \to \infty$.

(7.8A6) $\underline{f}(k, \underline{z}, \tau, \mu)$ obeys a Lipschitz condition in μ:

$$\|\underline{f}(k, \underline{z}, \tau, \mu) - \underline{f}(k, \underline{z}, \tau, \mu')\| \leq L|\mu - \mu|\|\underline{z}\|, \quad \|\underline{z}\| \leq h.$$

Proof. Notice the second Lipschitz condition in (7.8A5). The proof is much the same as that of Theorems 7.10, 7.11 with mainly notational changes. We discuss the differences. The discounted perturbation is defined as in (7.55) but with argument $\underline{z}$, τ

instead of $\underline{z}$. The Lyapunov function for (7.30) is $V(\underline{z}, \mu k)$ and has properties (7.4A1)–(7.4A4). In place of (7.57) we have, via (7.8A5)

$$\mu\|\underline{P}(k, \underline{z}, \tau) - \underline{P}(k, \underline{z}', \tau)\| \leq \xi(\mu)\|\underline{z} - \underline{z}'\| \tag{7.73a}$$

$$\mu\|\underline{P}(k, \underline{z}, \tau) - \underline{P}(k, \underline{z}, \tau')\| \leq \xi(\mu)\|\underline{z}\||\tau - \tau'|. \tag{7.73b}$$

In (7.60b) we find C_k is now

$$\begin{aligned} C_k &= \mu\left(\underline{P}(k, \underline{z}_k, \mu k) - \underline{P}(k, \underline{z}_{k-1}, \mu k - \mu)\right)^T V_z(\underline{z}_{k-1}, \mu k - \mu) \\ &= \mu \left\{ \begin{array}{c} \underline{P}(k, \underline{z}_k, \mu k) - \underline{P}(k, \underline{z}_{k-1}, \mu k) \\ +\underline{P}(k, \underline{z}_{k-1}, \mu k) - \underline{P}(k, \underline{z}_{k-1}, \mu k - \mu) \end{array} \right\}^T V_z(\underline{z}_{k-1}, \mu k - \mu) \\ &\leq \xi(\mu)(\|\underline{z}_k - \underline{z}_{k-1}\| + \mu\|\underline{z}_{k-1}\|)\|V_z(\underline{z}_{k-1}, \mu k - \mu)\| \quad \text{(by (7.73))} \end{aligned}$$

and this leads to a bound of the same size as before in (7.66). Other changes to be made are the same as the changes made in the proof of Theorem 7.11.

Now we turn to analyze the IOS of (7.26). Let us suppose that the unforced averaged system (7.29) is ES to $\underline{0}$. Then Theorem 7.14 tells us the unforced system (7.30) is ES to $\underline{0}$ and has an associated Lyapunov function $V(k, \underline{z}, \mu k)$.

We apply this Lyapunov function to the forced system (7.26). We have

$$V(k, \underline{z}_k, \mu k) = V(k, \underline{z}_{k-1} + \mu \underline{f}_{k-1} + \mu \underline{g}_{k-1} - \mu \underline{d}_{ok}, \mu k) \tag{7.74}$$

where

$$\underline{f}_{k-1} = \underline{f}(k, \underline{z}_{k-1}, \mu k, 0) \tag{7.75a}$$

$$\underline{g}_{k-1} = \underline{f}(k, \underline{z}_{k-1}, \mu k, \mu) - \underline{f}(k, \underline{z}_{k-1}, \mu k, 0) \tag{7.75b}$$

$$\underline{d}_{ok} = \delta \underline{z}_o(\mu k)/\mu. \tag{7.75c}$$

Note that from (7.8A6)

$$\|\underline{g}_{k-1}\| \leq L\mu\|\underline{z}_{k-1}\|. \tag{7.76a}$$

We further assume

(7.8A7) $\dot{\underline{z}}_o(\tau) = d\underline{z}_o(\tau)/d\tau$ obeys

$$\|\dot{\underline{z}}_o(\tau)\| \leq \dot{z}_o < \infty.$$

Then by the mean value theorem

$$\|\underline{d}_{ok}\| \leq \dot{z}_o. \tag{7.76b}$$

Also note that

$$\lim_{\mu \to 0} \|\underline{d}_{ok} - \dot{z}_o(\mu k)\| = 0. \tag{7.77}$$

Now applying a Taylor series (Appendix D13) to (7.74) gives

$$V(k, \underline{z}_k, k\mu) = V(k, \underline{z}_{k-1} + \mu \underline{f}_{k-1}, k\mu - \mu) + \mu(\underline{g}_{k-1} - \underline{d}_{ok})^T V_z(\underline{z}_k^*) \tag{7.78}$$

where (with $0 < \lambda_k < 1$)

$$\underline{z}_k^* = \underline{z}_{k-1} + \mu \underline{f}_{k-1} + \lambda_{k-1}\mu(\underline{g}_{k-1} - \underline{d}_{ok}). \tag{7.79}$$

However from (7.4A3), (7.75)

$$\begin{aligned}
&\mu|(\underline{g}_{k-1} - \underline{d}_{ok})^T V_z(\underline{z}_k^*)| \\
&\le \mu(\|\underline{g}_{k-1}\| + \|\underline{d}_{ok}\|)\alpha_4\|\underline{z}_k^*\| \\
&\le \mu\alpha_4(\|\underline{g}_{k-1}\| + \|\underline{d}_{ok}\|)(\|\underline{z}_{k-1}\| + \mu\|\underline{f}_{k-1}\| + \mu\|\underline{g}_{k-1}\| + \mu\|\underline{d}_{ok}\|).
\end{aligned}$$

But in view of (7.76), (7.8A1), and (7.8A3), this is further bounded as follows:

$$\begin{aligned}
&\mu|(\underline{g}_{k-1} - \underline{d}_{ok})^T V_z(\underline{z}_k^*)| \\
&\le \mu\alpha_4(\mu L\|\underline{z}_{k-1}\| + \|\underline{d}_{ok}\|)((1 + 2\mu L)\|\underline{z}_{k-1}\| + \mu\|\underline{d}_{ok}\|) \\
&\le \mu^2\alpha_4 L\|\underline{z}_{k-1}\|^2 + 2\mu a\|\underline{d}_{ok}\|\|\underline{z}_{k-1}\| + \mu^2\alpha_4\|\underline{d}_{ok}\|^2
\end{aligned}$$

for some constant a. Continuing, we have

$$\mu|(\underline{g}_{k-1} - \underline{d}_{ok})^T V_z(\underline{z}_k^*)| \le \frac{1}{2}\alpha_3\mu\|\underline{z}_{k-1}\|^2 + \mu b\|\underline{d}_{ok}\|^2$$

for some constant b (where $b \to$ positive constant as $\mu \to 0$).

Using this in (7.78) together with (7.4A2) gives

$$V(k, \underline{z}_k, \mu k) \le V(k-1, \underline{z}_{k-1}, \mu k - \mu) - \frac{1}{2}\alpha_3\mu\|\underline{z}_{k-1}\|^2 + \mu b\|\underline{d}_{ok}\|^2. \tag{7.80}$$

This is our fundamental bounding relation and we use it to develop two types of IOS result: (i) an amplitude bound on $\|\underline{z}_k\|$ and (ii) a power bound on $\|\underline{z}_k\|$.

From (7.4A1) we have (writing $V_k = V(k, \underline{z}_k, \mu k)$)

$$\begin{aligned}
V_k &\le (1 - \frac{\alpha_3}{2\alpha_2}\mu)V_{k-1} + \mu b\|\underline{d}_{ok}\|^2 \\
&\le \lambda V_{k-1} + \mu b\dot{z}_o^2, \quad \text{by (7.8A7)} \\
\lambda &= 1 - \alpha_3\mu/2\alpha_2.
\end{aligned}$$

Iterating gives

$$\begin{aligned}
V_k &\le \lambda^k V_0 + \sum_{s=1}^{k}\lambda^{k-s}\mu b\dot{z}_o^2 \\
&\le V_0 + \mu b\dot{z}_o^2/(1-\lambda) \\
&\le V_0 + 2b\dot{z}_o^2\alpha_2/\alpha_3.
\end{aligned} \tag{7.81}$$

From (7.4A1) we thus get the amplitude bound

$$\|\underline{z}_k\| \le (V_0/\alpha_1 + 2b\dot{z}_o^2\alpha_2/\alpha_3\alpha_1)^{\frac{1}{2}}. \tag{7.82}$$

Returning to (7.80), sum to find

$$\frac{1}{2}\mu\alpha_3 \sum_{s=k+1}^{k+N} \|\underline{z}_{s-1}\|^2 \le V_k + \mu b \sum_{s=k+1}^{k+N} \|\underline{d}_{os}\|^2$$

and in view of (7.82) we deduce

$$\lim_{N\to\infty} \frac{1}{N} \sum_{s=k+1}^{k+N} \|\underline{z}_{s-1}\|^2 \le \frac{2b}{\alpha_3} \lim_{N\to\infty} \frac{1}{N} \sum_{s=k+1}^{k+N} \|\underline{d}_{os}\|^2 \tag{7.83}$$

uniformly in k. This is the required power bound.

The results (7.82), (7.83) are not as precise as the result obtained in, say, Example 7.3 but do give valuable IOS results. We summarize these results as follows:

Theorem 7.15. Consider the forced system (7.26) and suppose (7.8A1)–(7.8A7) hold, then the amplitude bound (7.82) and the power bound (7.83) are valid.

7.9 THE "ODE" METHOD

7.9.1 Heuristics

Historically averaging methods were introduced into the adaptive area through the ordinary differential equation (ode) method (see Notes at the end of this chapter).

By stretching the time scale we can associate the averaged system (and hence the primary system) with an ode.

Let us first remind ourselves that in (7.1), (7.5), and (7.10), $\underline{z}_k$, $\bar{\underline{z}}_k$ are actually functions of μ (or sequences indexed by μ) and will be denoted in this section as $\underline{z}_k^{\mu}$, $\bar{\underline{z}}_k^{\mu}$. Now we introduce the stretched time scale functions

$$\bar{\underline{z}}_{\mu}(\tau) = \bar{\underline{z}}^{\mu}_{[\tau\mu^{-1}]}$$

$$\underline{z}_{\mu}(\tau) = \underline{z}^{\mu}_{[\tau\mu^{-1}]}$$

where $[x] =$ largest integer $\le x$. Then we can rewrite (7.10) as

$$[\bar{\underline{z}}_{\mu}(\tau+\mu) - \bar{\underline{z}}_{\mu}(\tau)]/\mu = \underline{f}_{av}(\bar{\underline{z}}_{\mu}(\tau)).$$

Proceeding heuristically we see that the left hand side looks like a derivative as $\mu \to 0$, and we are led to conjecture that if

$$\bar{\underline{z}}_{\mu}(0) = \bar{\underline{z}}_0^{\mu} \to \underline{\xi} \quad \text{as } \mu \to 0 \tag{7.84a}$$

then

$$\sup_{0\le\tau\le T} \|\bar{\underline{z}}_{\mu}(\tau) - \bar{\underline{z}}(\tau)\| \to 0 \quad \text{as } \mu \to 0 \tag{7.84b}$$

where $\bar{\underline{z}}(\tau)$ is the solution to the ode

$$d\bar{\underline{z}}(\tau)/d\tau = \underline{f}_{av}(\bar{\underline{z}}(\tau)), \quad \bar{\underline{z}}(0) = \underline{\xi}. \tag{7.85}$$

Also we can rewrite the conclusion of Theorem 7.1, 7.2 to read

$$\sup_{0\le\tau\le T} \|\underline{z}_\mu(\tau) - \bar{\underline{z}}_\mu(\tau)\| \to 0, \quad \text{as } \mu \to 0. \tag{7.86}$$

So putting (7.84), (7.86) together gives the conjecture: If $\underline{z}_\mu(0) = \underline{z}_0^\mu \to \underline{\xi}$ as $\mu \to 0$ then

$$\sup_{0\le\tau\le T} \|\underline{z}_\mu(\tau) - \bar{\underline{z}}(\tau)\| \to 0, \quad \text{as } \mu \to 0. \tag{7.87}$$

We may regard the ode (7.85) as a continuous time approximation of the discrete averaged system.

With a uniformity assumption (7.5A1) we will get a Hovering theorem

$$\sup_{\tau\ge 0} \|\underline{z}_\mu(\tau) - \bar{\underline{z}}(\tau)\| \to 0 \text{ as } \mu \to 0.$$

Again this is not a stability result. One expects to be able to develop a perturbed Lyapunov stability theory based on the ode but we do not pursue that here.

7.9.2 Rigor

Now we show how to make the ode idea rigorous. To begin with it is necessary to make $\bar{\underline{z}}_\mu(\tau)$ continuous in τ. So replace its jumps by straight line segments giving the piecewise continuous function

$$\hat{\underline{z}}_\mu(\tau) = \bar{\underline{z}}_{[\tau\mu^{-1}]} + (\tau\mu^{-1} - [\tau\mu^{-1}])(\bar{\underline{z}}_{[\tau\mu^{-1}+1]} - \bar{\underline{z}}_{[\tau\mu^{-1}]}). \tag{7.88}$$

Notice that because of the Lipschitz conditions (7.2A3), (7.2C3) on $\underline{f}_{av}(\bullet)$, the norm of the correction term in (7.88) is bounded as follows (using (7.2A4), (7.2C4)):

$$\|\text{correction term}\| \le 1 \times \mu\|\underline{f}_{av}(\bar{\underline{z}}_{[\tau\mu^{-1}]})\| \le \mu L_f h \to 0, \quad \text{as } \mu \to 0. \tag{7.89}$$

Now fix τ; then $\hat{\underline{z}}_\mu(\tau)$ is a sequence (indexed by μ) of bounded numbers and so by the Bolzano-Weierstrass theorem (Appendix D4) has at least one limit point. As τ varies we could hope to string those limit points together into a limit function which we would like to be continuous. To ensure this we need firstly uniform boundedness. Fix $\mu_0 > 0$; then we need

$$\sup_{0\le\tau\le T} \|\hat{\underline{z}}_\mu(\tau)\| \le h < \infty, \quad 0 \le \mu \le \mu_0. \tag{7.90}$$

This is just (7.2A4), (7.2C4) in another guise. But to string the limit points together into a continuous function we need something more, namely, equi-continuity. This brings us to Arzela's theorem (Appendix D, Section D11) which says that a sequence of uniformly bounded continuous functions on a finite interval has at least one limit function (which must be continuous) if the sequence is equi-continuous. That is

Given $\epsilon > 0$ there is $\delta(\epsilon) > 0$ so that

$$\sup_{|\tau_2-\tau_1|\le\delta(\epsilon)} \|\hat{\underline{z}}_\mu(\tau_2) - \hat{\underline{z}}_\mu(\tau_1)\| \le \epsilon, \quad 0 \le \mu \le \mu_0 \tag{7.91a}$$

and $\delta(\epsilon) \to 0$ as $\epsilon \to 0$.

Equivalently

Given $\delta > 0$ there is $\epsilon(\delta) > 0$ so that

$$\sup_{|\tau_2-\tau_1|\le\delta} \|\hat{\underline{z}}_\mu(\tau_2) - \hat{\underline{z}}_\mu(\tau_1)\| \le \epsilon(\delta), \quad 0 \le \mu \le \mu_0 \tag{7.91b}$$

and $\epsilon(\delta) \to 0$ as $\delta \to 0$.

We will show (7.91) below. So given (7.90) and (7.91) we can say there is a subsequence $\{\hat{\underline{z}}_{\mu_t}(\tau)\}$ of functions converging uniformly to a limit function, say $\underline{z}_e(\tau)$. From this we deduce via (7.90), (7.91) that

$$\sup_{0\le\tau\le T} \|\underline{z}_e(\tau)\| \le h < \infty \tag{7.92a}$$

$$\sup_{|\tau_2-\tau_1|\le\delta} \|\underline{z}_e(\tau_2) - \underline{z}_e(\tau_1)\| \le \epsilon(\delta). \tag{7.92b}$$

If we suppose (7.84a), then we have $\underline{z}_e(0) = \underline{\xi}$. If we can then show that the only possible limit function is the one satisfying (7.85), then we have established that the whole sequence converges to $\bar{\underline{z}}(\tau)$ (see Appendix D), i.e.,

$$\sup_{0\le\tau\le T} \|\hat{\underline{z}}_\mu(\tau) - \bar{\underline{z}}(\tau)\| \to 0, \quad \text{as } \mu \to 0. \tag{7.93}$$

In view of (7.89) this yields (7.84), and so via (7.86) we get (7.87).

To show that a limit function $\underline{z}_e(\tau)$ must obey (7.85), we proceed in two steps. We show that any limit function $\underline{z}_e(\tau)$ obeys

$$\underline{z}_e(\tau + \delta) - \underline{z}_e(\tau) = \int_\tau^{\tau+\delta} \underline{f}_{av}(\underline{z}_e(\sigma))d\sigma. \tag{7.94}$$

Note that the Lipschitz condition on $\underline{f}_{av}(\bullet)$ and the continuity of $\underline{z}_e(\tau)$ ensure that $\underline{f}_{av}(\underline{z}_e(\sigma))$ is continuous in σ and hence integrable.

Secondly it follows from (7.94), (7.2A3), (7.2C3), and (7.92) that $\underline{z}_e(\tau)$ is absolutely continuous (see Appendix D11) and so differentiable and in fact is the integral of its derivative (see Appendix D12). In particular, differentiating (7.94) we see that $\underline{z}_e(\tau)$ obeys (7.85). Now the Lipschitz condition also ensures the solution to (7.85) is unique (why?) and so there is only one possible limit function, namely, $\bar{\underline{z}}(\tau)$. Thus the whole sequence converges to $\bar{\underline{z}}(\tau)$ (why?).

We can summarize all this as follows.

Theorem 7.16. For the primary system (7.5) and the ode (7.85) with identical initial conditions, suppose (7.2A1)–(7.2A4) or (7.2C1)–(7.2C5) hold. Then for any fixed $T < \infty$, (7.87) holds.

Proof. We proceed in several stages.

Proof of (7.94). Let $\{\hat{\underline{z}}_{\mu_t}(\tau)\}$ be a convergent subsequence (t is a sequence of integers $\to \infty$ so that $\mu_t \to 0$) with limit function $\underline{z}_e(\tau)$, i.e.,

$$\sup_{0\le\tau\le T} \|\hat{\underline{z}}_{\mu_t}(\tau) - \underline{z}_e(\tau)\| \to 0, \quad \text{as } \mu_t \to 0.$$

Then in view of (7.88) and (7.89)

$$\sup_{0\le\tau\le T} \|\bar{\underline{z}}^{\mu_t}_{[\tau\mu_t^{-1}]} - \underline{z}_e(\tau)\| \to 0, \quad \text{as } \mu_t \to 0. \tag{7.95}$$

We now consider (7.94) and by adding and subtracting various terms split it into four pieces as follows:

$$\underline{z}_e(\tau+\delta) - \underline{z}_e(\tau) - \int_\tau^{\tau+\delta} \underline{f}_{av}(\underline{z}_e(\sigma))d\sigma = a_t + b_t + c_t + d_t$$

where, with $l_t = [(\tau+\delta)\mu_t^{-1}]$, $k_t = [\tau\mu_t^{-1}]$

$$a_t = \underline{z}_e(\tau+\delta) - \bar{\underline{z}}^{\mu_t}_{l_t} - (\underline{z}_e(\tau) - \bar{\underline{z}}^{\mu_t}_{k_t})$$

$$b_t = \bar{\underline{z}}^{\mu_t}_{l_t} - \bar{\underline{z}}^{\mu_t}_{k_t} - \mu_t \sum_{s=k_t+1}^{l_t+1} \underline{f}_{av}(\bar{\underline{z}}^{\mu_t}_{s-1})$$

$$c_t = \int_{\mu_t k_t}^{\mu_t l_t} \underline{f}_{av}(\underline{z}_e(\sigma))d\sigma - \mu_t \sum_{s=k_t+1}^{l_t+1} \underline{f}_{av}(\bar{\underline{z}}^{\mu_t}_{s-1})$$

$$d_t = -\int_{\mu_t l_t}^{\tau+\delta} \underline{f}_{av}(\underline{z}_e(\sigma))d\sigma + \int_{\mu_t k_t}^{\tau} \underline{f}_{av}(\underline{z}_e(\sigma))d\sigma.$$

Now by (7.95)

$$a_t \to 0 \quad \text{as } \mu_t \to 0, \quad \text{uniformly in } \tau.$$

Next

$$b_t = 0$$

while via (7.2A3), (7.2C3), and (7.92a)

$$\begin{aligned}\|d_t\| &\le (\tau+\delta-\mu_t l_t)L_f h + (\tau - \mu_t k_t)L_f h \\ &\le \mu_t L_f h\left([(\tau+\delta)\mu_t^{-1} - l_t] + [\tau\mu_t^{-1} - k_t]\right) \\ &\le 2\mu_t L_f h \to 0, \quad \text{as } \mu_t \to 0, \quad \text{uniformly in } \tau.\end{aligned}$$

So we are left with c_t. We write

$$c_t = c_{t1} + c_{t2}$$

$$c_{t1} = \int_{\mu_t k_t}^{\mu_t l_t} \underline{f}_{av}(\underline{z}_e(\sigma))d\sigma - \mu_t \sum_{s=k_t+1}^{l_t+1} \underline{f}_{av}(\underline{z}_e(\mu_t(s-1)))$$

$$c_{t2} = \mu_t \sum_{s=k_t+1}^{l_t+1} \left(\underline{f}_{av}(\underline{z}_e(\mu_t(s-1))) - \underline{f}_{av}(\bar{\underline{z}}^{\mu_t}_{s-1})\right).$$

However

$$c_{t1} = \sum_{s=k_t+1}^{l_t+1} \int_{\mu_t(s-1)}^{\mu_t s} [\underline{f}_{av}(\underline{z}_e(\sigma)) - \underline{f}_{av}(\underline{z}_e(\mu_t(s-1)))] d\sigma$$

$$\Rightarrow \qquad \|c_{t1}\| \leq L_f \mu_t \sum_{s=k_t+1}^{l_t+1} \sup_{\mu_t(s-1)\leq\sigma\leq\mu_t s} \|\underline{z}_e(\sigma) - \underline{z}_e(\mu_t(s-1))\|$$

$$\leq \mu_t L_f (l_t - k_t) \sup_{|\tau_2-\tau_1|\leq\mu_t} \|\underline{z}_e(\tau_2) - \underline{z}_e(\tau_1)\|. \tag{7.96}$$

So letting $\mu_t \to 0$ in (7.96) gives via (7.92b) that

$$\|c_{t1}\| \to 0, \quad \text{as } \mu_t \to 0.$$

For c_{t2} we have

$$\|c_{t2}\| \leq \mu_t \sum_{s=k_t+1}^{l_t+1} L_f \|\underline{z}_e(\mu_t(s-1)) - \bar{\underline{z}}_{s-1}^{\mu_t}\|$$

$$\leq \mu_t L_f \sum_{s=k_t+1}^{l_t+1} \sup_{0\leq\tau\leq T} \|\underline{z}_e(\tau) - \bar{\underline{z}}_{[\tau\mu_t^{-1}]}^{\mu_t}\|$$

$$\leq \mu_t L_f (2\mu_t + \delta) \sup_{0\leq\tau\leq T} \|\underline{z}_e(\tau) - \bar{\underline{z}}_{[\tau\mu_t^{-1}]}^{\mu_t}\|.$$

Thus by (7.95)

$$\|c_{t2}\| \to 0, \quad \text{as } \mu_t \to 0.$$

Putting these calculations together gives (7.94).

Proof of (7.91). Let $k_t = [\tau_1 \mu_t^{-1}]$, $l_t = [\tau_2 \mu_t^{-1}]$, $|\tau_2 - \tau_1| \leq \delta$. There are two cases: (i) $0 \leq \delta < \mu_t$ and (ii) $\delta \geq \mu_t$.

In case (i) $l_t = k_t$ so that the left hand side of (7.91b) is 0.

In case (ii)

$$\hat{\underline{z}}_{\mu_t}(\tau_2) - \hat{\underline{z}}_{\mu_t}(\tau_1) = \bar{\underline{z}}_{l_t}^{\mu_t} + ((\tau+\delta)\mu_t^{-1} - l_t)(\bar{\underline{z}}_{l_t+1}^{\mu_t} - \bar{\underline{z}}_{l_t}^{\mu_t})$$
$$- \bar{\underline{z}}_{k_t}^{\mu_t} - (\tau\mu_t^{-1} - k_t)(\bar{\underline{z}}_{k_t+1}^{\mu_t} - \bar{\underline{z}}_{k_t}^{\mu})$$

$$\Rightarrow \qquad \|\hat{\underline{z}}_{\mu_t}(\tau_2) - \hat{\underline{z}}_{\mu_t}(\tau_1)\| \leq \mu_t \sum_{s=k_t+1}^{l_t+1} \|\underline{f}_{av}(\bar{\underline{z}}_{s-1}^{\mu_t})\|$$

$$+ 1 \times \mu_t \times \|\underline{f}_{av}(\bar{\underline{z}}_{l_t}^{\mu_t})\| + 1 \times \mu_t \times \|\underline{f}_{av}(\bar{\underline{z}}_{k_t}^{\mu_t})\|$$

$$\leq \mu_t L_f h (l_t - k_t) + 2\mu_t L_f h, \quad \text{by (7.90)}$$

$$\leq L_f h(\delta + 4\mu_t)$$

$$\leq L_f h 5\delta.$$

Thus (7.91b) is established.

7.10 NOTES

Averaging methods have a long history in astronomy (see [S2] for historical notes), although they were only relatively recently introduced into the adaptive area. There is, needless to say, a large literature on averaging (including a sizeable Russian component) and we only offer a very selective list of references here. One of the earliest applications of (stochastic) averaging in the adaptive area is [D3] which considers finite time averaging and a Hovering theorem for single time scale problems. This approach was extended to mixed time scale problems by the work of Ljung (e.g., [L6], [L8]) in the context of long memory stochastic algorithms.

Rigorous development of averaging analysis begins only with the work of Bogoliuboff [B13] and the transformation technique introduced by Bogoliuboff has dominated the deterministic theory: See [B2], [B13], [H3], and [S3]. A fairly recent formulation of this technique treating systems more general than those treated in our Chapters 7 and 8 is in [B2]; they also give useful historical notes. A very clear exposé of the Bogoliuboff technique is available in [S3] (continuous time) and [B1] (discrete time).

There are three main approaches to the development of general averaging results. The first is the Bogoliuboff transformation technique just mentioned which allows development of both finite time and infinite time results (including stability). However compared to the technique used here it is rather cumbersome. Also the method used here delivers results under weaker regularity conditions. (This is discussed further below.)

Another technique is due originally to [G6] (in a single time scale, finite time averaging, stochastic setting) and in [K4] (where it is extended to mixed time scale problems) is called direct averaging: It also features in the work of [V1]. In these works finite time averaging theorems and Hovering theorems are developed. The method has recently been used by [S17] in the deterministic continuous time setting and in unpublished notes of Solo (the present author) extended to produce infinite time averaging and stability results.

The third method is that used here. The stochastic version is due apparently to [B12], and has been developed by [K4] (as his perturbed test function method). Blankenship and Papanicoloau used the perturbed Lyapunov function method on Markov problems [B12], and in [K4] it was extended to non-Markov settings. The specific discounted perturbation used here allows weaker regularity conditions than the discounted perturbation used by [B12] and [K4] allows.

With regard to regularity conditions there are several points worth noting. Firstly many traditional averaging results (based on the Bogoliuboff method) require signal periodicity. Our formulation makes it clear that such conditions are not needed. Secondly, note that for finite time averaging we do not require that the averaged function exist as a uniform limit. That is only needed for infinite time averaging. The Bogoliuboff technique on the other hand uses uniformity even for finite time averaging (e.g., see [S3]). Thirdly for finite time averaging we allow $\underline{f}(k, \underline{z}, \mu)$ to be discontinuous in $\underline{z}$. This is not allowed in the Bogoliuboff approach.

We should note here that the proof technique used by [S2] bears some similarity (for finite time averaging and the Hovering theorem) to ours. But they require continuity in $\underline{z}$ of $\underline{f}(k, \underline{z}, \mu)$ which we do not, and they do not realize that uniformity of averages is needed for the Hovering theorem. They do not prove any stability results. There seems to be confusion about the Hovering theorem in the literature, with some authors giving the impression it is a stability result. We have emphasized exponential stability in the book, but the Hovering theorem, for example, can be developed with only asymptotic stability [M4].

There is one other averaging method that is worth mentioning here (described in Bitmead and Johnson [B10]) although it has not been developed in full detail: the clearest exposé is in [B10], but it is used in an intuitive way in [R1] and [S3]. The linearization is done before averaging so that Theorem 7.7 is invoked and then averaging is applied to the linear system (7.41). In this way one only needs an averaging based stability theorem for linear time varying systems, and [B10] has its own relatively brief development of such a result. Note however that (7.40) may be quite hard to check or may not hold. Secondly, only stability results based on linearized approximation can be obtained so that a result such as that of Example 7.7 is not accessible. The method has further disadvantages in a stochastic setting. In its favor it may be said that aside from (7.40) it is very simple in conception and use although this is not true for the mixed time scale problems considered in Chapter 8.

EXERCISES

7.1 **(a)** Consider the EWLS algorithm of Section 4.7:

$$\delta \underline{\hat{w}}_k = H_k^{-1} \underline{x}_k e_k$$

$$H_k = \lambda H_{k-1} + \underline{x}_k \underline{x}_k^T$$

$$e_k = y_k - \underline{x}_k^T \underline{\hat{w}}_{k-1}$$

Assuming time invariant parameters, no noise, and deterministic $\underline{x}_k$, set $\lambda = 1 - \mu$ and find the corresponding averaged system. Carefully check any required regularity conditions.

(b) Find the corresponding ODE.

(c) Use a perturbed Lyapunov function calculation to give conditions for convergence of EWLS. Again carefully check any required regularity conditions.

Note: $\underline{\tilde{w}}_k$ converges, but H_k only hovers.

7.2 **(a)** Consider the relaxation LMS algorithm

$$\delta \underline{\hat{w}}_k = -\alpha \mu \underline{\hat{w}}_{k-1} + \mu \underline{x}_k e_k$$

$$e_k = y_k - \underline{x}_k^T \underline{\hat{w}}_{k-1}.$$

Assuming time invariant parameters, no noise, and deterministic $\underline{x}_k$, find the corresponding averaged system. Check carefully any required regularity conditions.

(b) Find the corresponding ODE.

(c) What are the equilibrium points of relaxation LMS? What can you say about convergence?

7.3 **(a)** Consider the LMS$_P$ algorithm

$$\delta \underline{\hat{w}}_k = \mu \underline{x}_k e_k / (1 + \mu \|\underline{x}_k\|^2)$$

$$e_k = \underline{y}_k - \underline{x}_k^T \underline{\hat{w}}_{k-1}.$$

Assuming time invariant parameters, no noise, and deterministic $\underline{x}_k$, find the averaged system. Be careful to check regularity conditions.

(b) Find the corresponding ODE.

(c) Discuss the convergence of LMS_P using a perturbed Lyapunov function argument. Be careful to check regularity conditions.

7.4 **(a)** Consider the Loop Filter LMS algorithm of Section 4.6:

$$\delta \underline{M}_k = \mu(A\underline{M}_{k-1} + \underline{b}\underline{x}_k^T e_k)$$
$$\delta \underline{\hat{w}}_k = \mu \underline{M}_k^T \underline{c} + \mu \underline{x}_k e_k$$
$$e_k = y_k - \underline{x}_k^T \underline{\hat{w}}_{k-1}.$$

Assuming time invariant parameters, no noise, and deterministic $\underline{x}_k$, find the averaged system. Be careful to check regularity conditions.

(b) Find the corresponding ODE.

(c) Use a perturbed Lyapunov function calculation to give conditions for convergence. Check regularity conditions carefully.

7.5 **(a)** Consider the signed LMS algorithm

$$\delta \underline{\hat{w}}_k = \mu \underline{s}_k e_k$$
$$e_k = y_k - \underline{x}_k^T \underline{\hat{w}}_{k-1}$$
$$\underline{s}_k = (s_{k1}\; s_{k2})^T$$
$$s_{ku} = \text{sgn}(x_{ku}), \quad 1 \le u \le 2$$
$$\text{sgn}(\xi) = \begin{cases} +1 & \xi > 0 \\ 0 & \xi = 0 \\ -1 & \xi < 0 \end{cases}$$
$$\underline{x}_k = (x_{k-1}\; x_{k-2})^T.$$

Suppose the auxiliary signal is

$$x_k = \sin(2\pi k/N), \quad N = 4.$$

Assuming time invariant parameters, no noise, and deterministic $\underline{x}_k$, exhibit the error system for the algorithm and then find the averaged system (check regularity conditions carefully).

(b) Find the corresponding ODE.

(c) Discuss the convergence of the algorithm using a perturbed Lyapunov function calculation.

7.6 Consider the Kalman filter gain equations for a Random Walk parameter model

$$\delta \underline{w}_{ok} = \underline{v}_k$$
$$y_k = \underline{w}_{ok}^T \underline{\phi}_k + \epsilon_k, \quad E(\underline{\phi}_k \underline{\phi}_k^T) = \underline{\Sigma}$$

where $\underline{v}_k$, ϵ_k are independent white noise of variances $\epsilon^2\sigma^2\underline{Q}$, σ^2; also $\epsilon \ll 1$; $\underline{\phi}_k$ is deterministic. The system matrix is $\underline{F}_k = I$, so from Appendix B9 the Kalman filter equations are (we write $\underline{P}_{k+1|k}$ as $\underline{P}_{k+1}$)

$$\underline{P}_{k+1} = \underline{P}_k - \frac{\underline{P}_k \underline{\phi}_k \underline{\phi}_k^T \underline{P}_k}{\sigma^2 + \underline{\phi}_k^T \underline{P}_k \underline{\phi}_k} + \epsilon^2 \underline{Q} \sigma^2$$

$$\delta \underline{\hat{w}}_{k+1} = \frac{\underline{P}_k \underline{\phi}_k}{\sigma^2 + \underline{\phi}_k^T \underline{P}_k \underline{\phi}_k} e_k$$

$$e_k = y_k - \underline{\phi}_k^T \underline{\hat{w}}_k.$$

Show that the averaged system for $\underline{P}_k / \epsilon\sigma^2$ is

$$\delta \underline{\prod}_{k+1} = -\epsilon \underline{\prod}_k \underline{\Sigma} \underline{\prod}_k + \epsilon \underline{Q} \tag{7.97}$$

where

$$\underline{\Sigma} = \lim_{N\to\infty} \frac{1}{N} \sum_{u=k+1}^{k+N} \underline{\phi}_u \underline{\phi}_u^T$$

and

$$\sup_{0 \le k \le T\epsilon^{-1}} \| \underline{P}_k / (\epsilon\sigma^2) - \underline{\prod}_k \| \to 0 \quad \text{as } \epsilon \to 0.$$

Note that (7.97) has a steady state solution obeying

$$\underline{\prod} \underline{\Sigma} \underline{\prod} = \underline{Q}.$$

If $\underline{\Sigma}^{\frac{1}{2}}$ (respectively $\underline{Q}^{\frac{1}{2}}$) is a positive definite symmetric square root of $\underline{\Sigma}$ (respectively $\underline{Q}$), then

$$\underline{\prod} = \underline{\Sigma}^{-\frac{1}{2}} \underline{Q}^{\frac{1}{2}}$$

so that

$$\underline{P}_k \simeq \sigma^2 \epsilon \underline{\Sigma}^{-\frac{1}{2}} \underline{Q}^{\frac{1}{2}}.$$

Thus for the Random Walk model the Kalman filter yields the optimal gain (see Section 5.3).

7.7 Consider the Kalman filter gain equations for a large amplitude slowly time varying model

$$y_k = \underline{\phi}_k^T \underline{w}_{ok} + \epsilon_k$$

$$\delta \underline{w}_{ok} = -\epsilon \underline{w}_{o,k-1} + \epsilon^{\frac{1}{2}} \underline{\nu}_k, \quad E(\underline{\phi}_k \underline{\phi}_k^T) = \underline{\Sigma}$$

where $\underline{\nu}_k$, ϵ_k are independent white noise of variances $\sigma_\nu^2 \underline{Q}$, σ^2; also $\epsilon \ll 1$; $\underline{\phi}_k$ is deterministic. Show that the Kalman filter matrix $\underline{P}_{k+1|k}$ has the same averaged system as in Exercise 7.6.

7.8 Show that when (7.6A5) holds, then

$$\lim_{\mu\to 0} \mu \sum_{s=k+1}^{\infty} (1-\mu)^{s-k} \underline{f}(s, \underline{z}) = \underline{f}_{av}(\underline{z}).$$

This exhibits a temporal average as a so-called Abel sum and explains to some extent the appearance of the discounted perturbation.

7.9 Consider the adaptive blind equalization algorithm of Section 4.5 given in (4.70). Find the averaged system associated with (4.70).

8

Deterministic Averaging: Mixed Time Scale

8.1 INTRODUCTION

The development of averaging methods in a mixed time scale setting is in principle much the same as in the single time scale case. However the way in which the averaged function is calculated is not quite so obvious. Further, the rigorous development must include consideration of the stability of the fast time scale auxiliary state equation.

It is worth reemphasizing that before engaging in an averaging analysis it is necessary to first assemble the adaptive algorithm in error system form.

The mixed time scale algorithm that we will consider has the general form

$$\delta \underline{z}_k = \mu \underline{f}(k, \underline{z}_{k-1}, \underline{y}_{k-1}, \mu) \tag{8.1a}$$

$$\underline{y}_k = \underline{A}(\underline{z}_{k-1})\underline{y}_{k-1} + \underline{h}(k, \underline{z}_{k-1}) + \mu \underline{g}(k, \underline{z}_{k-1}, \underline{y}_{k-1}, \mu). \tag{8.1b}$$

Initially, however, we will work with a simpler version of (8.1a), namely

$$\delta \underline{z}_k = \mu \underline{f}(k, \underline{z}_{k-1}, \underline{y}_{k-1}). \tag{8.2}$$

We call $\underline{z}_k$ the slow state and $\underline{y}_k$ the fast state.

Note that the first term in the auxiliary fast state equation is linear in the fast state $\underline{y}$; we call this semi-linear. We can certainly develop averaging analysis when (8.1b) is nonlinear, but it makes our subsequent calculations more tedious. To justify the structure (8.1) let us consider the following example.

Example 8.1. Mixed Time Scale LMS

The algorithm is given (see Section 4.3) as follows:

$$\delta\hat{\underline{\theta}}_k = \mu\hat{\underline{\varphi}}_k e_k \tag{8.3a}$$

$$e_k = y_k - \hat{\underline{\varphi}}_k^T\hat{\underline{\theta}}_{k-1} = y_k - \hat{s}_k \tag{8.3b}$$

$$\hat{\underline{\varphi}}_k = (-\hat{s}_{k-1}\cdots - \hat{s}_{k-p}\; x_{k-1}\cdots x_{k-m}). \tag{8.3c}$$

We form an error system by following the type of argument used in Section 6.2 where the stability of the posterior version of this algorithm was considered. We start with a modeling assumption:

(8.1a1)

$$y_k = s_k = \underline{\varphi}_k^T\underline{\theta}_o = -\sum_{r=1}^{p} s_{k-r}a_r + \sum_{r=1}^{m} x_{k-r}b_r$$

$$\underline{\varphi}_k = (-s_{k-1}\cdots - s_{k-p}\; x_{k-1}\cdots x_{k-m})$$

$$\underline{\theta}_o = (a_1\cdots a_p\; b_1\cdots b_m)$$

We rapidly find (much as in Section 6.2) that

$$(1 + a_o(q^{-1}))e_k = -\hat{\underline{\varphi}}_k^T\tilde{\underline{\theta}}_{k-1} \tag{8.4}$$

where, as usual

$$a_o(q^{-1}) = \sum_{r=1}^{p} a_r q^{-r}$$

and

$$\tilde{\underline{\theta}}_k = \hat{\underline{\theta}}_k - \underline{\theta}_o$$

is the parameter estimation error. If we write (8.4) in the form

$$e_k = -\frac{a_o(q^{-1})}{1 + a_o(q^{-1})}\hat{u}_k + \hat{u}_k \tag{8.5}$$

where $\hat{u}_k$ is the pseudo input

$$\hat{u}_k = -\hat{\underline{\varphi}}_k^T\tilde{\underline{\theta}}_{k-1} \tag{8.6}$$

then it is easy to see that we can express (8.4) in state space form as

$$\underline{\xi}_{k+1} = \underline{F}(\underline{a}_o)\underline{\xi}_k + \underline{b}\hat{u}_k \tag{8.7a}$$

$$e_k = \underline{c}^T\underline{\xi}_k + \hat{u}_k \tag{8.7b}$$

where

$$\underline{\xi}_k = (e_{k-1} \cdots e_{k-p})^T$$
$$\underline{b} = (1\ 0\ \cdots\ 0)^T$$
$$\underline{c} = -(a_1 \cdots a_p)^T = -\underline{a}_o$$
$$\underline{F} = \underline{F}(\underline{a}_o) = \begin{pmatrix} -a_1 \cdots & -a_p \\ I & \underline{0} \end{pmatrix}. \tag{8.7c}$$

Also from (8.3a), (8.3b), and (8.4) we can write

$$\begin{aligned} \underline{\hat{\varphi}}_k &= (-y_{k-1} + e_{k-1}, \cdots - y_{k-p} + e_{k-p},\ x_{k-1} \cdots x_{k-m})^T \\ &= \underline{\varphi}_k + (\underline{\xi}_k^T\ \underline{0}^T)^T \\ &= \underline{\varphi}_k + \underline{Q}\underline{\xi}_k \end{aligned} \tag{8.8}$$

where

$$\underline{Q} = \begin{pmatrix} I \\ \underline{0} \end{pmatrix}.$$

Putting (8.6), (8.8) into (8.7) gives

$$\begin{aligned} \underline{\xi}_{k+1} &= \underline{F}\underline{\xi}_k + \underline{b}(-\underline{\varphi}_k - \underline{Q}\underline{\xi}_k)^T \underline{\tilde{\theta}}_{k-1} \\ &= (\underline{F} - \underline{b}\underline{\tilde{\theta}}_{k-1}^T \underline{Q})\underline{\xi}_k - \underline{b}\underline{\varphi}_k^T \underline{\tilde{\theta}}_{k-1}. \end{aligned} \tag{8.9}$$

On the other hand, putting (8.6), (8.7b), (8.8) into (8.3) gives

$$\begin{aligned} \delta\underline{\tilde{\theta}}_k &= \mu\underline{\hat{\varphi}}_k(\underline{c}^T \underline{\xi}_k - \underline{\hat{\varphi}}_k^T \underline{\tilde{\theta}}_{k-1}) \\ &= \mu(\underline{\varphi}_k + \underline{Q}\underline{\xi}_k)(\underline{c}^T \underline{\xi}_k - (\underline{\varphi}_k + \underline{Q}\underline{\xi}_k)^T \underline{\tilde{\theta}}_{k-1}). \end{aligned} \tag{8.10}$$

The error system is then the pair (8.9), (8.10), and we see that it has the form (8.1), (8.2).

Note that in (8.9), (8.10), $\underline{\varphi}_k$ is made up of signals completely exogenous to the adaptive algorithm.

Before we turn to develop an averaging theory that will cover the pair (8.9), (8.10), it is useful to give a block diagram view of the pair (8.1a), (8.1b). We can view the pair as a feedback interconnection of two systems: a slow subsystem and a fast subsystem. This is illustrated in Fig. 8.1. Bearing in mind a similar view given in Section 6.2 we can look forward to a stability analysis involving Lyapunov functions for both the slow and fast systems.

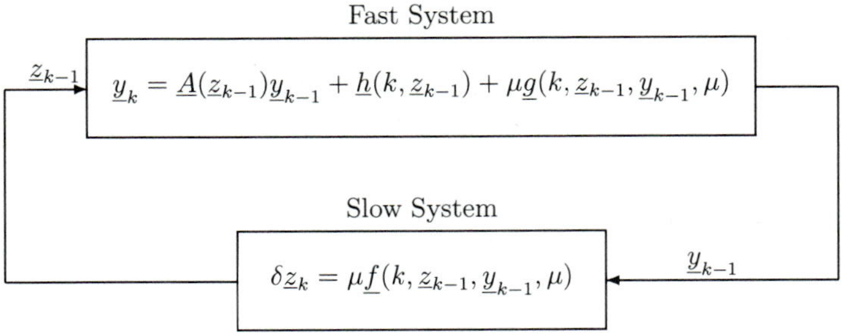

Figure 8.1 Mixed time scale error system, feedback as interconnected slow and fast systems.

8.2 FINITE TIME AVERAGING—TIME INVARIANT PARAMETERS

To see how to form an averaged system for the pair (8.1a), (8.1b) we pursue the heuristic argument used in Section 7.1. To keep the discussion clear we work initially with (8.2). Sum up (8.2) to find

$$\underline{z}_{k+N} - \underline{z}_k = \mu \sum_{s=k+1}^{k+N} \underline{f}(s, \underline{z}_{s-1}, \underline{y}_{s-1}).$$

As before, if μ is small, $\underline{z}_s$ changes slowly, so if N is not too large

$$\underline{z}_{k+N} - \underline{z}_k \simeq \mu \sum_{s=k+1}^{k+N} \underline{f}(s, \underline{z}_{k-1}, \underline{y}_{s-1}).$$

Now what of $\underline{y}_k$? Since $\underline{z}_s$ changes slowly, then over the interval $[k, k+N]$ it should be well approximated by $\underline{y}_s(\underline{z}_{k-1})$ where the so-called "frozen state" $\underline{y}_s(\underline{z})$ satisfies

$$\underline{y}_k(\underline{z}) = \underline{A}(\underline{z})\underline{y}_{k-1}(\underline{z}) + \underline{h}(k, \underline{z}). \tag{8.11a}$$

The other term in (8.1b) is $O(\mu)$ and will not affect the averaging process. Thus we have

$$\underline{z}_{k+N} - \underline{z}_k \simeq \mu \sum_{s=k+1}^{k+N} \underline{f}(s, \underline{z}_{k-1}, \underline{y}_{s-1}(\underline{z}_{k-1}))$$

and we are led to introduce the averaged function

$$\begin{aligned} \underline{f}_{av}(\underline{z}) &= \operatorname{av}(\underline{f}(s, \underline{z}, y_{s-1}(\underline{z}))) \\ &= \lim_{N\to\infty} \frac{1}{N} \sum_{s=k+1}^{k+N} \underline{f}(s, \underline{z}, \underline{y}_{s-1}(\underline{z})) \end{aligned} \tag{8.11b}$$

and the averaged system

$$\delta \bar{\underline{z}}_k = \mu \underline{f}_{av}(\bar{\underline{z}}_{k-1}). \tag{8.11c}$$

We claim that (8.11c) approximates (8.2). Before trying to justify this claim let us hasten to see how this looks on the pair (8.9), (8.10).

Example 8.2. Averaging of Mixed Time Scale LMS

From (8.9) we introduce the frozen state $\underline{\xi}_k(\underline{\tilde{\theta}})$ obeying

$$\underline{\xi}_{k+1}(\underline{\tilde{\theta}}) = (\underline{F} - \underline{b}\underline{\tilde{\theta}}^T \underline{Q})\underline{\xi}_k(\underline{\tilde{\theta}}) - \underline{b}\underline{\varphi}_k^T \underline{\tilde{\theta}}.$$

For subsequent calculation it is convenient to write this in operator form

$$\begin{aligned}\underline{\xi}_k(\underline{\tilde{\theta}}) &= \left[qI - (\underline{F} - \underline{b}\underline{\tilde{\theta}}^T \underline{Q})\right]^{-1}(-\underline{b}\underline{\varphi}_k^T)\underline{\tilde{\theta}} \\ &= (qI - \underline{F} + \underline{b}\underline{\tilde{\theta}}^T \underline{Q})^{-1}(-\underline{b}\underline{\varphi}_k^T)\underline{\tilde{\theta}}.\end{aligned}$$

Applying the Matrix Inversion lemma (Appendix A5) gives

$$\underline{\xi}_k(\underline{\tilde{\theta}}) = \left\{(qI - \underline{F})^{-1} - \frac{(qI - \underline{F})^{-1}\underline{b}\underline{\tilde{\theta}}^T \underline{Q}(qI - \underline{F})^{-1}}{1 + \alpha(q)}\right\}(-\underline{b}\underline{\varphi}_k^T \underline{\tilde{\theta}})$$

where

$$\alpha(q) = \underline{\tilde{\theta}}^T \underline{Q}^T (qI - \underline{F})^{-1}\underline{b}. \tag{8.12a}$$

Thus

$$\underline{\xi}_k(\underline{\tilde{\theta}}) = (qI - \underline{F})^{-1}\underline{b}(\underline{\varphi}_k^T \underline{\tilde{\theta}})/(1 + \alpha(q)). \tag{8.12b}$$

With (8.10) in mind, we then make the following calculations:

$$\begin{aligned}\underline{Q}\underline{\xi}_k(\underline{\tilde{\theta}}) &= \underline{Q}(qI - \underline{F})^{-1}\underline{b}(-\underline{\varphi}_k^T \underline{\tilde{\theta}})/(1 + \alpha(q)) \\ \underline{c}^T \underline{\xi}_k(\underline{\tilde{\theta}}) &= \underline{c}^T (qI - \underline{F})^{-1}\underline{b}(-\underline{\varphi}_k^T \underline{\tilde{\theta}})/(1 + \alpha(q)) \\ \underline{\varphi}_k + \underline{Q}\underline{\xi}_k(\underline{\tilde{\theta}}) &= \left\{I - \frac{\underline{Q}(qI - \underline{F})^{-1}\underline{b}\underline{\tilde{\theta}}^T}{1 + \alpha(q)}\right\}\underline{\varphi}_k \qquad (8.13) \\ \underline{\tilde{\theta}}^T (\varphi_k + \underline{Q}\underline{\xi}_k(\underline{\tilde{\theta}})) &= \underline{\tilde{\theta}}^T \underline{\varphi}_k/(1 + \alpha(q))\end{aligned}$$

Thus, we find

$$\underline{c}^T \underline{\xi}_k(\underline{\tilde{\theta}}) - \left(\underline{\varphi}_k + \underline{Q}\underline{\xi}_k(\underline{\tilde{\theta}})\right)^T \underline{\tilde{\theta}} = \frac{(\underline{c}^T (qI - \underline{F})^{-1}\underline{b} - 1)(-\underline{\tilde{\theta}}^T \underline{\varphi}_k)}{1 + \alpha(q)}$$

and in view of (8.5)–(8.7), this is

$$\underline{c}^T \underline{\xi}_k(\underline{\tilde{\theta}}) - \left(\underline{\varphi}_k + \underline{Q}\underline{\xi}_k(\underline{\tilde{\theta}})\right)^T \underline{\tilde{\theta}} = (1 + a_o(q^{-1}))^{-1}(1 + \alpha(q))^{-1}(-\underline{\tilde{\theta}}^T \underline{\varphi}_k). \tag{8.14}$$

Now we can use the pair (8.13), (8.14) to compute the averaged function from (8.10). To do this we introduce the following assumption:

(8.2a1) x_t is uniformly stationary

Then according to Appendix C8, y_k is uniformly stationary; thus, $\underline{\varphi}_k$ is uniformly stationary as is $\underline{\xi}_k(\tilde{\theta})$. Then according to the pseudo Parseval theorem in Appendix C8 and using (8.13) and (8.14) we have

$$\text{av}\left\{\left[\underline{\varphi}_k + \underline{Q}\underline{\xi}_k(\underline{\tilde{\theta}})\right]\left[\underline{c}^T\underline{\xi}_k(\underline{\tilde{\theta}}) - (\underline{\varphi}_k + \underline{Q}\underline{\xi}_k(\underline{\tilde{\theta}}))^T\underline{\tilde{\theta}}\right]\right\} = -\underline{R}(\underline{\tilde{\theta}})\underline{\tilde{\theta}} \tag{8.15a}$$

where

$$\underline{R}(\underline{\tilde{\theta}}) = \int_{-\pi}^{\pi}\left\{I - \frac{\underline{Q}(e^{j\omega}I - \underline{F})^{-1}\underline{b}\underline{\tilde{\theta}}^T}{1+\alpha(e^{j\omega})}\right\}\left\{\frac{1}{[1+a_o(e^{j\omega})]}\cdot\frac{1}{[1+\alpha(e^{-j\omega})]}\right\} d\underline{F}_\varphi(\omega)/2\pi \tag{8.15b}$$

and $\underline{F}_\varphi(\omega)$ is the spectral distribution function of $\underline{\varphi}_k$ (this holds provided $qI - \underline{F} - \underline{b}\underline{\tilde{\theta}}^T\underline{Q}$ is a stable matrix).

The averaged system is thus

$$\delta\underline{\tilde{\theta}}_k = -\mu\underline{R}(\underline{\tilde{\theta}}_{k-1})\underline{\tilde{\theta}}_{k-1}. \tag{8.16a}$$

A full stability analysis of this averaged system is more appropriately developed later, so here we just consider the linearized averaged system, which is

$$\delta\underline{\tilde{\theta}}_k = -\mu\underline{R}(0)\underline{\tilde{\theta}}_{k-1} \tag{8.16b}$$

where

$$\underline{R}(0) = \int_{-\pi}^{\pi}\frac{d\underline{F}_\varphi(\omega)/2\pi}{1+a_o(e^{-j\omega})}. \tag{8.17}$$

We see that if

(8.2a2) $1 + a_o(q^{-1})$ is strictly positive real (SPR) (i.e., $\text{Re}(1 + a_o(e^{-j\omega}))^{-1} \geq \rho > 0$) and so stable

then (see Appendix A1(vi)) $\underline{R}(0)$ is positive definite and

$$\underline{R}(0) > \underline{\Phi} = \rho\int_{-\pi}^{\pi} d\underline{F}_\varphi(\omega)/2\pi.$$

We see that if (see Section 6.1)

(8.2a3) $\underline{\varphi}_k$ is weakly PE

then $\underline{R}(0)$ is positive definite and so (8.16b) is ES to $\underline{0}$. As in Section 6.1 we can replace (8.2a3) with

(8.2a4) x_k is PE of order $2[(p+m)/2]$

This analysis may be compared with that in Section 6.2. This discussion is linked to the original system in Section 8.6.

Before concluding the example, we investigate the possibility of other equilibrium points of (8.16) aside from $\underline{\tilde{\theta}} = \underline{0}$. By definition an equilibrium point satisfies

$$\underline{R}(\underline{\tilde{\theta}})\underline{\tilde{\theta}} = \underline{0}$$

$$\Rightarrow \quad \underline{\tilde{\theta}}^T\underline{R}(\underline{\tilde{\theta}})\underline{\tilde{\theta}} = 0.$$

From (8.15) this yields

$$\underline{\tilde{\theta}}^T \underline{G}(\tilde{\theta})\underline{\tilde{\theta}} = 0 \tag{8.18a}$$

where

$$\underline{G}(\tilde{\theta}) = \int_{-\pi}^{\pi} d\underline{F}_{\varphi}|1+\alpha(e^{j\omega})|^{-2}(1+a_0(e^{-j\omega}))^{-1}d\omega/2\pi \tag{8.18b}$$

and under (8.2a2)

$$\underline{G}(\tilde{\theta}) \geq \rho \int_{-\pi}^{\pi} d\underline{F}_{\varphi}|1+\alpha(e^{j\omega})|^{-2}d\omega/2\pi.$$

Then under (8.2a3) we deduce from (8.18) that

$$\|\underline{\tilde{\theta}}\| = 0$$

i.e., $\underline{\tilde{\theta}} = \underline{0}$ is the only equilibrium point.

To justify the averaged system (8.11) we proceed as in Section 7.2. Sum (8.1), (8.11c) and subtract them to obtain

$$\underline{\Delta}_k = \underline{z}_k - \underline{\bar{z}}_k = \underline{\Delta}_0 + \mu \sum_{s=1}^{k} \underline{\tilde{f}}_s \tag{8.19}$$

where

$$\underline{\tilde{f}}_s = \underline{f}(s, \underline{z}_{s-1}, \underline{y}_{s-1}) - \underline{f}_{av}(\underline{\bar{z}}_{s-1}). \tag{8.20}$$

Now split (8.19) into two pieces:

$$\underline{\Delta}_k = \underline{\Delta}_0 + \mu \sum_{s=1}^{k} (\underline{f}_{av}(\underline{z}_{s-1}) - \underline{f}_{av}(\underline{\bar{z}}_{s-1})) + \mu \underline{J}_k \tag{8.21}$$

$$\underline{J}_k = \sum_{s=1}^{k} (\underline{f}(s, \underline{z}_{s-1}, \underline{y}_{s-1}) - \underline{f}_{av}(\underline{z}_{s-1})) \tag{8.22}$$

To continue, we fix an interval $[0, T]$ and introduce some regularity conditions:

(8.2A1)

$$\|\underline{f}(k, \underline{z}, \underline{y})\| \leq B_f, \qquad \|\underline{z}\| \leq h_0,\ \|\underline{y}\| \leq h_y$$
$$\|\underline{g}(k, \underline{z}, \underline{y}, \mu)\| \leq B_g, \qquad \|\underline{z}\| \leq h_0,\ \|\underline{y}\| \leq h_y$$

(8.2A2) Introduce the perturbation

$$\underline{p}(k, \underline{z}) = \sum_{s=1}^{k} (\underline{f}(s, \underline{z}, \underline{y}_{s-1}(\underline{z})) - \underline{f}_{av}(\underline{z}))$$

where $\underline{y}_s(\underline{z})$ is the frozen state (8.11a). Suppose

$$\|\underline{p}(k, \underline{z}) - \underline{p}(k, \underline{z}')\| \leq L_p \|\underline{z} - \underline{z}'\|$$
$$\|\underline{p}(k, \underline{0})\| \leq B_p$$

for $\|\underline{z}\|, \|\underline{z}'\| \leq h_0$.

(8.2A3) $\underline{f}_{av}(\underline{z})$ obeys a Lipschitz condition

$$\|\underline{f}_{av}(\underline{z}) - \underline{f}_{av}(\underline{z}')\| \leq L_f \|\underline{z} - \underline{z}'\|, \quad \|\underline{z}\|, \|\underline{z}'\| \leq h_0.$$

(8.2A4) $\|\bar{\underline{z}}_0\|$ is so small that $\|\bar{\underline{z}}_k\| \leq h/2$, $1 \leq k \leq T\mu^{-1}$, where $h < h_0$ is to be chosen.

(8.2A5) $\underline{f}(s, \underline{z}, \underline{y})$ obeys a Lipschitz condition in $\|\underline{y}\|$:

$$\|\underline{f}(s, \underline{z}, \underline{y}) - \underline{f}(s, \underline{z}, \underline{y}')\| \leq L_f \|\underline{y} - \underline{y}'\|, \quad \|\underline{y}\|, \|\underline{y}'\| \leq h_y, \quad \|\underline{z}\|, \|\underline{z}'\| \leq h_0$$

(8.2A6) $\underline{h}(t, \underline{0}) = \underline{0}$; $\underline{h}(t, \underline{z})$ obeys a Lipschitz condition:

$$\|\underline{h}(t, \underline{z}) - \underline{h}(t, \underline{z}')\| \leq L_f \|\underline{z} - \underline{z}'\|, \quad \|\underline{z}\|, \|\underline{z}'\| \leq h_0$$

We also require two regularity conditions that deal with the stability of the frozen state:

(8.2A7) $\underline{A}(\underline{z})$ is a stability matrix for all $\|\underline{z}\| \leq h$; i.e., all eigenvalues of $\underline{A}(\underline{z})$ have modulus which is $\leq \lambda < 1$. Also $\|A(\underline{z})\| \leq A$, $\|\underline{z}\| \leq h_0$.

(8.2A8) $\underline{A}(\underline{z})$ obeys a Lipschitz condition:

$$\|\underline{A}(\underline{z}) - \underline{A}(\underline{z}')\| \leq L_A \|\underline{z} - \underline{z}'\|, \quad \|\underline{z}\|, \|\underline{z}'\| \leq h_0$$

Let

$$T'_{\mu z} = \text{first time } k \text{ that } \|\underline{z}_k\| < h_0, \ \|\underline{z}_{k+1}\| \geq h_0$$

and similarly define $T'_{\mu y}$ based on h_y. Then let

$$T_\mu = \min\{T'_{\mu z}, T'_{\mu y}, T/\mu\}. \tag{8.23}$$

Below we will show that

$$\|\underline{J}_k\| \leq a + bT, \quad 1 \leq k \leq T_\mu. \tag{8.24}$$

Then returning to (8.21) use (8.2A3) to see that

$$\|\underline{\Delta}_k\| \leq \|\underline{\Delta}_0\| + \mu L_f \sum_{s=1}^{k} \|\underline{\Delta}_{s-1}\| + \mu(a + bT), \quad 1 \leq k \leq T_\mu.$$

Now apply the discrete Bellman-Gronwall lemma (see Appendix C2) to find that

$$\begin{aligned} \|\underline{\Delta}_k\| &\leq \left(\|\underline{\Delta}_0\| + \mu(a + bT)\right)(1 + \mu L_f)^{k-1}, \quad 1 \leq k \leq T_\mu \\ &\leq \left(\|\underline{\Delta}_0\| + c_T(\mu)\right) e^{\mu L_f k}, \quad 1 \leq k \leq T_\mu \end{aligned}$$

where

$$c_T(\mu) = \mu(a + bT) \downarrow 0 \text{ as } \mu \downarrow 0.$$

Thus we find

$$\begin{aligned} \max_{1 \leq k \leq T_\mu} \|\underline{\Delta}_k\| &\leq (\|\underline{\Delta}_0\| + c_T(\mu)) e^{\mu L_f T_\mu} \\ &\leq (\|\underline{\Delta}_0\| + c_T(\mu)) e^{T L_f}. \end{aligned} \tag{8.25}$$

Then we have our main result:

Theorem 8.1. For the primary system (8.2), (8.1b) and averaged system (8.11a)–(8.11c) with identical initial conditions, suppose (8.2A1)–(8.2A8) hold. Then given $T > 0$, there exist $b_T > 0$, $\mu_T > 0$ such that

$$\max_{1 \le k \le T/\mu} \|\underline{z}_k - \bar{\underline{z}}_k\| \le c_T(\mu) b_T, \quad 0 \le \mu \le \mu_T$$

and

$$c_T(\mu) \to 0 \text{ as } \mu \to 0.$$

Proof. Take μ_T so small that

$$c_T(\mu_T) e^{T L_f} < h/4, \quad \mu B_f < h/4.$$

Then from (8.2A4)

$$\begin{aligned}\|\underline{z}_k\| &\le \|\underline{\Delta}_k\| + \|\bar{\underline{z}}_k\|, \quad 1 \le k \le T_\mu \\ &< h/4 + h/2 = 3h/4 \\ \|\underline{z}_{T_\mu + 1}\| &\le \|\underline{z}_{T_\mu}\| + \mu B_f < h.\end{aligned}$$

But in view of (8.34) (which shows how h is chosen) and (8.35) below, this means $T_\mu = T/\mu$.

Proof of (8.24). We split $\underline{J}_k$ into two pieces by approximating $\underline{y}_s$ by $\underline{y}_s(\underline{z}_s)$:

$$\underline{J}_k = \underline{J}_{k,1} + \underline{J}_{k,2}$$

$$\underline{J}_{k,1} = \sum_{s=1}^{k} \left(\underline{f}(s, \underline{z}_{s-1}, \underline{y}_{s-1}) - \underline{f}(s, \underline{z}_{s-1}, \underline{y}_{s-1}(\underline{z}_{s-1})) \right)$$

$$\underline{J}_{k,2} = \sum_{s=1}^{k} \left(\underline{f}(s, \underline{z}_{s-1}, \underline{y}_{s-1}(\underline{z}_{s-1})) - \underline{f}_{av}(\underline{z}_{s-1}) \right).$$

From (8.2A5)

$$\|\underline{J}_{k,1}\| \le \sum_{s=1}^{k} L_f \|\underline{y}_{s-1} - \underline{y}_{s-1}(\underline{z}_{s-1})\| \tag{8.26}$$

and below we bound this further as

$$\|\underline{J}_{k,1}\| \le a_1 + b_1 T, \quad 1 \le k \le T_\mu. \tag{8.27}$$

Now we turn to $\underline{J}_{k,2}$ which we approximate by

$$\underline{P}_k = \underline{p}(k, \underline{z}_{k-1}).$$

We have that

$$\begin{aligned}\underline{P}_k - \underline{P}_{k-1} &= \underline{p}(k, \underline{z}_{k-1}) - \underline{p}(k-1, \underline{z}_{k-2}) \\ &= \underline{p}(k, \underline{z}_{k-1}) - \underline{p}(k-1, \underline{z}_{k-1}) + \underline{\eta}_k \\ &= [\underline{f}(k, \underline{z}_{k-1}, \underline{y}_{k-1}(\underline{z}_{k-1})) - \underline{f}_{av}(\underline{z}_{k-1})] + \underline{\eta}_k\end{aligned}$$

where

$$\underline{\eta}_k = \underline{p}(k-1, \underline{z}_{k-1}) - \underline{p}(k-1, \underline{z}_{k-2}).$$

Thus

$$\underline{P}_k - \underline{P}_{k-1} = \underline{J}_{k,2} - \underline{J}_{k-1,2} + \underline{\eta}_k.$$

Summing up gives

$$\underline{P}_k = \underline{J}_{k,2} + \sum_{s=1}^{k} \underline{\eta}_s. \tag{8.28}$$

However from (8.2A2)

$$\begin{aligned} \|\underline{\eta}_s\| &\leq L_p \|\underline{z}_{s-1} - \underline{z}_{s-2}\| \\ &= L_p \mu \|\underline{f}(s-1, \underline{z}_{s-2}, \underline{y}_{s-2})\| \\ &\leq \mu L_p B_f = \mu b_2 \end{aligned} \tag{8.29a}$$

while from (8.2A2) also

$$\begin{aligned} \|\underline{P}_k\| &\leq \|\underline{p}(k, \underline{z}_{k-1}) - \underline{p}(k, \underline{0})\| + \|\underline{p}(k, \underline{0})\| \\ &\leq L_p \|\underline{z}_{k-1}\| + B_p \\ &\leq L_p h + B_p. \end{aligned} \tag{8.29b}$$

So from (8.28)

$$\|\underline{J}_{k,2}\| \leq a_2 + \mu k b_2. \tag{8.29c}$$

Putting (8.27), (8.29) together gives (8.24).

Proof of (8.27). We begin with two fast state equations:

$$\underline{y}_t = \underline{A}(\underline{z}_{t-1})\underline{y}_{t-1} + \underline{h}(t, \underline{z}_{t-1}) + \mu \underline{g}(t, \underline{z}_{t-1}, \underline{y}_{t-1}, \mu) \tag{8.30}$$

$$\underline{y}_t(\underline{z}_{s-1}) = \underline{A}(\underline{z}_{s-1})\underline{y}_{t-1}(\underline{z}_{s-1}) + \underline{h}(t, \underline{z}_{s-1}), \quad \text{(think of } s \text{ fixed, } t \text{ varying)}$$

In the second equation set $t = s$; consider the resulting time varying system:

$$\underline{y}_s(\underline{z}_{s-1}) = \underline{A}(\underline{z}_{s-1})\underline{y}_{s-1}(\underline{z}_{s-2}) + \underline{h}(s, \underline{z}_{s-1}) - \underline{\rho}_s \tag{8.31}$$

where $\underline{\rho}_s = \underline{A}(\underline{z}_{s-1})(\underline{y}_{s-1}(\underline{z}_{s-2}) - \underline{y}_{s-1}(\underline{z}_{s-1}))$. Introduce $\underline{\delta}_s = \underline{y}_s - \underline{y}_s(\underline{z}_{s-1})$ and subtract (8.31) from (8.30) to find another time varying system

$$\underline{\delta}_s = \underline{A}(\underline{z}_{s-1})\underline{\delta}_{s-1} + \mu \underline{g}(s, \underline{z}_{s-1}, \underline{y}_{s-1}, \mu) + \underline{\rho}_s. \tag{8.32}$$

Below we show

$$\|\underline{y}_{s-1}(\underline{z}_{s-1}) - \underline{y}_{s-1}(\underline{z}_{s-2})\| \leq b\mu. \tag{8.33}$$

Next note that from (8.2A8)

$$\begin{aligned}\|\underline{A}(\underline{z}_s) - \underline{A}(\underline{z}_{s-1})\| &\le L_A\|\underline{z}_s - \underline{z}_{s-1}\| \\ &\le L_A\mu\|\underline{f}(s, \underline{z}_{s-1}, \underline{y}_{s-1})\| \\ &\le \mu L_A B_f, \quad 1 \le s \le T_\mu. \end{aligned} \tag{8.34a}$$

Then according to the slowly time varying stability result of Appendix C7(iv), if μ is small enough we have for some m $(1 < m < \infty)$ and λ $(0 < \lambda < 1)$

$$\left\|\underline{\prod}_{k,s}^{A}\right\| \le m\lambda^{k-s}, \quad 1 \le s \le k \le T_\mu \tag{8.34b}$$

where $\underline{\prod}_{k,s}^{A}$ is the transition matrix for the homogeneous time varying system $\underline{u}_s = \underline{A}(\underline{z}_{s-1})\underline{u}_{s-1}$ (see Appendix C6). Thus we find from (8.30), (8.2A6), as well as Appendix C6(v), that

$$\begin{aligned}\|\underline{y}_k\| &\le m\lambda^k\|\underline{y}_0\| + m\sum_{s=1}^{k}\lambda^{k-s}(L_f\|\underline{z}_{s-1}\| + \mu B_g) \\ &\le m\|\underline{y}_0\| + m(1-\lambda)^{-1}[L_f h + \mu B_g], \quad 1 \le k \le T_\mu + 1.\end{aligned}$$

So if we take $\|\underline{y}_0\|$, h and then μ small enough we can ensure

$$\|\underline{y}_k\| < h_y, \quad 1 \le k \le T_\mu + 1. \tag{8.35a}$$

From the definition of T_μ this means

$$T_\mu = \min\{T'_{\mu z}, T/\mu\}. \tag{8.35b}$$

Next, from (8.32), (8.33), and (8.2A7)

$$\begin{aligned}\|\underline{\delta}_k\| &\le m\lambda^k\|\underline{\delta}_0\| + \mu m(B_g + Ab)\sum_{s=1}^{k}\lambda^{k-s} \\ &\le m\lambda^k\|\underline{\delta}_0\| + \mu m(B_g + Ab)(1-\lambda)^{-1}.\end{aligned}$$

Then from (8.26)

$$\begin{aligned}\|\underline{J}_{k,1}\| &\le L_f\sum_{s=1}^{k}\|\underline{\delta}_{s-1}\| \\ &\le L_f m(1-\lambda)^{-1}[\|\underline{\delta}_0\| + \mu k(B_g + Ab)] \\ &\le L_f m(1-\lambda)^{-1}[\|\underline{\delta}_0\| + (B_g + Ab)T], \quad 1 \le k \le T_\mu\end{aligned}$$

which is the required result.

Proof of (8.33). Consider that we start two frozen state processes with identical initial conditions

$$\underline{y}_s(\underline{z}) = A(\underline{z})\underline{y}_{s-1}(\underline{z}) + \underline{h}(s, \underline{z})$$

$$\underline{y}_s(\underline{z}') = A(\underline{z}')\underline{y}_{s-1}(\underline{z}') + \underline{h}(s, \underline{z}').$$

So denoting the difference by $\underline{d}_s$ we get

$$\underline{d}_s = A(\underline{z})\underline{d}_{s-1} + \left(A(\underline{z}') - A(\underline{z})\right)\underline{y}_{s-1}(\underline{z}') + \underline{h}(s, \underline{z}) - \underline{h}(s, \underline{z}').$$

We deduce that

$$\|\underline{d}_k\| \leq m \sum_{s=1}^{k} \lambda^{k-s}(L_f\|\underline{z} - \underline{z}'\| + L_A\|\underline{z} - \underline{z}'\|h_y)$$

$$\leq m(1-\lambda)^{-1}(L_f\|\underline{z} - \underline{z}'\| + L_A h_y\|\underline{z} - \underline{z}'\|) = b'\|\underline{z} - \underline{z}'\|.$$

Thus from (8.2A1)

$$\|\underline{y}_s(\underline{z}_{s-1}) - \underline{y}_s(\underline{z}_{s-2})\| \leq b'\|\underline{z}_{s-1} - \underline{z}_{s-2}\| \leq b\mu. \tag{8.36}$$

Example 8.3. Checking Regularity Conditions

We continue discussion of mixed time scale LMS. The only regularity conditions requiring special consideration are (8.2A2) and (8.2A7). The reader is invited to check the others. We look at (8.2A7) first. From (8.9) we have

$$\underline{A}(\tilde{\underline{\theta}}) = \underline{F} - \underline{b}\tilde{\underline{\theta}}^T \underline{Q}.$$

Now the eigenvalues of $\underline{A}(\tilde{\underline{\theta}})$ are continuous functions of $\tilde{\underline{\theta}}$; also $\underline{A}(\underline{0}) = \underline{F}$. So given $\epsilon > 0$, there is a sufficiently small h_0 such that

$$\|\tilde{\underline{\theta}}\| \leq h_0 \Rightarrow |\lambda_{(r)}(\underline{A}(\tilde{\underline{\theta}})) - \lambda_{(r)}(\underline{F})\| \leq \epsilon$$

where $\lambda_{(r)}(\underline{M})$ is the rth largest eigenvalue of $\underline{M}$. Thus by (8.2a2) if ϵ is small enough, $\underline{A}(\tilde{\underline{\theta}})$ is a stability matrix.

Now we consider (8.2A2). The details are, in fact, fairly straightforward. In (8.13), (8.14) we see that $\underline{f}(k, \underline{z}, \underline{y}_{k-1}(\underline{z}))$ has the form $\underline{u}(k, \underline{z}, \underline{y}_{k-1}(\underline{z}))v(k, \underline{z}, \underline{y}_{k-1}(\underline{z}))$. Thus

$$\sum_{s=1}^{k}\left(\underline{f}(s, \underline{z}, y_{s-1}(\underline{z})) - \underline{f}(s, \underline{z}', y_{s-1}(\underline{z}'))\right)$$

$$= \sum_{s=1}^{k}\left(\underline{u}(s, \underline{z}, \underline{y}_{s-1}(\underline{z})) - \underline{u}(s, \underline{z}', \underline{y}_{s-1}(\underline{z}'))\right) v(s, \underline{z}, \underline{y}_{s-1}(\underline{z}))$$

$$+ \sum_{s=1}^{k}\underline{u}(s, \underline{z}', \underline{y}_{s-1}(\underline{z}'))\left(v(s, \underline{z}, \underline{y}_{s-1}(\underline{z})) - v(s, \underline{z}', \underline{y}_{s-1}(\underline{z}'))\right).$$

So it will do if (8.2A2) is satisfied by $\underline{u}(\bullet)$, $v(\bullet)$ separately. If we look now at (8.14), for example, we have (dropping the q^{-1} argument for convenience)

$$(1 + a_0)^{-1}(1 + \alpha)^{-1}\underline{\varphi}_k^T\tilde{\underline{\theta}} - (1 + a_0)^{-1}(1 + \alpha')^{-1}\underline{\varphi}_k^T\tilde{\underline{\theta}}' = d_k + f_k$$

where

$$d_k = (1 + a_0)^{-1}(1 + \alpha)^{-1}\underline{\varphi}_k^T(\underline{\tilde{\theta}} - \underline{\tilde{\theta}}')$$

$$f_k = (1 + a_0)^{-1}\left((1 + \alpha)^{-1} - (1 + \alpha')^{-1}\right)\underline{\varphi}_k^T\underline{\tilde{\theta}}'$$

$$= (1 + a_0)^{-1}(1 + \alpha)^{-1}(1 + \alpha')^{-1}\left(\underline{b}^T(qI - \underline{F})\right)^{-1}\underline{Q}(\underline{\tilde{\theta}}' - \underline{\tilde{\theta}})\underline{\varphi}_k^T\underline{\tilde{\theta}}'.$$

Now since $\underline{\varphi}_k$ is uniformly stationary and since this will be preserved under stable filtering (see Appendix C, section C8) then d_k and f_k will each satisfy (8.2A2) separately.

8.2.1 Extensions

Clearly the result is easily modified to cover (8.1a). The averaged function is now

$$\underline{f}_{av}(\underline{z}) = \text{av}(\underline{f}(s, \underline{z}, \underline{y}_{s-1}(\underline{z}), 0))$$

$$= \lim_{N\to\infty} \frac{1}{N}\sum_{s=k}^{k+N} \underline{f}(s, \underline{z}, \underline{y}_{s-1}(\underline{z}), 0). \tag{8.37}$$

We introduce the following regularity conditions:

(8.2B1) For $\|\underline{z}\| \le h_0$ and $\|\underline{y}\| \le h_y$,

$$\|\underline{f}(k, \underline{z}, \underline{y}, \mu)\| \le B_f$$

$$\|\underline{g}(k, \underline{z}, \underline{y}, \mu)\| \le B_g, \quad \|\underline{h}(k, \underline{z})\| \le B_h$$

(8.2B2) Introduce the perturbation

$$\underline{p}(k, \underline{z}) = \sum_{s=1}^{k}(\underline{f}(s, \underline{z}, \underline{y}_{s-1}(\underline{z}), 0) - \underline{f}_{av}(\underline{z}))$$

where $\underline{y}_s(\underline{z})$ is the frozen state. Suppose for $\|\underline{z}\|, \|\underline{z}'\| \le h_0$

$$\|\underline{p}(k, \underline{z}) - \underline{p}(k, \underline{z}')\| \le L_p\|\underline{z} - \underline{z}'\|$$

$$\|\underline{p}(k, \underline{0})\| \le B_p.$$

(8.2B3) Same as (8.2A3).

(8.2B4) Same as (8.2A4).

(8.2B5) $\underline{f}(s, \underline{z}, \underline{y}, \mu)$ obeys a Lipschitz condition in $\|\underline{y}\|$:

$$\|\underline{f}(s, \underline{z}, \underline{y}, \mu) - \underline{f}(s, \underline{z}, \underline{y}', \mu)\| \le L_y\|\underline{y} - \underline{y}'\|, \quad \|\underline{y}\|, \|\underline{y}'\| \le h_y, \quad \|\underline{z}\| \le h_0$$

(8.2B6) Same as (8.2A6).

(8.2B7) Same as (8.2A7).

(8.2B8) Same as (8.2A8).

(8.2B9) $\underline{f}(k, \underline{z}, \underline{y}, \mu)$ obeys a Lipschitz condition in μ:

$$\|\underline{f}(k, \underline{z}, \underline{y}, \mu) - \underline{f}(k, \underline{z}, \underline{y}, \mu')\| \le L|\mu - \mu'|, \quad \|\underline{z}\| \le h_0, \|\underline{y}\| \le y_0$$

We then have:

Theorem 8.2. For the primary system (8.1a), (8.1b) and the averaged system (8.11a), (8.11c), (8.37) with identical initial conditions, suppose (8.2B1)–(8.2B9) hold. Then given $T > 0$ there exist $b_T > 0$, $\mu_T > 0$ such that

$$\max_{1 \le k \le T/\mu} \|\underline{z}_k - \bar{\underline{z}}_k\| \le c_T(\mu) b_T, \quad 0 \le \mu \le \mu_T$$

and

$$c_T(\mu) \to 0 \text{ as } \mu \to 0.$$

Proof. Once we establish (8.24) the proof proceeds as before.

Proof of (8.24) for system (8.1a). We split $\underline{J}_k$ into three pieces:

$$\underline{J}_k = \underline{J}_{k,1} + \underline{J}_{k,2} + \underline{J}_{k,3}$$

$$\underline{J}_{k,1} = \sum_{s=1}^{k} \Big(\underline{f}(s, \underline{z}_{s-1}, \underline{y}_{s-1}, 0) - \underline{f}(s, \underline{z}_{s-1} \underline{y}_{s-1}(\underline{z}_{s-1}), 0) \Big)$$

$$\underline{J}_{k,2} = \sum_{s=1}^{k} \Big(\underline{f}(s, \underline{z}_{s-1}, \underline{y}_{s-1}(\underline{z}_{s-1}), 0) - \underline{f}_{av}(\underline{z}_{s-1}) \Big)$$

$$\underline{J}_{k,3} = \sum_{s=1}^{k} \Big(\underline{f}(s, \underline{z}_{s-1}, \underline{y}_{s-1}, \mu) - \underline{f}(s, \underline{z}_{s-1}, \underline{y}_{s-1}, 0) \Big)$$

We treat $\underline{J}_{k,1}$ and $\underline{J}_{k,2}$ as before. For $\underline{J}_{k,3}$ we find from (8.2B9) that

$$\begin{aligned} \|\underline{J}_{k,3}\| &\le Lk\mu, \quad 1 \le k \le T_\mu \\ &\le LT, \quad 1 \le k \le T_\mu. \end{aligned}$$

Adding this to the other two bounds gives (8.24) as required.

8.2.2 Weaker Regularity Conditions

As in the single time scale case we can weaken (8.2B2) to

(8.2C2) For $\|\underline{z}\|, \|\underline{z}'\| \le h_0$

$$\begin{aligned} \|\underline{p}(k, \underline{z}) - \underline{p}(k, \underline{z}')\| &\le k\gamma(k) L_p \|\underline{z} - \underline{z}'\| \\ \|\underline{p}(k, \underline{0})\| &\le k\gamma(k) B_p \end{aligned}$$

where $\gamma(k)$ is a non-increasing convergence function $\to 0$ as $k \to \infty$.

Details are much as in the single time scale case and are left to the reader. So in Theorems 8.1, 8.2, (8.2A2), (8.2B2) can be replaced by (8.2C2). We thus have:

Theorem 8.3. For the primary system (8.1a), (8.1b) and the averaged system (8.11a), (8.11c), (8.37) with identical initial conditions, suppose (8.2B1), (8.2C2), (8.2B3)–(8.2B9) hold. Then given $T > 0$ there exist $b_T > 0$, $\mu_T > 0$ such that

$$\max_{1 \le k \le T/\mu} \|\underline{z}_k - \bar{\underline{z}}_k\| \le c_T(\mu) b_T, \quad 0 \le \mu \le \mu_T$$

and

$$c_T(\mu) \to 0 \text{ as } \mu \to 0.$$

Proof. Much the same as before but modified by the same changes as made in the proof of Theorem 7.3.

8.3 FINITE TIME AVERAGING—TIME VARIANT PARAMETERS

To see what form the system takes with time variant parameters we return to Example 8.1, mixed time scale LMS. We begin by replacing (8.1a1) with a time varying specification:

(8.3a1)

$$y_k = s_k = \underline{\varphi}_k^T \underline{\theta}_{ok} = -\sum_{r=1}^{p} s_{k-r} a_{kr} + \sum_{r=1}^{m} x_{k-r} b_{kr}$$

$$\underline{\theta}_{ok} = (a_{k1} \ \cdots \ a_{kp}, b_{k1} \ \cdots \ b_{km})^T$$

With this model we find

$$(1 + a_{ok}(q^{-1}))e_k = -\hat{\underline{\varphi}}_k^T \tilde{\underline{\theta}}_{k-1} = \hat{u}_k$$

where now

$$\tilde{\underline{\theta}}_k = \hat{\underline{\theta}}_k - \underline{\theta}_{ok}$$

and

$$a_{ok}(q^{-1})e_k = a_{k1}e_{k-1} + a_{k2}e_{k-2} + \cdots + a_{kp}e_{k-p}.$$

This can be put into state space form as in Example 8.1, yielding

$$\underline{\xi}_{k+1} = \underline{F}(\underline{a}_{ok})\underline{\xi}_k + \underline{b}\hat{u}_k$$

$$e_k = \underline{c}_k^T \underline{\xi}_k + \hat{u}_k$$

where $\underline{F}(\underline{a})$ is defined in (8.7c) and $\underline{c}_k = \underline{c}(\underline{a}_{ok})$ with

$$\underline{c}(\underline{a}) = -(a_1 \ \cdots \ a_p)^T.$$

Proceeding as before we find, in place of (8.9)

$$\begin{aligned} \underline{\xi}_{k+1} &= \underline{F}(\underline{a}_{ok})\underline{\xi}_k + \underline{b}(-\underline{\varphi}_k - \underline{Q}\underline{\xi}_k)^T \tilde{\underline{\theta}}_{k-1} \\ &= (\underline{F}(\underline{a}_{ok}) - \underline{b}\tilde{\underline{\theta}}_{k-1}^T \underline{Q})\underline{\xi}_k - \underline{b}\underline{\varphi}_k^T \tilde{\underline{\theta}}_{k-1}. \end{aligned} \tag{8.38}$$

On the other hand (8.10) becomes

$$\begin{aligned}\delta\tilde{\underline{\theta}}_k &= \mu\hat{\underline{\varphi}}_k(\underline{c}_k^T\underline{\xi}_k - \hat{\underline{\varphi}}_k^T\tilde{\underline{\theta}}_{k-1}) - \delta\underline{\theta}_{ok} \\ &= \mu(\underline{\varphi}_k + \underline{Q}\underline{\xi}_k)\left(\underline{c}_k^T\underline{\xi}_k - (\underline{\varphi}_k + \underline{Q}\underline{\xi}_k)^T\tilde{\underline{\theta}}_{k-1}\right) - \delta\underline{\theta}_{ok}. \qquad (8.39)\end{aligned}$$

The pair (8.38), (8.39) is the error system. To put these into a form suitable for averaging we follow the large amplitude slowly varying specification of Section 7.2:

$$\underline{\theta}_{ok} = \underline{\theta}_o(\epsilon k); \quad \underline{\theta}(\bullet) \text{ is differentiable; } \epsilon \text{ is small.}$$

The error system now takes the general form

$$\begin{aligned}\delta\underline{z}_k &= \mu\underline{f}(k, \underline{z}_{k-1}, \underline{y}_{k-1}, \epsilon k, \mu) - \delta\underline{z}_o(\epsilon k) \\ \underline{y}_k &= \underline{A}(\epsilon k, \underline{z}_{k-1})\underline{y}_{k-1} + \underline{h}(k, \underline{z}_{k-1}, \epsilon k) + \mu\underline{g}(k, \underline{z}_{k-1}, \underline{y}_{k-1}, \epsilon k, \mu).\end{aligned}$$

As in Section 7.2 we note the appearance of the slow time ϵk. Following the discussion in Section 7.2 we turn immediately to consider the case of matched adaptation $\mu = \epsilon$. So the general system form is

$$\delta\underline{z}_k = \mu\underline{f}(k, \underline{z}_{k-1}, \underline{y}_{k-1}, \mu k, \mu) - \delta\underline{z}_o(\mu k) \qquad (8.40)$$

$$\underline{y}_k = \underline{A}(\mu k, \underline{z}_{k-1})\underline{y}_{k-1} + \underline{h}(k, \underline{z}_{k-1}, \mu k) + \mu\underline{g}(k, \underline{z}_{k-1}, \underline{y}_{k-1}, \mu k, \mu). \qquad (8.41)$$

To assemble an averaged system corresponding to this pair we follow once more our heuristic argument. For simplicity, we consider initially (8.40) with the μ argument missing from $\underline{f}(\bullet)$. Also, as in Section 7.3 we commence by considering the homogeneous or unforced system

$$\delta\underline{z}'_k = \mu\underline{f}(k, \underline{z}'_{k-1}, \underline{y}'_{k-1}, \mu k) \qquad (8.42)$$

$$\underline{y}'_k = \underline{A}(\mu k, \underline{z}'_{k-1})\underline{y}'_{k-1} + \underline{h}(k, \underline{z}'_{k-1}, \mu k) + \mu\underline{g}(k, \underline{z}'_{k-1}, \underline{y}'_{k-1}, \mu k). \qquad (8.43)$$

Sum up (8.42) to find

$$\underline{z}'_{k+N} - \underline{z}'_k = \mu \sum_{s=k+1}^{k+N} \underline{f}(s, \underline{z}'_{s-1}, \underline{y}'_{s-1}, \mu s).$$

If μ is small then $\underline{z}_s$ changes slowly and so does μs; then if N is not too large, we can replace $\underline{z}'_{s-1}$, μs by their values at the beginning of the interval

$$\underline{z}'_{k+N} - \underline{z}'_k = \mu \sum_{s=k+1}^{k+N} \underline{f}(s, \underline{z}'_{k-1}, \underline{y}'_{s-1}, \mu k).$$

Then $\underline{y}'_s$ will be well approximated by $\underline{y}_s(\underline{z}_{k-1}, \mu k)$ where the frozen state $\underline{y}'_k(\underline{z}, \tau)$ obeys

$$\underline{y}'_k(\underline{z}, \tau) = \underline{A}(\tau, \underline{z})\underline{y}'_{k-1}(\underline{z}, \tau) + \underline{h}(k, \underline{z}, \tau).$$

Thus we have

$$\underline{z}'_{k+N} - \underline{z}'_k \simeq \mu \sum_{s=k+1}^{k+N} \underline{f}(s, \underline{z}'_{k-1}, \underline{y}_{s-1}(\underline{z}'_{k-1}, \mu k), \mu k)$$

and we are led to introduce the averaged function

$$\begin{aligned}\underline{f}_{av}(\underline{z}, \tau) &= \text{av}\left\{\underline{f}(s, \underline{z}, \underline{y}_{s-1}(\underline{z}, \tau), \tau)\right\} \\ &= \lim_{N\to\infty} \frac{1}{N} \sum_{s=k+1}^{k+N} \underline{f}(s, \underline{y}_{s-1}(\underline{z}, \tau), \tau).\end{aligned}$$

The resultant averaged system is

$$\delta\bar{\underline{z}}_k = \mu \underline{f}_{av}(\bar{\underline{z}}_{k-1}, \mu k). \tag{8.44}$$

Note that the averaged system is time varying or non-autonomous but it is nevertheless simpler than the original system (8.42). For the more general system

$$\delta\underline{z}'_k = \mu \underline{f}(k, \underline{z}'_{k-1}, \underline{y}'_{k-1}, \mu k, \mu) \tag{8.45}$$

augmented by (8.43) the averaged system will be (8.44) with the averaged function now given by

$$\begin{aligned}\underline{f}_{av}(\underline{z}, \tau) &= \text{av}\left\{\underline{f}(s, \underline{z}, \underline{y}_{s-1}(\underline{z}, \tau), \tau, 0)\right\} \\ &= \lim_{N\to\infty} \frac{1}{N} \sum_{s=k+1}^{k+N} \underline{f}(s, \underline{z}, \underline{y}_{s-1}(\underline{z}, \tau), \tau, 0).\end{aligned} \tag{8.46}$$

Returning to the forced system, we claim, as in Section 7.3 that the partially averaged system is

$$\delta\bar{\underline{z}}_k = \mu \underline{f}_{av}(\bar{\underline{z}}_{k-1}, \mu k) - \delta\underline{z}_o(\mu k). \tag{8.47}$$

We now illustrate these calculations in our example.

Example 8.4. Averaging of Mixed Time Scale LMS with Time Varying Parameters

In fact the calculations are nearly the same as in Example 8.2. The partial averaged system is

$$\delta\bar{\underline{\theta}}_k = \mu \underline{R}(\bar{\underline{\theta}}_{k-1}, \mu k)\bar{\underline{\theta}}_{k-1} - \delta\underline{\theta}_o(\mu k)$$

where

$$\underline{R}(\underline{\theta}, \tau) = \int_{-\pi}^{\pi} \left\{\frac{I - \underline{Q}(e^{j\omega}I - \underline{F}_\tau)^{-1}\underline{b}\underline{\theta}^T}{1 + \alpha(e^{j\omega})}\right\}\left\{\frac{1}{(1 + a_{o\tau}(e^{-j\omega}))(1 + \alpha(e^{-j\omega}))}\right\} \frac{d\underline{F}_\varphi(\omega)}{2\pi}$$

where

$$a_{o\tau}(q) = a_1(\tau)q^{-1} + \cdots + a_p(\tau)q^{-p}$$
$$a_{r,k} = a_r(\mu k)$$
$$\underline{\theta}(\tau) = \left(a_1(\tau) \cdots a_p(\tau)\right)^T$$
$$\underline{F}_\tau = \underline{F}(\underline{a}_{o\tau}).$$

The linearized homogeneous or unforced system will be

$$\delta\underline{\bar{\theta}}_k = -\mu\underline{R}(\underline{0}, \mu k)\underline{\bar{\theta}}_{k-1}$$

where

$$\underline{R}(\underline{0}, \tau) = \int_{-\pi}^{\pi} \frac{d\underline{F}_\varphi(\omega)}{(1 + a_{o\tau}(e^{-j\omega}))} \frac{d\omega}{2\pi}.$$

So if we introduce a time varying version of (8.2a2), namely

(8.3a2) $1 + a_{o\tau}(q^{-1})$ is SPR for all τ, i.e., $\text{Re}(1 + a_{o\tau}(e^{j\omega})) \geq \rho > 0$.

Then $\underline{R}(\underline{0}, \tau)$ is positive definite if (8.2a2) holds and the linearized analysis can proceed as in Section 7.3. Because $\underline{R}(\underline{0}, \tau)$ is positive definite, a slow time varying condition is not needed although it holds anyway.

Now we look at a rigorous development of the results. The proof is much as in Section 7.3. Sum (8.47) (or (8.45)) and subtract it from (8.40) (respectively (8.42)) to find

$$\underline{\Delta}_k = \underline{z}_k - \underline{\bar{z}}_k = \underline{\Delta}_0 + \mu \sum_{s=1}^{k} \left(\underline{f}_{av}(\underline{z}_{s-1}, \mu s) - \underline{f}_{av}(\underline{\bar{z}}_{s-1}, \mu s)\right) + \mu \underline{J}_k$$

$$\underline{J}_k = \sum_{s=1}^{k} \left(\underline{f}(s, \underline{z}_{s-1}, \underline{y}_{s-1}, \mu s, \mu) - \underline{f}_{av}(\underline{z}_{s-1}, \mu s)\right).$$

The following regularity conditions are required:

(8.3A1) For $\|\underline{z}\| \leq h_0$, $\|\underline{y}\| \leq h_y$

$$\|\underline{f}(k, \underline{z}, \underline{y}, \tau, \mu)\| \leq B_f,$$
$$\|\underline{g}(k, \underline{z}, \underline{y}, \tau, \mu)\| \leq B_g.$$

(8.3A2) Introduce the perturbation

$$\underline{p}(k, \underline{z}, \tau) = \sum_{s=1}^{k} \left(\underline{f}(s, \underline{z}, \underline{y}_{s-1}(\underline{z}), \tau, 0) - \underline{f}_{av}(\underline{z}, \tau)\right)$$

and suppose that

$$\|\underline{p}(k, \underline{0}, \tau)\| \leq B_p k\gamma(k)$$

$$\|\underline{p}(k, \underline{z}, \tau) - \underline{p}(k, \underline{z}', \tau)\| \leq (L_{pa}\|\underline{z} - \underline{z}'\| + L_{pb}|\tau - \tau'|)k\gamma(k), \quad \|\underline{z}\|, \|\underline{z}'\| \leq h_0$$

where $\gamma(k)$ is a non-increasing convergence function $\to 0$ as $k \to \infty$.

(8.3A3) $\underline{f}_{av}(\underline{z}, \tau)$ obeys a Lipschitz condition:

$$\|\underline{f}_{av}(\underline{z}, \tau) - \underline{f}_{av}(\underline{z}', \tau)\| \leq L_f \|\underline{z} - \underline{z}'\|, \quad \|\underline{z}\|, \|\underline{z}'\| \leq h_0$$

(8.3A4) Same as (8.2A4).

(8.3A5) Same as (8.2A5) but for $\underline{f}(s, \underline{z}, \underline{y}, \tau)$.

(8.3A6) $\underline{h}(t, \underline{0}, \tau) = \underline{0}$, $\underline{h}(t, \underline{z}, \tau)$ obeys a Lipschitz condition:

$$\|\underline{h}(t, \underline{z}, \tau) - \underline{h}(t, \underline{z}', \tau)\| \leq L_f \|\underline{z} - \underline{z}'\|, \quad \|\underline{z}\|, \|\underline{z}'\| \leq h_0$$

(8.3A7) $\underline{A}(\underline{z}, \tau)$ is a stability matrix for all $\|\underline{z}\| \leq h$; i.e., all eigenvalues of $\underline{A}(\underline{z}, \tau)$ have modulus which is $\leq \lambda < 1$. Also $\|\underline{A}(\underline{z}, \tau)\| \leq A$, $\|\underline{z}\| \leq h_0$.

(8.3A8) $\underline{A}(\underline{z}, \tau)$ obeys a Lipschitz condition:

$$\|\underline{A}(\underline{z}, \tau) - \underline{A}(\underline{z}', \tau')\| \leq L_a \|\underline{z} - \underline{z}'\| + L_b |\tau - \tau'|, \quad \|\underline{z}\|, \|\underline{z}'\| \leq h_0$$

(8.3A9) $\underline{f}(k, \underline{z}, \underline{y}, \tau, \mu)$ obeys a Lipschitz condition in μ for $\|\underline{z}\| \leq h_0$, $\|\underline{y}\| \leq h_y$.

We come to the following result:

Theorem 8.4. For the primary forced system (8.40), (8.41) with associated partially averaged system (8.47) with identical initial conditions, suppose (8.3A1)–(8.3A9) hold. Then given $T > 0$, there exist $b_T > 0$, $\mu_T > 0$ such that

$$\max_{1 \leq k \leq T/\mu} \|\underline{z}_k - \bar{\underline{z}}_k\| \leq c_T(\mu) b_T, \quad 0 \leq \mu \leq \mu_T$$

and $c_T(\mu) \to 0$ as $\mu \to 0$.

Proof. Follows the same route as the proof of Theorems 8.1, 8.2, 8.3. We need only look at $\underline{J}_k$ in detail. We split it into three pieces:

$$\underline{J}_k = \underline{J}_{k,0} + \underline{J}_{k,1} + \underline{J}_{k,2}$$

$$\underline{J}_{k,0} = \sum_{s=1}^{k} \Big(\underline{f}(s, \underline{z}_{s-1}, \underline{y}_{s-1}, \mu s, \mu) - \underline{f}(s, \underline{z}_{s-1}, \underline{y}_{s-1}, \mu s, 0)\Big)$$

$$\underline{J}_{k,1} = \sum_{s=1}^{k} \Big(\underline{f}(s, \underline{z}_{s-1}, \underline{y}_{s-1}, \mu s, 0) - \underline{f}(s, \underline{z}_{s-1}, \underline{y}_{s-1}(\underline{z}_{s-1}), \mu s, 0)\Big)$$

$$\underline{J}_{k,2} = \sum_{s=1}^{k} \Big(\underline{f}(s, \underline{z}_{s-1}, \underline{y}_{s-1}(\underline{z}_{s-1}), \mu s, 0) - \underline{f}_{av}(\underline{z}_{s-1}, \mu s)\Big)$$

For $\underline{J}_{k,0}$ we have by (8.3A9)

$$\|\underline{J}_{k,0}\| \leq L\mu k \leq LT, \quad 0 \leq k \leq T_\mu.$$

Next, $\underline{J}_{k,2}$ is treated as $\underline{J}_{k,2}$ in the proof of Theorem 7.4. Continuing, $\underline{J}_{k,1}$ is treated as $\underline{J}_{k,1}$ in the proof of Theorem 8.1 with some minor changes. In place of (8.34) we get, from (8.3A8)

$$\begin{aligned}\|\underline{A}(\underline{z}_s, \mu s) - \underline{A}(\underline{z}_{s-1}, \mu s - \mu)\| &\leq L_a \|\underline{z}_s - \underline{z}_{s-1}\| + L_b \mu \\ &\leq L_a \mu \|\underline{f}(s, \underline{z}_{s-1}, \underline{y}_{s-1}, \mu)\| + L_b \mu \\ &\leq \mu(L_a B_f + L_b), \quad 1 \leq s \leq T/\mu\end{aligned}$$

which is of the appropriate size. The rest of the proof proceeds as before.

8.4 INFINITE TIME AVERAGING—HOVERING THEOREM

The Hovering theorem holds in the same way as it does in the single time scale case. We only need to make (8.2A2) or (8.2B2) or (8.2C2) uniform. So we introduce

(8.4A1) Consider the perturbation

$$\underline{p}(k, \underline{z}) = \sum_{s=1}^{k} \left(\underline{f}(s, \underline{z}, \underline{y}_{s-1}(\underline{z}), 0) - \underline{f}_{av}(\underline{z}) \right)$$

and suppose there is a non-increasing convergence function $\gamma(k) \geq 0$ so that for any r

$$\|\underline{p}_r(k + r, \underline{z}) - \underline{p}_r(k + r, \underline{z}')\| \leq k\gamma(k) L_p \|\underline{z} - \underline{z}'\|, \quad \|\underline{z}\|, \|\underline{z}'\| \leq h_0$$

where

$$\underline{p}_r(k, \underline{z}) = \underline{p}(k, \underline{z}) - \underline{p}(r, \underline{z}).$$

Then we have:

Theorem 8.5. (Hovering theorem). Consider the primary system (8.1) and the associated averaged system (8.11c), (8.4A1), (8.2B3)–(8.2B9) hold. Suppose $\underline{z}_e$ is an ES equilibrium point of the averaged system (but we do not assume $\underline{z}_e$ is an equilibrium point of the primary system) and the initial condition is in the stability region for $\underline{z}_e$. Then

$$\sup_{k \geq 1} \|\underline{z}_k - \bar{\underline{z}}_k\| \leq \xi(\mu)$$

where $\xi(\mu) \to 0$ as $\mu \to 0$.

Proof. Same as the proof of Theorem 7.9.

8.5 INFINITE TIME AVERAGING—STABILITY VIA PERTURBED LYAPUNOV FUNCTION

The ideas here are much the same as in the single time scale case of Section 7.6. Recall the block diagram representation of the mixed time scale algorithm as an interconnected feedback system and also the discussion in Section 6.2. There it was mentioned that a joint Lyapunov stability analysis was possible using Lyapunov functions for both the fast and slow subsystems. It is thus not surprising to find here that such a joint Lyapunov computation is needed.

In any case we start here, as in the single time scale case, with a Lyapunov computation based on the slow system. Soon we will find it necessary to add in a Lyapunov function for the fast system too. We introduce the following regularity conditions:

(8.5A1) $\underline{0}$ is an equilibrium point of (8.1a).

(8.5A2) $\underline{0}$ is an ES equilibrium point of (8.11c).

(8.5A3) $\underline{f}(k, \underline{z}, \underline{y})$, $\underline{f}_{av}(\underline{z})$ obey Lipschitz conditions

$$\|\underline{f}(k, \underline{z}, \underline{y}) - \underline{f}(k, \underline{z}', \underline{y})\| \le L\|\underline{z} - \underline{z}'\|$$

$$\|\underline{f}_{av}(\underline{z}) - \underline{f}_{av}(\underline{z}')\| \le L\|\underline{z} - \underline{z}'\|$$

for $\|\underline{z}\|, \|\underline{z}'\| \le h_0$.

(8.5A4) $\|\underline{f}_{av,z_i}(\underline{z})\| \le d$, $\|\underline{z}\| \le h_0$, $1 \le i \le p$; where

$$\underline{f}_{av,z_i}(\underline{z}) = d^2\underline{f}_{av}(\underline{z})/d\underline{z}^T dz_i.$$

(8.5A5) $\underline{f}(k, \underline{z}, \underline{y})$ obeys a Lipschitz condition in $\underline{y}$:

$$\|\underline{f}(k, \underline{z}, \underline{y}) - \underline{f}(k, \underline{z}, \underline{y}')\| \le L_f\|\underline{y} - \underline{y}'\|, \quad \|\underline{y}\|, \|\underline{y}'\| \le h_y, \quad \|\underline{z}\|, \|\underline{z}'\| \le h_0$$

(8.5A6) Let $\tilde{\underline{f}}(s, \underline{z}) = \underline{f}(s, \underline{z}, \underline{y}_{s-1}(\underline{z})) - \underline{f}_{av}(\underline{z})$ and then suppose

$$\|\sum_{s=k+1}^{k+N}\left(\tilde{\underline{f}}(s, \underline{z}) - \tilde{\underline{f}}(s, \underline{z}')\right)\| \le L_p\|\underline{z} - \underline{z}'\|\gamma(N)N, \quad \|\underline{z}\|, \|\underline{z}'\| \le h_0.$$

(8.5A7) Same as (8.2A7).

(8.5A8) Same as (8.2A8).

(8.5A9) $\underline{g}(k, \underline{0}, \underline{0}, \mu) = \underline{0}$; $\underline{g}(k, \underline{z}, \underline{y}, \mu)$ obeys a Lipschitz condition

$$\|\underline{g}(k, \underline{z}, \underline{y}, \mu) - \underline{g}(k, \underline{z}', \underline{y}', \mu)\| \le L_a\|\underline{z} - \underline{z}'\| + L_b\|\underline{y} - \underline{y}'\|$$

for $\|\underline{z}\|, \|\underline{z}'\| \le h_0$ and $\|\underline{y}\|, \|\underline{y}'\| \le h_y$.

(8.5A10) $\underline{h}(k, \underline{0}) = \underline{0}$; $\underline{h}(k, \underline{z})$ obeys a Lipschitz condition

$$\|\underline{h}(k, \underline{z}) - \underline{h}(k, \underline{z}')\| \le L_h\|\underline{z} - \underline{z}'\|, \quad \|\underline{z}\|, \|\underline{z}'\| \le h_0.$$

We now prove the following result:

Theorem 8.6. Consider the primary system (8.1) and the averaged system (8.11c). Suppose (8.5A1)–(8.5A10) hold. Then (8.1) is locally ES provided μ is sufficiently small.

Proof. From Theorem 7.6 there is a Lyapunov function $V(\underline{z})$ for (8.11c) obeying (C7a1)–(C7a4) of Appendix C7. We now proceed exactly as in Section 7.5 except for two points:

(i) Treatment of $\underline{y}_t$. We require

$$\|\underline{y}_0\| \le H_y, \text{ where } H_y \text{ is to be chosen.} \tag{8.48}$$

Then suppose inductively

$$\|\underline{y}_{t-1}\| \le h_y/(1+\mu L), \quad 1 \le t \le k. \tag{8.49}$$

Provided $H_y \le h_y/(1+\mu L)$, this holds for $\|\underline{y}_0\|$.

(ii) In the development of the perturbed Lyapunov function everything proceeds exactly as before, leading to equation (7.60) (note that, for the discounted perturbation (7.55), $\underline{\tilde{f}}(s, \underline{z})$ is defined in (8.5A6)), except that $\underline{\tilde{f}}_k$ in (7.61) is now given by

$$\underline{\tilde{f}}_k = \underline{f}(k, \underline{z}_{k-1}, \underline{y}_{k-1}) - \underline{f}_{av}(\underline{z}_{k-1}) \tag{8.50}$$

while in (7.62) $\underline{\tilde{f}}_k$ is replaced by $\underline{\hat{f}}_k$:

$$\underline{\hat{f}}_k = \underline{f}(k, \underline{z}_{k-1}, \underline{y}^*_{k-1}) - \underline{f}_{av}(\underline{z}_{k-1}) \tag{8.51}$$

$$\underline{y}^*_{k-1} = \underline{y}_{k-1}(\underline{z}_{k-1}). \tag{8.52}$$

Thus in the "$A_k + B_k$" inequality (7.65a) there is an extra term

$$(1-\mu)\mu(\underline{\hat{f}}_k - \underline{\tilde{f}}_k)^T V_z(\underline{z}_{k-1}).$$

In view of (8.5A5) ((C7a3) of Appendix C7) this is bounded as follows:

$$(1-\mu)\mu(\underline{\hat{f}}_k - \underline{\tilde{f}}_k)^T V_z(\underline{z}_{k-1}) \le \mu L \|\underline{\delta}_{k-1}\| \alpha_4 \|\underline{z}_{k-1}\| \tag{8.53}$$

$$\underline{\delta}_{k-1} = \underline{y}_{k-1} - \underline{y}^*_{k-1} \tag{8.54}$$

There is now another problem here because this bound is $O(\mu)$. This term can be handled by adding in the Lyapunov function for the fast system. In fact, the Lyapunov function will be assembled in two stages. Firstly a Lyapunov function for the fast error state in (8.32)

$$\underline{\delta}_s = \underline{A}(\underline{z}_{s-1})\underline{\delta}_{s-1} + \mu \underline{g}(s, \underline{z}_{s-1}, \underline{y}_{s-1}, \mu) + \underline{\rho}_s \tag{8.55a}$$

$$\underline{\rho}_s = \underline{A}(\underline{z}_{s-1})\left(\underline{y}_{s-1}(\underline{z}_{s-2}) - \underline{y}_{s-1}(\underline{z}_{s-1})\right)$$

and then one for the semi-frozen state (8.52) (cf. (8.31))

$$\underline{y}^*_k = \underline{A}(\underline{z}_{k-1})\underline{y}^*_{k-1} + \underline{h}(k, \underline{z}_{k-1}) - \underline{\rho}_k. \tag{8.55b}$$

Note that (8.5A7) and (8.5A8) ensure that the homogeneous time varying system

$$\underline{u}_s = \underline{A}_s \underline{u}_{s-1}, \quad \underline{A}_s = \underline{A}(\underline{z}_{s-1})$$

is ES to $\underline{0}$. We use, then, the Lyapunov function of Appendix C6(v), namely

$$V_{\delta,k} = \underline{\delta}_k^T \underline{P}_k \underline{\delta}_k \tag{8.56}$$

where

$$\underline{P}_k = \underline{P}(\underline{z}_{k-1}) = \sum_{r=0}^{\infty} \underline{A}_k^{Tr} \underline{A}_k^r$$

$$= \sum_{r=0}^{\infty} \underline{A}(\underline{z}_{k-1})^{Tr} \underline{A}(\underline{z}_{k-1})^r$$

and $\underline{P}_k$ has the following properties:

$$\underline{P}_k \geq I, \quad \|\underline{P}_k\| \leq P \tag{8.57a}$$

$$\|\underline{P}_k - \underline{P}_{k-1}\| \leq \mu c, \quad \text{for some } c \tag{8.57b}$$

$$\underline{A}_k^T \underline{P}_k \underline{A}_k = \underline{P}_k - I. \tag{8.57c}$$

Then applying $V_{\delta,k}$ to (8.55a) gives

$$V_{\delta,k} = (\underline{A}_k \underline{\delta}_{k-1} + \mu \underline{g}_k + \underline{\rho}_k)^T \underline{P}_k (\underline{A}_k \underline{\delta}_{k-1} + \mu \underline{g}_k + \underline{\rho}_k)$$

with

$$\underline{g}_k = \underline{g}(k, \underline{z}_{k-1}, \underline{y}_{k-1}, \mu).$$

Expanding gives, via (8.57c),

$$\begin{aligned} V_{\delta,k} &= \underline{\delta}_{k-1}^T \underline{A}_k^T \underline{P}_k \underline{A}_k \underline{\delta}_{k-1} + 2\mu(\underline{g}_k + \underline{\rho}_k)^T \underline{P}_k \underline{A}_k \underline{\delta}_{k-1} + \mu^2 (\underline{g}_k + \underline{\rho}_k)^T \underline{P}_k (\underline{g}_k + \underline{\rho}_k) \\ &= \underline{\delta}_{k-1}^T \underline{P}_k \underline{\delta}_{k-1} - \|\underline{\delta}_{k-1}\|^2 + 2\mu(\|\underline{g}_k\| + \|\underline{\rho}_k\|) B \|\underline{\delta}_{k-1}\| + 2\mu^2 P(\|\underline{g}_k\|^2 + \|\underline{\rho}_k\|^2). \end{aligned}$$

So for small μ we get for some a (via (8.57))

$$\begin{aligned} V_{\delta,k} &\leq V_{\delta,k-1} - \frac{3}{4}\|\underline{\delta}_{k-1}\|^2 + \frac{1}{8}\|\underline{\delta}_{k-1}\|^2 + \mu^2 a(\|\underline{g}_k\|^2 + \|\underline{\rho}_k\|^2) \\ &= V_{\delta,k-1} - \frac{5}{8}\|\underline{\delta}_{k-1}\|^2 + \mu^2 a(\|\underline{g}_k\|^2 + \|\underline{\rho}_k\|^2). \end{aligned}$$

From (8.5A9) we find

$$\begin{aligned} \|\underline{g}_k\| &\leq L_a \|\underline{z}_{k-1}\| + L_b \|\underline{y}_{k-1}\| \\ &\leq L_a \|\underline{z}_{k-1}\| + L_b \|\underline{\delta}_{k-1}\| + L_b \|\underline{y}_{k-1}^*\|. \end{aligned}$$

Notice that (8.36) is valid in the current infinite time setting so that from (8.5A3), (8.5A5)

$$\begin{aligned} \|\underline{\rho}_k\| &\leq Ab'\mu L \left(\|\underline{z}_{k-2}\| + \|\underline{y}_{k-1}\|\right) \\ &\leq \mu b' AL \left(\|\underline{z}_{k-2}\| + \|\underline{y}_{k-1}^*\| + \|\underline{\delta}_{k-1}\|\right) \\ &\leq \mu b' AL \left(\|\underline{z}_{k-1}\| + \|\underline{y}_{k-1}^*\| + \|\underline{\delta}_{k-1}\|\right)(1 - \mu L)^{-1} \\ &\leq 2\mu b' AL \left(\|\underline{z}_{k-1}\| + \|\underline{y}_{k-1}^*\| + \|\underline{\delta}_{k-1}\|\right), \quad \text{for small } \mu. \end{aligned}$$

Thus for some b

$$V_{\delta,k} \le V_{\delta,k-1} - \frac{5}{8}\|\underline{\delta}_{k-1}\|^2 + \mu^2 b\left(L_a^2\|\underline{z}_{k-1}\|^2 + L_b^2\|\underline{\delta}_{k-1}\|^2 + L_b^2\|\underline{y}^*_{k-1}\|^2\right).$$

For small μ we get

$$V_{\delta,k} \le V_{\delta,k-1} - \frac{9}{16}\|\underline{\delta}_{k-1}\|^2 + \mu^2 b\left(L_a^2\|\underline{z}_{k-1}\|^2 + L_b^2\|\underline{y}^*_{k-1}\|^2\right). \tag{8.58}$$

To deal with the $\underline{y}^*_k$ term we need the Lyapunov function $V_{y,k} = \underline{y}^{*T}_k \underline{P}_k \underline{y}^*_k$. As for (8.56) we find for some $b > 0$

$$V_{y,k} \le V_{y,k-1} - \frac{9}{16}\|\underline{y}^*_{k-1}\|^2 + \mu^2 b\|\underline{h}_k\|^2 \quad ; \quad \underline{h}_k = \underline{h}(k, \underline{z}_{k-1}).$$

Then from (8.5A10) we get

$$\|\underline{h}_k\| \le L_h\|\underline{z}_{k-1}\|.$$

Thus

$$V_{y,k} \le V_{y,k-1} - \frac{9}{16}\|\underline{y}^*_{k-1}\|^2 + \mu^2 b L_h^2\|\underline{z}_{k-1}\|^2. \tag{8.59}$$

Putting the two Lyapunov inequalities together gives for small μ and some $c > 0$

$$V_{\delta,k} + V_{y,k} \le V_{\delta,k-1} + V_{y,k-1} - \frac{1}{2}\|\underline{\delta}_{k-1}\|^2 - \frac{1}{2}\|\underline{y}^*_{k-1}\|^2 + \mu^2 c\|\underline{z}_{k-1}\|^2.$$

Returning to (7.60), (7.65a), and (8.53) we get

$$\left[V(k, \underline{z}_k) - V(k-1, \underline{z}_{k-1})\right] + \left[V_{y,k} - V_{y,k-1}\right] + \left[V_{\delta,k} - V_{\delta,k-1}\right] \le AB_k + C_k + D_k$$

with C_k, D_k as before but

$$\begin{aligned} AB_k &= A_k + B_k + \mu\alpha_4 L\|\underline{\delta}_{k-1}\|\|\underline{z}_{k-1}\| - \frac{1}{2}\|\underline{\delta}_{k-1}\|^2 - \frac{1}{2}\|\underline{y}^*_{k-1}\|^2 + \mu^2 c\|\underline{z}_{k-1}\|^2 \\ &\le A_k + B_k - \frac{1}{4}\|\underline{\delta}_{k-1}\|^2 - \frac{1}{4}\|\underline{y}^*_{k-1}\|^2 + \mu^2 d\|\underline{z}_{k-1}\|^2 \end{aligned}$$

for small μ and some d. In place of (7.63) we thus conclude

$$\delta W_k \le -\frac{1}{2}\mu(\alpha_3 - \psi(\mu))(\|\underline{z}_{k-1}\|^2 + \|\underline{\delta}_{k-1}\|^2 + \|\underline{y}^*_{k-1}\|^2)$$

where

$$W_k = V(k, \underline{z}_k) + V_{y,k} + V_{\delta,k}$$
$$\psi(\mu) \to 0 \text{ as } \mu \to 0.$$

The rest of the proof is completed as before, via (8.57).

8.6 INFINITE TIME AVERAGING—TIME INVARIANT PARAMETERS

We now apply the results of Section 8.5 to an example.

Example 8.5. Convergence of Mixed Time Scale LMS

We already did a linearized stability analysis of the averaged system (8.16) in Section 8.2. To connect that to ES of (8.16) we appeal to Theorem C7.1. To conclude ES of (8.16) from ES of (8.17) we have to show

$$\lim_{\|\underline{\tilde{\theta}}\|\to 0} \|(\underline{R}(\underline{\tilde{\theta}}) - \underline{R}(\underline{0}))\underline{\tilde{\theta}}\|/\|\underline{\tilde{\theta}}\| = 0.$$

This follows if $\underline{R}(\underline{\tilde{\theta}})$ is continuous at $\underline{0}$. From (8.15)

$$\underline{R}(\underline{\tilde{\theta}}) - \underline{R}(\underline{0}) = A(\tilde{\theta}) + B(\tilde{\theta})$$

$$A(\tilde{\theta}) = \frac{1}{2\pi}\int_{-\pi}^{\pi}\left(\frac{1}{(1+a_o(e^{j\omega}))(1+\alpha(e^{-j\omega}))} - \frac{1}{1+a_o(e^{j\omega})}\right) d\underline{F}_{\varphi}(\omega)$$

$$= \frac{1}{2\pi}\int_{-\pi}^{\pi}\frac{-\alpha(e^{-j\omega})}{(1+a_o(e^{j\omega}))(1+\alpha(e^{-j\omega}))} d\underline{F}_{\varphi}(\omega)$$

$$B(\tilde{\theta}) = \frac{1}{2\pi}\int_{-\pi}^{\pi}\frac{\underline{Q}(e^{j\omega}I - \underline{F})^{-1}\underline{b}\underline{\tilde{\theta}}^{T}}{|1+\alpha(e^{j\omega})|^2(1+a_o(e^{j\omega}))} d\underline{F}_{\varphi}(\omega).$$

We see easily that $A(\underline{\tilde{\theta}})$ and $B(\underline{\tilde{\theta}})$ are both continuous in $\underline{\tilde{\theta}}$ so the result follows. The reader is invited now to consider the regularity conditions in Section 8.5. They are all straightforward to check.

8.7 NOTES

The Bogoliuboff approach of course handles mixed time scale systems (see, e.g., [B1], [B2], [H3], and [S3]) but as before is more cumbersome than the one used here. Finite time averaging analysis of mixed time scale systems with a nonlinear fast state (but no slow time) was developed by [V5]. Infinite time results (including, as well, systems with a slow time) were developed by [R3] for systems with a semi-linear fast state. These results were extended to cover systems with a nonlinear fast state by [B2]. In a long memory context, mixed time scale systems with a semi-linear fast state were introduced into the adaptive area in [L6]. The first application of deterministic averaging results for slowly time varying systems to adaptive algorithms is in [S17]. Power and amplitude bounds can be found in an infinite time averaging time variant parameter setting much as in Section 7.8, and there is little new to be learned about this in the mixed time scale setting. Of course slow system and fast system Lyapunov functions are used. Similarly an ode associated with (8.11c), namely (7.85), is obtained, as in Section 7.9; again there is nothing new, given the earlier sections of this chapter.

EXERCISES

8.1 Show that the momentum algorithm described in Section 4.6 can be formulated as a mixed time scale algorithm. Discuss convergence of

(a) the momentum LMS algorithm.

(b) the SD time delay estimation algorithm sketched in Section 4.6.

8.2 Consider adaptive estimation of the parameter a_o in the single pole model

$$s_k = a_o s_{k-1} + x_k.$$

A pseudo SD algorithm will have the form

$$\begin{aligned} \delta\hat{a}_k &= \mu\hat{s}_{k-1}e_k \\ e_k &= s_k - \hat{s}_k \\ \hat{s}_k &= \hat{a}_{k-1}\hat{s}_{k-1} + x_k. \end{aligned}$$

Now suppose a projection feature is used to ensure stability of the fast state. So the $\hat{s}_k$ equation is replaced with

$$\hat{s}_k = f(\hat{a}_{k-1})\hat{s}_{k-1} + x_k$$

$$f(\hat{a}) = \begin{cases} \hat{a}, & \text{if } |\hat{a}| \le \rho < 1 \\ \rho/\hat{a}, & \text{if } |\hat{a}| > \rho. \end{cases}$$

Find the averaged system associated with this algorithm. See if you can develop a perturbed Lyapunov based stability analysis that yields ES if μ is small enough.

8.3 Consider the SD_2 algorithm developed in Section 4.4 for an IIR filter—namely (4.29)–(4.34). Find the averaged system and develop a perturbed Lyapunov based stability analysis.

PART III

STOCHASTIC AVERAGING

In this last part we develop a stochastic version of averaging theory. The ideas have much similarity to the deterministic case. However the background needed is much greater and the technical detail increases. Still, we feel that the ideas retain an essential simplicity. Even so we have not attempted to provide full rigor or treatment since that would require a monograph in itself. We have however given enough detail to ensure that the diligent reader can retain a "grip" on the discussion. At many places in the next two chapters we do indeed reach the research frontier existing at the time of writing.

9

Stochastic Averaging: Single Time Scale

9.1 INTRODUCTION

In the deterministic setting we confined attention to stability, while in the stochastic setting performance becomes an issue. Stochastic stability is typically addressed by means of a first order analysis, while questions of performance (e.g., involving misadjustment) are illuminated by second order calculations. Note that, as with stability, performance determination requires an infinite time analysis.

9.2 FINITE TIME AVERAGING: TIME INVARIANT PARAMETERS

To motivate the system to be considered here, let us again refer to the LMS algorithm and recall the error system (5.4), which describes LMS in the presence of noise:

$$\delta \underline{\tilde{w}}_k = -\mu \underline{x}_k \underline{x}_k^T \underline{\tilde{w}}_{k-1} + \mu \underline{x}_k \epsilon_k \tag{9.1}$$

where now $\underline{x}_k$, ϵ_k are stochastic, but not supposed to be Gaussian or white noise. With this example in mind we are led to consider the following general primary system:

$$\delta \underline{z}_k = \mu \underline{f}(k, \underline{z}_{k-1}), \quad \underline{f}(k, \underline{z}) \text{ is stochastic.} \tag{9.2}$$

The primary system (9.2) is sometimes written in the form

$$\delta \underline{z}_{k-1} = \mu \underline{f}(\underline{\xi}_k, \underline{z}_{k-1}) \tag{9.3}$$

where $\underline{\xi}_k$ denotes a driving stochastic signal. Now we claim that the first order averaged system associated with (9.2) is

$$\delta \underline{\bar{z}}_k = \mu \underline{f}_{av}(\underline{\bar{z}}_{k-1}) \tag{9.4a}$$

$$\underline{f}_{av}(\underline{z}) = E(\underline{f}(k, \underline{z})) \tag{9.4b}$$

where we have assumed

(9.2A1) $E(\underline{f}(k, \underline{z}))$ is time invariant.

9.2.1 First Order Analysis

An analysis can be developed to justify the association of (9.4) with (9.2) in a manner similar to the deterministic case of Section 7.2. Sum (9.2) and (9.4a) and subtract them to find

$$\underline{\Delta}_k = \underline{z}_k - \underline{\bar{z}}_k = \underline{\Delta}_0 + \mu \sum_{s=1}^{k} \left(\underline{f}(s, \underline{z}_{s-1}) - \underline{f}_{av}(\underline{\bar{z}}_{s-1}) \right)$$

$$= \underline{\Delta}_0 + \mu \sum_{s=1}^{k} \left(\underline{f}(s, \underline{z}_{s-1}) - \underline{f}(s, \underline{\bar{z}}_{s-1}) \right) + \mu \underline{J}_k \tag{9.5a}$$

$$\underline{J}_k = \sum_{s=1}^{k} \left(\underline{f}(s, \underline{\bar{z}}_{s-1}) - \underline{f}_{av}(\underline{\bar{z}}_{s-1}) \right). \tag{9.5b}$$

Notice that the split here is a little different to that in (7.13), (7.14): The reason will soon be clear. Now we introduce some regularity conditions. Fix an interval $[0, T]$ and suppose

(9.2A2) $\underline{f}(s, \underline{z})$ obeys a global stochastic Lipschitz condition

$$\|\underline{f}(s, \underline{z}) - \underline{f}(s, \underline{z}')\| \le L_s \|\underline{z} - \underline{z}'\|, \quad \text{for all } \underline{z}, \underline{z}'$$

where L_s is a random sequence which obeys a strong law of large number (SLLN): as $N \to \infty$

$$\frac{1}{N} \sum_{s=1}^{N} L_s \to L \quad \text{w.p.1.}$$

Then in (9.5a) we find

$$\|\underline{\Delta}_k\| \le \|\underline{\Delta}_0\| + \mu \sum_{s=1}^{k} L_s \|\underline{\Delta}_{s-1}\| + \mu \|\underline{J}_k\|. \tag{9.6}$$

As in the deterministic case we will approximate $\underline{J}_k$ by a perturbation $\underline{P}_k = \underline{p}(k, \underline{\bar{z}}_{k-1})$ where

$$\underline{p}(k, \underline{z}) = \sum_{s=1}^{k} \left(\underline{f}(s, \underline{z}) - \underline{f}_{av}(\underline{z}) \right).$$

We suppose then

(9.2A3) $\underline{p}(k, \underline{z})$ obeys a global Lipschitz condition,

$$\|\underline{p}(k, \underline{z}) - \underline{p}(k, \underline{z}')\| \le M_k \|\underline{z} - \underline{z}'\|, \quad \text{for all } \underline{z}, \underline{z}'$$
$$\|\underline{p}(k, \underline{0})\| \le p_k$$

where, as $k \to \infty$

$$M_k/k \to 0 \quad \text{w.p.1}$$
$$p_k/k \to 0 \quad \text{w.p.1.}$$

It follows from (9.2A3) that $N^{-1} \sum_{s=1}^{N} \underline{f}(s, \underline{z}) \to \underline{f}_{av}(\underline{z})$ w.p.1. Then (see Exercise 9.1) (9.2A2) ensures that $\underline{f}_{av}(\underline{z})$ obeys a Lipschitz condition.

(9.2A4) $\|\bar{\underline{z}}_k\| \le h$, for $1 \le k \le T/\mu$
(9.2A5) $\|\underline{f}_{av}(\underline{z})\| \le B_a$, for $\|\underline{z}\| \le h$

Much as in the deterministic case we find

$$\underline{P}_k - \underline{P}_{k-1} = \underline{J}_k - \underline{J}_{k-1} + \underline{\eta}_k \tag{9.7}$$

$$\underline{\eta}_k = \underline{p}(k-1, \bar{\underline{z}}_{k-1}) - \underline{p}(k-1, \bar{\underline{z}}_{k-2})$$

$$\Rightarrow \quad \|\underline{\eta}_k\| \le M_{k-1}\|\bar{\underline{z}}_{k-1} - \bar{\underline{z}}_{k-2}\|, \quad \text{by (9.2A3)}$$
$$\le M_{k-1}\mu\|\underline{f}_{av}(\bar{\underline{z}}_{k-2})\|$$
$$\le \mu B_a M_{k-1} \quad \text{(by (9.2A5))}. \tag{9.8}$$

On the other hand it also follows from (9.2A3) that

$$\|\underline{P}_k\| \le \|\underline{p}(k, \bar{\underline{z}}_{k-1}) - \underline{p}(k, \underline{0})\| + \|\underline{p}(k, \underline{0})\|$$
$$\le M_k\|\bar{\underline{z}}_{k-1}\| + p_k$$
$$\le M_k h + p_k.$$

Thus from (9.7), (9.8) we find

$$\mu\|\underline{J}_k\| \le \mu\|\underline{P}_k\| + \mu\|\sum_{s=1}^{k} \underline{\eta}_s\|$$

$$\le \mu M_k h + \mu p_k + \mu^2 B_a \sum_{s=1}^{k} M_{s-1}. \tag{9.9a}$$

From (9.2A3) we have that, as $\mu \to 0$ (see Exercise 9.2)

$$\mu \max_{1 \le k \le T/\mu} M_k \to 0 \quad \text{w.p.1}$$

$$\mu \max_{1 \le k \le T/\mu} p_k \to 0 \quad \text{w.p.1}$$

so that

$$\mu^2 \max_{1\le k\le T/\mu} \sum_{s=1}^{k} M_{s-1} \le T\mu \max_{1\le k\le T/\mu} M_k \to 0 \quad \text{w.p.1.}$$

Thus we have, as $\mu \to 0$

$$\mu\|\underline{J}_k\| \le \mu \max_{1\le k\le T/\mu} \|\underline{J}_k\| = d_T(\mu) \to 0 \quad \text{w.p.1.} \tag{9.9b}$$

Finally from (9.6) and the Bellman-Gronwall lemma it follows that

$$\|\underline{\Delta}_k\| \le (\|\underline{\Delta}_0\| + d_T(\mu))e^{\mu \sum_{s=1}^{k} L_s}$$

$$\le (\|\underline{\Delta}_o\| + d_T(\mu))e^{TL + \mu \max_{1\le k\le T/\mu} \sum_{s=1}^{k}(L_s - L)}$$

but in view of (9.2A2) (cf. Exercise 9.2)

$$\mu \max_{1\le k\le T/\mu} \sum_{s=1}^{k} (L_s - L) \to 0 \quad \text{w.p.1,} \quad \text{as } \mu \to 0.$$

Thus the following result has been established:

Theorem 9.1. (First order finite time averaging). For the primary system (9.2) and averaged system (9.4) with identical initial condition, suppose (9.2A1)–(9.2A5) hold. Then given $T > 0$, there exists $c_T(\mu)$ so that

$$\max_{1\le k\le T/\mu} \|\underline{z}_k - \bar{\underline{z}}_k\| \le c_T(\mu) \quad \text{w.p.1} \tag{9.10a}$$

$$c_T(\mu) \to 0 \quad \text{w.p.1} \quad \text{as } \mu \to 0 \tag{9.10b}$$

This result is now illustrated with the LMS algorithm.

Example 9.1. LMS Algorithm

Let us firstly consider the noise-free case where

$$\underline{f}(k, \underline{w}) = -\underline{x}_k \underline{x}_k^T \underline{w}.$$

Then (9.2A1) requires

(9.2a1) $E(\underline{x}_k \underline{x}_k^T) = \underline{R}_x$

For (9.2A2) we have $L_k = \|\underline{x}_k\|^2$ and so require

(9.2a2) $N^{-1} \sum_{k=1}^{N} \|\underline{x}_k\|^2 \to r_x$ w.p.1.

For (9.2A3) we find

$$\underline{p}(k, \underline{w}) - \underline{p}(k, \underline{w}') = -\sum_{s=1}^{k} (\underline{x}_s \underline{x}_s^T - \underline{R}_x)(\underline{w} - \underline{w}')$$

so that $\underline{p}(k, \underline{0}) = 0$ while

$$M_k = \| \sum_{s=1}^{k} (\underline{x}_s \underline{x}_s^T - \underline{R}_x) \|.$$

Thus we need

(9.2a3) $N^{-1} \sum_{k=1}^{N} \underline{x}_k \underline{x}_k^T \to \underline{R}_x$ w.p.1.

Turning to the noisy case we have

$$\underline{f}(k, \underline{w}) = -\underline{x}_k \underline{x}_k^T \underline{w} + \underline{x}_k \epsilon_k$$

so that we require

(9.2a4)

$$E(\underline{x}_k \epsilon_k) = \underline{0}. \tag{9.11}$$

Next (9.2A2) is satisfied if (9.2a2) holds, while in (9.2A3) we now find

$$\underline{p}(k, \underline{0}) = \sum_{s=1}^{k} \underline{x}_s \epsilon_s.$$

Thus we require

(9.2a5) $N^{-1} \sum_{k=1}^{N} \underline{x}_k \epsilon_k \to 0$ w.p.1.

If $(\underline{x}_k, \epsilon_k)$ is strictly stationary then the ergodic theorem and (9.9) ensure (9.2a2), (9.2a5). Other conditions could be allowed too. Thus we have established that the first order averaged system associated with (9.1) is

$$\delta \underline{\bar{w}}_k = -\mu \underline{R}_x \underline{\bar{w}}_{k-1}.$$

9.2.2 Extensions

It is straightforward to extend the analysis to deal with systems of the form

$$\delta \underline{z}_k = \mu \underline{f}(k, \underline{z}_{k-1}, \mu). \tag{9.12}$$

Under suitable conditions the associated averaged system will be (9.4a) where now

$$\underline{f}_{av}(\underline{z}) = E(\underline{f}(k, \underline{z}, 0)) \tag{9.13}$$

and we have supposed

(9.2B1) $E(\underline{f}(k, \underline{z}, 0))$ is time invariant.

We further suppose

(9.2B2) $\underline{f}(s, \underline{z}, \mu)$ obeys a global Lipschitz condition

$$\|\underline{f}(s, \underline{z}, \mu) - \underline{f}(s, \underline{z}', \mu)\| \le L_s \|\underline{z} - \underline{z}'\|, \quad \text{for all } \underline{z}, \underline{z}'$$

where L_s is a random sequence which obeys an SLLN.

(9.2B3) Introduce the perturbation

$$\underline{p}(k, \underline{z}) = \sum_{s=1}^{k} \left(\underline{f}(s, \underline{z}, 0) - \underline{f}_{av}(\underline{z}) \right)$$

and then suppose (9.2A3) holds for this new $\underline{p}(k, \underline{z})$.

(9.2B4) As for (9.2A4).

(9.2B5) As for (9.2A5).

(9.2B6) $\underline{f}(k, \underline{z}, \mu)$ obeys a global Lipschitz condition in μ:

$$\|\underline{f}(k, \underline{z}, \mu) - \underline{f}(k, \underline{z}, \mu')\| \le B_k|\mu - \mu'|, \quad \text{for all } \underline{z}$$

where B_k obeys an SLLN: $N^{-1} \sum_{k=1}^{N} B_k \to B$ w.p.1.

Again the argument is developed much as in the deterministic case. Proceeding as before we obtain (9.5a) where now

$$\underline{J}_k = \sum_{s=1}^{k} \left(\underline{f}(s, \bar{\underline{z}}_{s-1}, \mu) - \underline{f}_{av}(\bar{\underline{z}}_{s-1}) \right).$$

This is next split into two pieces:

$$\underline{J}_k = \underline{J}_{k1} + \underline{J}_{k2}$$

$$\underline{J}_{k1} = \sum_{s=1}^{k} \left(\underline{f}(s, \bar{\underline{z}}_{s-1}, \mu) - \underline{f}(s, \bar{\underline{z}}_{s-1}, 0) \right)$$

$$\underline{J}_{k2} = \sum_{s=1}^{k} \left(\underline{f}(s, \bar{\underline{z}}_{s-1}, 0) - \underline{f}_{av}(\bar{\underline{z}}_{s-1}) \right)$$

Now $\underline{J}_{k2}$ can be treated exactly as before. For $\underline{J}_{k1}$ we use (9.2B6) to find

$$\|\underline{J}_{k1}\| \le \mu \sum_{s=1}^{k} B_s = \mu \sum_{s=1}^{k} (B_s - B) + \mu k B$$

so that

$$\max_{1 \le k \le T/\mu} \|\underline{J}_{k1}\| \le BT + \mu \max_{1 \le k \le T/\mu} \sum_{s=1}^{k} (B_s - B).$$

Adding this to the bound (9.8) gives again (9.9b), which is all that is needed. We have thus established

Theorem 9.2. (First order finite time averaging). Consider the primary system (9.12) and averaged system (9.4a) with $\underline{f}_{av}(\bullet)$ given in (9.13) and with identical initial conditions. Suppose (9.2B1)– (9.2B6) hold. Then given $T > 0$

$$\max_{1 \le k \le T/\mu} \|\underline{z}_k - \bar{\underline{z}}_k\| \le c_T(\mu) \tag{9.14a}$$

$$c_T(\mu) \to 0 \quad \text{w.p.1} \quad \text{as } \mu \to 0 \tag{9.14b}$$

9.2.3 Second Order Analysis

In the stochastic setting one can obtain detailed information about the fluctuations of the system trajectory from a second order analysis.

Considering (9.2), (9.4) let us introduce the fluctuation

$$\begin{aligned}\underline{U}_k &= \mu^{-1/2}(\underline{z}_k - \bar{\underline{z}}_k)\\ &= \mu^{-1/2}\underline{\Delta}_k.\end{aligned} \tag{9.15}$$

Then subtracting (9.4a) from (9.2) we can write

$$\delta\underline{\Delta}_k = \mu(\underline{f}(k, \underline{z}_{k-1}) - \underline{f}_{av}(\bar{\underline{z}}_{k-1})). \tag{9.16}$$

Now proceeding heuristically, apply a Taylor series to find

$$\delta\underline{\Delta}_k = \mu\left(\underline{f}(k, \bar{\underline{z}}_{k-1}) - \underline{f}_{av}(\bar{\underline{z}}_{k-1})\right) + \mu\nabla\underline{f}(k, \bar{\underline{z}}_{k-1})\underline{\Delta}_{k-1} + O(\|\underline{\Delta}_{k-1}\|^2)\mu.$$

Norm up by $\mu^{-\frac{1}{2}}$ and rewrite this as

$$\delta\underline{U}_k = \mu\nabla\underline{f}(k, \bar{\underline{z}}_{k-1})\underline{U}_{k-1} + \mu^{\frac{1}{2}}\tilde{\underline{f}}_k + O(\mu^{\frac{3}{2}}) \tag{9.17a}$$

where

$$\tilde{\underline{f}}_k = \underline{f}(k, \bar{\underline{z}}_{k-1}) - \underline{f}_{av}(\bar{\underline{z}}_{k-1}) \tag{9.17b}$$

and $\tilde{\underline{f}}_k$ has zero mean.

Now considering (9.17) carefully, two things are apparent. Firstly one suspects that $\nabla\underline{f}(k, \bar{\underline{z}}_{k-1})$ can be replaced by $\nabla\underline{f}_{av}(\bar{\underline{z}}_{k-1}) = d\underline{f}_{av}(\underline{z})/d\underline{z}^T|_{\underline{z}=\bar{\underline{z}}_{k-1}}$. Secondly and somewhat more subtly we notice from (9.7), (9.8) that $\sum_{s=1}^{k}\tilde{\underline{f}}_u$ and $\underline{P}_k = \underline{p}(k, \bar{\underline{z}}_{k-1})$ differ by a term which $\to 0$ w.p.1 as $\mu \to 0$. So we are led to approximate (9.17a) by

$$\delta\bar{\underline{U}}_k = \mu\nabla\underline{f}_{av}(\bar{\underline{z}}_{k-1})\bar{\underline{U}}_{k-1} + \mu^{\frac{1}{2}}\tilde{\underline{f}}_k \tag{9.18}$$

or

$$\delta\bar{\underline{U}}_k = \mu\nabla\underline{f}_{av}(\bar{\underline{z}}_{k-1})\bar{\underline{U}}_{k-1} + \mu^{\frac{1}{2}}\delta\underline{P}_k. \tag{9.19}$$

Now (9.18) or (9.19) is a linear time varying stochastic difference equation with a stochastic forcing term (the $O(\mu^{\frac{1}{2}})$ term, called the diffusion term) and a drift term (the $O(\mu)$ term). It is straightforward to use (9.18) or (9.19) to compute a fluctuation variance.

Example 9.2. LMS Algorithm Continued

From (9.1) we have

$$\underline{f}(k, \underline{w}) = -\underline{x}_k\underline{x}_k^T\underline{w} + \underline{x}_k\epsilon_k$$

so that via (9.2a4)

$$\underline{f}_{av}(\underline{w}) = -\underline{R}_x\underline{w}.$$

Thus

$$\underline{\tilde{f}}(k, \underline{w}) = \underline{f}(k, \underline{w}) - \underline{f}_{av}(\underline{w})$$
$$= -(\underline{x}_k \underline{x}_k^T - \underline{R}_x)\underline{w} + \underline{x}_k \epsilon_k$$

so that

$$\underline{p}(k, \underline{w}) = -\sum_{s=1}^{k} (\underline{x}_s \underline{x}_s^T - \underline{R}_x)\underline{w} + \sum_{s=1}^{k} \underline{x}_s \epsilon_s. \tag{9.20}$$

The averaged system is

$$\delta \underline{\bar{w}}_k = -\mu \underline{R}_x \underline{\bar{w}}_{k-1}$$

and if we suppose

(9.2a6) $0 \leq \mu\lambda_{max} < 2$, where λ_{max} is the largest eigenvalue of $\underline{R}_x$

then the averaged system is clearly exponentially stable at $\underline{0}$. So we are led to (9.19), which becomes

$$\delta \underline{\bar{U}}_k = -\mu \underline{R}_x \underline{\bar{U}}_{k-1} + \mu^{\frac{1}{2}} \underline{x}_k \epsilon_k - \mu^{\frac{1}{2}} (\underline{x}_k \underline{x}_k^T - \underline{R}_x)\underline{\bar{w}}_{k-1} - \mu^{\frac{1}{2}} \sum_{s=1}^{k} (\underline{x}_s \underline{x}_s^T - \underline{R}_x)\delta \underline{\bar{w}}_k. \tag{9.21}$$

It is left to the reader to show that the last term in (9.21) can be dropped. To put it another way, $\delta \underline{P}_k$ in (9.19) can be replaced by $\underline{\tilde{f}}_k$ in (9.18). In Exercise 9.7 it is asked to show that if $\underline{x}_k$ is strictly stationary and (9.2a6) holds, then as $\mu \to 0$

$$\mu^{\frac{1}{2}} \max_{1 \leq k \leq T\mu^{-1}} \left\| \sum_{s=1}^{k} (\underline{x}_s \underline{x}_s^T - \underline{R}_x)\underline{\bar{w}}_{s-1} \right\| \to 0 \quad \text{w.p.1.}$$

Thus the third term in (9.21) can be neglected too.

To keep the discussion simple, take $\underline{\bar{U}}_0 = \underline{0}$ (i.e., identical initial conditions for $\underline{z}_k$, $\underline{\bar{z}}_k$). The variance of $\underline{\bar{U}}_k$ due to the remaining terms in (9.21) is then

$$\text{var}(\underline{\bar{U}}_k) = \mu \sum_{s=1}^{k} \sum_{t=1}^{k} (I - \mu \underline{R}_x)^{k-t} \underline{\Sigma}_{t,s} (I - \mu \underline{R}_x)^{k-s}$$
$$\underline{\Sigma}_{t,s} = \text{cov}(\underline{x}_t \epsilon_t, \underline{x}_s \epsilon_s)$$
$$\underline{\bar{U}}_k = \mu^{-1/2}(\underline{\tilde{w}}_k - \underline{\bar{w}}_k).$$

To permit compact computation, now suppose $(\underline{x}_k, \epsilon_k)$ is stationary with zero mean and autocovariance matrix $\underline{\Sigma}_u$. Then

$$var(\underline{\bar{U}}_k) = \mu \sum_{s=1}^{k} \sum_{t=1}^{k} (I - \mu \underline{R}_x)^{k-t} \underline{\Sigma}_{t-s} (I - \mu \underline{R}_x)^{k-s}.$$

If $\underline{F}(\omega)$ is the spectrum associated with $\underline{\Sigma}_u$, then

$$\underline{\Sigma}_u = \int_{-\pi}^{\pi} e^{j\omega u} \underline{F}(\omega) d\omega / 2\pi.$$

Although we have not discussed steady state behavior, we look at it briefly here. It is easily seen that, as $k \to \infty$

$$var(\bar{\underline{U}}_k) \to \Gamma^{\mu}$$

where

$$\Gamma^{\mu} = \frac{\mu}{2\pi}\int_{-\pi}^{\pi}\left(I - e^{j\omega}(I - \mu\underline{R}_x)\right)^{-1}\underline{F}(\omega)\left(I - e^{-j\omega}(I - \mu\underline{R}_x)\right)^{-1} d\omega. \tag{9.22}$$

To see what happens to Γ^{μ} as $\mu \to 0$, change variables to $\omega = \mu\lambda$ so that

$$\Gamma^{\mu} = \frac{1}{2\pi}\int_{-\pi/\mu}^{\pi/\mu}\left(j\lambda I + \underline{R}_x + O(\mu)\right)^{-1}\underline{F}(\mu\lambda)\left(-j\lambda I + \underline{R}_x + O(\mu)\right)^{-1} d\lambda$$

$$\Rightarrow \quad \Gamma = \frac{1}{2\pi}\int_{-\infty}^{\infty}\left(j\lambda I + \underline{R}_x\right)^{-1}\underline{F}(0)\left(-j\lambda I + \underline{R}_x\right)^{-1} d\lambda, \quad \text{as } \mu \to 0 \tag{9.23a}$$

$$= \int_{0}^{\infty} e^{-R_x t}\underline{F}(0)e^{-R_x t}dt \tag{9.23b}$$

where Parseval's theorem has been used. To put it another way, Γ solves the Lyapunov equation

$$\underline{R}_x\Gamma + \Gamma\underline{R}_x = \underline{F}(0). \tag{9.23c}$$

If ϵ_k, $\underline{x}_t$ are independent white noises, then the Lyapunov equation becomes

$$\underline{R}_x\Gamma + \Gamma\underline{R}_x = \sigma_\epsilon^2\underline{R}_x \quad \Rightarrow \quad \Gamma = \sigma_\epsilon^2 I/2.$$

(This agrees with a Taylor series expansion in Exercise 5.5). Loosely speaking we have that for large k and small μ, $\mu^{-1/2}(\underline{z}_k - \bar{\underline{z}}_k)$ is close in distribution (as $k \to \infty$) to a random variable whose variance is Γ. This limiting steady state notion is pursued more carefully in Sections 9.6–9.7.

9.2.4 Rigorous Development of Second Order Analysis

In this subsection we develop a rigorous justification of the replacement of (9.16) by (9.18) leading to Theorem 9.3 stated at the end of the subsection. To justify the replacement of (9.16) by (9.18) we proceed by breaking (9.16) into several pieces as follows:

$$\delta\underline{\Delta}_k = \mu\nabla\underline{f}(k, \bar{\underline{z}}_{k-1})\underline{\Delta}_{k-1} + \mu\tilde{\underline{f}}_k + \mu\underline{a}_k \tag{9.24}$$

where ($\tilde{\underline{f}}_k$ is defined in (9.17b))

$$\underline{a}_k = \underline{f}(k, \underline{z}_{k-1}) - \underline{f}(k, \bar{\underline{z}}_{k-1}) - \nabla\underline{f}(k, \bar{\underline{z}}_{k-1})\underline{\Delta}_{k-1}. \tag{9.25}$$

Norming up by $\mu^{-\frac{1}{2}}$ gives

$$\delta\underline{U}_k = \mu\nabla\underline{f}(k, \bar{\underline{z}}_{k-1})\underline{U}_{k-1} + \mu^{\frac{1}{2}}\tilde{\underline{f}}_k + \mu^{\frac{1}{2}}\underline{a}_k. \tag{9.26}$$

We will work firstly with (9.18). Subtracting (9.18) from (9.26) and introducing the derivation $\underline{d}_k = \underline{U}_k - \bar{\underline{U}}_k$ gives

$$\delta \underline{d}_k = \mu \nabla \underline{f}(k, \bar{\underline{z}}_{k-1}) \underline{d}_{k-1} + \mu^{\frac{1}{2}} \underline{a}_k + \mu \nabla \tilde{\underline{f}}_k \bar{\underline{U}}_{k-1} \qquad (9.27)$$

$$\nabla \tilde{\underline{f}}_k = \nabla \underline{f}(k, \bar{\underline{z}}_{k-1}) - \nabla \underline{f}_{av}(\bar{\underline{z}}_{k-1}). \qquad (9.28)$$

To deal with the $\underline{a}_k$ term in (9.27) we suppose

(9.2C1)

$$\|\nabla \underline{f}(k, \underline{z}) - \nabla \underline{f}(k, \underline{z}')\| \le c_k \|\underline{z} - \underline{z}'\|/2$$

where c_k obeys an SLLN: $N^{-1} \sum_{k=1}^{N} c_k \to c$ w.p.1.

Using (9.2C1) we find

$$\mu^{\frac{1}{2}} \|\underline{a}_k\| \le \mu^{\frac{1}{2}} c_k \|\underline{\Delta}_{k-1}\|^2$$
$$= \mu \|\underline{\Delta}_{k-1}\| c_k \|\underline{U}_{k-1}\|$$
$$\le \mu \|\underline{\Delta}_{k-1}\| c_k (\|\underline{d}_{k-1}\| + \|\bar{\underline{U}}_{k-1}\|). \qquad (9.29)$$

Now apply Theorem 9.1 or 9.2 to find

$$\mu^{\frac{1}{2}} \|\underline{a}_k\| \le \mu c_T(\mu) c_k (\|\underline{d}_{k-1}\| + \|\bar{\underline{U}}_{k-1}\|), \quad 1 \le k \le T\mu^{-1}. \qquad (9.30)$$

To deal with the last term in (9.27) we are led to introduce a new perturbation (this need for a second perturbation is a standard occurrence in second order analysis).

(9.2C2) Introduce

$$\underline{p}_z(k, \underline{z}) = \nabla \underline{p}(k, \underline{z}) = \sum_{s=1}^{k} \left(\nabla \underline{f}(s, \underline{z}) - \nabla \underline{f}_{av}(\underline{z}) \right)$$

and suppose

$$\|\underline{p}_z(k, \underline{z}) - \underline{p}_z(k, \underline{z}')\| \le M_{zk} \|\underline{z} - \underline{z}'\|$$
$$\|\underline{p}_z(k, \underline{0})\| \le p_{zk}$$

where, as $k \to \infty$

$$M_{zk}/k \to 0 \quad \text{w.p.1}$$
$$p_{zk}/k \to 0 \quad \text{w.p.1.}$$

Now introduce

$$\underline{P}_{z,k} = \underline{p}_z(k, \bar{\underline{z}}_k)$$

and observe, as usual, that

$$\underline{P}_{z,k} - \underline{P}_{z,k-1} = \nabla \tilde{\underline{f}}_k + \underline{\eta}_{z,k} \qquad (9.31)$$

where

$$\underline{\eta}_{z,k} = \underline{p}_z(k-1, \bar{\underline{z}}_{k-1}) - \underline{p}_z(k-1, \bar{\underline{z}}_{k-2}).$$

Also note that, from (9.2C3),

$$\|\underline{\eta}_{z,k}\| \le \mu B h M_{zk}. \tag{9.32}$$

Continuing, we use (9.31) as follows:

$$\begin{aligned}
\sum_{s=1}^{k} \nabla \tilde{\underline{f}}_s \bar{\underline{U}}_{s-1} &= \sum_{s=1}^{k} \left(\delta \underline{P}_{z,s} + \underline{\eta}_{z,s}\right) \bar{\underline{U}}_{s-1} \\
&= \sum_{s=1}^{k} (\underline{P}_{z,s}\bar{\underline{U}}_s - \underline{P}_{z,s-1}\bar{\underline{U}}_{s-1}) + \sum_{s=1}^{k} \underline{P}_{z,s}(\bar{\underline{U}}_{s-1} - \bar{\underline{U}}_s) + \sum_{s=1}^{k} \underline{\eta}_{z,s}\bar{\underline{U}}_{s-1} \\
&= \underline{P}_{z,k}\bar{\underline{U}}_k - \sum_{s=1}^{k} \underline{P}_{z,s}\left(\mu \nabla \underline{f}_{av}(\bar{\underline{z}}_{s-1})\bar{\underline{U}}_{s-1} + \mu^{\frac{1}{2}} \delta \underline{P}_s\right) + \sum_{s=1}^{k} \underline{\eta}_{z,s}\bar{\underline{U}}_{s-1}.
\end{aligned}$$

Thus we find, via (9.2C3), (9.32)

$$\begin{aligned}
\mu \left\| \sum_{s=1}^{k} \nabla \tilde{\underline{f}}_s \bar{\underline{U}}_{s-1} \right\| \le \mu (M_{zk} h + p_{z,k}) \|\bar{\underline{U}}_{k-1}\| &+ \mu^2 L \sum_{s=1}^{k} \|\bar{\underline{U}}_{s-1}\| (M_{z,s-1} h + p_{z,s-1}) \\
&+ \mu^2 B h \sum_{s=1}^{k} \|\bar{\underline{U}}_{s-1}\| M_{z,s-1} + \mu^{\frac{3}{2}} \|\underline{T}_k\|
\end{aligned} \tag{9.33}$$

$$\underline{T}_k = \sum_{s=1}^{k} \underline{P}_{z,s} \delta \underline{P}_s. \tag{9.34}$$

Below we show

$$\max_{1 \le k \le T\mu^{-1}} \|\bar{\underline{U}}_k\| \le \bar{U} \quad \text{w.p.1} \tag{9.35}$$

and under certain additional regularity conditions we show

$$\mu^{\frac{3}{2}} \max_{1 \le k \le T\mu^{-1}} \|\underline{T}_k\| \to 0 \quad \text{w.p.1.} \tag{9.36}$$

Putting these in (9.30), (9.33), and then using these together with (9.28) in (9.27)—once it is summed, gives

$$\|\underline{d}_k\| \le \|\underline{d}_0\| + \mu \sum_{s=1}^{k} (L_s + c_T(\mu) c_s) \|\underline{d}_{s-1}\| + d_T(\mu)$$

where

$$d_T(\mu) \to 0 \quad \text{w.p.1}, \qquad \text{as } \mu \to 0.$$

Now using the Bellman-Gronwall lemma gives

$$\|\underline{d}_k\| \le (\|\underline{d}_0\| + d_T(\mu))e^{\mu \sum_{s=1}^{k}(L_s + c_T(\mu)c_s)}. \tag{9.37}$$

So provided $\underline{U}_k$ and $\bar{\underline{U}}_k$ have the same initial condition, we have

$$\max_{1\le k\le T\mu^{-1}} \|\underline{d}_k\| \to 0 \quad \text{w.p.1,} \quad \text{as } \mu \to 0.$$

Proof of (9.35). From (9.18), (9.2A2)

$$\|\bar{\underline{U}}_k\| \le \|\bar{\underline{U}}_0\| + \mu \sum_{s=1}^{k} L_s \|\bar{\underline{U}}_{s-1}\| + \mu^{\frac{1}{2}} \|\delta \underline{P}_k\|$$

$$\le \left(\|\bar{\underline{U}}_0\| + \mu^{\frac{1}{2}} \max_{1\le k\le T\mu^{-1}} \|\delta \underline{P}_k\|\right) + \mu \sum_{s=1}^{k} L_s \|\bar{\underline{U}}_{s-1}\|.$$

Now apply the Bellman-Gronwall lemma to find

$$\max_{1\le k\le T\mu^{-1}} \|\bar{\underline{U}}_k\| \le \left(\|\bar{\underline{U}}_0\| + \mu^{\frac{1}{2}} \max_{1\le k\le T\mu^{-1}} \|\delta \underline{P}_k\|\right) e^{\mu \max_{1\le k\le T\mu^{-1}}(L_s - L) + TL}.$$

From (9.7), (9.8) and recalling Exercise 9.2 we find

$$\mu^{\frac{1}{2}} \max_{1\le k\le T\mu^{-1}} \|\delta \underline{P}_k\| \to 0 \quad \text{w.p.1,} \quad \text{as } \mu \to 0$$

provided

$$\|\tilde{\underline{f}}_k\|^2/k \to 0 \quad \text{w.p.1,} \quad \text{as } k \to \infty. \tag{9.38}$$

Since

$$\|\tilde{\underline{f}}_k\| \le \|\tilde{\underline{f}}(k, \underline{0})\| + L_k h$$
$$\le \|\underline{f}(k, \underline{0})\| + \|\underline{f}_{av}(\underline{0})\| + L_k h$$

(9.38) follows from

(9.2C3) $\|\underline{f}(k, \underline{0})\|^2/k,\ L_k^2/k \to 0$ w.p.1, as $k \to \infty$.

Thus (9.35) is established.

Proof of (9.36). The proof is somewhat tedious but entirely straightforward. The chief feature of note is the use of the Martingale (MG) approximation method of Appendix E12.

To begin we can replace $\underline{T}_k$ by

$$\underline{T}_{k,1} = \sum_{s=1}^{k} \underline{P}_{z,s} \tilde{\underline{f}}_s$$

provided we dispose of the difference

$$\underline{D}_{k,1} = \sum_{s=1}^{k} \underline{P}_{z,s}\underline{\eta}_s.$$

However by (9.8)

$$\begin{aligned}\|\underline{D}_{k,1}\| &\le \mu \sum_{s=1}^{k} \|\underline{P}_{z,s}\| M_{s-1} h \\ &\le \mu h \sum_{s=1}^{k} \left(h M_{z,s} + p_{z,s}\right) M_{s-1}\end{aligned}$$

thus we can conclude

$$\mu^{\frac{3}{2}} \max_{1 \le k \le T/\mu} \|\underline{D}_{k,1}\| \to 0 \quad \text{w.p.1}, \qquad \text{as } \mu \to 0$$

provided we suppose

(9.2C4) $(p_{z,s}, M_{z,s}, M_s)/s^{3/4} \to 0$ w.p.1, as $s \to \infty$.

Continuing, we split $\underline{T}_{k,1}$ into two pieces:

$$\begin{aligned}\underline{T}_{k,1} &= \underline{T}_{k,2} + \underline{D}_{k,2} \\ \underline{T}_{k,2} &= \sum_{s=1}^{k} \underline{P}_{z,s-1}\underline{\tilde{f}}_s \\ \underline{D}_{k,2} &= \sum_{s=1}^{k} \nabla \underline{\tilde{f}}_s \underline{\tilde{f}}_s\end{aligned}$$

Below we show that under certain mild conditions, the following is true:

$$\mu^{\frac{3}{2}} \max_{1 \le k \le T/\mu} \|\underline{D}_{k,2}\| \to 0 \quad \text{w.p.1}. \tag{9.39}$$

Now we deal with $\underline{T}_{k,2}$ by using the MG approximation method of Appendix E12. Thus we decompose $\underline{\tilde{f}}_s$ as

$$\underline{\tilde{f}}_s = \underline{\nu}_s + \underline{\upsilon}_{s-1} - \underline{\upsilon}_s \tag{9.40}$$

where $\underline{\nu}_s$ is an MG difference with respect to increasing σ-fields $\mathcal{F}_k$ generated by the underlying stochastic signal $\underline{\xi}_k$ of (9.3). This leads to a breakup of $\underline{T}_{k,2}$ into two pieces:

$$\underline{T}_{k,2} = \underline{T}_{k,3} + \underline{D}_{k,3} \tag{9.41a}$$

$$\underline{T}_{k,3} = \sum_{s=1}^{k} \underline{P}_{z,s-1}\underline{v}_s \tag{9.41b}$$

$$\underline{D}_{k,3} = \sum_{s=1}^{k} \underline{P}_{z,s-1}(\underline{v}_{s-1} - \underline{v}_s) \tag{9.41c}$$

We deal with $\underline{D}_{k,3}$ first. Summation by parts gives

$$\underline{D}_{k,3} = -\underline{P}_{z,k}\underline{v}_k + \sum_{s=1}^{k} \nabla\underline{\tilde{f}}_s\underline{v}_s.$$

Next, by (9.2C2)

$$\|\underline{P}_{z,k}\underline{v}_k\| \leq (M_{z,k}h + p_{z,k})\|\underline{v}_k\|.$$

So provided

$$\|\underline{v}_k\|/k^{\frac{1}{2}} \to 0 \quad \text{w.p.1,} \quad \text{as } k \to \infty \tag{9.42}$$

then (9.2C2) ensures

$$\mu^{\frac{3}{2}} \max_{1\leq k\leq T/\mu} \|\underline{P}_{z,k}\underline{v}_k\| \to 0 \quad \text{w.p.1,} \quad \text{as } \mu \to 0.$$

For the second term in $\underline{D}_{k,3}$

$$\mu^{\frac{3}{2}} \max_{1\leq k\leq T/\mu} \left\| \sum_{s=1}^{k} \nabla\underline{\tilde{f}}_s\underline{v}_s \right\| \leq \mu^{\frac{1}{2}} \max_{1\leq k\leq T/\mu} \|\underline{v}_k\| \mu \max_{1\leq k\leq T/\mu} \sum_{s=1}^{k} \|\nabla\underline{\tilde{f}}_s\|. \tag{9.43}$$

However, by (9.2C1)

$$\|\nabla\underline{\tilde{f}}_s\| \leq \|\nabla\underline{f}(s, \underline{0})\| + c_s h$$

so if we suppose

(9.2C5) $\|\nabla\underline{f}(s, \underline{0})\|$ obeys an SLLN

$$\frac{1}{N}\sum_{s=1}^{N} \|\nabla\underline{f}(s, \underline{0})\| \to \Psi \quad \text{w.p.1}$$

then (9.42), (9.2C2) ensure (9.43) $\to 0$ w.p.1, as $\mu \to 0$ so that

$$\mu^{\frac{3}{2}} \max_{1\leq k\leq T/\mu} \|\underline{D}_{k,3}\| \to 0 \quad \text{w.p.1,} \quad \text{as } \mu \to 0.$$

Now we turn to $\underline{T}_{k,3}$ in (9.41). It will do to show

$$k^{-\frac{3}{2}}\underline{T}_{k,3} \to 0 \quad \text{w.p.1,} \quad \text{as } k \to \infty$$

which will follow, by Kronecker's lemma if

$$\sum_{s=1}^{n} \underline{P}_{z,s-1}\underline{\nu}_s/s^{\frac{3}{2}} \quad \text{converges w.p.1.}$$

By the conditional MG convergence theorem (Theorem E7.5) this occurs if

$$\sum_{s=1}^{\infty} \|\underline{P}_{z,s-1}\|^2 E(\|\underline{\nu}_s\|^2|\mathcal{F}_{s-1})/s^3 < \infty \quad \text{w.p.1}$$

which occurs, via (9.2C2) if

$$2\sum_{s=1}^{\infty}(M^2_{z,s-1}h^2 + p^2_{z,s-1})E(\|\underline{\nu}_s\||\mathcal{F}_{s-1})/s^3 < \infty \quad \text{w.p.1.}$$

Via (9.2C4) this occurs if

$$\sum_{s=1}^{\infty} E(\|\underline{\nu}_s\|^2|\mathcal{F}_{s-1})/s^{\frac{3}{2}} < \infty \quad \text{w.p.1}$$

which holds, via the Monotone Convergence theorem if

$$\sum_{s=1}^{\infty} E(\|\underline{\nu}_s\|^2)/s^{\frac{3}{2}} < \infty$$

which follows if

$$\sup_{1\le s\le\infty} E(\|\underline{\nu}_s\|^2) \le \infty. \tag{9.44}$$

We prove this below via (9.40) by showing $\sup_s E\|\underline{\tilde{f}}_s\|^2 < \infty$ and $\sup_s E\|\underline{\nu}_s\|^2 \le \infty$.

Proof of (9.39). We can split $\underline{D}_{k,2}$ as follows:

$$\underline{D}_{k,2} = \sum_{s=1}^{k}\left(\nabla\underline{\tilde{f}}_s - \nabla\underline{\tilde{f}}(s,\underline{0})\right)\underline{\tilde{f}}_s + \sum_{s=1}^{k}\nabla\underline{\tilde{f}}(s,\underline{0})(\underline{\tilde{f}}_s - \underline{\tilde{f}}(s,\underline{0})) + \sum_{s=1}^{k}\nabla\underline{\tilde{f}}(s,\underline{0})\underline{\tilde{f}}(s,\underline{0})$$

$$\Rightarrow \quad \|\underline{D}_{k,2}\| \le \sum_{s=1}^{k} hc_s\|\underline{\tilde{f}}_s\| + \sum_{s=1}^{k}\|\nabla\underline{\tilde{f}}(s,\underline{0})\|L_s h + \sum_{s=1}^{k}\|\nabla\underline{f}(s,\underline{0})\|\|\underline{\tilde{f}}(s,\underline{0})\|$$

where (9.2A2) and (9.2C1) have been used. Continuing,

$$\|\underline{D}_{k,2}\| \le h^2\sum_{s=1}^{k} c_s L_s + h\sum_{s=1}^{k} c_s\|\underline{\tilde{f}}(s,\underline{0})\|$$
$$+\sum_{s=1}^{k}\|\nabla\underline{\tilde{f}}(s,\underline{0})\|L_s h + \sum_{s=1}^{k}\|\nabla\underline{\tilde{f}}(s,\underline{0})\|\|\underline{\tilde{f}}(s,\underline{0})\|. \tag{9.45}$$

Since, for instance

$$\mu^{\frac{3}{2}} \max_{1\le k\le T/\mu} \sum_{s=1}^{k} c_s L_s \le \mu^{\frac{1}{2}} \left(\max_{1\le k\le T/\mu} L_s \right) \mu \max_{1\le k\le T/\mu} \sum_{s=1}^{k} c_s$$

then the SLLN for c_k in (9.2C1) together with (9.2C3) ensures the whole term $\to 0$ w.p.1 as $\mu \to 0$. Similarly (9.2C3), (9.2C5) then ensures (9.39).

Proof of (9.44). From Appendix E12 we see, via (9.40), that (9.44) is ensured if the following condition holds

$$\underline{\xi}_k, \text{ the underlying driving stochastic signal of (9.3) is strong mixing with} \tag{9.46}$$

$$\sum_{n=0}^{\infty} \alpha_n^{1/4} < 0. \tag{9.47}$$

Also

$$\sup_{1\le s\le\infty} E\|\underline{\tilde{f}}_s\|^4 < \infty. \tag{9.48}$$

Since $\|\underline{\tilde{f}}_s\| \le \|\underline{\tilde{f}}(s, \underline{0})\| + hL_s$, (9.48) follows if

$$\sup_{1\le s\le\infty} E\|\underline{\tilde{f}}(s,\underline{0})\|^4 < \infty, \qquad \sup_{1\le s\le\infty} E(L_s^4) < \infty. \tag{9.49}$$

Proof of (9.42). This follows if $\sum_{k=1}^{\infty} P(\|\underline{v}_k\|/k^{\frac{1}{2}} > \epsilon) < \infty$, which follows via Chebyshev inequality if for some $\delta > 0$

$$\sup_{1\le k\le\infty} E\|\underline{v}_k\|^{2+\delta} < \infty.$$

From Appendix E12 this occurs if

$$\sup_{1\le s\le\infty} E\|\underline{\tilde{f}}_s\|^{4+2\delta} \le \infty \tag{9.50}$$

$$\sum_{n=0}^{\infty} \alpha_n^{1/(4+2\delta)} < \infty. \tag{9.51}$$

And so (9.47)–(9.51) follow from

(9.2C6) (9.46), (9.51), and (9.52) hold, and also

$$\sup_{1\le s\le\infty} E\|\underline{\tilde{f}}(s,\underline{0})\|^{4+2\delta} < \infty, \qquad \sup_{1\le s\le\infty} E(L_s^{4+2\delta}) < \infty. \tag{9.52}$$

We have thus established the following result.

Theorem 9.3. (Second order finite time averaging). For the primary system (9.2) and averaged system (9.4), consider the fluctuation $\underline{U}_k$ in (9.15). Then under (9.2A1)–(9.2A5) and (9.2C1)–(9.2C6) we have

$$\max_{1 \le k \le T\mu^{-1}} \|\underline{U}_k - \bar{\underline{U}}_k\| \to 0 \quad \text{w.p.1,} \qquad \text{as } \mu \to 0$$

where $\bar{\underline{U}}_k$ is generated from (9.18) with the same initial condition as $\underline{U}_k$.

It is left to the reader to show that the result also holds with (9.19): see Exercise 9.8.

9.3 FINITE TIME AVERAGING: TIME VARIANT PARAMETERS

To motivate the issues here it is useful, once again, to return to the LMS algorithm with time variant parameters. The error system (5.35) is

$$\delta\tilde{\underline{w}}_k = -\mu \underline{x}_k \underline{x}_k^T \tilde{\underline{w}}_{k-1} - \delta \underline{w}_{ok} + \mu \underline{x}_k \epsilon_k.$$

From this an initial general specification is apparent:

$$\delta \underline{z}_k = \mu \underline{f}(k, \underline{z}_{k-1}) - \delta \underline{z}_{ok}.$$

As in Section 7.3 it is necessary to be specific about the parameter variation, and there are several options. In Section 5.3 all the parameter variation was placed in the stochastic part of the model. In Section 7.3 the parameter variation was purely deterministic. Here we look at a purely stochastic specification, and there are two options, both of which are large amplitude slowly time variant parameter models.

(9.3S1) Smooth signal model

$$\underline{z}_{ok} = \underline{z}_o(\varepsilon k) + \underline{z}_o, \qquad \varepsilon \ll 1 \text{ and } \underline{z}_o(\bullet) \text{ is differentiable.}$$

(9.3S2) Rough signal model

$$\underline{z}_{ok} = \underline{z}_o(\varepsilon k) + \underline{z}_o, \qquad \varepsilon \ll 1 \text{ and}$$
$$\underline{z}_o(\bullet) \text{ is not differentiable but is continuous w.p.1.}$$

An example of (9.3S1) is $\underline{z}_o(\tau)$, a continuous time second order autoregression. An example of (9.3S2) is $\underline{z}_o(\tau)$, a continuous time first order autoregression. Another example that essentially fits (9.3S2) is

$$\delta \underline{z}_{ok} = -\varepsilon \underline{z}_{ok} + \varepsilon^{\frac{1}{2}} \underline{\nu}_k \tag{9.53}$$

where $\underline{\nu}_k$ is a white noise: roughly, $\underline{z}_{ok} = \underline{z}_o(\epsilon k)$ where $\underline{z}_o(\tau)$ is a first order autoregression.

The parameter model provides, then, the following general specification:

$$\delta \underline{z}_k = \mu \underline{f}(k, \underline{z}_{k-1}) - \delta \underline{z}_o(\varepsilon k). \tag{9.54}$$

As in Section 7.3 it is necessary to consider the relation between the time scales μ^{-1} and ε^{-1}. The stochastic case is somewhat richer than the deterministic case now and there are four possibilities:

(i) Slow adaptation: $\mu \ll \varepsilon$, e.g., $\mu = \varepsilon^r$, $r > 1$

(ii) Matched adaptation:

(a) First order: $\mu =$ constant $\times\, \varepsilon$

(b) Second order: $\mu =$ constant $\times\, \varepsilon^{\frac{1}{2}}$

(iii) Fast adaptation: $\mu \gg \varepsilon$, e.g., $\mu = \varepsilon^r$, $r < \frac{1}{2}$

In case (i) adaptation fails, and case (iii) is not treated here, so we concentrate on case (ii).

9.3.1 First Order Matched Adaptation

In the first order case we claim the associated partially averaged system is

$$\delta \bar{\underline{z}}_k = \mu \underline{f}_{av}(\bar{\underline{z}}_{k-1}) - \delta \underline{z}_o(\mu k) \tag{9.55}$$

where $\underline{f}_{av}(\bullet)$ is given in (9.4b) and (9.2A1) is assumed. To see this, subtract (9.55) from (9.54) to obtain (9.5) again. The formal justification of (9.55) then proceeds as in Section 9.2.

9.3.2 Second Order Matched Adaptation

First Order Analysis. In the second order matched adaptation case we need to look at both a first order and a second order analysis. The associated first order averaged system is

$$\delta \bar{\underline{z}}_k = \mu \underline{f}_{av}(\bar{\underline{z}}_{k-1}) \tag{9.56}$$

where $\underline{f}_{av}(\bullet)$ is given in (9.4b) and (9.2A1) is assumed. To see this subtract (9.56) from (9.54) and sum to obtain

$$\begin{aligned}\underline{\Delta}_k &= \underline{z}_k - \bar{\underline{z}}_k \\ &= \underline{\Delta}_0 + \mu \sum_{s=1}^{k} \left(\underline{f}(s, \underline{z}_{s-1}) - \underline{f}_{av}(\bar{\underline{z}}_{s-1}) \right) + \mu \underline{J}_k + \underline{z}_o(\mu^2 k) - \underline{z}_o\end{aligned}$$

where $\underline{J}_k$ is given in (9.5b). Now using the regularity conditions (9.2A2)–(9.2A5) proceed as in Section 9.2 to see that (9.10) still holds provided we can show that

$$\max_{1 \le k \le T/\mu} \|\underline{z}_o(\mu^2 k) - \underline{z}_o(0)\| \to 0 \quad \text{w.p.1}, \quad \text{as } \mu \to 0. \tag{9.57}$$

This is equivalent to

$$\sup_{0 \le \tau \le \mu T} \|\underline{z}_o(\tau) - \underline{z}_o(0)\| \to 0 \quad \text{w.p.1}, \quad \text{as } \mu \to 0$$

which is just the requirement of w.p.1 continuity. This holds for either signal model (9.3S1) or (9.3S2). To sum up, the following result has been established.

Theorem 9.4. Consider the primary system (9.54) and averaged system (9.56) and suppose (9.2A1)–(9.2A5) hold as well as (9.3S1) or (9.3S2). Also suppose second order matched adaptation, $\mu =$ constant $\times\, \varepsilon^{1/2}$. Then the first order averaging result (9.4) holds.

The perceptive reader will have noticed that Theorem 9.4 applies also to case (iii), fast adaptation. In Exercise 9.13 the reader is asked to show that the signal model (9.53) satisfies (9.57).

Second Order Analysis. The primary system is

$$\delta \underline{z}_k = \mu \underline{f}(k, \underline{z}_{k-1}) - \delta \underline{z}_o(\mu^2 k) \tag{9.58}$$

and the first order averaged system is

$$\delta \bar{\underline{z}}_k = \mu \underline{f}_{av}(\bar{\underline{z}}_{k-1}). \tag{9.59}$$

So we must consider the fluctuation

$$\underline{U}_k = (\underline{z}_k - \bar{\underline{z}}_k)/\mu^{\frac{1}{2}}.$$

Subtracting (9.59) from (9.58) gives

$$\delta \underline{U}_k = \mu^{\frac{1}{2}} \left(\underline{f}(k, \underline{z}_{k-1}) - \underline{f}_{av}(\bar{\underline{z}}_{k-1}) \right) - \mu^{-\frac{1}{2}} \delta \underline{z}_o(\mu^2 k). \tag{9.60}$$

Now proceeding as in Section 9.2 we are led to approximate this by

$$\delta \bar{\underline{U}}_k = \mu \nabla \underline{f}_{av}(\bar{\underline{z}}_{k-1}) \bar{\underline{U}}_{k-1} + \mu^{\frac{1}{2}} \delta \underline{P}_k - \mu^{-\frac{1}{2}} \delta \underline{z}_o(\mu^2 k) \tag{9.61}$$

where $\delta \underline{P}_k$ could be replaced by $\tilde{\underline{f}}_k$ as in (9.18). The justification follows as in Section 9.2 because the new forcing term $\mu^{-1/2} \delta \underline{z}_o(\mu^2 k)$ does not depend on any other system signals.

Example 9.3. LMS with Time Varying Parameters

Continuing Examples 9.1, 9.2, a time variant parameter is now inserted into the LMS error system. We treat a variant of the slowly time varying model of Section 5.3 (see (5.3A4(ii)), (5.65))

$$\delta \underline{w}_{o,k} = -\epsilon \underline{w}_{o,k-1} + \epsilon^{\frac{1}{2}} \underline{\nu}_k$$

where now $\underline{\nu}_k$ is a colored noise of spectrum $\underline{F}_\nu(\omega)$. We also assume for simplicity (not necessity),

(9.3a1) The signals $\{\underline{x}_k, \epsilon_k\}$ and $\{\underline{w}_{o,k}\}$ (or $\{\underline{\nu}_k\}$) are statistically independent.

With second order matched adaptation we take $\mu = \epsilon^{1/2}$ and the approximating system (9.61) becomes (following the discussion in Example 9.2)

$$\begin{aligned}\delta \bar{\underline{U}}_k &= -\mu \underline{R}_x \bar{\underline{U}}_{k-1} + \mu^{\frac{1}{2}} \underline{x}_k \epsilon_k - \mu^{-\frac{1}{2}} \delta \underline{w}_{o,k} \\ &= -\mu \underline{R}_x \bar{\underline{U}}_{k-1} + \mu^{\frac{1}{2}} \underline{x}_k \epsilon_k + \mu^{\frac{3}{2}} \underline{w}_{o,k-1} - \mu^{\frac{1}{2}} \underline{\nu}_k.\end{aligned}$$

Since $\text{var}(\underline{w}_{o,k}) = O(1)$, it is easy to show that the third term may be neglected and we are left with

$$\delta \bar{\underline{U}}_k = -\mu \underline{R}_x \bar{\underline{U}}_{k-1} + \mu^{\frac{1}{2}} \underline{x}_k \epsilon_k - \mu^{\frac{1}{2}} \underline{\nu}_k.$$

A variance calculation as in Example 9.2 leads to

$$\mathrm{var}(\underline{\bar{U}}_k) \to \Gamma_a^\mu + \Gamma_b^\mu$$

where Γ_a^μ is given in (9.22) while Γ_b^μ is the same but with $\underline{F}(\omega)$ replaced by $\underline{F}_\nu(\omega)$. As in Example 9.2 we find $\Gamma_a^\mu \to \Gamma_a$ given in (9.23b) or (9.23c) and $\Gamma_b^\mu \to \Gamma_b$ given as in (9.23b) or (9.23c) but with $\underline{F}(0)$ replaced by $\underline{F}_\nu(0)$.

The result $\Gamma_a + \Gamma_b$ may be compared with the result obtained in (5.76) which applies only in a fully white noise setting.

Example 9.4. Time Delay Estimation

Consider the idealized adaptive time delay estimation problem of Section 4.5, but allow all signals to be stochastic and allow additive noise. The algorithm becomes

$$\delta\hat{\theta}_k = -\mu\left(s'_{k\Delta-\hat{\theta}_{k-1}} + n'_{k\Delta-\hat{\theta}_{k-1}}\right)\left(s_{k\Delta-\theta_o} - s_{k\Delta-\hat{\theta}_{k-1}} + m_{k\Delta-\theta_o} - n_{k\Delta-\hat{\theta}_{k-1}}\right)$$

where $m_{k\Delta} = m(k\Delta)$ and $m(t)$ is a continuous time noise statistically independent of $n(t)$ with $n_{k\Delta} = n(k\Delta)$. The averaged function is

$$\begin{aligned} f_{av}(\tilde{\theta}) &= -E\left(s'_{k\Delta-\tilde{\theta}-\theta_o}(s_{k\Delta-\theta_o} - s_{k\Delta-\tilde{\theta}-\theta_o})\right) - E(n'_{k\Delta-\tilde{\theta}-\theta_o} n_{k\Delta-\tilde{\theta}-\theta_o}) \\ &= -R_s'(-\tilde{\theta}) + R_s'(0) - R_n'(0) \\ &= R_s'(\tilde{\theta}) \end{aligned}$$

where $R_s(\bullet)$, $R_n(\bullet)$ are the autocorrelation functions of $s(t)$, $n(t)$ respectively. So the averaged system is

$$\delta\bar{\theta}_k = \mu R_s'(\bar{\theta}_{k-1})$$

which has already been discussed (see Section 7.7). Continuing, to assemble the terms in (9.18) we find

$$\nabla f_{av}(\tilde{\theta}) = R_s''(\tilde{\theta})$$

and

$$\begin{aligned} \tilde{f}(k,\tilde{\theta}) &= -\left(s'_{k\Delta-\tilde{\theta}-\theta_o}(s_{k\Delta-\theta_o} - s_{k\Delta-\tilde{\theta}-\theta_o}) - R_s'(\tilde{\theta})\right) \\ &\quad - n'_{k\Delta-\tilde{\theta}-\theta_o} e_k - s'_{k\Delta-\tilde{\theta}-\theta_o}(m_{k\Delta-\theta_o} - n_{k\Delta-\tilde{\theta}-\theta_o}) \\ e_k &= (s_{k\Delta-\theta_o} - s_{k\Delta-\tilde{\theta}-\theta_o} + m_{k\Delta-\theta_o} - n_{k\Delta-\tilde{\theta}-\theta_o}) \end{aligned}$$

while

$$\tilde{f}(k,0) = -s'_{k\Delta-\theta_o}(m_{k\Delta-\theta_o} - n_{k\Delta-\theta_o}).$$

Under suitable regularity conditions (see Section 7.7)

$$\begin{aligned} |\tilde{f}(k,\tilde{\theta}) - \tilde{f}(k,0)| &= \left|\int_0^{\tilde{\theta}} \frac{d\tilde{f}(k,u)}{du} du\right| \\ &\le \text{const.} \times |\tilde{\theta}| \end{aligned}$$

and so if the initial condition for $\tilde{\theta}_k$ is in the domain of attraction of the equilibrium point $\tilde{\theta} = 0$, then (with $\tilde{\theta}_k = \hat{\theta}_k - \theta_o$)

$$\mu^{\frac{1}{2}} \max_{1 \le k \le T\mu^{-1}} |\tilde{f}(k, \tilde{\theta}_k) - \tilde{f}(k, 0)| \le \mu^{\frac{1}{2}} \times \text{const.} \to 0 \quad \text{w.p.1}, \quad \text{as } \mu \to 0.$$

So we are left with ($\bar{U}_k = \mu^{-1/2}(\tilde{\theta}_k - \bar{\theta}_k)$)

$$\delta \bar{U}_k = \mu R_s''(\tilde{\theta}_{k-1})\bar{U}_{k-1} + \mu^{\frac{1}{2}} \tilde{f}(k, 0).$$

Again continuity ensures, as $\mu \to 0$

$$\mu \max_{1 \le k \le T\mu^{-1}} \left| (R_s''(\tilde{\theta}_{k-1}) - R_s''(0))\bar{U}_{k-1} \right| \to 0 \quad \text{w.p.1}$$

so we end up with

$$\delta \bar{U}_k = \mu R_s''(0)\bar{U}_{k-1} + \mu^{\frac{1}{2}} \tilde{f}(k, 0).$$

A variance calculation as in Example 9.3 gives

$$\text{var}(\bar{U}_k) \to \Gamma^{\mu}, \quad \text{as } k \to \infty$$
$$\Gamma^{\mu} \to \Gamma, \quad \text{as } \mu \to 0$$
$$\Gamma = \frac{F(0)}{2R_s''(0)}$$

where $F(\omega)$ is the spectrum of $\tilde{f}(k, 0)$. Since

$$E\left(\tilde{f}(k, 0)\tilde{f}(0, 0)\right) = R_s''(k\Delta)\left(R_m(k\Delta) - R_n(k\Delta)\right)$$

$$\Rightarrow \qquad F(\omega) = \frac{1}{2\pi}\int_{-\pi}^{\pi} F_{s2}(\lambda)F_v(\omega - \lambda)d\lambda$$

where $F_{s2}(\lambda)$ is the spectrum of $s'_{k\Delta}$ and $F_v(\lambda) = F_m(\lambda) - F_n(\lambda)$ where $F_m(\lambda)$ and $F_n(\lambda)$ are the spectra of $m_{k\Delta}$ and $n_{k\Delta}$. We could re-express these in terms of the spectra of $s(\bullet)$, $m(\bullet)$, and $n(\bullet)$ of course. Finally then

$$F(0) = \frac{1}{2\pi}\int_{-\pi}^{\pi} F_{s2}(\lambda)\left(F_m(\lambda) - F_n(\lambda)\right) d\lambda.$$

9.4 INFINITE TIME AVERAGING: HOVERING THEOREM

We can extend the finite time averaging results of Section 9.2 to infinite time, in much the way it was done in Section 7.5. Some care is needed, however, because the uniformity condition used in Section 7.5 can only hold in an average sense in the stochastic case. Of course, exponential stability of the first order averaged system is needed as before.

The closest we can come to a (first order) stochastic version of the Hovering theorem seems to be the following result. There is an RV, $W(\mu)$ such that for any $A > 0$

$$\sup_{k \ge 1} P(\|\underline{z}_k - \bar{\underline{z}}_k\| > A) < P(W(\mu) > A) \tag{9.62}$$

and

$$W(\mu) \to 0 \quad \text{w.p.1}, \qquad \text{as } \mu \to 0. \tag{9.63}$$

From this it follows that

$$\sup_{k \geq 1} P\left(\|\underline{z}_k - \bar{\underline{z}}_k\| > \epsilon\right) \to 0, \qquad \text{as } \mu \to 0. \tag{9.64}$$

To obtain these sort of results some type of property such as strict stationarity or uniform mixing will be needed. To keep our presentation as clear as possible, strict stationarity is used. So we suppose:

(9.4A1) in (9.2A2), (9.2A3) the bounding sequences L_s, M_s, and p_s are strictly stationary.

We also need a slightly modified version of (9.2A3)

(9.4A3) There are matrices $\{\underline{M}_u, \underline{p}_u\}$ ($\underline{M}_0 = \underline{0} = \underline{p}_0$) so that

$$\left\| \sum_{s=r+1}^{k} \left(\tilde{\underline{f}}(s, \underline{z}) - \tilde{\underline{f}}(s, \underline{z}')\right) \right\| \leq \|\underline{M}_k - \underline{M}_r\| \|\underline{z} - \underline{z}'\|, \quad \text{for all } \underline{z}, \underline{z}'$$

$$\|\underline{M}_k\|/k \to 0 \quad \text{w.p.1}, \qquad \text{as } k \to \infty$$

$$\left\| \sum_{s=r+1}^{k} \tilde{\underline{f}}(s, \underline{0}) \right\| \leq \|\underline{p}_k - \underline{p}_r\|$$

$$\|\underline{p}_k\|/k \to 0 \quad \text{w.p.1}, \qquad \text{as } k \to \infty.$$

This leads to the following result.

Theorem 9.5. (First order stochastic Hovering theorem). Consider the primary system (9.2) and averaged system (9.4a) with identical initial conditions and suppose (9.2A1)–(9.2A5), (9.4A1) hold. Suppose $\underline{z}_e$ is an exponentially stable equilibrium point of the averaged system (9.4a) (it may not be an equilibrium point of the primary system), and the initial condition is in the stability region for $\underline{z}_e$. Then there is a random variable $W(\mu)$ such that (9.62), (9.63) hold. A consequence of this is (9.64).

Proof. We have already shown (9.64) follows from (9.62) and (9.63). The proof of these is similar to the deterministic proof in Section 7.5 with some necessary but subtle changes. The bound $c_T(\mu)$ used in Section 7.5 is now stochastic and depends on the interval $I_{n,n+1}$: Call it $c_{T,n}(\mu)$. Then proceeding as in Section 7.5, (7.45) is replaced by

$$\Delta_n \leq \lambda \left(\Delta_{n-1} + c_{T,n-1}(\mu)\right)$$

where $0 < \lambda = m\alpha^{T\mu^{-1}} < 1$. Iterating this gives (since $\Delta_0 = 0$)

$$\Delta_n \leq \lambda \sum_{s=1}^{n} \lambda^{n-s} c_{T,s-1}(\mu).$$

Thus in place of (7.48) we get

$$\max_{nT\mu^{-1}\le k<(n+1)T\mu^{-1}} \|\underline{z}_k - \bar{\underline{z}}_k\| \le c_{T,n}(\mu) + \sum_{r=0}^{n-1} \lambda^{n-r} c_{T,r}(\mu)$$

$$= \sum_{r=0}^{n} \lambda^{n-r} c_{T,r}(\mu) = W_n(\mu).$$

Thus

$$P\left(\max_{nT\mu^{-1}\le k\le(n+1)T\mu^{-1}} \|\underline{z}_k - \bar{\underline{z}}_k\| > A\right) \le P(W_n(\mu) > A)$$

$$= P\left(\sum_{r=0}^{n} \lambda^{n-r} c_{T,r-n}(\mu) > A\right)$$

by strict stationarity. Continuing,

$$P\left(\sum_{r=0}^{n} \lambda^{n-r} c_{T,r-n}(\mu) > A\right) = P\left(\sum_{u=0}^{n} \lambda^{u} c_{T,-u}(\mu) > A\right)$$

$$\le P(W(\mu) > A) \tag{9.65}$$

where

$$W(\mu) = \sum_{u=0}^{\infty} \lambda^{u} c_{T,-u}(\mu). \tag{9.66}$$

Two things now have to be shown:

$$W(\mu) < \infty \quad \text{w.p.1} \tag{9.67a}$$

$$W(\mu) \to 0 \quad \text{w.p.1}, \quad \text{as } \mu \to 0. \tag{9.67b}$$

To check (9.67a) we use the ratio test which requires

$$\limsup_{r\to\infty} \left(\lambda^r c_{T,-r}(\mu)\right)^{\frac{1}{r}} < 1.$$

This will follow if

$$\limsup_{r\to\infty} \left(c_{T,-r}(\mu)\right)^{\frac{1}{r}} \le 1. \tag{9.68}$$

Now $c_{T,-r}(\mu)$ has the form

$$c_{T,-r}(\mu) = d_{T,-r}(\mu) e^{v_{-r}(\mu)+TL}$$

$$d_{T,-r}(\mu) = \mu \max_{rT\mu^{-1}\le k\le(r+1)T\mu^{-1}} N_{-k,r}$$

$$v_{-r}(\mu) = \mu \sum_{rT\mu^{-1}}^{(r+1)T\mu^{-1}-1} \delta_{-s}, \quad \delta_{-s} = L_{-s} - L$$

where $N_{-k,r}$ is a sum of terms of the form

$$\|\underline{M}_{-k} - \underline{M}_{-rT\mu^{-1}}\|, \quad \|\underline{p}_{-k} - \underline{p}_{-rT\mu^{-1}}\|.$$

Consider firstly that by (9.2B2)

$$(rT)^{-1}v_{-r}(\mu) = (rT)^{-1}\mu \sum_{s=1}^{(r+1)T/\mu} \delta_{-s} - (rT)^{-1}\mu \sum_{s=1}^{rT/\mu-1} \delta_{-s}$$
$$\to 0 - 0 = 0 \quad \text{w.p.1}, \quad \text{as } \mu \to 0.$$

Next,

$$(d_{T,-r})^{\frac{1}{r}} \le 2rT\left(\max_{rT\mu^{-1}\le k\le (r+1)T\mu^{-1}} \frac{N_{-k,r}}{k}\right)^{\frac{1}{r}}$$
$$\le 4rT\left(\sup_{k\ge rT/\mu} \frac{N_{-k,0}}{k}\right)^{\frac{1}{r}}.$$

However $r^{1/r} \to 1$ as $r \to \infty$ while

$$\sup_{k\ge rT/\mu} \frac{N_{-k,0}}{k} \to 0 \quad \text{w.p.1}, \quad \text{as } k \to \infty.$$

Thus $\lim\sup(d_{T,-r})^{1/r} \le 1$ w.p.1 as $r \to \infty$. Thus (9.68) is established. From (9.65), (9.67a) we deduce (9.62).

To establish (9.67b) we proceed as follows. Introduce $\theta = \lambda^{1/2}$ and note

$$W(\mu) \le \sup_{u\ge 0} \theta^u c_{T,-u}(\mu) \sum_{u=0}^{\infty} \theta^u$$
$$\le (1-\theta)^{-1} \sup_{u\ge 0} \theta^u c_{T,-u}(\mu).$$

Again we consider $d_{T,-r}(\mu)$, $v_{-r}(\mu)$ separately. We have, for $r \ge 1$

$$d_{T,-r}(\mu) \le (1+r)T \max_{rT\mu^{-1}\le k\le (r+1)T\mu^{-1}} N_{-k,r}/k$$
$$\le 4rT\bar{m}(\mu)$$
$$\bar{m}(\mu) = \sup_{k\ge T\mu^{-1}} N_{-k,0}/k.$$

In view of (9.2A3)

$$\bar{m}(\mu) \to 0 \quad \text{w.p.1}, \quad \text{as } \mu \to 0. \tag{9.69}$$

Continuing, we have that, as $\mu \to 0$

$$\theta = \left(m\alpha^{T\mu^{-1}}\right)^{\frac{1}{2}} \to e^{-\alpha T/2}$$

so that, for some $\mu_* > 0, r \geq 1$

$$\theta^r c_{T,-r}(\mu) \leq \bar{m}(\mu) 2rTe^{-arT/2} e^{v_{-r}(\mu)+TL}, \quad \mu \leq \mu_*.$$

Now we look at $v_{-r}(\mu)$. We have by (9.2A2)

$$\bar{\delta}_n = \frac{1}{n}\sum_{s=1}^{n} \delta_{-s} \to 0 \quad \text{w.p.1}, \quad \text{as } n \to \infty.$$

Consequently, there is a non-decreasing sequence $\psi(n)$ with $\psi(n) \to \infty$ as $n \to \infty$ such that

$$\psi(n)\bar{\delta}_n \to 0 \quad \text{w.p.1}, \quad \text{as } n \to \infty$$

e.g., it will do to take

$$\psi(n) = \left(\sup_{k \geq n} \bar{\delta}_{-k}\right)^{-\frac{1}{2}}.$$

Next, for $\mu \leq \mu_*$

$$\begin{aligned}\left|\mu \sum_{s=1}^{rT\mu^{-1}} \delta_{-s}\right| &\leq \frac{r}{\psi(Tr\mu^{-1})}\psi(Tr\mu^{-1})|\bar{\delta}_{Tr\mu^{-1}}| \\ &\leq \frac{r}{\psi(Tr\mu_*^{-1})}\psi(Tr\mu^{-1})|\bar{\delta}_{Tr\mu^{-1}}|.\end{aligned}$$

Now take r_0 so that for all $r \geq r_0$

$$\frac{r}{\psi(Tr\mu_*^{-1})} \leq ar.$$

Also, there is k_0 so that for $Tr\mu^{-1} \geq k_0$

$$\psi(Tr\mu^{-1})|\bar{\delta}_{Tr\mu^{-1}}| < 1/4.$$

So when $r \geq r_0$ we have

$$\psi(Tr\mu^{-1})|\bar{\delta}_{Tr\mu^{-1}}| < 1/4 \quad \text{when} \quad \mu \leq \mu_1 = Tr_0/k_0.$$

Thus we have

$$|v_{-r}(\mu)| \leq \frac{1}{4}aTr, \quad r \geq r_o \text{ and } \mu \leq \min(\mu_*, \mu_1).$$

Consequently, we can write

$$\begin{aligned}\sup_{r\geq 1}\theta^r c_{T,-r}(\mu) &\leq 4\bar{m}(\mu)\left(\max_{1\leq r<r_0} rTe^{-aTr/2}e^{v_{-r}(\mu)+TL} + \sup_{r\geq r_0} rTe^{-aTr/2}e^{v_{-r}(\mu)+TL}\right)\\ &\leq 4\bar{m}(\mu)e^{TL}\left(\max_{1\leq r<r_0} rTe^{-aTr/2+v_{-r}(\mu)} + \sup_{r\geq r_0} a^{-1}(rTa/4)e^{-aTr/2}\right)\\ &\leq 16a^{-1}e^{-1}\bar{m}(\mu)e^{TL}\left(\max_{1\leq r<r_0} e^{v_{-r}(\mu)}+1\right).\end{aligned}$$

Continuing,

$$\max_{1\leq r<r_0} v_{-r}(\mu) \leq \mu\sum_{s=1}^{r_0T\mu^{-1}}|\delta_{-s}|.$$

However as $\mu\to 0$

$$\mu\sum_{s=1}^{r_0T\mu^{-1}}|\delta_{-s}| \to r_0TE|\delta_{-0}| \leq 2Lr_0T \quad \text{w.p.1.}$$

Thus there is an integer N_* so that

$$\sup_{\mu\leq N_*^{-1}}\mu\sum_{s=1}^{r_0T\mu^{-1}}|\delta_{-s}| \leq 4Lr_0T \quad \text{w.p.1.}$$

Thus, for $\mu \leq \mu_0 = \min(\mu_*, \mu_1, N_*^{-1})$

$$\begin{aligned}\sup_{r\geq 1}\theta^r c_{T,-r}(\mu) &\leq \bar{m}(\mu)e^{TL}8a^{-1}e^{-1}\left(e^{4Lr_0T}-1\right)\\ &= b\bar{m}(\mu), \qquad \text{for example.}\end{aligned}$$

Thus we have established

$$\sup_{k\geq 1} P\left(\|\underline{z}_k - \bar{\underline{z}}_k\| > A\right) \leq P\left(\max(c_{T,0}(\mu), b\bar{m}(\mu)) > A(1-\theta)\right)$$

and since as $\mu\to 0$, $\theta\to e^{-aT/2}$, (9.62) is established, while (9.63) follows easily from (9.69) and the fact that $c_{T,0}(\mu)\to 0$ w.p.1 as $\mu\to 0$.

9.5 INFINITE TIME AVERAGING—STABILITY

In connecting the stability of the averaged system to that of the primary system, it is useful to separate out two cases:

Noise-free time invariant parameters

$$\begin{aligned}\delta\underline{z}_k &= \mu\underline{f}(k, \underline{z}_{k-1})\\ \underline{f}(k, \underline{0}) &= \underline{0}.\end{aligned} \tag{9.70}$$

Noisy with possibly time variant parameters

$$\delta \underline{z}_k = \mu \underline{f}(k, \underline{z}_{k-1}) - \delta \underline{z}_o(\epsilon k)$$

$$\underline{f}(k, \underline{0}) \text{ is stochastic.} \tag{9.71}$$

In the first case w.p.1 convergence of $\underline{z}_k$ is possible and can be investigated using a perturbed Lyapunov function approach. In the second case performance analysis (e.g., an asymptotic variance for $\underline{z}_k$) is of interest, and this can be accomplished by determining "steady state" behavior.

In this section we consider stability, while steady state behavior is investigated in Section 9.6. The development here is similar to the deterministic one, although the reader will find it useful to consult Appendix E for review material.

Suppose then that the averaged system (9.4) is exponentially stable so that by Theorem C7.2 it has a Lyapunov function $V(\underline{z})$. The idea is to add a stochastic perturbation to $V(\underline{z})$ to yield a Lyapunov function suited to (9.70). Of course, we must suppose:

(9.5A1) $\underline{0}$ is an equilibrium point of (9.70).

(9.5A2) $\underline{0}$ is an exponentially stable equilibrium point of (9.4).

Consider then a stochastic version of the perturbation (7.55), namely, the conditional discounted perturbation

$$\underline{P}(k, \underline{z}) = \sum_{s=k+1}^{\infty} (1-\mu)^{s-k} E\left(\underline{\tilde{f}}(s, \underline{z})|\mathcal{F}_k\right) \tag{9.72a}$$

$$= \sum_{r=1}^{\infty} (1-\mu)^r E\left(\underline{\tilde{f}}(k+r, \underline{z})|\mathcal{F}_k\right)$$

where

$$\underline{\tilde{f}}(s, \underline{z}) = \underline{f}(s, \underline{z}) - \underline{f}_{av}(\underline{z})$$

and

$\mathcal{F}_k$ are the increasing σ-fields generated by the underlying random sequence $\underline{\xi}_u$, $u \le k$ (see (9.3)). (9.72b)

Note that $\underline{\tilde{f}}(s, \underline{z})$ is measurable with respect to $\mathcal{F}_s$ and that $\underline{P}(k, \underline{0}) = \underline{0}$.

The basic perturbation identity (7.56) now becomes (see Exercise 9.5)

$$\delta \underline{P}(k, \underline{z}) = -\underline{\tilde{f}}(k, \underline{z}) + \mu\left(\underline{\tilde{f}}(k, \underline{z}) + \underline{P}(k, \underline{z})\right) + \underline{\nu}_k(\underline{z}) \tag{9.72c}$$

where $\underline{\nu}_k(\underline{z})$ is an MG difference sequence

$$\underline{\nu}_k(\underline{z}) = (1-\mu)\sum_{r=0}^{\infty}(1-\mu)^r \left(E(\underline{\tilde{f}}(k+r, \underline{z})|\mathcal{F}_k) - E(\underline{\tilde{f}}(k+r, \underline{z})|\mathcal{F}_{k-1})\right)$$

so

$$E(\underline{\nu}_k(\underline{z})|\mathcal{F}_{k-1}) = 0, \quad \text{if } \underline{z} \text{ is adapted to } \mathcal{F}_{k-1}. \tag{9.72d}$$

We now place a mixing condition on $\{\xi_u\}$, the underlying stochastic driving signal of (9.3):

(9.5A3) $\underline{\xi}_u$ obeys a uniform mixing condition with mixing coefficient φ_s which (without loss of generality) is non-increasing, and $\varphi_s \to 0$ as $s \to \infty$.

By the fundamental uniform mixing lemma of Appendix E12 we have

$$\|E(\underline{u}_{k+r}|\mathcal{F}_k)\| \le \sqrt{2}\varphi_r^{1/2}\left[E(\|\underline{u}_{k+r}\|^2|\mathcal{F}_k) + E(\|\underline{u}_{k+r}\|^2)\right]^{\frac{1}{2}} \tag{9.73}$$

where

$$\underline{u}_{k+r} = \underline{\tilde{f}}(k+r, \underline{z}) - \underline{\tilde{f}}(k+r, \underline{z}').$$

Next we assume:

(9.5A4) $\underline{\tilde{f}}(k, \underline{z}), \underline{f}_{av}(\underline{z})$ obey uniform Lipschitz conditions so that

$$\|\underline{\tilde{f}}(s, \underline{z}) - \underline{\tilde{f}}(s, \underline{z}')\| \le L\|\underline{z} - \underline{z}'\|, \quad \text{for all } \underline{z}, \underline{z}'.$$

Then we find

$$\|E(\underline{u}_{k+r}|\mathcal{F}_k)\| \le 2\varphi_r^{1/2} L\|\underline{z} - \underline{z}'\|.$$

Thus from (9.5A1) and (9.72) we get

$$\mu\|\underline{P}(k, \underline{z}) - \underline{P}(k, \underline{z}')\| \le \xi(\mu)\|\underline{z} - \underline{z}'\| \tag{9.74a}$$

$$\xi(\mu) = 2L\mu\sum_{r=1}^{\infty}(1-\mu)^r\varphi_r^{1/2}.$$

This also ensures that $\|\underline{v}_k(\underline{z})\|$ has a finite first moment.

We now investigate the behavior of $\xi(\mu)$ as $\mu \to 0$ by using the Toeplitz lemma of Appendix D7. We have

$$\mu(1-\mu)^r \to 0, \quad \text{as } \mu \to 0, \ \ r \text{ fixed}$$

$$\mu\sum_{r=1}^{\infty}(1-\mu)^r = 1 - \mu \le 1$$

$$\mu\sum_{r=1}^{\infty}(1-\mu)^r = 1 - \mu \to 1 \quad \text{as } \mu \to 0.$$

Thus since $\varphi_r^{\frac{1}{2}} \to 0$ as $r \to \infty$ we deduce

$$\xi(\mu) \to 0 \quad \text{as } \mu \to 0. \tag{9.74b}$$

Now we introduce the perturbed Lyapunov function

$$V(k, \underline{z}) = V(\underline{z}) + \mu V_1(k, \underline{z}) \tag{9.75a}$$

$$V_1(k, \underline{z}) = \underline{P}^T(k, \underline{z})V_z(\underline{z}) \tag{9.75b}$$

and develop the stability argument very much as in the deterministic case (Section 7.6). We find, in view of (9.74), that (7.59) holds, that is, for μ small enough

$$0 \leq [\alpha_1 - \alpha_4 \xi(\mu)] \|\underline{z}\|^2 \leq V(k, \underline{z}) \leq [\alpha_2 + \alpha_4 \xi(\mu)] \|\underline{z}\|^2. \tag{9.76}$$

Now we split the increment in $V(k, \underline{z}_k)$ into pieces A, B, C, and D as in (7.60). We find that the A, C, and D bounds (7.61), (7.66), and (7.67) still hold and only the B_k calculation differs. In (7.62) there is an extra term

$$\mu \underline{\nu}_k(\underline{z}_{k-1})^T V_z(\underline{z}_{k-1}).$$

This term will be carried all the way through and so (7.63) becomes

$$V(k, \underline{z}_k) - V(k-1, \underline{z}_{k-1}) \leq -\mu(\alpha_3 - \psi(\mu)) \|\underline{z}_{k-1}\|^2 + \mu \underline{\nu}_k(\underline{z}_{k-1})^T V_z(\underline{z}_{k-1}).$$

However because of (9.72c) $\underline{\nu}_k(\underline{z}_{k-1})$ is also an MG difference sequence. Thus taking conditional expectations yields

$$E(V_k | \mathcal{F}_{k-1}) \leq V_{k-1} - \frac{1}{2}\mu\,(\alpha_3 - \psi(\mu))\,\|\underline{z}_{k-1}\|^2$$

where $V_k = V(k, \underline{z}_k)$ and $\psi(\mu) \to 0$ as $\mu \to 0$.

Taking μ small enough we now find via (9.76) that (with $b > 0$, $\mu b < 1$)

$$\begin{aligned} & E(V_k | \mathcal{F}_{k-1}) \leq (1 - \mu b) V_{k-1} \\ \Rightarrow \quad & E(V_k \lambda^{-k} | \mathcal{F}_{k-1}) \leq V_{k-1} \lambda^{-(k-1)} \\ & \lambda = 1 - \mu b. \end{aligned}$$

Thus by the submartingale convergence theorem (Appendix E) we find

$$V_k \lambda^{-k} \to R_\infty (< \infty) \quad \text{w.p.1}$$

or

$$V_k \sim R_\infty \lambda^k \quad \text{w.p.1, as } k \to \infty.$$

This establishes (w.p.1) exponential stability of $\underline{z}_k$ via (9.75). The following result has thus been obtained.

Theorem 9.6. Consider the noise-free primary system (9.70) and averaged system (9.4). Suppose that (9.5A1)–(9.5A4) hold. Then (9.70) is exponentially stable w.p.1.

In Exercise 9.6 the reader is asked to show that L in (9.5A4) can be replaced by an RV L_r that obeys an SLLN.

Example 9.5. Time Delay Estimation

We consider the noise-free idealized adaptive time delay estimation problem of Section 4.5, but now allow the underlying signal to be stochastic. The error system is

$$\delta\tilde{\theta}_k = -\mu s'_{k\Delta - \tilde{\theta} - \theta_o}(s_{k\Delta - \theta_o} - s_{k\Delta - \tilde{\theta}_{k-1} - \theta_o}). \tag{9.77}$$

The averaged function is, as in Example 9.4,

$$f_{av}(\tilde{\theta}) = R'_s(\tilde{\theta})$$

where $R_s(\bullet)$ is the autocovariance of $s(t)$. The corresponding averaged system is

$$\delta\bar{\theta}_k = \mu R_s'(\bar{\theta}_{k-1}).$$

We have already analyzed its behavior in Example 7.7. We now assume:

(9.5a1) $s_{k\Delta}$ obeys the uniform mixing assumption (9.5A3).

Next, the Lipschitz condition (9.5A4) holds if (cf. (7.7C1))

(9.5a2) $s_{k\Delta}$, $s'_{k\Delta}$, and $s''_{k\Delta}$ are uniformly bounded.

Thus we find via Theorem 9.6 that if the initial condition $\tilde{\theta}_0$ is in the region of attraction of the equilibrium point $\tilde{\theta} = 0$, then we get exponential convergence of (9.77) to 0 w.p.1 for μ that is small enough.

9.6 STEADY STATE—PRELIMINARIES

Here we discuss the noisy time variant parameter system (9.71). The aim is to do a performance analysis, e.g., to get an approximation to the weight error power ($\lim_{N\to\infty} N^{-1} \sum_{k=1}^{N} \|\underline{\tilde{z}}_k\|^2$) and related quantities.

To understand the issues, let us return to the LMS algorithm with time invariant parameters. The error system (5.5) is

$$\delta\underline{\tilde{w}}_k = -\mu \underline{x}_k \underline{x}_k^T \underline{\tilde{w}}_{k-1} + \mu \underline{x}_k \epsilon_k. \tag{9.78}$$

Let us for the moment consider the case treated in Section 5.2 where $\{\underline{x}_k, \epsilon_k\}$ are independent white noises with variance $(\underline{R}_x, \sigma_\epsilon^2)$ respectively. We look at the normalized weight error variance

$$\underline{G}_k^\mu = E(\underline{\tilde{w}}_k \underline{\tilde{w}}_k^T)/\mu$$

which, from (5.8), obeys

$$\delta\underline{G}_k^\mu = -\mu \underline{R}_x \underline{G}_{k-1}^\mu - \mu \underline{G}_{k-1}^\mu \underline{R}_x + \mu\sigma_\epsilon^2 \underline{R}_x + \mu^2 \underline{\psi}_x \tag{9.79}$$

where

$$\underline{\psi}_x = 2\underline{R}_x \underline{G}_{k-1}^\mu \underline{R}_x + \mathrm{tr}(\underline{G}_{k-1}^\mu \underline{R}_x)\underline{R}_x.$$

Under a stability condition (namely (5.12)) there is a steady state or limit variance $\underline{G}_\infty^\mu$ obeying (from (9.79))

$$\underline{R}_x \underline{G}_\infty^\mu + \underline{G}_\infty^\mu \underline{R}_x = \sigma_\epsilon^2 \underline{R}_x + \mu \underline{\psi}_x. \tag{9.80}$$

On the other hand, observe that $\underline{G}_k^\mu$, considered as a sequence of signals indexed by μ, can be approximated to order μ by a "limit" $\underline{\pi}_k^\mu$ obeying

$$\delta\underline{\pi}_k^\mu = -\mu \underline{R}_x \underline{\pi}_{k-1}^\mu - \mu \underline{\pi}_{k-1}^\mu \underline{R}_x + \mu\sigma_\epsilon^2 \underline{R}_x. \tag{9.81}$$

The statement is justified by subtracting (9.81) from (9.79) to get, with $\underline{E}_k^\mu = \underline{G}_k^\mu - \underline{\pi}_k^\mu$,

$$\delta\underline{E}_k^\mu = -\mu \underline{R}_x \underline{E}_{k-1}^\mu - \mu \underline{E}_{k-1}^\mu \underline{R}_x + \mu^2 \underline{\psi}_x.$$

So under the stability condition (5.12) we see that (see Exercise 9.3)

$$\sup_{k\geq 1} \|\underline{G}_k^\mu - \underline{\pi}_k^\mu\| = o(\mu) \to 0 \quad \text{as } \mu \to 0. \tag{9.82}$$

Let us further note that under the stability condition (5.12), $\underline{\pi}_k^\mu$ itself has a steady state value $\underline{\pi}_\infty$ satisfying

$$\underline{R}_x \underline{\pi}_\infty + \underline{\pi}_\infty \underline{R}_x = \sigma_\epsilon^2 \underline{R}_x. \tag{9.83}$$

Finally we can close a circle by observing from (9.80) that the sequence of steady state variance matrices $\underline{G}_\infty^\mu$ has a limit as $\mu \to 0$ which is $\underline{\pi}_\infty$!

These observations can be conveniently represented in a rectangular diagram as shown in Fig. 9.1.

$$\begin{array}{ccc} \underline{G}_k^\mu & \sim & \underline{G}_\infty^\mu \\ \Downarrow & & \Downarrow \\ \underline{\pi}_k^\mu & \sim & \underline{\pi}_\infty \end{array}$$

Figure 9.1 Invariant or steady state or limiting variance matrices. $\sim$ means "has steady state," $\Downarrow$ means the difference converges to $\underline{0}$.

The observations we have made on this special case allow us to perceive how to formulate a performance analysis in more complicated situations. There are several points to be made. We are interested in knowing $\underline{G}_k^\mu$ but it can only be computed in simple cases. If we cannot calculate $\underline{G}_k^\mu$, perhaps we can find $\underline{\pi}_k^\mu$, which is an approximation to $\underline{G}_k^\mu$ that discards terms of order $o(\mu)$. However from $\underline{\pi}_k^\mu$ we can find a steady state value $\underline{\pi}_\infty$, and we need to be sure it is the same as $\lim_{\mu\to 0} \underline{G}_\infty^\mu$. If this is so, then we have a 0th order approximation to $\underline{G}_k^\mu$.

These ideas lead to the following strategy:

(i) Find the averaged system with associated variance $\underline{\pi}_k^\mu$ and prove a Hovering theorem

$$\sup_{k\geq 1} \|\underline{G}_k^\mu - \underline{\pi}_k^\mu\| \leq c(\mu), \qquad \mu \leq \mu_0$$

where

$$c(\mu) \to 0 \quad \text{as } \mu \to 0.$$

Clearly, some stability results will be needed to achieve this.

(ii) Show that $\underline{\pi}_k^\mu$ has a steady state value $\underline{\pi}_\infty^\mu$. This is done in two steps. First, using a stability result show that $\underline{\pi}_k^\mu$ is bounded in k, uniformly in μ. Thus it must have a limit point. Then show somehow that there is only one possible limit point. For example from (9.81) any limit point must obey (9.83), the solution to which is unique if $\underline{R}_x$ is positive definite.

(iii) From (i) and (ii) it follows that $\underline{G}_k^\mu$ has a limit point. And from (i) it then follows that any such limit obeys

$$\|\underline{G}_\infty^\mu - \underline{\pi}_\infty^\mu\| \to 0 \quad \text{as } \mu \to 0.$$

Since, however, $\underline{\pi}_\infty^\mu$ is unique, then so is $\underline{\pi}_\infty^0$, and hence so is $\underline{G}_\infty^0$, and $\underline{G}_\infty^0 = \underline{\pi}_\infty^0$.

For general use, this strategy is better expressed (in a somewhat weaker way) in terms of $\mu^{-1/2}\underline{\tilde{w}}_k$ which we now denote as $\underline{U}_k^{\mu}$ to keep the μ-dependence in mind. The strategy becomes

(i) Show that there is a partially averaged trajectory $\underline{\bar{U}}_k^{\mu}$ which has a steady state or invariant measure with associated RV $\underline{\bar{U}}_{\infty}^{\mu}$. This follows if $\underline{\bar{U}}_k^{\mu}$ is tight with respect to k, uniformly in μ. This can be proved using Lyapunov methods.

(ii) Show that the deviation $\underline{d}_k = \underline{U}_k^{\mu} - \underline{\bar{U}}_k^{\mu}$ is tight in k, uniformly in μ, i.e.,

$$\lim_{A\to\infty} \sup_{k\geq 1} P(\|\underline{\bar{U}}_k^{\mu} - \underline{U}_k^{\mu}\| > A) = 0, \quad \text{uniformly in } \mu. \tag{9.84}$$

Also show

$$\sup_{k\geq 1} P(\|\underline{U}_k^{\mu} - \underline{\bar{U}}_k^{\mu}\| > \epsilon) \to 0, \quad \text{as } \mu \to 0. \tag{9.85}$$

This is a second order stochastic Hovering theorem. It can be proved by a perturbed Lyapunov method.

(iii) It follows from (ii) that $\underline{U}_k^{\mu}$ has a steady state or invariant measure with associated RV, $\underline{U}_{\infty}^{\mu}$.

(iv) It then follows from (iii) and (9.85) that

$$\|\underline{U}_{\infty}^{\mu} - \underline{\bar{U}}_{\infty}^{\mu}\| \xrightarrow{P} 0, \quad \text{as } \mu \to 0.$$

If we can show $\underline{\bar{U}}_{\infty}^{0}$ is unique then we have a 0th order approximation to $\underline{U}_{\infty}^{\mu}$.

Now turning to a more general setting there are several cases to cover:

(i) Time Invariant Parameters

$$\delta \underline{z}_k = \mu \underline{f}(k, \underline{z}_{k-1})$$
$$\underline{f}(k, \underline{0}) \text{ is stochastic.}$$

Here a second order analysis is appropriate.

(ii) Time Variant Parameters

$$\delta \underline{z}_k = \mu \underline{f}(k, \underline{z}_{k-1}) - \delta \underline{z}_o(\varepsilon k)$$
$$\underline{f}(k, \underline{0}) \text{ is stochastic.}$$

Depending on the matching between ε and μ, a first order analysis or a second order analysis will be relevant.

These cases are pursued in the next two sections.

9.7 STEADY STATE ANALYSIS—TIME INVARIANT PARAMETERS

Following the second order analysis of Section 9.2 we have to consider the following pair of systems: (9.26) and (9.18), which we reproduce here for convenience:

$$\delta \underline{U}_k^{\mu} = \mu \nabla \underline{f}(k, \bar{\underline{z}}_{k-1}) \underline{U}_{k-1}^{\mu} + \mu^{\frac{1}{2}} \tilde{\underline{f}}_k + \mu^{\frac{1}{2}} \underline{a}_k \tag{9.86}$$

$$\delta \bar{\underline{U}}_k^{\mu} = \mu \nabla \underline{f}_{av}(\bar{\underline{z}}_{k-1}) \bar{\underline{U}}_{k-1}^{\mu} + \mu^{\frac{1}{2}} \tilde{\underline{f}}_k. \tag{9.87}$$

The idea is that steady state properties of (9.87) are relatively easy to compute and can be related well to those of (9.86). We already observed that $\mu^{1/2} \tilde{\underline{f}}_k$ in (9.87) could be replaced by $\mu^{1/2} \delta \underline{P}_k$ as in (9.19). Furthermore, since generally $\bar{\underline{z}}_k$ will be ES at $\underline{0}$, we will hope to replace (9.87) by

$$\delta \underline{U}_{o,k}^{\mu} = \mu \nabla \underline{f}_{av}(\underline{0}) \underline{U}_{o,k-1}^{\mu} + \mu^{\frac{1}{2}} \delta \underline{p}_k \tag{9.88}$$

$$\underline{p}_k = \underline{p}(k, \underline{0}).$$

The justification of these replacements is covered in Exercises 9.9, and 9.10 and is left to the reader. We will work with (9.87). Also the tightness in k, uniform in μ, of $\bar{\underline{U}}_k^{\mu}$ is easily established from (9.87) by using ES and showing $\sup_k E(\|\bar{\underline{U}}_k^{\mu}\|^2) \leq$ const.

As indicated in the last section we need a second order stochastic Hovering theorem that establishes (9.84), (9.85). The development of such a second order result is much harder than the development of the first order result in Section 9.4. Since the original system will have an ES property, Lyapunov methods can be used throughout. The resultant argument is long and tedious although quite clear. For this reason and to retain clarity we will only give a sketch of the development. We will outline the form of the argument clearly and indicate special regularity conditions, but we will not develop the argument in intricate detail or with complete rigor.

Our argument begins with a difference equation for the deviation $\underline{d}_k = \underline{U}_k - \bar{\underline{U}}_k$ (namely (9.27)) (we now drop the μ superscript):

$$\delta \underline{d}_k = \mu \nabla \underline{f}(k, \bar{\underline{z}}_{k-1}) \underline{d}_{k-1} + \mu^{\frac{1}{2}} \underline{a}_k + \mu \nabla \tilde{\underline{f}}_k \bar{\underline{U}}_{k-1}. \tag{9.89}$$

We shall need ES of the unforced system

$$\delta \hat{\underline{\zeta}}_k = \mu \nabla \underline{f}(k, \bar{\underline{z}}_{k-1}) \hat{\underline{\zeta}}_{k-1}. \tag{9.90}$$

This will follow, via (9.2C1) if we have ES of

$$\delta \underline{\zeta}_k^* = \mu \nabla \underline{f}(k, \underline{0}) \underline{\zeta}_{k-1}^*. \tag{9.91}$$

By means of a perturbed Lyapunov argument (as in Section 9.5) this will follow from (9.7A1):

(9.7A1)

    (i) $\underline{0}$ is an ES equilibrium point of the averaged system (9.4a) with decay rate $O(\mu)$.

(ii) The linearized system

$$\delta \underline{\zeta}_k = \mu \nabla \underline{f}_{av}(\underline{0}) \underline{\zeta}_{k-1} \tag{9.92}$$

is ES to $\underline{0}$ with decay rate $O(\mu)$.

Of course if $\{\partial^2 \underline{f}_{av}(\underline{z})/\partial z_i \partial z_j\}$ is continuous at $\underline{0}$ then Theorem 7.8 tells us that (ii) $\Rightarrow$ (i).

The idea now is to show tightness (in k) of $\underline{d}_k$ (that is, to show (9.84)) by finding a perturbed Lyapunov function for (9.89). The uniform convergence in probability of (9.85) is discussed later.

9.7.1 Development of Tightness: (9.84)

The difficulty and subtlety in (9.89) is due to the $\underline{a}_k$ term, which is state dependent. Our subsequent argument hinges on reorganizing $\underline{a}_k$ in a judicious way as follows. From (9.25)

$$\begin{aligned}
\underline{a}_k &= \underline{f}(k, \bar{\underline{z}}_{k-1} + \underline{\Delta}_{k-1}) - \underline{f}(k, \bar{\underline{z}}_{k-1}) - \nabla \underline{f}(k, \bar{\underline{z}}_{k-1}) \underline{\Delta}_{k-1} \\
&= \underline{f}(k, \bar{\underline{z}}_{k-1} + \mu^{\frac{1}{2}} \underline{U}_{k-1}) - \underline{f}(k, \bar{\underline{z}}_{k-1}) - \nabla \underline{f}(k, \bar{\underline{z}}_{k-1}) \mu^{\frac{1}{2}} \underline{U}_{k-1} \\
&= \underline{f}(k, \bar{\underline{z}}_{k-1} + \mu^{\frac{1}{2}} \bar{\underline{U}}_{k-1} + \mu^{\frac{1}{2}} \underline{d}_{k-1}) - \underline{f}(k, \bar{\underline{z}}_{k-1}) - \nabla \underline{f}(k, \bar{\underline{z}}_{k-1})(\mu^{\frac{1}{2}} \bar{\underline{U}}_{k-1} + \mu^{\frac{1}{2}} \underline{d}_{k-1}) \\
&= \underline{q}_k + \underline{t}_k
\end{aligned} \tag{9.93}$$

where

$$\underline{q}_k = \underline{f}(k, \bar{\underline{z}}_{k-1} + \mu^{\frac{1}{2}} \bar{\underline{U}}_{k-1} + \mu^{\frac{1}{2}} \underline{d}_{k-1}) - \underline{f}(k, \bar{\underline{z}}_{k-1} + \mu^{\frac{1}{2}} \bar{\underline{U}}_{k-1}) - \nabla \underline{f}(k, \bar{\underline{z}}_{k-1}) \mu^{\frac{1}{2}} \underline{d}_{k-1}$$

$$\underline{t}_k = \underline{f}(k, \bar{\underline{z}}_{k-1} + \mu^{\frac{1}{2}} \bar{\underline{U}}_{k-1}) - \underline{f}(k, \bar{\underline{z}}_{k-1}) - \nabla \underline{f}(k, \bar{\underline{z}}_{k-1}) \mu^{\frac{1}{2}} \bar{\underline{U}}_{k-1}.$$

We note that from (9.2C1)

$$\mu^{\frac{1}{2}} \|\underline{t}_k\| \le \mu^{\frac{3}{2}} c_k \|\bar{\underline{U}}_{k-1}\|^2 \tag{9.94a}$$

$$\mu^{\frac{1}{2}} \|\underline{q}_k\| \le 2L\mu \|\underline{d}_{k-1}\|. \tag{9.94b}$$

This bound on the state dependent term $\|\underline{q}_k\|$ is too big, however. So a perturbation calculation is needed to handle $\underline{q}_k$. Introduce

$$\underline{q}(k, \bar{\underline{z}}, \bar{\underline{U}}, \underline{d}) = \underline{f}(k, \bar{\underline{z}} + \mu^{\frac{1}{2}} \bar{\underline{U}} + \mu^{\frac{1}{2}} \underline{d}) - \underline{f}(k, \bar{\underline{z}} + \mu^{\frac{1}{2}} \bar{\underline{U}}) - \nabla \underline{f}(k, \bar{\underline{z}}) \mu^{\frac{1}{2}} \underline{d}$$

so that

$$\underline{q}_k = \underline{q}(k, \bar{\underline{z}}_{k-1}, \bar{\underline{U}}_{k-1}, \underline{d}_{k-1}).$$

Also introduce the conditional discounted perturbation (cf. (9.72a), but notice the $\mu^{\frac{1}{2}}$)

$$\underline{Q}(k, \bar{\underline{z}}, \bar{\underline{U}}, \underline{d}) = \sum_{r=1}^{\infty} (1 - \mu^{\frac{1}{2}})^r E\left(\underline{q}(k + r, \bar{\underline{z}}, \bar{\underline{U}}, \underline{d}) | \mathcal{F}_k\right).$$

Then as in Section 9.5 there follows a perturbation decomposition (here • does duty for three arguments)

$$\delta \underline{Q}(k, \bullet) = -\underline{q}(k, \bullet) + \mu^{\frac{1}{2}} \Big(\underline{q}(k, \bullet) + \underline{Q}(k, \bullet)\Big) + \underline{\nu}_k(\bullet) \tag{9.95}$$

where $\underline{\nu}_k(\bar{\underline{z}}, \bar{\underline{U}}, \underline{d})$ is an MG difference sequence. Under the mixing condition (9.5A3) we obtain the uniform mixing inequality (9.73) with $\underline{u}_{k+r}$ replaced by $\underline{q}(k+r, \bullet)$. Next we require a uniform version of (9.2C1), namely,

(9.7A2) As for (9.2C1) but $c_k = c$, for all k.

Then (9.7A2) and the uniform Lipschitz condition (9.5A4) yield the following Lipschitz properties (cf. derivation of (9.74a))

$$\mu^{\frac{1}{2}} \|\underline{Q}(k, \bar{\underline{z}}, \bar{\underline{U}}, \underline{d}) - \underline{Q}(k, \bar{\underline{z}}', \bar{\underline{U}}, \underline{d})\| \leq \|\bar{\underline{z}} - \bar{\underline{z}}'\| \xi(\mu)(1 + \mu^{\frac{1}{2}} \|\underline{d}\|) \tag{9.96a}$$

$$\mu^{\frac{1}{2}} \|\underline{Q}(k, \bar{\underline{z}}, \bar{\underline{U}}, \underline{d}) - \underline{Q}(k, \bar{\underline{z}}, \bar{\underline{U}}', \underline{d})\| \leq \mu^{\frac{1}{2}} \xi(\mu) \|\bar{\underline{U}} - \bar{\underline{U}}'\| \tag{9.96b}$$

$$\mu^{\frac{1}{2}} \|\underline{Q}(k, \bar{\underline{z}}, \bar{\underline{U}}, \underline{d}) - \underline{Q}(k, \bar{\underline{z}}, \bar{\underline{U}}, \underline{d}')\| \leq \mu^{\frac{1}{2}} \xi(\mu) \|\underline{d} - \underline{d}'\| \tag{9.96c}$$

where

$$\xi(\mu) \to 0 \quad \text{as } \mu \to 0.$$

It is important to note that since

$$\underline{q}(k, \bar{\underline{z}}, \bar{\underline{U}}, \underline{0}) = \underline{0}, \quad \text{for any finite } \bar{\underline{z}}, \bar{\underline{U}}$$

then

$$\underline{Q}(k, \bar{\underline{z}}, \bar{\underline{U}}, \underline{0}) = \underline{0}$$

so that

$$\mu^{\frac{1}{2}} \|\underline{Q}(k, \bar{\underline{z}}, \bar{\underline{U}}, \underline{d})\| \leq \mu^{\frac{1}{2}} \xi(\mu) \|\underline{d}\|. \tag{9.97}$$

Now we turn to a Lyapunov analysis of (9.89) and are led to construct the following three-term perturbed Lyapunov function

$$V(k, \bar{\underline{z}}, \bar{\underline{U}}, \underline{d}) = V_0(\underline{d}) + V_1(k, \underline{d}) + V_2(k, \bar{\underline{z}}, \bar{\underline{U}}, \underline{d})$$

where $V_0(\underline{d})$ is the Lyapunov function for (9.92) and $V_0(\underline{d}) + V_1(k, \underline{d})$ is a perturbed Lyapunov function for (9.91), which is constructed as in Section 9.5 and so needs no spelling out. Finally the important term is

$$V_2(k, \bar{\underline{z}}, \bar{\underline{U}}, \underline{d}) = \mu^{\frac{1}{2}} \underline{Q}^T(k, \bar{\underline{z}}, \bar{\underline{U}}, \underline{d}) V_0'(\underline{d}) \tag{9.98}$$

$$V_0'(\underline{d}) = \frac{dV_0(\underline{d})}{d\underline{d}}.$$

In view of (9.97) the positivity of $V(k, \bullet)$ is ensured, i.e., for some α_1, α_2 and all small enough μ

$$\alpha_1 \|\underline{d}\|^2 \leq V(k, \bullet) \leq \alpha_2 \|\underline{d}\|^2. \tag{9.99}$$

Now we commence the Lyapunov computation. Let

$$V_k = V(k, \bar{\underline{z}}_k, \bar{\underline{U}}_k, \underline{d}_k)$$

and define $V_{0,k}$, $V_{1,k}$, and $V_{2,k}$ similarly. Our aim is to derive an inequality of the form

$$E(V_k | \mathcal{F}_{k-1}) \leq V_{k-1}(1 - a\mu) + \mu |b_k| \tag{9.100a}$$

where

$$E|b_k| \leq \text{constant}$$

and $\lim_{\mu \to 0}$ constant $< \infty$. Then it follows easily that

$$\sup_k E(V_k) \leq \text{constant} \times a^{-1}. \tag{9.100b}$$

In view of (9.99) this of course yields tightness since

$$\sup_k P(\|\underline{d}_k\| > A) \leq \sup_k E(\|d_k\|^2)/A^2.$$

In our subsequent argument we thus need only keep track of terms that are $O(\mu)|b_k|$ and ensure that state dependent terms are $o(\mu)V_{k-1}$ or $o(\mu)\|\underline{d}_{k-1}\|^2$. Continuing with the Lyapunov argument compute

$$\delta V_k = \delta V_{0,k} + \delta V_{1,k} + \delta V_{2,k}$$

by using Taylor series expansions as in Sections 7.6, 9.5. The term $\delta V_{1,k}$ deals with ES of (9.91) (or (9.90)) based on ES of (9.92) and so we do not discuss it any further. Of course $\delta V_{0,k}$ will provide the crucial term—$\mu\alpha\|\underline{d}_{k-1}\|^2$. We will discuss $\delta V_{2,k}$ in detail shortly. The other main term of interest in $\delta V_{0,k}$ will be

$$\mu^{\frac{1}{2}} \underline{a}_k^T V_0'(\underline{d}_k^*) \tag{9.101}$$

where

$$\underline{d}_k^* = \underline{d}_{k-1} + \mu \nabla \underline{f}_{av}(\underline{0})\underline{d}_{k-1}.$$

Other terms (which are easily bounded) are

$$\mu \underline{a}_k^T \underline{V}_0'' \underline{a}_k, \quad \mu(\nabla \tilde{\underline{f}}_k \underline{\bar{U}}_{k-1})^T V_0'(\underline{d}_k^*) \tag{9.102}$$

where, since $V_0(\underline{d})$ is quadratic, $\underline{V}_0''$ is positive definite. In view of (9.93) and (9.94) we can replace (9.101) by $\mu^{\frac{1}{2}} \underline{q}_k^T V_0'(\underline{d}_k^*)$, and since $\underline{d}_k^* - \underline{d}_{k-1}$ is $O(\mu)$ we can replace this by

$$\mu^{\frac{1}{2}} \underline{q}_k^T V_0'(\underline{d}_{k-1}). \tag{9.103}$$

Now turn to $\delta V_{2,k}$ which can be split (as usual) into three pieces

$$\delta V_{2,k} = B_k + C_k + D_k$$

$$B_k = \mu^{\frac{1}{2}} \left(\underline{Q}(k, []_{k-1}) - \underline{Q}(k-1, []_{k-1})\right)^T V_0'(\underline{d}_{k-1}) \tag{9.104a}$$

$$C_k = \mu^{\frac{1}{2}} \left(\underline{Q}(k, []_k) - \underline{Q}(k, []_{k-1})\right)^T V_0'(\underline{d}_{k-1}) \tag{9.104b}$$

$$D_k = \mu^{\frac{1}{2}} \left(\underline{Q}(k, []_k)\right)^T \left(V_0'(\underline{d}_k) - V_0'(\underline{d}_{k-1})\right) \tag{9.104c}$$

where $[]_k = \bar{z}_k, \underline{\bar{U}}_k, \underline{d}_k$.

For B_k we use (9.95) to see that (9.103) is cancelled and we are left with

$$\mu \left(\underline{q}_k + \underline{Q}(k, []_{k-1})\right)^T V_0'(\underline{d}_{k-1}) + \underline{\nu}_k^T V_0'(\underline{d}_{k-1}) \tag{9.105}$$

where $\underline{\nu}_k$ is an MG difference. Once we take conditional expectations, the last term will vanish. From (9.94), the first term in (9.105) is bounded by const. $\times\ \mu^{3/2}\|\underline{d}_{k-1}\|^2$ which is $o(\mu)\|\underline{d}_{k-1}\|^2$ as required. By (9.97) the second term in (9.105) is bounded by $\mu\xi(\mu) \times$ const. $\times\ \|\underline{d}_{k-1}\|^2$ which is $o(\mu)\|\underline{d}_{k-1}\|^2$ as required. Turning to C_k we find from (9.96) that

$$C_k \leq \mu^{\frac{1}{2}}\xi(\mu)\|\underline{d}_{k-1}\|(\|\underline{\bar{U}}_k - \underline{\bar{U}}_{k-1}\| + \|\underline{d}_k - \underline{d}_{k-1}\|)$$
$$+ \mu^{\frac{1}{2}}\xi(\mu)\|\underline{d}_{k-1}\|^2\|\bar{z}_k - \bar{z}_{k-1}\| + \xi(\mu)\|\bar{z}_k - \bar{z}_{k-1}\|\|\underline{d}_{k-1}\|$$

and since $\|\bar{z}_k - \bar{z}_{k-1}\| = O(\mu)$, $\|\underline{\bar{U}}_k - \underline{\bar{U}}_{k-1}\| = O(\mu^{1/2})$, $\|\underline{d}_k - \underline{d}_{k-1}\| = O(\mu^{1/2})$, all the terms bounding C_k are $o(\mu)\|\underline{d}_{k-1}\|^2$ or $o(\mu)|b_k|$ as required. Continuing, for D_k,

$$|D_k| \leq \mu^{\frac{1}{2}}\xi(\mu)\|\underline{d}_{k-1}\|\|\underline{d}_k - \underline{d}_{k-1}\|$$

which is bounded by terms that are $o(\mu)\|\underline{d}_{k-1}\|^2$ or $o(\mu)$. So we are led to conclude that (9.100a) and hence (9.100b) is valid as required.

9.7.2 Development of (9.85): Uniform Convergence in Probability

If we view $\underline{U}_k$ as an $O(\mu^{1/2})$ perturbation of z_k (i.e., $z_k = \bar{z}_k + \mu^{1/2}\underline{U}_k$) about $\bar{z}_k$ then the idea is to take this expansion one step further. First divide (9.89) throughout by $\mu^{1/2}$ to give (with $\underline{\delta}_k = \underline{d}_k\mu^{-1/2}$)

$$\delta\underline{\delta}_k = \mu\nabla \underline{f}(k, \bar{z}_{k-1})\underline{\delta}_{k-1} + \underline{a}_k + \mu^{\frac{1}{2}}\nabla\underline{\tilde{f}}_k\underline{\bar{U}}_{k-1}. \tag{9.106}$$

We are led now to introduce $\underline{\delta}_{o,k}$, satisfying

$$\delta\underline{\delta}_{o,k} = \mu\nabla \underline{f}_{av}(\bar{z}_{k-1})\underline{\delta}_{o,k-1} + \mu^{\frac{1}{2}}\nabla\underline{\tilde{f}}_k\underline{\bar{U}}_{k-1}. \tag{9.107}$$

It is straightforward to show that $\underline{\delta}_{o,k}$ is tight in k uniformly in μ by showing (see Exercise 9.11)

$$\sup_k E\|\underline{\delta}_{o,k}\|^2 \leq \text{constant} \tag{9.108}$$

Now subtract (9.107) from (9.106) to find (with $\underline{\eta}_k = \underline{\delta}_k - \underline{\delta}_{o,k}$)

$$\delta\underline{\eta}_k = \mu\nabla\underline{f}(k, \bar{\underline{z}}_{k-1})\underline{\eta}_{k-1} + \underline{a}_k + \mu\nabla\tilde{\underline{f}}_k\underline{\delta}_{o,k-1}. \tag{9.109}$$

If we can show

$$\sup_k E\|\underline{\eta}_k\|^2 \leq \text{constant} \tag{9.110}$$

then from this and (9.108) we have

$$\sup_k E\|\underline{d}_k\|^2 \leq \mu \times \text{constant}$$

which delivers (9.85) as required. In analyzing (9.109) the last term causes no problems. So again an analysis of (9.109) depends on judicious treatment of $\underline{a}_k$. And again we seek to establish a result such as (9.100a), (9.100b).

As before we break $\underline{a}_k$ into several pieces. From (9.110) (cf. (9.93)) we write

$$\begin{aligned}
\underline{a}_k &= \underline{f}(k, \bar{\underline{z}}_{k-1} + \mu^{\frac{1}{2}}\bar{\underline{U}}_{k-1} + \mu\underline{\delta}_{o,k-1} + \mu\underline{\eta}_{k-1}) - \underline{f}(k, \bar{\underline{z}}_{k-1}) \\
&\quad - \nabla\underline{f}(k, \bar{\underline{z}}_{k-1})(\mu^{\frac{1}{2}}\bar{\underline{U}}_{k-1} + \mu\underline{\delta}_{o,k-1} + \mu\underline{\eta}_{k-1}) \\
&= \underline{r}_k + \underline{t}_{k1} + \underline{t}_{k2} \\
\underline{r}_k &= \underline{f}(k, \bar{\underline{z}}_{k-1} + \mu^{\frac{1}{2}}\bar{\underline{U}}_{k-1} + \mu\underline{\delta}_{o,k-1} + \mu\underline{\eta}_{k-1}) \\
&\quad - \underline{f}(k, \bar{\underline{z}}_{k-1} + \mu^{\frac{1}{2}}\bar{\underline{U}}_{k-1} + \mu\underline{\delta}_{o,k-1}) - \nabla\underline{f}(k, \bar{\underline{z}}_{k-1})\mu\underline{\eta}_{k-1} \\
\underline{t}_{k1} &= \underline{f}(k, \bar{\underline{z}}_{k-1} + \mu^{\frac{1}{2}}\bar{\underline{U}}_{k-1} + \mu\underline{\delta}_{o,k-1}) - \underline{f}(k, \bar{\underline{z}}_{k-1} + \mu^{\frac{1}{2}}\bar{\underline{U}}_{k-1}) - \nabla\underline{f}(k, \bar{\underline{z}}_{k-1})\mu\underline{\delta}_{o,k-1} \\
\underline{t}_{k2} &= \underline{f}(k, \bar{\underline{z}}_{k-1} + \mu^{\frac{1}{2}}\bar{\underline{U}}_{k-1}) - \underline{f}(k, \bar{\underline{z}}_{k-1}) - \nabla\underline{f}(k, \bar{\underline{z}}_{k-1})\mu^{\frac{1}{2}}\bar{\underline{U}}_{k-1}.
\end{aligned}$$

Note that

$$\|\underline{r}_k\| \leq \mu\|\underline{\eta}_{k-1}\| \times \text{constant} \tag{9.111a}$$

$$\|\underline{t}_{k1}\| \leq \mu^2\|\underline{\delta}_{o,k-1}\|^2 \times \text{constant} \tag{9.111b}$$

$$\|\underline{t}_{k2}\| \leq \mu\|\bar{\underline{U}}_{k-1}\|^2 \times \text{constant} \tag{9.111c}$$

Again the bound on the state dependent term $\|\underline{r}_k\|$ is too large. So we look to a perturbation computation. Introduce then

$$\underline{r}(k, \bar{\underline{z}}, \bar{\underline{U}}, \underline{\delta}_o, \underline{\eta}) = \underline{f}(k, \bar{\underline{z}} + \mu^{\frac{1}{2}}\bar{\underline{U}} + \mu\underline{\delta}_o + \mu\underline{\eta}) - \underline{f}(k, \bar{\underline{z}} + \mu^{\frac{1}{2}}\bar{\underline{U}} + \mu\underline{\delta}_o) - \nabla\underline{f}(k, \bar{\underline{z}})\mu\underline{\eta}$$

so that

$$\underline{r}_k = \underline{r}(k, \bar{\underline{z}}_{k-1}, \bar{\underline{U}}_{k-1}, \underline{\delta}_{o,k-1}, \underline{\eta}_{k-1}).$$

As usual we introduce a discounted perturbation $\underline{R}(k, \bullet)$ with discount $(1 - \mu^{1/2})^r$ and find

$$\delta\underline{R}(k, \bullet) = -\underline{r}(k, \bullet) + \mu^{\frac{1}{2}}\left(\underline{r}(k, \bullet) + \underline{R}(k, \bullet)\right) + \underline{\nu}_k(\bullet) \tag{9.112}$$

where $\underline{v}_k(\bullet)$ is an MG difference sequence. There are Lipschitz properties analogous to (9.96) and (9.97). Indeed, using (9.7A2),

$$\|\underline{R}(k, \underline{\eta}, \bullet) - \underline{R}(k, \underline{\eta}', \bullet)\| \leq \mu^{\frac{1}{2}}\xi(\mu)\|\underline{\eta} - \underline{\eta}'\|$$

$$\|\underline{R}(k, \underline{\delta}_o, \bullet) - \underline{R}(k, \underline{\delta}_o', \bullet)\| \leq \mu^{\frac{1}{2}}\xi(\mu)\|\underline{\delta}_o - \underline{\delta}_o'\|$$

$$\|\underline{R}(k, \underline{\bar{U}}, \bullet) - \underline{R}(k, \underline{\bar{U}}', \bullet)\| \leq \xi(\mu)\left(\mu\|\underline{\eta}\|\|\underline{\bar{U}} - \underline{\bar{U}}'\| + \mu^{\frac{1}{2}}\|\underline{\bar{U}} - \underline{\bar{U}}'\|^2\right)$$

$$\|\underline{R}(k, \underline{\bar{z}}, \bullet) - \underline{R}(k, \underline{\bar{z}}', \bullet)\| \leq \mu^{-\frac{1}{2}}\xi(\mu)\left(\mu\|\underline{\bar{z}} - \underline{\bar{z}}'\|\|\underline{\eta}\| + \|\underline{\bar{z}} - \underline{\bar{z}}'\|^2\right)$$

where $\xi(\mu) \to 0$ as $\mu \to 0$. The third inequality follows because, via (9.7A2) and a Taylor series, we have for some λ, λ' with $0 \leq \lambda, \lambda' \leq 1$,

$$\begin{aligned}
&\|\underline{r}(k, \underline{\bar{U}}, \bullet) - \underline{r}(k, \underline{\bar{U}}', \bullet)\| \\
&= \Big\|\Big(\nabla \underline{f}(k, \underline{\bar{z}} + \mu^{\frac{1}{2}}\underline{\bar{U}} + \mu^{\frac{1}{2}}\underline{\delta}_o + \mu\underline{\eta} + \lambda\mu^{\frac{1}{2}}(\underline{\bar{U}} - \underline{\bar{U}}'))\mu^{\frac{1}{2}}(\underline{\bar{U}} - \underline{\bar{U}}') \\
&\quad - \nabla \underline{f}(k, \underline{\bar{z}} + \mu^{\frac{1}{2}}\underline{\bar{U}} + \mu\underline{\delta}_o + \lambda'\mu^{\frac{1}{2}}(\underline{\bar{U}} - \underline{\bar{U}}'))\mu^{\frac{1}{2}}(\underline{\bar{U}} - \underline{\bar{U}}')\Big)\Big\| \\
&\leq c\mu^{\frac{1}{2}}\|\underline{\bar{U}} - \underline{\bar{U}}'\|\left(\mu\|\underline{\eta}\| + 2\mu^{\frac{1}{2}}\|\underline{\bar{U}} - \underline{\bar{U}}'\|\right).
\end{aligned}$$

The fourth inequality follows similarly. As expected we also see that

$$\|\underline{R}(k, \underline{\bar{z}}, \underline{\bar{U}}, \underline{\delta}_o, \underline{\eta})\| \leq \mu^{\frac{1}{2}}\xi(\mu)\|\underline{\eta}\|. \tag{9.113}$$

The perturbed Lyapunov function for (9.109) has three perturbation terms, namely, $V_1(\bullet)$, $V_2(\bullet)$ as before (but now functions of $\underline{\eta}$), and

$$V_3(\bullet) = \underline{R}^T(k, \bullet)V_0'(\underline{\eta})$$

so that in view of (9.113), positivity with respect to $\underline{\eta}$ will be retained. As before there will be a problem term

$$\underline{r}_k^T V_0'(\underline{\eta}_{k-1}) \tag{9.114}$$

and a three-term breakup of $\delta V_3(k, \bullet)$ as in (9.104). (Note that, from (9.109), (9.111), $\|\delta\underline{\eta}_k\| = O(\mu)$.) The B_k term will cancel (9.114) and leave terms of size $o(\mu)$, $o(\mu)\|\underline{\eta}_{k-1}\|$ as required. Similarly the C_k, D_k terms are appropriately bounded. So (9.110) will be delivered as required.

Example 9.6.

We consider the LMS algorithm. In this case we saw in Example 9.2 that (9.18) is

$$\delta\underline{\bar{U}}_k = -\mu\underline{R}_x\underline{\bar{U}}_{k-1} + \mu^{\frac{1}{2}}\left(\underline{x}_k\epsilon_k - (\underline{x}_k\underline{x}_k^T - \underline{R}_x)\underline{\bar{w}}_{k-1}\right).$$

We see now that (9.88) becomes

$$\delta\underline{\bar{U}}_k = -\mu\underline{R}_x\underline{\bar{U}}_{k-1} + \mu^{\frac{1}{2}}\underline{x}_k\epsilon_k$$

and we have already computed the 0th order steady state variance in (9.22). That is

$$\lim_{\mu\to 0}\lim_{k\to\infty} E(\underline{\bar{U}}_k \underline{\bar{U}}_k^T) = \underline{\Gamma}.$$

Our results are slightly stronger than (9.84), (9.85). In fact we have shown

$$\lim_{\mu\to 0}\sup_k E\|\underline{U}_k - \underline{\bar{U}}_k\|^2 = 0$$

so we can conclude

$$\lim_{\mu\to 0}\lim_{k\to\infty} E(\underline{U}_k \underline{U}_k^T) = \underline{\Gamma}$$

i.e., $\underline{\Gamma}$ is the 0th order approximation to the steady state variance of $\underline{U}_k$.

9.8 STEADY STATE ANALYSIS—TIME VARYING PARAMETERS

There is not much to add here. In Section 9.3 we saw that first order matched adaptation led to a partially averaged system (9.55). Thus the second order analysis of the fluctuation $\underline{U}_k = \mu^{-1/2}\underline{\Delta}_k$ proceeds as in Section 9.7.

For second order matched adaptation we are led to the pair (9.60), (9.61). The only new feature is the additional forcing term $\mu^{-1/2}\delta\underline{z}_o(\mu^2 k)$ in each equation. This term then does not appear in the deviation equation and so analysis proceeds as in Section 9.7. The only difference is that the variance of $\underline{\bar{U}}_k$ obtained from (9.61) has an added term, but this is a minor feature of the analysis.

9.9 NOTES

Finite time first order stochastic averaging results are developed in a number of references such as [B7], [F4], and [K4], although none of these references provides a result as strong as Theorem 9.2. An early result is [G6] and in the adaptive area, [D3].

A Hovering theorem is given in [B7]. However stochastic stability results as in Section 9.5 are hard to find. The perturbed Lyapunov method used here has its origin in [B12] as further developed by [K4]. But our particular choice of perturbation differs from these and allows stronger results. In the special case of LMS and EWLS, special methods have been used by [C4], [G14], and [S15].

Finite time second order analysis is provided by [B7] and [F4] and has its origin in [K4].

Steady state computations are made in a number of places in the literature but rarely are justified. In fact only [B12] and [K4] seem to have done this. They both use perturbed Lyapunov function methods, but our argument is in fact quite different from theirs.

The procedure in Section 9.7 can be used to generate higher order (in μ) approximations to the steady state distribution. The details of this for LMS are discussed in [S16] although the analysis there does not use a perturbed Lyapunov function approach.

With regard to stochastic averaging in the presence of time varying parameters, there is only [B6], [B7], [K5], [M5] [S15], and [S16].

EXERCISES

9.1 Prove that (9.2A2) ensures

$$\|\underline{f}_{av}(\underline{z}) - \underline{f}_{av}(\underline{z}')\| \le L\|\underline{z} - \underline{z}'\|.$$

9.2 Prove that if

$$N_k/k \to 0 \quad \text{w.p.1, as } k \to \infty$$

then

$$\mu \max_{1\le k\le T\mu^{-1}} N_k \to 0 \quad \text{w.p.1, as } \mu \to 0.$$

9.3 **(a)** Prove (9.82).
(b) Prove that $\underline{\pi}_k^{\mu}$ has a steady state value satisfying (9.83); i.e., show (9.82) holds with $\underline{G}_k^{\mu}$ replaced by $\underline{\pi}_{\infty}$.

9.4 Prove, using Appendix C6(v), that (9.7A1), (9.7A2) ensure (9.87) is ES at $\underline{0}$.

9.5 Derive the perturbation identity (9.72c).

9.6 Substantiate the remark after Theorem 9.6.

9.7 In Example 9.2 show that if $\underline{x}_k$ is strictly stationary and (9.2a6) holds, then, as $\mu \to 0$

$$\mu^{\frac{1}{2}} \max_{1\le k\le T\mu^{-1}} \left\| \sum_{s=1}^{k} (\underline{x}_s\underline{x}_s^T - \underline{R}_x)\underline{\bar{w}}_{s-1} \right\| \to 0 \quad \text{w.p.1.}$$

(*Hint:* use the MG approximation method in Appendix E12.)

9.8 Show that under the conditions of Theorem 9.3

$$\mu^{\frac{1}{2}} \left\| \sum_{s=1}^{k} \underline{\tilde{f}}_s - \delta \underline{P}_k \right\| \le \mu^{\frac{3}{2}} B_a \sum_{s=1}^{k} M_{s-1}.$$

Hence show that (9.19) can be used in Theorem 9.3.

9.9 Show that $\mu^{1/2}\underline{\tilde{f}}_k$ in (9.87) can be replaced by $\mu^{1/2}\delta\underline{P}_k$ of (9.19) with regard to steady state behavior.

9.10 Show that (9.19) can be replaced by (9.88) with regard to steady state behavior.

9.11 Show that if $\nabla \underline{f}(k, \underline{0})$ is stationary then (9.108) holds.
Hint: The computation involves the cross-covariance of $\underline{\tilde{f}}_k$ and $\nabla\underline{\tilde{f}}_k$.

9.12 Find the averaged system corresponding to the EWLS algorithm (4.87), (4.89). Assume $\underline{x}_k$ is stationary and take $\lambda = 1 - \mu$.

9.13 Show that (9.57) holds for (9.53), i.e., with $\mu = \epsilon^{\frac{1}{2}}$:

$$\max_{1\le k\le T\mu^{-1}} \|\underline{z}_{o,k} - \underline{z}_{o,0}\| \to 0 \quad \text{w.p.1,} \quad \text{as } \mu \to 0.$$

10

Stochastic Averaging: Mixed Time Scale

10.1 INTRODUCTION

As in the deterministic case, mixed time scale stochastic averaging has much similarity to single time scale stochastic averaging. The chief differences are the computation of the averaged function and the treatment of stability of the fast state.

A full rigorous development of mixed time scale stochastic averaging would require more space than we can devote here, and so our aims are rather to lay out the ideas.

10.2 FINITE TIME AVERAGING—TIME INVARIANT PARAMETERS

Following Section 8.1 we start with the pair (8.1b), (8.2)

$$\delta \underline{z}_k = \mu \underline{f}(k, \underline{z}_{k-1}, \underline{y}_{k-1}) \tag{10.1}$$

$$\underline{y}_k = \underline{A}(\underline{z}_{k-1})\underline{y}_{k-1} + \underline{h}(k, \underline{z}_{k-1}) + \mu \underline{g}(k, \underline{z}_{k-1}, \underline{y}_{k-1}, \mu) \tag{10.2}$$

where now $\underline{f}(k, \bullet)$, $\underline{h}(k, \bullet)$, $\underline{g}(k, \bullet)$ are stochastic. We now have to deal with the stability of the frozen state (8.11a)

$$\underline{y}_k(\underline{z}) = \underline{A}(\underline{z})\underline{y}_{k-1}(\underline{z}) + \underline{h}(k, \underline{z}). \tag{10.3}$$

We suppose:

(10.2A1) Conditions (8.2A7), (8.2A8) hold.

It must be immediately pointed out that conditions (8.2A7), (8.2A8) are very strong in a stochastic setting and rule out many interesting cases. But at the time of writing, little improvement over them has been found.

From (10.2A1) one can show $\{\underline{y}_k(\underline{z})\}$ is tight in k (this is discussed later) and so has an invariant measure. We thus suppose:

(10.2A2) $\underline{y}_k(\underline{z})$ is strictly stationary.

The averaged function will be

$$\underline{f}_{av}(\underline{z}) = E\left(\underline{f}(k, \underline{z}, \underline{y}_{k-1}(\underline{z}))\right) \tag{10.4}$$

where we have assumed:

(10.2A3) $E\left(\underline{f}(k, \underline{z}, \underline{y}_{k-1}(\underline{z}))\right)$ is time invariant.

The averaged system is then

$$\delta\bar{\underline{z}}_k = \mu \underline{f}_{av}(\bar{\underline{z}}_{k-1}). \tag{10.5}$$

The justification of this averaged system proceeds along lines similar to those in Section 8.2. It leads to the Theorem stated at the end of this subsection.

10.2.1 First Order Analysis

Firstly we require random sequence, global versions of (8.2A1)–(8.2A6):

(10.2A4) Conditions (8.2A1)–(8.2A6) hold with the following changes:

(i) Constraints $\|\underline{z}\| \le h_0$, $\|\underline{y}\| \le h_y$, etc. are replaced by: for all $\underline{z}$, for all $\underline{y}$, etc.
(ii) Bounds such as L_p, B_p, L_f are replaced by sequences L_{pk}, B_{pk}, L_{fk} where

$$(L_{pk}, B_{pk}, L_{fk})/k \to 0 \quad \text{w.p.1}, \qquad \text{as } k \to \infty.$$

We also need additional Lipschitz conditions:

(10.2A5) $\underline{f}(s, \underline{z}, \underline{y})$ obeys a Lipschitz condition in $\underline{z}$,

$$\|\underline{f}(s, \underline{z}, \underline{y}) - \underline{f}(s, \underline{z}', \underline{y})\| \le L_{fs}\|\underline{z} - \underline{z}'\|, \quad \text{for all } \underline{z}, \underline{z}'.$$

(10.2A6) Lipschitz conditions (8.5A9), (8.5A10) hold but $\underline{h}_k = \underline{h}(k, \underline{0})$, $\underline{g}_k = \underline{g}(k, \underline{0}, \underline{0}, \mu)$ need not be 0.

We could proceed as in Section 8.2, but we vary the argument here to deal more easily with the issue of stochastic boundedness.

Subtract (10.5) from (10.1) and sum to get (8.19), (8.20). Now split (8.19) (but differently from the way it was split in (8.21)) as

$$\underline{\Delta}_k = \underline{\Delta}_0 + \mu \sum_{s=1}^{k} \left(\underline{f}(s, \underline{z}_{s-1}, \underline{y}_{s-1}) - \underline{f}(s, \bar{\underline{z}}_{s-1}, \bar{\underline{y}}_{s-1}) \right) + \mu \underline{J}_k \tag{10.6}$$

$$\underline{J}_k = \sum_{s=1}^{k} \left(\underline{f}(s, \bar{\underline{z}}_{s-1}, \bar{\underline{y}}_{s-1}) - \underline{f}_{av}(\bar{\underline{z}}_{s-1}) \right) \tag{10.7}$$

where $\bar{\underline{y}}_s$ obeys (10.2) but with $\underline{z}_{k-1}$ replaced by $\bar{\underline{z}}_{k-1}$. Below we show

$$\mu \max_{1 \le k \le T\mu^{-1}} \|\underline{J}_k\| \le d_T(\mu) \tag{10.8a}$$

$$d_T(\mu) \to 0 \quad \text{w.p.1}, \quad \text{as } \mu \to 0. \tag{10.8b}$$

Applying the Lipschitz conditions gives (with L_s denoting a generic Lipschitz sequence, and $\underline{\Delta}_{y,s} = \underline{y}_s - \bar{\underline{y}}_s$)

$$\|\underline{\Delta}_k\| \le \|\underline{\Delta}_0\| + \mu \sum_{s=1}^{k} \left(L_s \|\underline{\Delta}_{s-1}\| + L_s \|\underline{\Delta}_{y,s-1}\| \right) + d_T(\mu), \quad 1 \le k \le T/\mu. \tag{10.9}$$

To continue we need to bound $\|\underline{\Delta}_{y,s}\|$. From (10.2) and the definition of $\bar{\underline{y}}_k$ we find

$$\begin{aligned}\underline{\Delta}_{y,k} = {} & \underline{A}(\underline{z}_{k-1})\underline{\Delta}_{y,k-1} + \underline{h}(k, \underline{z}_{k-1}) - \underline{h}(k, \bar{\underline{z}}_{k-1}) + \left(\underline{A}(\underline{z}_{k-1}) - \underline{A}(\bar{\underline{z}}_{k-1})\right) \bar{\underline{y}}_{k-1} \\ & + \mu \left(\underline{g}(k, \underline{z}_{k-1}, \underline{y}_{k-1}, \mu) - \underline{g}(k, \bar{\underline{z}}_{k-1}, \bar{\underline{y}}_{k-1}, \mu) \right)\end{aligned}$$

so using the ES bound (8.34b) as well as the Lipschitz conditions gives

$$\begin{aligned}\|\underline{\Delta}_{y,k}\| &\le m\lambda^k \|\underline{\Delta}_{y,0}\| + m \sum_{s=1}^{k} \lambda^{k-s} \left(L_s \|\underline{\Delta}_{s-1}\| + \mu L_s \|\underline{\Delta}_{y,s-1}\| + L_s \|\underline{\Delta}_{s-1}\| \|\bar{\underline{y}}_{s-1}\| \right) \\ &\le m\lambda^k \|\underline{\Delta}_{y,0}\| + \xi(\mu) \sum_{s=1}^{k} \lambda^{k-s} \|\underline{\Delta}_{y,s-1}\| + \sum_{s=1}^{k} \lambda^{k-s} L_s \|\underline{\Delta}_{s-1}\| \left(\|\bar{\underline{y}}_{s-1}\| + 1 \right)\end{aligned}$$

where

$$\xi(\mu) = m\mu \max_{1 \le k \le T\mu^{-1}} L_k \to 0 \quad \text{w.p.1}, \quad \text{as } \mu \to 0.$$

Now apply the Bellman-Gronwall lemma to find

$$\|\underline{\Delta}_{y,k}\| \le m\lambda_\mu^k \|\underline{\Delta}_{y,0}\| + m \sum_{s=1}^{k} \lambda_\mu^{k-s} L_s \|\underline{\Delta}_{s-1}\| \left(\|\bar{\underline{y}}_{s-1}\| + 1 \right), \quad 1 \le k \le T/\mu$$

$$\lambda_\mu = \lambda + \xi(\mu) \le \lambda_e = \frac{1}{2}(1 + \lambda), \quad \text{if } \mu \text{ is small.}$$

Using this in (10.9) gives

$$\|\underline{\Delta}_k\| \le \|\underline{\Delta}_0\| + \mu \sum_{s=1}^{k} L_s \|\underline{\Delta}_{s-1}\| + d_T(\mu) + \mu m \|\underline{\Delta}_{y,0}\| \sum_{s=1}^{k} L_s \lambda_\mu^s$$

$$+ \mu m \sum_{s=1}^{k} L_s \sum_{u=1}^{s} \lambda_\mu^{s-u} L_u \|\underline{\Delta}_{u-1}\| \left(\|\underline{\bar{y}}_{u-1}\| + 1 \right)$$

$$\le \|\underline{\Delta}_0\| + d_T(\mu) + \|\underline{\Delta}_0\| \xi(\mu)(1-\lambda_\mu)^{-1} + \mu \sum_{s=1}^{k} L_s \|\underline{\Delta}_{s-1}\|$$

$$+ \mu \sum_{u=1}^{k} L_u \|\underline{\Delta}_{u-1}\| \left(\|\underline{\bar{y}}_{k-1}\| + 1 \right) \sum_{s=u}^{T/\mu} L_s \lambda_\mu^{s-u}.$$

Now apply the Bellman-Gronwall lemma to find for $1 \le k \le T/\mu$

$$\|\underline{\Delta}_k\| \le \left(\|\underline{\Delta}_0\| + d_T(\mu) + \|\underline{\Delta}_{y,0}\| \xi(\mu)(1-\lambda_\mu)^{-1} \right) \times$$

$$\exp\left(\mu \sum_{s=1}^{T/\mu} L_s + \mu \sum_{u=1}^{T/\mu} L_u \left(\|\underline{\bar{y}}_{u-1}\| + 1 \right) \sum_{s=u}^{T/\mu} L_s \lambda_\mu^{s-u} \right)$$

$$= \left(\|\underline{\Delta}_0\| + d_T(\mu) + \|\underline{\Delta}_{y,0}\| \xi(\mu)(1-\lambda_\mu)^{-1} \right) \times \exp\left(\mu \sum_{s=1}^{T/\mu} L_s + \mu \sum_{s=1}^{T/\mu} L_s v_s \right)$$

$$v_s = \sum_{u=1}^{s} \lambda_\mu^{s-u} L_u \left(\|\underline{\bar{y}}_{u-1}\| + 1 \right)$$

from which we can conclude

$$\sup_{1 \le k \le T\mu^{-1}} \|\underline{\Delta}_k\| \le c_T(\mu) \to 0 \quad \text{w.p.1,} \quad \text{as } \mu \to 0 \tag{10.10}$$

if we ensure

$$\limsup_{\mu \to 0} \mu \sum_{s=1}^{T/\mu} L_s v_s < \infty. \tag{10.11}$$

Now

$$\mu \sum_{s=1}^{T/\mu} L_s v_s \le \left(\mu \sum_{s=1}^{T/\mu} L_s^2 \mu \sum_{s=1}^{T/\mu} v_s^2 \right)^{1/2}. \tag{10.12a}$$

It is left to the reader to show

$$\mu \sum_{s=1}^{T/\mu} v_s^2 \le 2(1-\lambda_\mu)^{-1} \mu \sum_{s=1}^{T/\mu} L_s^2 \left(\|\underline{\bar{y}}_{s-1}\|^2 + 1 \right) \tag{10.12b}$$

$$\le 2(1-\lambda_\mu)^{-1} \left(\mu \sum_{s=1}^{T/\mu} L_s^4 \mu \sum_{s=1}^{T/\mu} \|\underline{\bar{y}}_{s-1}\|^4 \right)^{\frac{1}{2}} + 2(1-\lambda_\mu)^{-1} \mu \sum_{s=1}^{T/\mu} L_s^2. \tag{10.12c}$$

So we assume:

(10.2A7) L_s^4 obeys a SLLN.

We must deal with the $\|\underline{\bar{y}}_{s-1}\|$ term. From (10.2), (8.34b)

$$\|\underline{\bar{y}}_k\| \le m\lambda^k \|\underline{\bar{y}}_0\| + m \sum_{s=1}^{k} \lambda^{k-s} \left(L_s \|\underline{\bar{z}}_{s-1}\| + \|\underline{h}_s\| + \|\underline{g}_s\| + \mu L_s \|\underline{\bar{z}}_{s-1}\| + \mu L_s \|\underline{\bar{y}}_{s-1}\| \right)$$

$$\le m\lambda^k \|\underline{\bar{y}}_0\| + \xi(\mu) \sum_{s=1}^{k} \lambda^{k-s} \|\underline{\bar{y}}_{s-1}\| + m \sum_{s=1}^{k} \lambda^{k-s} \left(L_s h(1+\mu) + \|\underline{g}_s\| + \|\underline{h}_s\| \right).$$

Now apply the Bellman-Gronwall lemma to get

$$\|\underline{\bar{y}}_k\| \le m\lambda_\mu^k \|\underline{\bar{y}}_0\| + m \sum_{s=1}^{k} a_s \lambda_\mu^{k-s} \tag{10.13a}$$

$$a_s = L_s h(1+\mu) + \|\underline{g}_s\| + \|\underline{h}_s\|. \tag{10.13b}$$

The first term in (10.13a), a transient, causes no trouble, so we neglect it. For the second we have

$$\|\underline{\bar{y}}_k\|^2 \le \sum_{s=1}^{k} a_s^2 \lambda_\mu^{k-s} (1-\lambda_\mu)^{-1}.$$

From this follows (cf. (10.12))

$$\mu \sum_{s=1}^{T/\mu} \|\underline{\bar{y}}_s\|^4 \le (1-\lambda_\mu)^{-2} \mu \sum_{s=1}^{T/\mu} a_s^4.$$

So we assume:

(10.2A8) $\|\underline{g}_s\|^4$, $\|\underline{h}_s\|^4$ obey SLLNs.

Thus via (10.12) we have established (10.11) and hence (10.10).

Proof of (10.8a). As in the proof of (8.24), we split $\underline{J}_k$ into two pieces:

$$\underline{J}_k = \underline{J}_{k,1} + \underline{J}_{k,2}$$

$$\underline{J}_{k,1} = \sum_{s=1}^{k} \left(\underline{f}(s, \bar{\underline{z}}_{s-1}, \bar{\underline{y}}_{s-1}) - \underline{f}(s, \bar{\underline{z}}_{s-1}, \underline{y}_{s-1}(\bar{\underline{z}}_{s-1})) \right)$$

$$\underline{J}_{k,2} = \sum_{s=1}^{k} \left(\underline{f}(s, \bar{\underline{z}}_{s-1}, \underline{y}_{s-1}(\bar{\underline{z}}_{s-1})) - \underline{f}_{av}(\bar{\underline{z}}_{s-1}) \right).$$

We deal with $\underline{J}_{k,2}$ first. Using the perturbation (8.2A2) we find (8.29) and are led to

$$\|\underline{\eta}_k\| \le \mu L_{pk} B_{fk}$$

$$\|\underline{P}_k\| \le L_{pk} h + B_{pk}$$

so that

$$\mu \max_{1 \le k \le T\mu^{-1}} \|\underline{J}_{k,2}\| \le \mu \max_{1 \le k \le T\mu^{-1}} \left(L_{pk} h + B_{pk} + \mu \sum_{s=1}^{k} L_{ps} B_{fs} \right).$$

We assume:

(10.2A9) L_{ps}^2, B_{fs}^2 each obey an SLLN.

Then we can deduce

$$\mu \max_{1 \le k \le T\mu^{-1}} \|\underline{J}_{k,2}\| \to 0 \quad \text{w.p.1,} \quad \text{as } \mu \to 0.$$

Now we turn to $\underline{J}_{k,1}$. From the Lipschitz condition (and with $\underline{\delta}_s$ defined essentially as in (8.32))

$$\|\underline{J}_{k-1}\| \le \sum_{s=1}^{k} L_s \|\underline{\delta}_s\|$$

$$\underline{\delta}_s = \|\bar{\underline{y}}_{s-1} - \underline{y}_{s-1}(\bar{\underline{z}}_{s-1})\|.$$

Now $\underline{\delta}_s$ is treated as in the proof of (8.27) except that $\bar{\underline{y}}_k$ is handled as in (10.13). Next, (8.33) still holds with $\underline{z}$ replaced by $\bar{\underline{z}}$, but with a bound of the form μb_k. So from (8.32)

$$\|\underline{\delta}_k\| \le m\lambda^k \|\underline{\delta}_0\| + \mu m \sum_{s=1}^{k} \lambda^{k-s} \left(L_s \|\bar{\underline{z}}_{s-1}\| + L_s \|\bar{\underline{y}}_{s-1}\| + \|\underline{h}_s\| + \|\underline{g}_s\| + b_s \right).$$

The first term, a transient, causes no trouble, and the second term is $O(\mu)$. It is thus left to the reader to show, along lines similar to the way (10.11) was treated, that

$$\mu \max_{1 \le k \le T\mu^{-1}} \|\underline{J}_{k,1}\| \to 0 \quad \text{w.p.1,} \quad \text{as } \mu \to 0$$

as required. We have thus established the following result.

Theorem 10.1. Consider the primary system (10.1), (10.2) and averaged system (10.5) with identical initial conditions. Suppose (10.2A1)–(10.2A9) hold then (10.10) holds.

10.2.2 Second Order Analysis

Now we turn to look at a sketch of second order results following Section 9.2.

Introduce the fluctuation

$$\underline{U}_k = \mu^{-\frac{1}{2}} \underline{\Delta}_k$$

and subtract (10.5) from (10.1) to get

$$\delta \underline{\Delta}_k = \mu \left(\underline{f}(k, \underline{z}_{k-1}, \underline{y}_{k-1}) - \underline{f}_{av}(\bar{\underline{z}}_{k-1}) \right).$$

Now proceeding in a heuristic way, use a Taylor series to find

$$\begin{aligned}
\delta \underline{\Delta}_k &= \mu \left(\underline{f}(k, \bar{\underline{z}}_{k-1}, \underline{y}_{k-1}(\bar{\underline{z}}_{k-1})) - \underline{f}_{av}(\bar{\underline{z}}_{k-1}) \right) \\
&\quad + \mu \left(\underline{f}_z(k, \bar{\underline{z}}_{k-1}, \underline{y}_{k-1}(\bar{\underline{z}}_{k-1}))(\underline{z}_{k-1} - \bar{\underline{z}}_{k-1}) \right. \\
&\quad \left. + \underline{f}_y(k, \bar{\underline{z}}_{k-1}, \underline{y}_{k-1}(\bar{\underline{z}}_{k-1}))(\underline{y}_{k-1} - \underline{y}_{k-1}(\bar{\underline{z}}_{k-1})) \right) + \mu^2 O(\|\underline{\Delta}_{k-1}\|^2)
\end{aligned}$$

where $\underline{f}_z(\bullet) = \partial \underline{f} / \partial \underline{z}^T$, etc. Continuing (and denoting the first term as $\tilde{\underline{f}}_k$),

$$\begin{aligned}
\delta \underline{\Delta}_k &= \mu \tilde{\underline{f}}_k + \mu \underline{f}_z(k, \bar{\underline{z}}_{k-1}, \underline{y}_{k-1}(\bar{\underline{z}}_{k-1})) \underline{\Delta}_{k-1} \\
&\quad + \mu \underline{f}_y(k, \bar{\underline{z}}_{k-1}, \underline{y}_{k-1}(\bar{\underline{z}}_{k-1}))(\underline{y}_{k-1}(\underline{z}_{k-1}) - \underline{y}_{k-1}(\bar{\underline{z}}_{k-1})) + \mu^2 O(\|\underline{\Delta}_{k-1}\|^2) + O(\mu^2) \\
&= \mu \tilde{\underline{f}}_k + \mu \underline{f}_z(k, \bar{\underline{z}}_{k-1}, \underline{y}_{k-1}(\bar{\underline{z}}_{k-1})) \underline{\Delta}_{k-1} \\
&\quad + \mu \underline{f}_y(k, \bar{\underline{z}}_{k-1}, \underline{y}_{k-1}(\bar{\underline{z}}_{k-1})) \left. \frac{\partial \underline{y}_{k-1}(\underline{z})}{\partial \underline{z}^T} \right|_{\underline{z}=\bar{\underline{z}}_{k-1}} \underline{\Delta}_{k-1} + O(\mu^2).
\end{aligned}$$

Norming up by $\mu^{-1/2}$ leads us to rewrite this as

$$\begin{aligned}
\delta \underline{U}_k &= \mu \left(\underline{f}_z(k, \bar{\underline{z}}_{k-1}, \underline{y}_{k-1}(\bar{\underline{z}}_{k-1})) + \underline{f}_y(k, \bar{\underline{z}}_{k-1}, \underline{y}_{k-1}(\bar{\underline{z}}_{k-1})) \left. \frac{\partial \underline{y}_{k-1}(\underline{z})}{\partial \underline{z}^T} \right|_{\underline{z}=\bar{\underline{z}}_{k-1}} \right) \underline{U}_{k-1} \\
&\quad + \mu^{\frac{1}{2}} \tilde{\underline{f}}_k + O(\mu^{\frac{3}{2}})
\end{aligned}$$

and as in Section 9.2 we arrive at the approximation

$$\begin{aligned}\delta\underline{\bar{U}}_k &= \mu\underline{A}(\underline{\bar{z}}_{k-1})\underline{\bar{U}}_{k-1} + \mu^{\frac{1}{2}}\underline{\tilde{f}}_k \\ \underline{A}(\underline{z}) &= \underline{G}(\underline{z}) + \underline{H}(\underline{z}) \\ \underline{G}(\underline{z}) &= \text{av}\left(\underline{f}_z(k, \underline{z}, \underline{y}_{k-1}(\underline{z}))\right) = \underline{f}_{av,z}(\underline{z}) \\ \underline{H}(\underline{z}) &= \text{av}\left(\underline{f}_y(k, \underline{z}, \underline{y}_{k-1}(\underline{z}))\frac{\partial \underline{y}_{k-1}(\underline{z})}{\partial \underline{z}^T}\right).\end{aligned}$$

Example 10.1. Mixed Time Scale LMS

We pursue Example 8.2. The error system is as given in (8.9), (8.10) and the averaged system in (8.15a), (8.15b). We consider that $\underline{\varphi}_k$ is stochastic, however. For the second order analysis we simplify the computation by considering only the locally ES situation and only calculate $\underline{A}(\underline{0})$. We have firstly from (8.10), (8.15a)

$$\underline{G}(\underline{0}) = \underline{f}_{av,z}(\underline{0}) = -\underline{R}(\underline{0})$$

where $\underline{R}(\underline{0})$ is given in (8.17). Next (in the notation of Example 8.2)

$$\begin{aligned}\underline{H}(\underline{0}) &= \text{av}\left(\underline{f}_\xi(k, \underline{\tilde{\theta}}, \underline{\xi}_{k-1}(\underline{\tilde{\theta}}))\frac{\partial \underline{\xi}_{k-1}(\underline{\tilde{\theta}})}{\partial \underline{\tilde{\theta}}^T}\right)\Bigg|_{\underline{\tilde{\theta}}=\underline{0}} \\ &= \text{av}\left(\underline{f}_\xi(k, \underline{0}, \underline{0})\left.\frac{\partial \underline{\xi}_{k-1}(\underline{\tilde{\theta}})}{\partial \underline{\tilde{\theta}}^T}\right|_{\underline{\tilde{\theta}}=\underline{0}}\right).\end{aligned}$$

From (8.9), (8.10)

$$\begin{aligned}\underline{f}_\xi(k, \underline{0}, \underline{0}) &= \underline{\varphi}_k\underline{c}^T \\ \left.\frac{\partial \underline{\xi}_k(\underline{\tilde{\theta}})}{\partial \underline{\tilde{\theta}}^T}\right|_{\underline{\tilde{\theta}}=\underline{0}} &= \underline{F}\left.\frac{\partial \underline{\xi}_{k-1}(\underline{\tilde{\theta}})}{\partial \underline{\tilde{\theta}}^T}\right|_{\underline{\tilde{\theta}}=\underline{0}} - \underline{b}\underline{\varphi}_k^T\end{aligned}$$

so that

$$\underline{H}(\underline{0}) = -\text{av}\left(\underline{\varphi}_k\underline{c}^T(qI - \underline{F})^{-1}\underline{b}\underline{\varphi}_k^T\right)$$

which, in view of (8.5), (8.7) is

$$-\text{av}\left(\underline{\varphi}_k\left(\frac{-a_0(q^{-1})}{1 + a_0(q^{-1})}\right)\underline{\varphi}_k^T\right) = \frac{1}{2\pi}\int_{-\pi}^{\pi}\frac{a_0(e^{-j\omega})}{1 + a_0(e^{-j\omega})}d\underline{F}_\varphi(\omega).$$

Then we find

$$\underline{f}_{av,\tilde{\theta}}(\underline{0}) + \underline{H} = \underline{A}$$

$$\underline{A} = \frac{1}{2\pi}\int_{-\pi}^{\pi}\left(1 - 2\text{Re}\left(\frac{1}{1 + a_0(e^{-j\omega})}\right)\right)d\underline{F}_\varphi(\omega) = \underline{R}_\varphi^*.$$

So the second order approximation is

$$\delta\underline{\bar{U}}_k = -\mu \underline{R}_\varphi^* \underline{\bar{U}}_{k-1} + \mu^{\frac{1}{2}} \underline{\tilde{f}}_{k0}$$

$$\underline{\tilde{f}}_{k0} = \underline{f}(k, \underline{0}, \underline{0}).$$

If $\underline{\tilde{f}}_{k0}$ is stationary, then a limit variance is easily calculated as in Example 9.2.

10.3 NOTES

It is straightforward but tedious to develop a Hovering theorem analogous to that in Section 9.4. A treatment of time varying parameters can also be developed along the lines of Section 9.3. Steady state analysis can similarly be developed following Section 9.7.

As mentioned earlier, mixed time scale stochastic averaging was applied to long-memory adaptive algorithm analysis in [L5], [L6], and [L8]: However only stability analysis is developed. The perturbation based approach used here has its origin in [B12]. The discussion there is for continuous Markov processes: First and second order finite time averaging, stability, and a partial steady state analysis are developed. The approach was developed further in [K4] where first order finite time averaging and steady state analysis are discussed for both continuous and discrete time non-Markovian processes.

In the context of long memory algorithms the perturbation approach of [B12] has been applied in a discrete time Markovian setting in [B7] and [M5] under the name Poisson equation method: First and second order finite time averaging and some treatment of stability are given. Also in [B7, Chapter I.4 and II.4.6] short memory algorithms receive some treatment.

A

Review of Matrix Analysis

A1 BASIC CONCEPTS

(i) We denote a p-dimensional vector by

$$\underline{a} = \begin{pmatrix} a_1 \\ \vdots \\ a_p \end{pmatrix}.$$

An $m \times n$ matrix is an array of numbers

$$\underline{A} = [a_{ij}]_{m,n} = \begin{bmatrix} a_{11} & a_{12} & \cdots & a_{1n} \\ a_{21} & a_{22} & \cdots & a_{2n} \\ \vdots & \vdots & \ddots & \vdots \\ a_{m1} & a_{m2} & \cdots & a_{mn} \end{bmatrix}.$$

Matrices that can be multiplied together are called conformable.

(ii) A matrix is square if $m = n$. A diagonal matrix has its only non-zero entries on the main diagonal and is often denoted as $\text{diag}(a_1 \cdots a_m)$.

(iii) We introduce two products:

$$\text{inner product: } \underline{a}^T\underline{b} = \underline{a}\cdot\underline{b} = <\underline{a},\underline{b}> = \sum_{s=1}^{p} a_s b_s$$

$$\text{outer product or dyadic product: } \underline{ab}^T = [a_i b_j]_{m,n}$$

(iv) Matrix or vector transpose is defined as

$$\underline{A}^T = ([a_{ij}]_{m,n})^T = [a_{ji}]_{n,m}.$$

(v) A symmetric matrix has $\underline{A}^T = \underline{A}$.

(vi) A positive definite matrix $\underline{A}$ has the property

$$\underline{x}^T\underline{Ax} > 0 \quad \text{for any real } \underline{x} \neq \underline{0}.$$

A positive semi-definite matrix $\underline{A}$ has the property

$$\underline{x}^T\underline{Ax} \geq 0 \quad \text{for any real } \underline{x} \neq \underline{0}.$$

If $\underline{A}$, $\underline{B}$ are positive definite we say $\underline{A} \geq \underline{B}$ if $\underline{A} - \underline{B}$ is positive semi-definite.

(vii) Two vectors $\underline{a}$, $\underline{b}$ are perpendicular or orthogonal if $\underline{a}^T\underline{b} = 0$.

(viii) Trace of a square matrix is $\text{tr}(\underline{A}) = \sum_{i=1}^{m} a_{ii}$. Thus if $\underline{A}$, $\underline{B}$ are two matrices and $\underline{x}$ is a vector

$$\text{tr}(\underline{AB}) = \text{tr}(\underline{BA})$$
$$\text{tr}(\underline{xx}^T) = \text{tr}(\underline{x}^T\underline{x}) = \underline{x}^T\underline{x}.$$

(ix) Norm of a vector

$$\|\underline{x}\| = (\underline{x}^T\underline{x})^{\frac{1}{2}}.$$

Matrix norm is defined in section A2(vii).

(x) The rank of a matrix is the minimum of the number of linearly independent rows or columns. An $m \times m$ square matrix $\underline{A}$ is of full rank if its rank $= m$. In that case $\underline{A}$ is called nonsingular. If a square matrix fails to have full rank, it is called singular. If it is nonsingular it has a unique inverse $\underline{A}^{-1}$ such that

$$\underline{A}^{-1}\underline{A} = \underline{AA}^{-1} = \underline{I}.$$

(xi) Determinant for a square matrix $\underline{A}$

$$|\underline{A}| = \det \underline{A} = \sum_{i=1}^{m} a_{ij} D_{ij}$$
$$D_{ij} = (-1)^{i+j} M_{ij}$$

where M_{ij} is the minor of a_{ij}; that is the determinant of the submatrix of $\underline{A}$ obtained by deleting the ith row and j^{th} column. In the 2×2 case

$$\begin{vmatrix} a & b \\ c & d \end{vmatrix} = ad - bc.$$

For any two conformable square matrices $|\underline{A}\underline{B}| = |\underline{A}||\underline{B}|$.

(xii) $\underline{A}$ is singular if and only if $\det \underline{A} = 0$. This is easily seen in the 2×2 case, because the two columns are linearly dependent since

$$d\begin{pmatrix} a \\ c \end{pmatrix} - c\begin{pmatrix} b \\ d \end{pmatrix} = \begin{pmatrix} 0 \\ 0 \end{pmatrix}.$$

(xiii) A $p \times p$ matrix $\underline{O}_p$ is orthogonal if

$$\underline{O}_p^T \underline{O}_p = [\delta_{ij} a_i]_{p,p} = \det[a_1 \cdots a_p]$$

where $\delta_{ij} = 0$ for all $i \neq j$ and $\delta_{ij} = 1$ if $i = j$. If we also have $a_i = 1$ for all i, i.e.,

$$\underline{O}_p^T \underline{O}_p = \underline{I}$$

then the matrix $\underline{O}_p$ is called orthonormal. From this we see

$$\underline{O}_p^T = \underline{O}_p^{-1}, \qquad \det \underline{O}_p = \pm 1$$

so that we also have

$$\underline{O}_p \underline{O}_p^T = \underline{I}.$$

If we write $\underline{O}_p$ in column vector form

$$\underline{O}_p = [\underline{c}_1 \cdots \underline{c}_p]$$

then

$$\underline{O}_p^T \underline{O}_p = [\underline{c}_i^T \underline{c}_j] = [\delta_{ij}]$$

where δ_{ij} is the Kronecker delta, i.e., $\delta_{ij} = 1$ if $i = j$, and $\delta_{ij} = 0$ otherwise.

As an example consider that

$$\underline{O}_2 = \begin{bmatrix} \cos\theta & \sin\theta \\ -\sin\theta & \cos\theta \end{bmatrix}$$

is orthonormal. Also if we apply $\underline{O}_2$ to a 2×1 vector in the plane, the vector is rotated through an angle θ. Thus orthonormal matrices are also called rotation matrices.

A2 EIGENVALUES AND EIGENVECTORS

(i) A $p \times p$ matrix $\underline{G}$ has a possibly complex eigenvector $\underline{\xi} \neq \underline{0}$ and an eigenvalue λ if

$$\underline{G}\underline{\xi} = \lambda \underline{\xi}. \tag{A2.1}$$

Since also for any $a \neq 0$

$$\underline{G} a \underline{\xi} = \lambda a \underline{\xi}$$

to fix the scale we specify that $\underline{\xi}^H\underline{\xi} = 1$ (where the superscript H for Hermitian denotes complex conjugate transpose) and then call $\underline{\xi}$ a unit eigenvector. In two-dimensional space multiplication by a matrix rotates and scales a vector. If $\underline{G}$ is orthonormal, only rotation occurs. If $\underline{\xi}$ is an eigenvector of $\underline{G}$, it only gets scaled, not rotated, when multiplied by $\underline{G}$.

(ii) How many eigenvectors, eigenvalues can a $p \times p$ matrix have? From (A2.1) we have

$$(\lambda \underline{I} - \underline{G})\underline{\xi} = \underline{0}$$

and for this to have solutions $\underline{\xi} \neq \underline{0}$ we need that $\lambda \underline{I} - \underline{G}$ be singular, i.e.,

$$\det(\lambda \underline{I} - \underline{G}) = 0.$$

But from A1(xi) this equation is a pth order polynomial in λ, so it has p roots, which are then the p eigenvalues of $\underline{G}$. If some roots are repeated, then there are less than p distinct eigenvalues. As an example we find the eigenvalues of

$$\underline{A} = \begin{pmatrix} 2 & 1 \\ 1 & 1 \end{pmatrix}.$$

We have

$$|\lambda \underline{I} - \underline{A}| = \begin{vmatrix} \lambda - 2 & -1 \\ 1 & \lambda - 1 \end{vmatrix} = (\lambda - 2)(\lambda - 1) + 1 = 0$$

$$= \lambda^2 - 3\lambda + 3 = 0$$

$$\Rightarrow \qquad \lambda = (3 \pm \sqrt{9 - 12})/2 = (3 \pm j\sqrt{3})/2.$$

(iii) If eigenvalues are real, we will use the ordering convention

$$\lambda_1 \geq \lambda_2 \geq \cdots \geq \lambda_p.$$

If $\underline{A}$ is symmetric, then its eigenvalues are real. Let $\underline{\xi}$ be a possibly complex eigenvector and λ its corresponding possibly complex eigenvalue. Then

$$\underline{A}\underline{\xi} = \lambda \underline{\xi} \qquad \text{(A2.2)}$$

$$\Rightarrow \qquad \underline{A}\underline{\xi}^* = \lambda^* \underline{\xi}^*$$

$$\Rightarrow \qquad \underline{\xi}^H \underline{A} = \lambda^* \underline{\xi}^H \qquad \text{(A2.3)}$$

where superscript * denotes complex conjugate. From (A2.2)

$$\underline{\xi}^H \underline{A}\underline{\xi} = \lambda \underline{\xi}^H \underline{\xi}.$$

From (A2.3)

$$\underline{\xi}^H \underline{A}\underline{\xi} = \lambda^* \underline{\xi}^H \underline{\xi}.$$

Thus we must have $\lambda^* = \lambda$, i.e., λ is real.

(iv) Eigenvectors corresponding to distinct eigenvalues are orthogonal if $\underline{A}$ is symmetric. Suppose

$$\begin{aligned}
\underline{A}\underline{\xi}_i &= \lambda_i \underline{\xi}_i \\
\underline{A}\underline{\xi}_j &= \lambda_j \underline{\xi}_j, \qquad \lambda_i \neq \lambda_j \\
\Rightarrow \qquad \underline{\xi}_i^T \underline{A}\underline{\xi}_j &= \underline{\xi}_i^T \lambda_j \underline{\xi}_j = \lambda_j \underline{\xi}_i^T \underline{\xi}_j \\
\underline{\xi}_j^T \underline{A}\underline{\xi}_i &= \underline{\xi}_j^T \lambda_i \underline{\xi}_i = \lambda_i \underline{\xi}_j^T \underline{\xi}_i = \lambda_i \underline{\xi}_i^T \underline{\xi}_j.
\end{aligned}$$

Since $\underline{\xi}_i^T \underline{A}\underline{\xi}_j = (\underline{\xi}_i^T \underline{A}\underline{\xi}_j)^T = \underline{\xi}_j^T \underline{A}\underline{\xi}_i$, the above two equations lead to $(\lambda_i - \lambda_j)\underline{\xi}_i^T \underline{\xi}_j = 0$, which in turn implies

$$\underline{\xi}_i^T \underline{\xi}_j = 0, \quad \text{or } \lambda_i = \lambda_j (\text{a contradiction}).$$

(v) Eigenvalue Decomposition

Suppose $\underline{A}$ is full rank, symmetric with distinct eigenvalues $\lambda_1 \ldots \lambda_p$ and unit eigenvectors $\underline{\xi}_1 \ldots \underline{\xi}_p$, which we assemble into an orthonormal matrix

$$\underline{O}_p = (\underline{\xi}_1 \cdots \underline{\xi}_p).$$

Then

$$\begin{aligned}
\underline{A}\underline{O}_p &= \underline{A}(\underline{\xi}_1 \cdots \underline{\xi}_p) \\
&= (\lambda_1 \underline{\xi}_1 \cdots \lambda_p \underline{\xi}_p) \\
&= \underline{O}_p \underline{\Lambda} \\
\underline{\Lambda} &= \text{diag}[\lambda_1 \ldots \lambda_p].
\end{aligned}$$

Thus we get the spectral decomposition

$$\begin{aligned}
\underline{A} &= \underline{O}_p \underline{\Lambda} \underline{O}_p^T \\
&= \sum_{u=1}^{p} \lambda_u \underline{\xi}_u \underline{\xi}_u^T.
\end{aligned}$$

Also

$$\underline{O}_p^T \underline{A} \underline{O}_p = \underline{\Lambda}.$$

(vi) If $\underline{A}$ is symmetric then $\underline{A}$ is positive definite if and only if $\lambda_u > 0$, $u = 1 \ldots p$. This follows because if $\underline{A}$ is positive definite then

$$0 < \underline{\xi}_u^T \underline{A}\underline{\xi}_u = \lambda_u \underline{\xi}_u^T \underline{\xi}_u = \lambda_u.$$

On the other hand if $\lambda_u > 0$, $u = 1 \ldots p$, then for any $\underline{x} \neq \underline{0}$

$$\underline{x}^T \underline{A}\underline{x} = \underline{x}^T \left(\sum_{u=1}^{p} \lambda_u \underline{\xi}_u \underline{\xi}_u^T\right)\underline{x}$$

$$= \sum_{u=1}^{p} \lambda_u (\underline{\xi}^T \underline{x})^2 > 0.$$

(vii) Matrix Norm

$$\|\underline{A}\| = (\text{largest eigenvalue of } \underline{A}^T \underline{A})^{\frac{1}{2}}.$$

Thus

$$|\underline{x}^T \underline{A}\underline{y}| \leq (\underline{x}^T \underline{A}^T \underline{A}\underline{x}\underline{y}^T \underline{y})^{\frac{1}{2}}$$

$$\leq \|\underline{x}\| \|\underline{y}\| \|\underline{A}\|.$$

A3 MATRIX DIFFERENTIATION

If $\underline{w}$ is a p-vector, define

$$\partial f/\partial \underline{w} = \partial f(\underline{w})/\partial \underline{w} = (\partial f/\partial w_1 \cdots \partial f/\partial w_p)^T$$

$$\partial \underline{f}/\partial \underline{w}^T = (\partial \underline{f}/\partial w_1 \cdots \partial \underline{f}/\partial w_p).$$

Then

$$\partial(\underline{a}^T \underline{w})/\partial \underline{w} = \underline{a} \quad (\text{since } \partial(\underline{a}^T \underline{w})/\partial w_i = a_i\ )$$

$$\partial(\underline{w}^T \underline{A}\underline{w})/\partial \underline{w} = (\underline{A} + \underline{A}^T)\underline{w}$$

$$\partial(\underline{A}\underline{w})/\partial \underline{w}^T = \underline{A}.$$

Next define the matrix

$$\partial^2 f/\partial \underline{w}\partial \underline{w}^T = [\partial^2 f/\partial w_i \partial w_j] = \text{matrix with } i, j \text{ element } \partial^2 f/\partial w_i \partial w_j.$$

Then check that

$$\partial^2/\partial \underline{w}\partial \underline{w}^T (\underline{w}^T \underline{A}\underline{w}) = \underline{A} + \underline{A}^T.$$

A4 PARTITIONED MATRIX INVERSION

If $\underline{A}$, $\underline{B}$ are square matrices and the indicated inverses exist, then

$$\begin{pmatrix} \underline{A} & \underline{D} \\ \underline{B} & \underline{C} \end{pmatrix}^{-1} = \begin{bmatrix} \underline{A}^{-1} + \underline{A}^{-1}\underline{D}\underline{\Delta}_C^{-1}\underline{B}\underline{A}^{-1} & -\underline{A}^{-1}\underline{D}\underline{\Delta}_C^{-1} \\ -\underline{\Delta}_C^{-1}\underline{B}\underline{A}^{-1} & \underline{\Delta}_C^{-1} \end{bmatrix}$$

where

$$\underline{\Delta}_C = \underline{C} - \underline{B}\underline{A}^{-1}\underline{D}.$$

The result is easily checked by direct multiplication. In a similar way show

$$\begin{pmatrix} \underline{A} & \underline{D} \\ \underline{B} & \underline{C} \end{pmatrix}^{-1} = \begin{bmatrix} \underline{\Delta}_A^{-1} & -\underline{\Delta}_A^{-1}\underline{D}\underline{C}^{-1} \\ -\underline{C}^{-1}\underline{B}\underline{\Delta}_A^{-1} & \underline{C}^{-1} + \underline{C}^{-1}\underline{B}\underline{\Delta}_A^{-1}\underline{D}\underline{B}^{-1} \end{bmatrix}$$

where

$$\underline{\Delta}_A = \underline{A} - \underline{D}\underline{C}^{-1}\underline{B}.$$

A5 MATRIX INVERSION LEMMA

Equating the two inverses in section A3 gives

$$(\underline{A} - \underline{D}\underline{C}^{-1}\underline{B})^{-1} = \underline{A}^{-1} + \underline{A}^{-1}\underline{D}\underline{\Delta}_C^{-1}\underline{B}\underline{A}^{-1}$$

or in a more usual notation (change $-\underline{C}^{-1}$ with $\underline{C}$ and interchange $\underline{B}$ and $\underline{D}$)

$$(\underline{A} + \underline{B}\underline{C}\underline{D})^{-1} = \underline{A}^{-1} - \underline{A}^{-1}\underline{B}\underline{\Delta}^{-1}\underline{D}\underline{A}^{-1}$$

$$\underline{\Delta} = \underline{C}^{-1} + \underline{D}\underline{A}^{-1}\underline{B}.$$

A6 DETERMINANT IDENTITIES

Expand in rows to see that

$$\det \begin{bmatrix} \underline{A} & \underline{D} \\ \underline{0} & \underline{C} \end{bmatrix} = \det \underline{A} \det \underline{C} = \det \begin{bmatrix} \underline{A} & \underline{0} \\ \underline{B} & \underline{C} \end{bmatrix}.$$

Now use the identity

$$\begin{pmatrix} \underline{A} & \underline{D} \\ \underline{B} & \underline{C} \end{pmatrix} = \begin{pmatrix} \underline{A} & \underline{0} \\ \underline{B} & \underline{I} \end{pmatrix} \begin{pmatrix} \underline{I} & \underline{A}^{-1}\underline{D} \\ \underline{0} & \underline{\Delta}_C \end{pmatrix}$$

to find that (with $\underline{\Delta}_C$ as given in section A3)

$$\det \begin{pmatrix} \underline{A} & \underline{D} \\ \underline{B} & \underline{C} \end{pmatrix} = \det \underline{A} \det \underline{\Delta}_C.$$

Similarly show that this also is

$$\det \begin{pmatrix} \underline{A} & \underline{D} \\ \underline{B} & \underline{C} \end{pmatrix} = \det \underline{C} \det \underline{\Delta}_A.$$

Now set $\underline{A} = \underline{I}_m$, $\underline{C} = \underline{I}_n$ to see from the last two identities that

$$\det(\underline{I}_n - \underline{B}\underline{D}) = \det(\underline{I}_m - \underline{D}\underline{B}).$$

In particular, if $\underline{b}$, $\underline{d}$ are vectors,

$$\begin{aligned} \det(\underline{A} + \underline{b}\underline{d}^T) &= \det[(\underline{A})(\underline{I} + \underline{A}^{-1}\underline{b}\underline{d}^T)] \\ &= \det(\underline{A}) \det(\underline{I} + \underline{A}^{-1}\underline{b}\underline{d}^T) \\ &= (1 + \underline{d}^T\underline{A}^{-1}\underline{b}) \det \underline{A}. \end{aligned}$$

A7 PROPERTIES OF TOEPLITZ MATRICES

Suppose $\underline{\Sigma}$ is the autocovariance matrix of a stationary time series

$$\underline{\Sigma} = \begin{bmatrix} \gamma_0 & \gamma_1 & \cdots & \gamma_{p-1} \\ \gamma_1 & \gamma_0 & \cdots & \gamma_{p-2} \\ \vdots & \vdots & \ddots & \vdots \\ \gamma_{p-1} & \gamma_{p-2} & \cdots & \gamma_0 \end{bmatrix}$$

then $\underline{\Sigma}$ is called a Toeplitz matrix because its cross-diagonal entries are identical. Suppose $\underline{\Sigma}$ has eigenvectors $\underline{\xi}_1^{(p)} \ldots \underline{\xi}_p^{(p)}$ and eigenvalues $\lambda_1^{(p)} \ldots \lambda_p^{(p)}$. Then as $p \to \infty$

$$\lambda_k^{(p)} \sim F(2\pi k/p)$$
$$\underline{\xi}_u^{(p)} \sim [e^{-j2\pi(u/p)0}, e^{-j2\pi(u/p)}, \ldots, e^{-j2\pi(u/p)(p-1)}]/\sqrt{p}$$

where $F(\omega)$ is the spectrum of the time series and $\sim$ means that the norm of the difference between the two terms goes to 0 as $p \to \infty$. For a proof see [B14].

A8 CHOLESKY DECOMPOSITION/FACTORIZATION

If Γ is symmetric positive definite then there exists a factorization or decomposition of Γ as

$$\Gamma = LDL^T$$

where L is lower triangular with 1s on the main diagonal, and D is diagonal. The decomposition can be computed recursively as follows.

Call the $n \times n$ top left corner submatrix of Γ as Γ_n. Given the Cholesky Decomposition of Γ_{n-1} as

$$\Gamma_{n-1} = L_{n-1} D_{n-1} L_{n-1}^T$$

we can get the order n factors as follows. We write the equation

$$\Gamma_n = L_n D_n L_n^T$$

in partitioned form

$$\begin{pmatrix} \Gamma_{n-1} & \underline{g}_n \\ \underline{g}_n^T & g_{nn} \end{pmatrix} = \begin{pmatrix} L_{n-1} & 0 \\ \underline{l}_n^T & 1 \end{pmatrix} \begin{pmatrix} D_{n-1} & 0 \\ 0 & d_n \end{pmatrix} \begin{pmatrix} L_{n-1}^T & \underline{l}_n \\ 0 & 1 \end{pmatrix}$$
$$= \begin{pmatrix} L_{n-1} D_{n-1} L_{n-1}^T & L_{n-1} D_{n-1} \underline{l}_n \\ \underline{l}_n^T D_{n-1} L_{n-1}^T & \underline{l}_n^T D_{n-1} \underline{l}_n + d_n \end{pmatrix}.$$

Equating terms gives

$$L_{n-1} D_{n-1} \underline{l}_n = \underline{g}_n$$
$$g_{nn} = \underline{l}_n^T D_{n-1} \underline{l}_n + d_n.$$

Thus

$$\underline{l}_n = D_{n-1}^{-1} L_{n-1}^{-1} \underline{g}_n \tag{A8.1}$$

$$d_n = g_{nn} - \underline{l}_n^T D_{n-1} \underline{l}_n.$$

The inversion of D_{n-1} requires just component-wise division. It is straightforward to derive a recursive algorithm to invert a lower triangular matrix. So the computation can be made fully recursive. The Cholesky algorithm is popular because of its good numerical properties and its recursive nature.

A9 VEC CALCULUS

If $\underline{M}$ is a matrix with columns $\underline{c}_1 \cdots \underline{c}_m$, we define vec($\underline{M}$) or $\underline{\vec{m}}$ as a vector consisting of the columns of $\underline{M}$ stacked one underneath another

$$\underline{\vec{m}} = \text{vec}(\underline{M}) = \begin{bmatrix} \underline{c}_1 \\ \vdots \\ \underline{c}_m \end{bmatrix}.$$

If $\underline{A}$ and $\underline{B}$ are two matrices, the Kronecker product $\underline{A} \otimes \underline{B}$ is defined as

$$\underline{A} \otimes \underline{B} = \text{block matrix whose } i, j \text{ block element is } a_{i,j}\underline{B}.$$

The following relationships are basic:

$$\begin{aligned} \text{vec}(\underline{ABC}) &= \underline{C}^T \otimes \underline{A}\, \text{vec}(\underline{B}) \\ \text{tr}(\underline{AB}) &= \text{vec}^T(\underline{A})\text{vec}(\underline{B}^T) \\ \text{tr}(\underline{ABCD}) &= \text{vec}^T(\underline{ABC})\text{vec}(\underline{D}^T) \\ &= \text{vec}^T(\underline{B})\, \underline{C} \otimes \underline{A}\, \text{vec}(\underline{D}^T) \end{aligned}$$

Derivations are left to the reader.

B

Stochastic Signals and Systems Review

B1 MOMENTS

If a random variable X has a probability density function (pdf) $p_X(x)$, then its mean is defined as

$$\mu_X = E(X) = \int x p_X(x) dx.$$

Note that $E(X)$ is linear in X, i.e., $E(aX + b) = aE(X) + b$, for constants a and b. The second moment is

$$R_X = E(X^2).$$

The variance is

$$\begin{aligned}\gamma_X = \sigma_X^2 &= E(X - \mu_X)^2 \\ &= E(X^2 - 2\mu_X X + \mu_X^2) \\ &= R_X - \mu_X^2.\end{aligned}$$

B2 CORRELATION

(i) If X, Y are two random variables with means μ_X, μ_Y, moments R_X, R_Y, and variances σ_X^2, σ_Y^2, then the co-moment or correlation is

$$R_{XY} = E(XY).$$

The covariance is

$$\begin{aligned}\gamma_{XY} = \sigma_{XY} &= E(X - \mu_X)(Y - \mu_Y) \\ &= E(XY - \mu_X Y - \mu_Y X + \mu_X \mu_Y) \\ &= E(XY) - \mu_X \mu_Y \\ &= R_{XY} - \mu_X \mu_Y.\end{aligned}$$

The normalized correlation is

$$\rho_{XY} = \sigma_{XY}/\sigma_X \sigma_Y = \gamma_{XY}/\sqrt{\gamma_X \gamma_Y}.$$

X and Y are (linearly) uncorrelated if $\rho_{XY} = 0$. It is left to the reader to show that $|\rho_{XY}| \leq 1$.

The normalized correlation should really be called the *linear* normalized correlation because it measures the amount of linear relation between X and Y. For example if $Y = X^2$ then Y, X are perfectly (nonlinearly) correlated, but not perfectly linearly correlated.

(ii) Correlation Matrix. If $\underline{Y} = (Y_1 \ldots Y_p)$ is a random vector, its correlation matrix is defined as

$$\begin{aligned}\underline{R} &= [E(Y_i Y_j)] \\ &= \begin{bmatrix} E(Y_1Y_1) & E(Y_1Y_2) & \cdots & E(Y_1Y_p) \\ E(Y_2Y_1) & E(Y_2Y_2) & \cdots & E(Y_2Y_p) \\ \vdots & \vdots & \ddots & \vdots \\ E(Y_pY_1) & E(Y_pY_2) & \cdots & E(Y_pY_p) \end{bmatrix} \\ &= E\begin{bmatrix} Y_1Y_1 & Y_1Y_2 & \cdots & Y_1Y_p \\ Y_2Y_1 & Y_2Y_2 & \cdots & Y_2Y_p \\ \vdots & \vdots & \ddots & \vdots \\ Y_pY_1 & Y_pY_2 & \cdots & Y_pY_p \end{bmatrix} \\ &= E(\underline{Y}\underline{Y}^T).\end{aligned}$$

Similarly, the covariance matrix is

$$\underline{\Gamma} = E[(\underline{Y} - \underline{\mu})(\underline{Y} - \underline{\mu})^T]$$

where $\underline{\mu}$ is the vector mean

$$\underline{\mu} = E(\underline{Y}).$$

If $\underline{X}$ and $\underline{Y}$ have a joint multivariate density, we may partition the correlation matrix as

$$\underline{R}_{X,Y} = \begin{pmatrix} \underline{R}_X & \underline{R}_{XY} \\ \underline{R}_{YX} & \underline{R}_Y \end{pmatrix} = \begin{bmatrix} E(\underline{X}\underline{X}^T) & E(\underline{X}\underline{Y}^T) \\ E(\underline{Y}\underline{X}^T) & E(\underline{Y}\underline{Y}^T) \end{bmatrix}.$$

The covariance matrix may be similarly partitioned.

B3 CONDITIONAL EXPECTATION

The conditional mean of X given Y is defined as

$$\mu_{X|Y} = E(X|Y) = \int xp(x|y)dx$$

where $p(x|y)$ is the conditional pdf of X given Y. The conditional (second) moment of X given Y is

$$R_{X|Y} = E(X^2|Y).$$

The conditional variance is

$$\gamma^2_{X|Y} = \text{var}(X|Y).$$

It is straightforward to show that

$$R_{X|Y} = \gamma^2_{X|Y} + \mu^2_{X|Y}.$$

The following two iterated conditional expectation identities are useful:

$$\begin{aligned} E(E(X|Y)) &= E(X) \\ \text{var}(X) &= E(\text{var}(X|Y)) + \text{var}(E(X|Y)). \end{aligned}$$

To show the first equation, let $p(x, y)$ be the joint pdf of X and Y, then

$$\begin{aligned} E(E(X|Y)) &= \int \left[\int xp(x|y)dx \right] p_Y(y)dy \\ &= \int \int x(p(x, y)/p_Y(y))dxp_Y(y)dy \\ &= \int \int xp(x, y)dxdy \\ &= \int xp_X(x)dx = E(X). \end{aligned}$$

The second equality is left to the reader.

B4 PROPERTIES OF THE MULTIVARIATE GAUSSIAN DENSITY

(i) A scalar random variable X with pdf $p(x)$ has a Gaussian density (GD) when

$$p(x) = \frac{1}{\sqrt{2\pi}\sigma} e^{-\frac{1}{2}(x-\mu)^2/\sigma^2}, \quad -\infty < x < \infty.$$

The mean is then

$$\begin{aligned} E(X) &= \int_{-\infty}^{\infty} x p(x) dx \\ &= \mu + \int_{-\infty}^{\infty} (x-\mu) p(x) dx \\ &= \mu + \frac{1}{\sqrt{2\pi}\sigma} \int_{-\infty}^{\infty} t e^{-\frac{1}{2}t^2/\sigma^2} dt \\ &= \mu \end{aligned}$$

The integral vanishes since t is an odd function while the rest of the integrand is an even function.

The variance is

$$\begin{aligned} \mathrm{var}(X) = \sigma_X^2 = \gamma_X &= E(X-\mu)^2 \\ &= \int_{-\infty}^{\infty} (x-\mu)^2 p(x) dx \\ &= \frac{1}{\sqrt{2\pi}\sigma} \int_{-\infty}^{\infty} t^2 e^{-\frac{1}{2}t^2/\sigma^2} dt. \end{aligned}$$

Now change variables ($u = t/\sigma$) to find

$$\begin{aligned} \mathrm{var}(X) &= \frac{\sigma^2}{\sqrt{2\pi}} \int_{-\infty}^{\infty} u^2 e^{-u^2/2} du \\ &= \sigma^2. \qquad \text{(Prove it!)} \end{aligned}$$

(ii) Show as in (i) that

$$E(X-\mu)^4 = 3\sigma^4.$$

(iii) A random p-vector $\underline{X}$ has a multivariate Gaussian density $p(\underline{x})$ (MVGD) when $\underline{\Gamma}$ is positive definite and

$$p(\underline{x}) = \frac{1}{(2\pi |\underline{\Gamma}|)^{p/2}} e^{-\frac{1}{2}(\underline{x}-\underline{\mu})^T \underline{\Gamma}^{-1} (\underline{x}-\underline{\mu})}.$$

To calculate the mean we orthogonalize. From (A1)(v) $\underline{\Gamma}$ has a spectral decomposition

$$\underline{\Gamma} = \underline{Q} \underline{\Lambda} \underline{Q}^T$$

where $\underline{Q}$ is orthonormal. Thus, by (A0)(xiii) we have

$$|\underline{\Gamma}| = |\underline{\Lambda}|.$$

Now introduce the transformed random vector

$$\underline{Y} = \underline{Q}(\underline{X} - \underline{\mu}) = (Y_1 \ldots Y_p)^T.$$

Since $\underline{Q}$ is orthonormal, the Jacobian of this transformation is 1, so the pdf of $\underline{Y}$ is

$$
\begin{aligned}
p_Y(\underline{y}) &= \frac{1}{(2\pi|\underline{\Lambda}|)^{p/2}} e^{-\frac{1}{2}\underline{y}^T\underline{\Lambda}^{-1}\underline{y}} \\
&= \prod_{u=1}^{p} \left\{ \frac{e^{-\frac{1}{2}y_u^2/\lambda_u}}{(2\pi\lambda_u)^{\frac{1}{2}}} \right\} \\
&= \text{product of pdfs of } p\text{-independent GDs.}
\end{aligned}
$$

Thus

$$
\begin{aligned}
& E(Y_u) = 0, \qquad u = 1 \ldots p \\
\Rightarrow \quad & E(\underline{Y}) = \underline{0} = E(\underline{Q}(\underline{X} - \underline{\mu})) \\
\Rightarrow \quad & E(\underline{X}) = \underline{\mu}.
\end{aligned}
$$

Now we turn to the variance calculation. First note that

$$
\begin{aligned}
& \operatorname{var}(Y_u) = \lambda_u \\
\Rightarrow \quad & E(\underline{Y}\underline{Y}^T) = \underline{\Lambda}, \quad \text{by independence} \\
\Rightarrow \quad & \underline{Q}E(\underline{X} - \underline{\mu})(\underline{X} - \underline{\mu})^T\underline{Q}^T = \underline{\Lambda} \\
\Rightarrow \quad & E(\underline{X} - \underline{\mu})(\underline{X} - \underline{\mu})^T = \underline{Q}^T\underline{\Lambda}\underline{Q} = \underline{\Gamma}.
\end{aligned}
$$

(iv) Independence. If $\underline{X}$ and $\underline{Y}$ have jointly an MVGD and $\underline{X}$, $\underline{Y}$ are uncorrelated, then $\underline{X}$ and $\underline{Y}$ are independent.

Proof. Take $\underline{\mu}_X = \underline{0} = \underline{\mu}_Y$ without loss of generality. Since $\underline{X}$ and $\underline{Y}$ are uncorrelated, the joint covariance matrix must have the form

$$
\begin{aligned}
& \underline{\Gamma} = \begin{bmatrix} \underline{\Gamma}_X & \underline{0} \\ \underline{0} & \underline{\Gamma}_Y \end{bmatrix} \\
\Rightarrow \quad & |\underline{\Gamma}| = |\underline{\Gamma}_X||\underline{\Gamma}_Y|. \\
& \text{Also } [\underline{X}^T\underline{Y}^T]\underline{\Gamma}^{-1}\begin{bmatrix} \underline{X} \\ \underline{Y} \end{bmatrix} = \underline{X}^T\underline{\Gamma}_X^{-1}\underline{X} + \underline{Y}^T\underline{\Gamma}_Y^{-1}\underline{Y}.
\end{aligned}
$$

From these two equalities we find

$$
p(\underline{x}, \underline{y}) = p_x(\underline{x})p_Y(\underline{y})
$$

where $p(\underline{x}, \underline{y})$ is the joint density and $p_X(\underline{x})$, $p_Y(\underline{y})$ are marginal densities. Thus independence follows.

(v) Conditional Densities. If $\underline{X}$, $\underline{Y}$ have a joint MVGD then

$$\underline{\mu}_{X|Y} = E(X|Y) = \underline{\Gamma}_{XY}\underline{\Gamma}_Y^{-1}\underline{Y} \tag{B4.1}$$

$$\underline{\Gamma}_{X|Y} = \text{var}(\underline{X}|\underline{Y}) = \underline{\Gamma}_X - \underline{\Gamma}_{XY}\Gamma_Y^{-1}\underline{\Gamma}_{YX}. \tag{B4.2}$$

Proof. From the definition of conditional density we have

$$p(\underline{x}|\underline{y}) = p(\underline{x}, \underline{y})/p_Y(\underline{y}).$$

Now the exponent of the left hand side (lhs) is

$$\text{const.} - \frac{1}{2}(\underline{x} - \underline{\mu}_{X|Y})^T \underline{\Gamma}_{X|Y}^{-1}(\underline{x} - \underline{\mu}_{X|Y}) \tag{B4.3}$$

and equals the exponent on the right hand side (rhs) which is

$$\text{const.} - \frac{1}{2}(\underline{x}^T \underline{y}^T)\begin{pmatrix} \underline{\Gamma}_X & \underline{\Gamma}_{XY} \\ \underline{\Gamma}_{YX} & \underline{\Gamma}_Y \end{pmatrix}^{-1}\begin{pmatrix} \underline{x} \\ \underline{y} \end{pmatrix} - \frac{1}{2}\underline{y}^T\underline{\Gamma}_Y^{-1}\underline{y}.$$

Now apply the partitioned matrix inversion formulae of Appendix A, section A4 to find this is

$$\text{const.} - \frac{1}{2}(\underline{x}^T \underline{y}^T)\begin{pmatrix} \underline{\Delta}_X^{-1} & -\underline{\Delta}_X^{-1}\underline{\Gamma}_{XY}\underline{\Gamma}_Y^{-1} \\ -\underline{\Gamma}_Y^{-1}\underline{\Gamma}_{YX}\underline{\Delta}_X^{-1} & \underline{\Delta}_Y^{-1} \end{pmatrix}\begin{pmatrix} \underline{x} \\ \underline{y} \end{pmatrix} - \frac{1}{2}\underline{y}^T\underline{\Gamma}_Y^{-1}\underline{y} \tag{B4.4}$$

where

$$\underline{\Delta}_X = \underline{\Gamma}_X - \underline{\Gamma}_{XY}\underline{\Gamma}_Y^{-1}\underline{\Gamma}_{YX}$$

$$\underline{\Delta}_Y = \underline{\Gamma}_Y - \underline{\Gamma}_{YX}\underline{\Gamma}_X^{-1}\underline{\Gamma}_{XY}.$$

Differentiate (B4.3) and (B4.4) with respect to $\underline{x}$ to find via Appendix A, section A3 that

$$-\Gamma_{X|Y}^{-1}(\underline{x} - \underline{\mu}_{X|Y}) = -\underline{\Delta}_X^{-1}\underline{x} - \underline{\Delta}_X^{-1}\underline{\Gamma}_{XY}\Gamma_Y^{-1}\underline{y}.$$

Equating functions of $\underline{x}$ here gives (B4.2). Equating functions of $\underline{y}$ then gives (B4.1).

(vi) Fourth Moments. If $\underline{X}$ has an MVGD with 0 mean and $\underline{\alpha}$ and $\underline{\beta}$ are fixed constant vectors, then

$$E[(\underline{\alpha}^T\underline{X})^2(\underline{\beta}^T\underline{X})^2] = \underline{\alpha}^T\underline{\Gamma}_X\underline{\alpha}\underline{\beta}^T\underline{\Gamma}_X\underline{\beta} + 2(\underline{\alpha}^T\underline{\Gamma}_X\underline{\beta})^2.$$

Since $\underline{\alpha}$ is arbitrary this can also be written

$$E[\underline{X}\underline{X}^T(\underline{\beta}^T\underline{X})^2] = \underline{\Gamma}_X\underline{\beta}^T\underline{\Gamma}_X\underline{\beta} + 2\underline{\Gamma}_X\underline{\beta}\underline{\beta}^T\underline{\Gamma}_X.$$

Proof. Let $U = \underline{\alpha}^T\underline{X}$, $W = \underline{\beta}^T\underline{X}$. Then U, W have a joint MVGD with 0 mean and with covariance matrix

$$\begin{pmatrix} \sigma_U^2 & \sigma_{UW} \\ \sigma_{WU} & \sigma_W^2 \end{pmatrix} = \begin{pmatrix} \underline{\alpha}^T\underline{\Gamma}_X\underline{\alpha} & \underline{\alpha}^T\underline{\Gamma}_X\underline{\beta} \\ \underline{\beta}^T\underline{\Gamma}_X\underline{\alpha} & \underline{\beta}^T\underline{\Gamma}_X\underline{\beta} \end{pmatrix}.$$

Now we proceed via a conditioning argument to calculate

$$\begin{aligned} E(U^2W^2) &= E(E(U^2W^2|W)) \\ &= E(W^2E(U^2|W)) \\ &= E(W^2[\text{var}(U|W) + E^2(U|W)]) \\ &= E(W^2)\sigma^2_{U|W} + E(W^2(\sigma_{UW}\sigma_W^{-2}W)^2) \\ &= \sigma^2_W(\sigma^2_U - \sigma_{UW}\sigma_W^{-2}\sigma_{UW}) + E(W^4)\sigma^2_{UW}\sigma_W^{-4} \end{aligned}$$

where (B4)(v) has been used. However from (B4)(ii) this is

$$\begin{aligned} E(U^2W^2) &= \sigma^2_W\sigma^2_U - \sigma^2_{UW} + 3\sigma^2_{UW} \\ &= \sigma^2_W\sigma^2_U + 2\sigma^2_{UW} \end{aligned}$$

which is the quoted result.

B5 CONVERGENCE OF SEQUENCES OF RANDOM VARIABLES

(i) Let $\{Y_n\}$, $n = 1, 2, \cdots$ be a sequence of random variables. We say Y_n converges weakly or in distribution to Y,

$$Y_n \Rightarrow Y$$

if, at each continuous point of $F(y) = P(Y \le y)$

$$P(Y_n \le y) \to P(Y \le y), \quad \text{as } n \to \infty.$$

(ii) We say Y_n converges in probability to Y,

$$Y_n \xrightarrow{p} Y$$

if for any $\epsilon > 0$

$$P(|Y_n - Y| > \epsilon) \to 0, \quad \text{as } n \to \infty.$$

(iii) We say Y_n converges in mean square to Y,

$$Y_n \xrightarrow{ms} Y$$

if

$$E|Y_n - Y|^2 \to 0, \quad \text{as } n \to \infty.$$

(iv) Chebyshev-Biename inequality. If X is a positive random variable with $E(X) < \infty$ and tail distribution function

$$G(x) = P(X \ge x)$$

then

$$\begin{aligned}
E(X) &= -\int_0^\infty x(dG/dx)dx \\
&= -xG(x)\Big|_0^\infty + \int_o^\infty G(x)dx \\
&= \int_0^\infty G(x)dx \\
&\geq \int_0^\alpha G(x)dx, \quad \text{for any } \alpha > 0 \\
&\geq G(\alpha)\int_0^\alpha dx, \quad \text{why?} \\
&= \alpha G(\alpha) \\
&= \alpha P(X \geq \alpha).
\end{aligned}$$

(v) From the Chebyshev-Biename inequality

$$\begin{aligned}
P(|Y_n - Y| > \epsilon) &\leq E|Y_n - Y|/\epsilon \\
&\leq E^{\frac{1}{2}}(Y_n - Y)^2/\epsilon.
\end{aligned}$$

So we have

convergence in mean square $\Longrightarrow$ convergence in probability.

The reader is invited to show that

convergence in probability $\Longrightarrow$ convergence in distribution.

B6 DISCRETE RANDOM SIGNALS

(i) Autocovariance and Cross-Covariance. A discrete random signal or time series is just a sequence X_t, $t = 0, 1, \cdots$ of random variables. If the mean is time invariant, i.e.,

$$E(X_t) = \text{constant} = \mu_X = \mu$$

we say X_t is first order stationary. If also the auto-moment or autocorrelation is time invariant, i.e.,

$$E(X_t X_s) = \text{function of } t - s = R_{t-s} = R_{X,t-s}$$

we say X_t is second order stationary. This may be equivalently expressed in terms of the autocovariance

$$E(X_t - \mu)(X_s - \mu) = \gamma_{t-s} = \gamma_{X,t-s}.$$

If X_t and Y_t are both second order stationary and if their cross-covariance is also time invariant, so that

$$\gamma_{XY,t,s} = E(X_t - \mu_X)(Y_s - \mu_Y) = \gamma_{XY,t-s}$$

then we say that X_t, Y_t are jointly second order stationary. Note that then

$$\gamma_{YX,r} = \gamma_{XY,-r}.$$

(ii) Autocovariance Matrix. If we collect p successive values of a stationary time series X_t into a vector $\underline{X}_t = (X_{t-1} \dots X_{t-p})^T$ then the autocovariance matrix of $\underline{X}_t$ is

$$\begin{aligned}
\underline{\Gamma}_X &= E(\underline{X}_t \underline{X}_t^T) \\
&= [E(X_{t-r}X_{t-s})] \\
&= [\gamma_{r-s}] \\
&= \begin{bmatrix} \gamma_0 & \gamma_1 & \cdots & \gamma_{p-1} \\ \gamma_1 & \gamma_0 & \cdots & \gamma_{p-2} \\ \vdots & \vdots & \ddots & \vdots \\ \gamma_{p-1} & \gamma_{p-2} & \cdots & \gamma_0 \end{bmatrix}.
\end{aligned}$$

Notice that the cross-diagonal entries in this matrix are all the same. Such a matrix is called a Toeplitz matrix. Note that $\underline{\Gamma}_X$ is positive definite.

(iii) For manipulating time series models the backwards operator proves useful:

$$\begin{aligned}
q^{-1}X_t &= X_{t-1} \\
q^{-2}X_t &= q^{-1}(q^{-1}X_t) = q^{-1}X_{t-1} = X_{t-2}, \quad \text{etc.}
\end{aligned}$$

(iv) White Noise. A white noise is a sequence of independent zero mean random variables usually with constant variance.

(v) Stationary Time Series are easily generated by filtering white noise. A moving average time series of order m (MA(m)) is generated by passing white noise through an FIR filter with delay m:

$$\begin{aligned}
X_t &= (1 + \sum_{s=1}^{m} c_s q^{-s})\epsilon_t \\
&= \epsilon_t + \sum_{s=1}^{m} c_s \epsilon_{t-s}.
\end{aligned}$$

This is also called an all zero filter.

(vi) An Autoregressive Time Series of order p (AR(p)) is generated by passing a white noise through an IIR all pole filter of order p.

$$X_t = [1 + a(q^{-1})]^{-1}\epsilon_t, \quad a(q^{-1}) = \sum_{s=1}^{p} a_s q^{-s}$$

$$\Rightarrow \qquad [1 + a(q^{-1})]X_t = \epsilon_t$$

$$\Rightarrow \qquad X_t = -\sum_{s=1}^{p} a_s q^{-s} X_t + \epsilon_t$$

$$= -\sum_{s=1}^{p} a_s X_{t-s} + \epsilon_t. \tag{B6.1}$$

It is necessary that $1 + a(q^{-1})$ be a stable polynomial (see Appendix C4) in order that X_t be stationary. Also the recursive computation in (B6.1) must be initialized properly.

(vii) An autoregressive moving average time series (ARMA(p, m)) is generated by passing a white noise through a pole zero filter as follows:

$$X_t = [1 + a(q^{-1})]^{-1}(1 + c(q^{-1}))\epsilon_t$$

$$\Rightarrow \qquad X_t = -\sum_{s=1}^{p} a_s X_{t-s} + \epsilon_t + \sum_{s=1}^{m} c_s \epsilon_{t-s}.$$

Again to generate a stationary time series careful initialization is needed.

(viii) Autocovariance of AR Time Series. If in (B6.1) ϵ_t has zero mean (and variance σ_ϵ^2), then clearly so does X_t. Multiply through (B6.1) by X_{t-k} and take expectations to find

$$\gamma_{X,k} = \gamma_k = -\sum_{s=1}^{p} a_s \gamma_{k-s}, \quad k \geq 1 \tag{B6.2}$$

$$\gamma_{X,0} = \gamma_0 = -\sum_{s=1}^{p} a_s \gamma_s, \quad k = 1. \tag{B6.3}$$

Equations (B6.2) are the so-called Yule-Walker equations that generate the acf. Multiply (B6.1) through by ϵ_t and take expectations to find

$$E(\epsilon_t X_t) = 0 + \sigma_\epsilon^2.$$

Thus (B6.3) becomes

$$\gamma_0 = -\sum_{s=1}^{p} a_s \gamma_s + \sigma_\epsilon^2. \tag{B6.4}$$

We now look at two particular cases:

AR(1) Note that stability (and so stationarity) requires $|a| < 1$. Then (B6.4) becomes

$$\gamma_0 = -a_1\gamma_1 + \sigma_\epsilon^2$$

while (B6.2) gives

$$\begin{aligned} & \gamma_1 = -a_1\gamma_0 \\ \Rightarrow \quad & \gamma_0 = a_1^2\gamma_0 + \sigma_\epsilon^2 \\ \Rightarrow \quad & \gamma_0 = \sigma_\epsilon^2/(1 - a_1^2). \end{aligned} \tag{B6.5}$$

Using (B6.2) again gives

$$\begin{aligned} \gamma_k &= -a_1\gamma_{k-1} \\ &= (-a_1)^k\gamma_0 \\ \Rightarrow \quad \rho_k &= \gamma_k/\gamma_0 = (-a_1)^k. \end{aligned} \tag{B6.6}$$

AR(2) Stationarity requires that both roots of $q^2 + a_1 q + a_2 = 0$ have modulus < 1. Equivalently the coefficients must be in the triangular region of Fig. B.1. Now again if ϵ_t has zero mean so does X_t. Next (B6.4) becomes

$$\gamma_0 = -a_1\gamma_1 - a_2\gamma_2 + \sigma_\epsilon^2 \tag{B6.7}$$

while (B6.2) gives

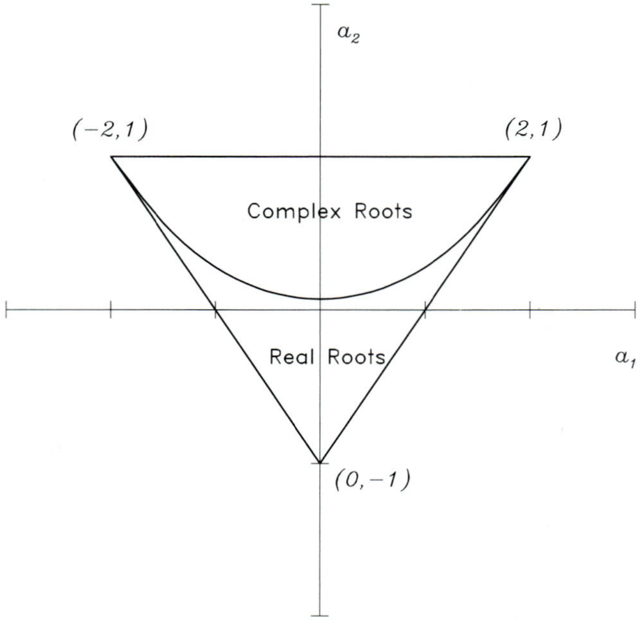

Figure B.1 Stability/Stationarity region for AR(2) process.

$$\gamma_1 = -a_1\gamma_0 - a_2\gamma_1$$

$$\Rightarrow \quad \gamma_1 = -\frac{a_1}{1+a_2}\gamma_0 \tag{B6.8}$$

$$\Rightarrow \quad \rho_1 = -\frac{a_1}{1+a_2}. \tag{B6.9}$$

Also

$$\gamma_2 = -a_1\gamma_1 - a_2\gamma_0$$

$$= \left(\frac{a_1^2}{1+a_2} - a_2\right)\gamma_0. \tag{B6.10}$$

Using (B6.8), (B6.10) in (B6.7) gives

$$\gamma_0 = \frac{a_1^2}{1+a_2}\gamma_0 - a_2\left(\frac{a_1^2}{1+a_2} - a_2\right)\gamma_0 + \sigma_\epsilon^2$$

$$\Rightarrow \quad \gamma_0 = \frac{(1+a_2)}{(1-a_2)}\frac{\sigma_\epsilon^2}{[(1+a_2)^2 - a_1^2]}. \tag{B6.11}$$

To generate the acf for lags $k \geq 2$ use (B6.2) with initial conditions (B6.9), (B6.10). AR(p) $p \geq 2$. To generate autocovariances for $p \geq 2$ the reader is referred to the algorithm of [H11]. That algorithm also generates autocovariances for ARMA(p,m) models.

B7 STOCHASTIC FOURIER ANALYSIS

(i) The Spectrum for a Finite Energy Signal. If X_t is a deterministic, finite energy, discrete, aperiodic signal, then its Fourier transform is defined by

$$X(\omega) = \sum_{t=-\infty}^{\infty} X_t e^{-j\omega t}$$

and X_t can be recovered from $X(\omega)$:

$$X_t = \int_{-\pi}^{\pi} X(\omega)e^{j\omega t}d\omega/2\pi.$$

Also Parseval's theorem shows how the total signal energy is distributed in the frequency domain

$$\sum_{t=-\infty}^{\infty} X_t^2 = \int_{-\pi}^{\pi} |X(\omega)|^2 d\omega/2\pi$$

and $|X(\omega)|^2$ is called the spectrum of X_t.

(ii) The Spectrum for an Infinite Energy Signal. If X_t is an infinite energy, finite power signal, a Fourier analysis is still possible as follows: Introduce the finite time Fourier transform

$$X_N(\omega) = \sum_{t=1}^{N} X_t e^{-j\omega t}$$

and define the total power spectrum as

$$P(\omega) = \lim_{N\to\infty} \frac{1}{N}|X_N(\omega)|^2 = \lim_{N\to\infty} \mathcal{P}_N(\omega) \tag{B7.1}$$

where $\mathcal{P}_N(\omega)$ is called the total periodogram. If X_t has a "dc" component μ_X, we can define the "ac" power spectrum as

$$F_X(\omega) = \lim_{N\to\infty} \frac{1}{N}|\tilde{X}_N(\omega)|^2$$

$$\tilde{X}_N(\omega) = \sum_{t=1}^{N}(X_t - \mu_X)e^{-j\omega t}.$$

Then we find

$$P(\omega) = F_X(\omega) + \mu_X^2\delta(\omega).$$

Because the total power spectrum thus has a "dc" spike in it, we prefer to work with the ac power spectrum and also drop the ac qualifier.

(iii) The Spectrum for a Stationary Time Series. When X_t is a stationary stochastic time series the limit (B7.1) does not exist as a limit with probability 1. Rather it exists as a limit in distribution. Under certain regularity conditions we can define the power spectrum as

$$F_X(\omega) = \lim_{N\to\infty} \frac{1}{N}E|\tilde{X}_N(\omega)|^2 = \lim_{N\to\infty} E(I_N(\omega))$$

where $I_N(\omega)$ is the (ac) periodogram. Then it turns out that

$$\frac{I_N(\omega)}{F_X(\omega)} = \frac{|X_N(\omega)|^2}{NF_X(\omega)} \Rightarrow \chi_2^2, \quad \text{as } N \to \infty$$

where χ_2^2 is a chi-square with two degrees of freedom, i.e., an exponentially distributed random variable. For a full discussion of these ideas see [B14].

(iv) Wiener-Khinchin relations. From

$$I_N(\omega) = \frac{1}{N}\sum_{s=1}^{N}\sum_{t=1}^{N} \tilde{X}_t\tilde{X}_s e^{-j\omega(t-s)}, \quad \tilde{X}_t = X_t - \mu_X$$

$$= \sum_{s=-(N-1)}^{(N-1)} \tilde{C}_{Ns}e^{-j\omega s}$$

$$\tilde{C}_{Ns} = \frac{1}{N}\sum_{t=1}^{N-|s|} \tilde{X}_t\tilde{X}_{t+s}$$

we find that

$$E(I_N(\omega)) = \sum_{s=-(N-1)}^{N-1} \gamma_{X,s}(1 - |s|/N)e^{-j\omega s}.$$

It is not surprising then to find that, under certain regularity conditions

$$F_X(\omega) = \lim_{N\to\infty} E(I_N(\omega))$$
$$= \sum_{s=-\infty}^{\infty} \gamma_{X,s}e^{-j\omega s} \tag{B7.2}$$

so that the spectrum is the Fourier series of the autocovariance sequence. Naturally the Fourier series can be inverted to yield

$$\gamma_{X,s} = \frac{1}{2\pi}\int_{-\pi}^{\pi} e^{j\omega s}F_X(\omega)d\omega. \tag{B7.3}$$

The pair (B7.2), (B7.3) are the Wiener-Khinchin pair.

(v) Cross Spectrum. If X_t, Y_t are jointly stationary, the cross spectrum is defined as the following limit:

$$F_{XY}(\omega) = \lim_{N\to\infty} \frac{1}{N} E(\tilde{X}_N(\omega)\tilde{Y}_N^*(\omega)).$$

By repeating the type of calculation made in (iv) we find the following Fourier series relations

$$F_{XY}(\omega) = \sum_{s=-\infty}^{\infty} \gamma_{XY,s}e^{-j\omega s}$$

$$\gamma_{XY,s} = \frac{1}{2\pi}\int_{-\pi}^{\pi} e^{j\omega s}F_{XY}(\omega)d\omega$$

where $\gamma_{XY,s}$ is the cross-covariance

$$\gamma_{XY,s} = E(\tilde{X}_t\tilde{Y}_{t+s}), \quad s = 0, \pm 1, \cdots.$$

Note that

$$F_{XY}(\omega) = F_{YX}^*(\omega). \tag{B7.4}$$

(vi) Effect of Linear Filtering on Spectra. Suppose X_t is filtered to produce Y_t

$$Y_t = h(q^{-1})X_t = \sum_{s=-\infty}^{\infty} h_s X_{t-s}$$

$$\Rightarrow \qquad \mu_Y = \sum_{s=-\infty}^{\infty} h_s\mu_X$$

$$\Rightarrow \qquad \tilde{Y}_t = \sum_{s=-\infty}^{\infty} h_s\tilde{X}_{t-s}, \quad \tilde{Y}_t = Y_t - \mu_Y, \text{ etc.}$$

Thus

$$\gamma_{YX,r} = E(\tilde{Y}_t \tilde{X}_{t+s})$$
$$= \sum_{s=-\infty}^{\infty} h_s \gamma_{X,r+s}.$$

Taking Fourier series gives

$$F_{YX}(\omega) = \sum_{r=-\infty}^{\infty} e^{-j\omega r} \gamma_{YX,r}$$
$$= \sum_{r=-\infty}^{\infty} e^{-j\omega r} \sum_{s=-\infty}^{\infty} h_s \gamma_{X,r+s}$$
$$= \sum_{s=-\infty}^{\infty} h_s e^{j\omega s} \sum_{r=-\infty}^{\infty} e^{-j\omega(r+s)} \gamma_{X,r+s}$$
$$= \sum_{r=-\infty}^{\infty} h_s e^{j\omega s} \sum_{t=-\infty}^{\infty} e^{-j\omega t} \gamma_{X,t}$$
$$= H^*(\omega) F_X(\omega) \tag{B7.5}$$

where $H(\omega) = \sum_{s=-\infty}^{\infty} h_s e^{-j\omega s}$. Similarly we find

$$F_Y(\omega) = H^*(\omega) F_{XY}(\omega). \tag{B7.6}$$

We then deduce from (B7.4)–(B7.6) that

$$F_Y(\omega) = |H(\omega)|^2 F_X(\omega). \tag{B7.7}$$

It is left as an exercise for the reader to show that if

$$Y_t = h_1(q^{-1}) X_t$$
$$Z_t = h_2(q^{-1}) X_t$$

then

$$F_{YZ}(\omega) = H_1(\omega) H_2^*(\omega) F_X(\omega).$$

(vii) Spectra of ARMA Processes. Using the results of B7(vi) and the definitions in B6(v)–(vii) we find:

Spectrum of white noise. The acv is

$$\gamma_{X,s} = \delta_s \sigma^2, \quad \delta_s = \begin{cases} 1 & s = 0 \\ 0 & s \neq 0 \end{cases}$$

$$\Rightarrow \quad F_X(\omega) = \sigma^2.$$

Spectrum of an ARMA process

$$F_X(\omega) = \left| \frac{1 + c(e^{-j\omega})}{1 + a(e^{-j\omega})} \right|^2 \sigma^2.$$

B8 MINIMUM VARIANCE ESTIMATION

Let $\underline{Y}$, $\underline{Z}$ be two zero mean, correlated random variables. Then the minimum variance estimator of $\underline{Y}$ given $\underline{Z}$ is $E(\underline{Y}|\underline{Z})$. To prove this denote

$$\hat{\underline{Y}} = E(\underline{Y}|\underline{Z}).$$

Let $\tilde{\underline{Y}}$ = some function of $\underline{Z}$ be any other estimator. Note that by definition of conditional expectation, for any function of $\underline{Z}$, say $\Phi(\underline{Z})$

$$E[(\underline{Y} - \hat{\underline{Y}})\Phi^T] = \underline{0}. \tag{B8.1}$$

This holds because, by iterated conditional expectation

$$\begin{aligned}
&E[(\underline{Y} - \hat{\underline{Y}})\Phi^T] \\
&= E(\underline{Y}\Phi^T) - E(\hat{\underline{Y}}\Phi^T) \\
&= E(E(\underline{Y}\Phi^T|Z)) - E(\hat{\underline{Y}}\Phi^T) \\
&= E(E(\underline{Y}|\underline{Z})\Phi^T) - E(\hat{\underline{Y}}\Phi^T) \\
&= E(\hat{\underline{Y}}\Phi^T) - E(\hat{\underline{Y}}\Phi^T) \\
&= 0.
\end{aligned}$$

Now use (B8.1) to see that

$$\begin{aligned}
E\|\underline{Y} - \tilde{\underline{Y}}\|^2 &= E\|\underline{Y} - \hat{\underline{Y}} + \hat{\underline{Y}} - \tilde{\underline{Y}}\|^2 \\
&= E\|\underline{Y} - \hat{\underline{Y}}\|^2 + E\|\hat{\underline{Y}} - \tilde{\underline{Y}}\|^2 \\
&\geq E\|\underline{Y} - \hat{\underline{Y}}\|^2.
\end{aligned}$$

Equality occurs if and only if $\tilde{\underline{Y}} = \hat{\underline{Y}}$ as is required.

B9 DISCRETE CONDITIONAL KALMAN FILTER

Consider the state estimation problem for the state space model

$$\underline{z}_{k+1} = \underline{F}_k \underline{z}_k + \underline{b}_k x_k + \underline{v}_k \tag{B9.1a}$$

$$y_k = \underline{c}_k^T \underline{z}_k + d_k x_k + n_k \tag{B9.1b}$$

where $(\underline{v}_k^T, n_k)$ is a zero mean, white noise with covariance

$$\text{var}\begin{pmatrix} \underline{v}_k \\ n_k \end{pmatrix} = \begin{bmatrix} \underline{Q}_k & \underline{0} \\ \underline{0}^T & \sigma_k^2 \end{bmatrix}. \tag{B9.2}$$

Also x_k is an external measured signal. Note that we allow $\underline{F}_k$, $\underline{b}_k$, $\underline{c}_k$, d_k, and x_k to be functions of past data $y_1, y_2, \ldots, y_{k-1}$. The state estimator is supplied by the Kalman filter, which can be developed in a number of ways, but for our purposes a Gaussian setting will do. We suppose $(\underline{v}_k^T, n_k)$ is Gaussian and that $\underline{z}_0$ is Gaussian, so that $(\underline{z}_k, y_k)$ is too (if $\underline{F}_k$, $\underline{b}_k$, $\underline{c}_k$, d_k, and x_k are deterministic). Given data $y_1, y_2, \ldots, y_{k-1} = y_1^{k-1}$, the aim is to predict the state $\underline{z}_k$. According to Appendix B8 and B4(v) the minimum variance prediction is

$$\begin{aligned} \underline{z}_{k|k-1} &= E(\underline{z}_k | y_1^{k-1}) \\ &= E(\underline{z}_k y_1^{k-1}) \underline{R}_{k-1}^{-1} y_1^{k-1} \end{aligned}$$

where

$$\underline{R}_{k-1} = E(y_1^{k-1} y_1^{(k-1)T}).$$

However this formula is useless for real time computation. Rather we seek a recursive algorithm that has finite memory, so that limited data need only be stored, so that the algorithm can be used in real time.

The natural idea is to use a Cholesky factorization of $\underline{R}_k$ (which is recursive – see Appendix A8), and indeed the Kalman filter can be shown to be a specialization of the Cholesky algorithm (when $\underline{F}_k$, $\underline{b}_k$, $\underline{c}_k$, d_k, and x_k are deterministic). However specializing the Cholesky algorithm is rather tedious, so instead we develop an independent derivation. Our derivation allows $\underline{F}_k$, $\underline{b}_k$, $\underline{c}_k$, d_k, and x_k to be functions of past data y_1^{k-1}.

We start with conditional densities. We have

$$\begin{aligned} p(\underline{z}_k | y_1^k) &= p(y_1^k, \underline{z}_k)/p(y_1^k) \\ &= p(y_k, y_1^{k-1}, \underline{z}_k)/p(y_1^k) \\ &= p(y_k | y_1^{k-1}, \underline{z}_k) p(y_1^{k-1}, \underline{z}_k)/p(y_1^k) \\ &= p(y_k | y_1^{k-1}, \underline{z}_k) p(\underline{z}_k | y_1^{k-1})/p(y_k | y_1^{k-1}). \end{aligned} \tag{B9.3}$$

We proceed now by induction. If $\underline{z}_k | y_1^{k-1}$ is conditionally Gaussian, then from (B9.1) $y_k | y_1^{k-1}$ is conditionally Gaussian. Also it follows directly anyway that $y_k | y_1^{k-1}, \underline{z}_k$ is conditionally Gaussian. Thus $\underline{z}_k | y_1^k$ is conditionally Gaussian. Then from (B9.1) $\underline{z}_{k+1} | y_1^k$ is conditionally Gaussian, and the inductive step is complete. Thus all the conditional densities in (B9.3) are Gaussian, so equating exponents gives

$$\begin{aligned} &(\underline{z}_k - \underline{z}_{k|k})^T \underline{P}_{k|k}^{-1} (\underline{z}_k - \underline{z}_{k|k}) \\ &= (y_k - d_k x_k - \underline{c}_k^T \underline{z}_k)^2 \sigma_k^{-2} + (\underline{z}_k - \underline{z}_{k|k-1})^T \underline{P}_{k|k-1}^{-1} (\underline{z}_k - \underline{z}_{k|k-1}) - e_k^2 V_k^{-1} \end{aligned} \tag{B9.4}$$

where we have introduced

$$\begin{aligned} \underline{z}_{k|k} &= E(\underline{z}_k | y_1^k), & \underline{P}_{k|k} &= \mathrm{var}(\underline{z}_k | y_1^k) \\ \underline{z}_{k|k-1} &= E(\underline{z}_k | y_1^{k-1}), & \underline{P}_{k|k-1} &= \mathrm{var}(\underline{z}_k | y_1^{k-1}) \\ e_k &= y_k - y_{k|k-1}, & V_k &= \mathrm{var}(e_k | y_1^{k-1}). \end{aligned}$$

Note from (B9.1) that taking conditional expectations gives

$$e_k = y_k - d_k x_k - \underline{c}_k^T \underline{z}_{k|k-1}. \tag{B9.5}$$

Now differentiate (B9.4) with respect to $\underline{z}_k$ to find

$$\underline{P}_{k|k}^{-1}(\underline{z}_k - \underline{z}_{k|k}) = -\underline{c}_k \sigma_k^{-2}(y_k - d_k x_k - \underline{c}_k^T \underline{z}_k) + \underline{P}_{k|k-1}^{-1}(\underline{z}_k - \underline{z}_{k|k-1}). \quad \text{(B9.6)}$$

Differentiating again gives

$$\underline{P}_{k|k}^{-1} = \underline{c}_k \underline{c}_k^T \sigma_k^{-2} + \underline{P}_{k|k-1}^{-1}. \quad \text{(B9.7a)}$$

On the other hand putting $\underline{z}_k = \underline{0}$ in (B9.6) gives

$$\underline{P}_{k|k}^{-1}\underline{z}_{k|k} = \underline{c}_k \sigma_k^{-2} y_k + \underline{P}_{k|k-1}^{-1}\underline{z}_{k|k-1}. \quad \text{(B9.7b)}$$

From (B9.7) we derive the conditional Kalman filter equations. Using (B9.7a) in (B9.7b) gives, via (B9.5),

$$\underline{P}_{k|k}^{-1}\underline{z}_{k|k} = \underline{P}_{k|k}^{-1}\underline{z}_{k|k-1} + \underline{c}_k \sigma_k^{-2} e_k$$

$$\Rightarrow \qquad \underline{z}_{k|k} = \underline{z}_{k|k-1} + \underline{P}_{k|k}\underline{c}_k \sigma_k^{-2} e_k. \quad \text{(B9.8a)}$$

Taking conditional expectations in (B9.1) gives

$$\underline{z}_{k+1|k} = \underline{F}_k \underline{z}_{k|k} + \underline{b}_k x_k. \quad \text{(B9.8b)}$$

Now return to (B9.7a) and use the matrix inversion lemma (Appendix A4) to find

$$\underline{P}_{k|k} = \underline{P}_{k|k-1} - \frac{\underline{P}_{k|k-1}\underline{c}_k \underline{c}_k^T \underline{P}_{k|k-1}}{\sigma_k^2 + \underline{c}_k^T \underline{P}_{k|k-1}\underline{c}_k}. \quad \text{(B9.9a)}$$

Subtract (B9.8b) from (B9.1) and take variances to find

$$\underline{P}_{k+1|k} = \underline{F}_k \underline{P}_{k|k} \underline{F}_k^T + \underline{Q}_k. \quad \text{(B9.9b)}$$

The Kalman filter is the triplet (B9.5), (B9.8), (B9.9).

It is worth noting that these equations can be collapsed into a smaller set, the so-called innovations form as follows. Put (B9.1b) into (B9.5) and take conditional variances to find

$$V_k = \sigma_k^2 + \underline{c}_k^T \underline{P}_{k|k-1}\underline{c}_k. \quad \text{(B9.10)}$$

Next, from (B9.9a):

$$\underline{P}_{k|k}\underline{c}_k \sigma_k^{-2} = \underline{P}_{k|k-1}\underline{c}_k V_k^{-1}. \quad \text{(B9.11)}$$

Using also (B9.8b), (B9.11) in (B9.8a) as well as (B9.9a) in (B9.9b) gives the innovations form

$$\underline{z}_{k+1|k} = \underline{F}_k \underline{z}_{k|k-1} + \underline{b}_k x_k + \underline{F}_k \underline{P}_{k|k-1}\underline{c}_k V_k^{-1} e_k \quad \text{(B9.12a)}$$

$$\underline{P}_{k+1|k} = \underline{F}_k(\underline{P}_{k|k-1} - \underline{P}_{k|k-1}\underline{c}_k \underline{c}_k^T \underline{P}_{k|k-1} V_k^{-1})\underline{F}_k^T + \underline{Q}_k \quad \text{(B9.12b)}$$

$$V_k = \sigma_k^2 + \underline{c}_k^T \underline{P}_{k|k-1}\underline{c}_k \quad \text{(B9.13a)}$$

$$e_k = y_k - \underline{c}_k^T \underline{z}_{k|k-1} - d_k x_k. \quad \text{(B9.13b)}$$

Finally to exhibit the filtering aspect fully, use (B9.13b) in (B9.12a) to obtain

$$\underline{z}_{k+1|k} = (\underline{F}_k - \underline{K}_k \underline{c}_k^T)\underline{z}_{k|k-1} + \underline{b}_k x_k + \underline{K}_k y_k \tag{B9.14a}$$

where we have introduced the Kalman gain

$$\underline{K}_k = \underline{F}_k \underline{P}_{k|k-1} \underline{c}_k V_k^{-1}. \tag{B9.14b}$$

The stability of the time varying filter (B9.14a) is discussed in [A3] as are other derivations of the Kalman filter. Stability is also discussed briefly in Appendix B10 below. Special cases of the conditional filter developed here have been given by [G3] and [M3]. The reader is invited to show that $\underline{Q}_k$, σ_k^2 in (B9.2) may be a function of y_1^{k-1} too. For other derivations of the Kalman filter see [A3].

B10 STEADY STATE KALMAN FILTER AND SPECTRAL FACTORIZATION

(i) Consider a time invariant state space model

$$\underline{z}_{k+1} = \underline{F}\underline{z}_k + \underline{b}v_k \tag{B10.1a}$$

$$y_k = \underline{c}^T \underline{z}_k + n_k \tag{B10.1b}$$

where v_k is a white noise of variance σ_v^2; n_k is an independent white noise of variance σ_n^2; $\underline{z}_{k+1}$ is a p-vector. Under certain regularity conditions (see below) the Kalman filter has a steady state form in which $\underline{P}_{k|k-1}$ converges to a limit value $\underline{P}$ (and so of course $\underline{P}_{k|k}$, V_k have limit values too). The steady state innovations form can be found from (B9.11)

$$\underline{z}_{k+1|k} = \underline{F}\underline{z}_{k|k-1} + \underline{F}\underline{P}\underline{c}V^{-1}e_k \tag{B10.2a}$$

$$y_k = \underline{c}^T \underline{z}_{k|k-1} + e_k \tag{B10.2b}$$

and from (B9.9) (where substituting (B9.9a) into (B9.9b) leads to)

$$\underline{P} = \underline{F}(\underline{P} - \underline{P}\underline{c}\underline{c}^T\underline{P}V^{-1})\underline{F}^T + \underline{b}\underline{b}^T\sigma_v^2 \tag{B10.3a}$$

$$V = \sigma_n^2 + \underline{c}^T\underline{P}\underline{c}. \tag{B10.3b}$$

If we introduce the gain vector

$$\underline{g} = \underline{F}\underline{P}\underline{c}V^{-1} \tag{B10.4}$$

and assume $\underline{F}$ is a stability matrix (see Appendix C6) then (B10.2) can be written

$$\begin{aligned} y_k &= [\underline{c}^T(qI - \underline{F})^{-1}\underline{g} + 1]e_k \\ &= [g(q^{-1})/F(q^{-1}) + 1]e_k \end{aligned}$$

where

$$q^p F(q^{-1}) = \det(qI - \underline{F}).$$

This implies the spectrum of y_k (see Appendix B7(vi)) is

$$F_y(\omega) = |1 + g(e^{-j\omega})/F(e^{-j\omega})|^2 V. \tag{B10.5}$$

On the other hand from (B10.1)

$$\begin{aligned} y_k &= \underline{c}^T(qI - F)^{-1}\underline{b}v_k + n_k \\ &= (b(q^{-1})/F(q^{-1}))v_k + n_k \end{aligned}$$

so that the spectrum is

$$F_y(\omega) = \left|\frac{b(e^{-j\omega})}{F(e^{-j\omega})}\right|^2 \sigma_v^2 + \sigma_n^2. \tag{B10.6}$$

Equating (B10.5), (B10.6) we see that the steady state Kalman filter yields a spectral factorization of the spectrum of y_k.

(ii) This can be linked back to (B10.3) as follows: Use (B10.4) to rewrite (B10.3a) as

$$\underline{P} = \underline{FPF^T} - \underline{gg}^T V + \underline{bb}^T.$$

Then

$$\begin{aligned} (qI - \underline{F})^{-1}\underline{P}(q^{-1}I - \underline{F}^T)^{-1} &= (qI - \underline{F})^{-1}\underline{FPF}^T(q^{-1}I - \underline{F}^T)^{-1} + \underline{E} \\ \underline{E} &= -(qI - \underline{F})^{-1}\underline{gg}^T(q^{-1}I - \underline{F}^T)^{-1}V \\ &\quad + (qI - \underline{F})^{-1}\underline{bb}^T(q^{-1}I - \underline{F}^T)^{-1}. \end{aligned}$$

But

$$q(qI - \underline{F})^{-1} = I + (qI - \underline{F})^{-1}\underline{F} = I + \underline{F}(qI - \underline{F})^{-1}$$

so that

$$\begin{aligned} & (qI - \underline{F})^{-1}\underline{P}(q^{-1}I - \underline{F}^T)^{-1} \\ &= (-I + q(qI - \underline{F})^{-1})\underline{P}(-I + q^{-1}(q^{-1}I - \underline{F}^T)^{-1}) + \underline{E} \\ &= \underline{P} - q^{-1}\underline{P}(q^{-1}I - \underline{F}^T)^{-1} - q(qI - \underline{F})^{-1}\underline{P} \\ &\quad + (qI - \underline{F})^{-1}\underline{P}(q^{-1}I - \underline{F}^T)^{-1} + \underline{E} \\ \Rightarrow \quad 0 &= \underline{P} - (I + (qI - \underline{F})^{-1}\underline{F})\underline{P} \\ &\quad - \underline{P}(I + \underline{F}^T(q^{-1}I - \underline{F}^T)^{-1}) + \underline{E} \\ &= \underline{P} - \underline{P} - (qI - \underline{F})^{-1}\underline{FP} - \underline{P} - \underline{PF}^T(q^{-1}I - \underline{F}^T)^{-1} + \underline{E} \\ \Rightarrow \quad 0 &= -\underline{c}^T\underline{Pc} - \underline{c}^T(qI - \underline{F})^{-1}\underline{g}V - V\underline{g}^T(q^{-1}I - \underline{F}^T)^{-1}\underline{c} + \underline{c}^T\underline{Ec}. \end{aligned}$$

Using (B10.3b) this is

$$\begin{aligned} 0 = \sigma_n^2 - V\{&1 + \underline{c}^T(qI - \underline{F})^{-1}\underline{g} + \underline{g}^T(q^{-1}I - \underline{F}^T)^{-1}\underline{c} \\ &+ \underline{c}^T(qI - \underline{F})^{-1}\underline{gg}^T(q^{-1}I - \underline{F}^T)^{-1}\underline{c}\} \\ &+ \underline{c}^T(qI - \underline{F})^{-1}\underline{bb}^T(q^{-1}I - \underline{F}^T)^{-1}\underline{c} \end{aligned}$$

which gives

$$\sigma_n^2 + \frac{b(q^{-1})}{F(q^{-1})}\frac{b(q)}{F(q)} = V\left(1 + \frac{g(q^{-1})}{F(q^{-1})}\right)\left(1 + \frac{g(q)}{F(q)}\right) \qquad \text{(B10.7)}$$

which is the spectral factorization again. The spectral factorization can be regarded as a limiting form of the Cholesky factorization for a Markov model.

(iii) Regularity conditions that ensure a steady state Kalman filter (see [A3]) are: $(\underline{F}, \underline{b})$ is controllable and $(\underline{c}^T, \underline{F})$ is observable, i.e.,

$$\mathcal{C}(\underline{F}, \underline{b}) = [\underline{b}\ \underline{F}\underline{b}\ \cdots\ \underline{F}^{p-1}\underline{b}] \text{ is full rank}$$

$$\mathcal{O}(\underline{c}, \underline{F}) = \begin{bmatrix} \underline{c}^T \\ \underline{c}^T\underline{F} \\ \ddots \\ \underline{c}^T\underline{F}^{p-1} \end{bmatrix} \text{ is full rank.}$$

B11 SPECTRAL FACTORIZATION BY WILSON'S ALGORITHM

Here we consider directly solving the spectral factorization problem by a Newton algorithm. We will give a transfer function derivation of the algorithm due to [W7]. A state space derivation can be found in [H8], [H9].

Consider (B10.7) which can be rewritten

$$\sigma_n^2 F(q)F(q^{-1}) + b(q^{-1})b(q) = d(q)d(q^{-1})V. \qquad \text{(B11.1)}$$

$$d(q^{-1}) = g(q^{-1}) + F(q^{-1}).$$

From this we can formulate the spectral factorization problem as follows. Given a moving average polynomial

$$\gamma(q) = \sum_{s=-p}^{p} \gamma_s q^{-s} \qquad \text{(B11.2a)}$$

which is positive definite, i.e.,

$$\gamma(e^{-j\omega}) \geq 0, \quad |\omega| \leq \pi \qquad \text{(B11.2b)}$$

find a one-sided polynomial

$$\phi(q^{-1}) = \sum_{s=0}^{p} \phi_s q^{-s} \qquad \text{(B11.2c)}$$

so that

$$\gamma(q) = \phi(q^{-1})\phi(q). \qquad \text{(B11.2d)}$$

Once we have found $\phi(q^{-1})$ we can write

$$\phi(q^{-1}) = d(q^{-1})V^{\frac{1}{2}} \tag{B11.3a}$$

$$d(q^{-1}) = 1 + \sum_{s=1}^{p} d_s q^{-s} \tag{B11.3b}$$

$$V = \phi_0^2, \quad d_s = \phi_s/\phi_0, \, 1 \le s \le p \tag{B11.3c}$$

to agree with (B11.1).

The derivation of the Newton algorithm proceeds as follows. Given the kth iterate $\phi_k(q^{-1})$, we make a correction $\eta_k(q^{-1})$ to it, giving $\phi_{k+1}(q^{-1}) = \phi_k(q^{-1}) + \eta_k(q^{-1})$ so as to try to ensure that

$$\phi_{k+1}(q^{-1})\phi_{k+1}(q) = \gamma(q).$$

Expanding the left hand side and dropping the second order term $\eta_k(q)\eta_k(q^{-1})$ suggests defining $\eta_k(q)$ to satisfy

$$\phi_k(q)\phi_k(q^{-1}) + \eta_k(q)\phi_k(q^{-1}) + \eta_k(q^{-1})\phi_k(q) = \gamma(q) \tag{B11.4a}$$

$$\Rightarrow \qquad \frac{\hat{\phi}_k(q)}{\phi_k(q)} + \frac{\hat{\phi}_k(q^{-1})}{\phi_k(q^{-1})} = \frac{\gamma(q)}{\phi_k(q)\phi_k(q^{-1})} \tag{B11.4b}$$

where we have put

$$\hat{\phi}_k(q) = \frac{1}{2}\phi_k(q) + \eta_k(q)$$

so that

$$\phi_{k+1}(q) = \frac{1}{2}\phi_k(q) + \hat{\phi}_k(q). \tag{B11.5}$$

Equating coefficients in (B11.4) gives (see also [H11])

$$\underline{\hat{\phi}}_k = \underline{T}^{-1}(\underline{\phi}_k)\underline{\gamma} \tag{B11.6}$$

$$\underline{\hat{\phi}}_k = (\hat{\phi}_{0k} \cdots \hat{\phi}_{pk})^T$$

$$\underline{\gamma} = (\gamma_0 \cdots \gamma_p)^T$$

$$\underline{T}(\underline{\phi}) = \begin{bmatrix} \phi_0 & \phi_1 & \cdots & \phi_{p-1} & \phi_p \\ \phi_1 & \phi_2 & \cdots & \phi_p & 0 \\ \vdots & \vdots & \ddots & \vdots & \vdots \\ \phi_{p-1} & \phi_p & \cdots & 0 & 0 \\ \phi_p & 0 & \cdots & 0 & 0 \end{bmatrix} + \begin{bmatrix} \phi_0 & \phi_1 & \cdots & \phi_{p-1} & \phi_p \\ 0 & \phi_0 & \cdots & \phi_{p-2} & \phi_{p-1} \\ \vdots & \vdots & \ddots & \vdots & \vdots \\ 0 & 0 & \cdots & \phi_0 & \phi_1 \\ 0 & 0 & \cdots & 0 & \phi_0 \end{bmatrix}. \tag{B11.7}$$

Then (B11.5) gives

$$\underline{\phi}_{k+1} = \frac{1}{2}\underline{\phi}_k + \underline{\hat{\phi}}_k. \tag{B11.8}$$

Now we look at the convergence of this iterative algorithm. We find that it has the remarkable property that if initiated with a stable iteration (e.g., $\phi_0(q) = 1$ will do), then all subsequent iterations are stable and the algorithm converges to $\phi(q)$, solving (B11.2a).

From (B11.4a), on $|q| = 1$

$$\begin{aligned} |\phi_{k+1}(q)|^2 &= |\phi_k(q) + \eta_k(q)|^2 \\ &= |\phi_k(q)|^2 + \phi_k(q)\eta_k(q^{-1}) + \phi_k(q^{-1})\eta_k(q) + |\eta_k(q)|^2 \\ \Rightarrow \quad |\phi_{k+1}(q)|^2 &= \gamma(q) + |\phi_{k+1}(q) - \phi_k(q)|^2 \geq |\phi_{k+1}(q) - \phi_k(q)|^2. \end{aligned} \tag{B11.9}$$

Thus by Rouche's theorem [R4], $\phi_k(q)$, $\phi_{k+1}(q)$ have the same number of zeroes inside $|q| = 1$. Since $\phi_0(q) = 1$ has no zeroes inside $|q| = 1$, all iterations are stable (note here that $\phi_k(q)$ having all roots outside the unit circle $|q| = 1$ is the same as $\phi_k(q^{-1})$ having all roots inside the unit circle). For a proof of convergence of the algorithm the reader is directed to [W7].

B12 EXTENDED KALMAN FILTER

Consider state estimation for the nonlinear state space model

$$\underline{z}_{k+1} = \underline{f}(k, \underline{z}_k) + \underline{v}_k \tag{B12.1a}$$

$$y_k = h(k, \underline{z}_k) + n_k \tag{B12.1b}$$

where $(\underline{v}_k^T, n_k)$ is a zero mean white noise with covariance

$$\begin{bmatrix} \underline{Q}_k & \underline{0} \\ \underline{0}^T & \sigma_k^2 \end{bmatrix}.$$

The idea behind the Extended Kalman filter (EKF) state estimator is to produce an approximate linear filter via linearization of (B12.1). At stage k we let

$$\underline{z}_{k|k-1} = \text{estimate of } \underline{z}_k \text{ based on data } y_1^{k-1} = \{y_{k-1}, y_{k-2}, \ldots, y_1\}$$

$$\underline{P}_{k|k-1} = \text{approximate covariance of } \underline{z}_k - \underline{z}_{k|k-1}.$$

Now expand (B12.1) in a Taylor series about $\underline{z}_{k|k-1}$ to obtain

$$\underline{z}_{k+1} \approx \underline{f}(k, \underline{z}_{k|k-1}) + \underline{F}_k(\underline{z}_k - \underline{z}_{k|k-1}) + \underline{v}_k \tag{B12.2a}$$

$$y_k = h(k, \underline{z}_{k|k-1}) + \underline{H}_k^T(\underline{z}_k - \underline{z}_{k|k-1}) + n_k \tag{B12.2b}$$

where

$$\underline{F}_k = \left. \frac{\partial \underline{f}(k, \underline{z})}{\partial \underline{z}^T} \right|_{\underline{z} = \underline{z}_{k|k-1}} \tag{B12.3a}$$

$$\underline{H}_k = \left. \frac{\partial h(k, \underline{z})}{\partial \underline{z}} \right|_{\underline{z} = \underline{z}_{k|k-1}}. \tag{B12.3b}$$

Now rewrite (B12.2) to have the form of (B9.1), (B9.2)

$$z_{k+1} = \underline{F}_k z_k + \left[\underline{f}(k, z_{k|k-1}) - \underline{F}_k z_{k|k-1}\right] + \underline{v}_k$$

$$y_k = \underline{H}_k^T z_k + \left[h(k, z_{k|k-1}) - \underline{H}_k^T z_{k|k-1}\right] + n_k.$$

It is now appropriate to apply the Kalman filter (B9.12), (B9.13) to this pair: e.g., $\underline{F}_k$ and $\underline{H}_k$ are functions of y_1^{k-1}. We thus obtain

$$\begin{aligned} z_{k+1|k} &= \underline{F}_k z_{k|k-1} + \left[\underline{f}(k, z_{k|k-1}) - \underline{F}_k z_{k|k-1}\right] + \underline{K}_k e_k \\ &= \underline{f}(k, z_{k|k-1}) + \underline{K}_k e_k \end{aligned} \qquad \text{(B12.4a)}$$

$$\underline{P}_{k+1|k} = \underline{F}_k \underline{P}_{k|k-1} \underline{F}_k^T - \underline{K}_k V_k \underline{K}_k^T + \underline{Q}_k \qquad \text{(B12.4b)}$$

$$\underline{K}_k = \underline{F}_k \underline{P}_{k|k-1} \underline{H}_k V_k^{-1} \qquad \text{(B12.4c)}$$

$$\begin{aligned} e_k &= y_k - \underline{H}_k^T z_{k|k-1} - \left[h(k, z_{k|k-1}) - \underline{H}_k^T z_{k|k-1}\right] \\ &= y_k - h(k, z_{k|k-1}) \end{aligned} \qquad \text{(B12.5a)}$$

$$V_k = \sigma_k^2 + \underline{H}_k^T \underline{P}_{k|k-1} \underline{H}_k. \qquad \text{(B12.5b)}$$

The EKF is the set (B12.4), (B12.5). Further discussion of the EKF is available in [J1] and [M2].

C

Deterministic Signals and Systems Review

C1 SUM AND PRODUCT LEMMA

Suppose $h_t \geq 0$, $t = 1, 2, \ldots$ and define

$$\prod_t = \prod_{u=1}^{t}(1 + h_u) = (1 + h_t)(1 + h_{t-1}) \cdots (1 + h_1)$$
$$\prod_0 = 1 + h_0 = 1.$$

Then

$$\prod_t = 1 + \prod_t \sum_{s=1}^{t} h_s / \prod_s; \qquad \prod_t = 1 + \sum_{s=1}^{t} h_s \prod_{s-1}.$$

Proof.

$$\prod_t = (1 + h_t)\prod_{t-1} = \prod_{t-1} + h_t \prod_{t-1}$$
$$\Rightarrow \qquad \prod_{t-1}^{-1} = \prod_t^{-1} + h_t / \prod_t.$$

Sum the first line from $t = 1$ to T to get the second result. Now sum the second line from $t = 1$ to T to get

$$1 = \prod_T^{-1} + \sum_{t=1}^{T} h_t / \prod_t$$

$$\Rightarrow \qquad \prod_T = 1 + \prod_T \sum_{t=1}^{T} h_t / \prod_t.$$

C2 DISCRETE BELLMAN GRONWALL LEMMA

Suppose

$$h_t \geq 0, \xi_t \geq 0, \text{ for } t = 0, 1, \ldots$$

$$\xi_t \leq g + \sum_{s=1}^{t} h_{s-1} \xi_{s-1}, \quad t \geq 1$$

$$\xi_0 \leq g, \quad h_0 = 0.$$

Then

$$\xi_t \leq g \prod_{t-1}.$$

Proof. Use induction. It is true for $t = 0$. Then

$$\begin{aligned} \xi_t &\leq g + \sum_{s=1}^{t} h_{s-1} g \prod_{s-2} \\ &= g + g \sum_{u=1}^{t-1} h_u \prod_{u-1}. \end{aligned}$$

Now use the sum and product lemma to find

$$\begin{aligned} \xi_t &\leq g + g(\prod_{t-1} - 1) \\ &= g \prod_{t-1} \end{aligned}$$

so the inductive step is complete.

C3 EXTENDED DISCRETE BELLMAN GRONWALL LEMMA

Suppose

$$g_t \geq 0, h_t \geq 0, \xi_t \geq 0, \text{ for } t = 0, 1, \ldots$$

$$\xi_t \leq g_t + \sum_{s=1}^{t} h_{s-1} \xi_{s-1}, \quad t \geq 1$$

$$\xi_0 \leq g_0, \quad h_0 = 0.$$

Then

$$\xi_t \le g_t + \prod_{t-1} \sum_{s=1}^{t-1} g_s h_s / \prod_{s-1}$$

or alternatively

$$\xi_t \le g_0 \prod_{t-1} + \prod_{t-1} \sum_{s=1}^{t} (g_s - g_{s-1}) \prod_{s-1}^{-1}.$$

Proof. Let $s_t = \sum_{u=1}^{t} h_u \xi_u$ so that

$$\begin{aligned}
& & \xi_t &\le g_t + s_{t-1} \\
&\Rightarrow & h_t \xi_t &\le g_t h_t + h_t s_{t-1} \\
&\Rightarrow & s_t - s_{t-1} &\le g_t h_t + h_t s_{t-1} \\
&\Rightarrow & s_t &\le (1 + h_t) s_{t-1} + g_t h_t \\
&\Rightarrow & s_t / \prod_t &\le s_{t-1} / \prod_{t-1} + g_t h_t / \prod_t .
\end{aligned}$$

Sum from $t = 1$ to n to find

$$\frac{s_n}{\prod_n} \le \sum_{s=1}^{n} \frac{g_s h_s}{\prod_s} \quad \Rightarrow \quad s_n \le \prod_n \sum_{s=1}^{n} \frac{g_s h_s}{\prod_s}$$

and this gives the first result. Continuing,

$$\begin{aligned}
& \prod_{t-1}^{-1} - \prod_t^{-1} = h_t / \prod_t \\
\Rightarrow \quad & s_t / \prod_t - s_{t-1} / \prod_{t-1} \le g_t \left(\prod_{t-1}^{-1} - \prod_t^{-1} \right) \\
& \qquad = g_{t-1} \prod_{t-1}^{-1} - g_t \prod_t^{-1} + (g_t - g_{t-1}) \prod_{t-1}^{-1}.
\end{aligned}$$

Sum from $t = 1$ to n (recall $h_0 = 0$)

$$\Rightarrow \quad s_n / \prod_n - h_0 \xi_o / \prod_0 \le g_0 / \prod_0 - g_n / \prod_n + \sum_{t=1}^{n} (g_t - g_{t-1}) \prod_{t-1}^{-1}$$

$$\Rightarrow \quad s_n \le g_0 \prod_n - g_n + \prod_n \sum_{t=1}^{n} (g_t - g_{t-1}) \prod_{t-1}^{-1}.$$

Finally

$$\begin{aligned}
\xi_t &\le g_t + s_{t-1} \\
&\le g_t + g_0 \prod_{t-1} - g_{t-1} + \prod_{t-1} \sum_{s=1}^{t-1} (g_s - g_{s-1}) \prod_{s-1}^{-1} \\
&= g_0 \prod_{t-1} + \prod_{t-1} \sum_{s=1}^{t} (g_s - g_{s-1}) \prod_{s-1}^{-1}.
\end{aligned}$$

Corollary C3.1. If $g_t = g_0 + \alpha t$; $h_t = h$ then

$$\xi_t \le (g_0 + \alpha + \alpha h^{-1})\prod_{t-1} = (g_0 + \alpha + \alpha h^{-1})(1+h)^{t-1}.$$

Proof.

$$\xi_t \le g_0 \prod_{t-1} + \alpha \prod_{t-1} \sum_{s=1}^{t} \prod_{s-1}^{-1}.$$

Use the sum and product lemma

$$\begin{aligned}
\xi_t &\le g_0 \prod_{t-1} + \alpha h^{-1}\Big(\prod_{t-1} \sum_{s=1}^{t-1} h \prod_s^{-1} + \prod_{t-1} h\Big) \\
&= g_0 \prod_{t-1} + \alpha h^{-1}\Big(\prod_{t-1} - 1 + \prod_{t-1} h\Big) \\
&\le g_0 \prod_{t-1} + \alpha h^{-1} \prod_{t-1} + \alpha \prod_{t-1} \\
&= (g_0 + \alpha h^{-1} + \alpha) \prod_{t-1}.
\end{aligned}$$

C4 STABILITY OF LINEAR TIME INVARIANT SYSTEMS

(i) A linear time invariant system with input x_t and output s_t is characterized by its impulse response $\{h_t\}$ as follows:

$$s_t = \sum_{r=-\infty}^{\infty} h_r x_{t-r}. \tag{C4.1}$$

If $x_t = \delta_{t-t_0}$ (Kronecker delta), an impulse at time t_0, then (C4.1) gives $s_t = h_{t-t_0}$.

(ii) Such a system is called bounded input, bounded output stable if

$$|x_t| \le b \Rightarrow |s_t| \le B < \infty.$$

Theorem C4.1. A linear time invariant system is bounded input, bounded output stable if and only if

$$\sum_{t=-\infty}^{\infty} |h_t| < \infty. \tag{C4.2}$$

Proof. If (C4.2) holds, then from (C4.1)

$$|s_t| \le b \sum_{s=-\infty}^{\infty} |h_s| < \infty.$$

If stability holds, let

$$x_t = \operatorname{sgn}(h_{-t})$$

so $|x_t| \le 1$. Then we must have

$$\infty > |s_0| = |\sum_{s=-\infty}^{\infty} h_s x_{-s}|$$
$$= |\sum_{s=-\infty}^{\infty} h_s \operatorname{sgn}(h_s)| = \sum_{s=-\infty}^{\infty} |h_s|.$$

(iii) If we have a causal all pole system

$$s_t = [1 + a(q^{-1})]^{-1} x_t$$
$$a(q^{-1}) = \sum_{s=1}^{p} a_s q^{-s}$$

then comparing with (C4.1) we have

$$[1 + a(q^{-1})]^{-1} = \sum_{s=0}^{\infty} h_s q^{-s}.$$

To investigate stability, suppose $q^p(1 + a(q^{-1}))$ has roots $\alpha_1 \ldots \alpha_p$, then (suppose they are distinct for simplicity) a partial fraction expansion tells us that for some constants $A_i, i = 1 \ldots p$

$$[1 + a(q^{-1})]^{-1} = [\textstyle\prod_{i=1}^{p}(1 - \alpha_i q^{-1})]^{-1} = \sum_{i=1}^{p} A_i (1 - \alpha_i q^{-1})^{-1}$$
$$= \sum_{s=0}^{\infty}\sum_{i=1}^{p} A_i \alpha_i^s q^{-s}$$
$$\Rightarrow \qquad h_s = \sum_{i=1}^{p} A_i \alpha_i^s.$$

From this it is clear that (C4.2) can hold if and only if

$$|\alpha_i| < 1, \quad i = 1 \cdots p \tag{C4.3}$$

i.e., all roots or poles have modulus < 1. If (C4.3) holds we call $1 + a(q^{-1})$ a stability polynomial or loosely a stable polynomial.

(iv) Minimum Phase. We say $h(q^{-1}) = \sum_{s=0}^{\infty} h_s q^{-s}$ is minimum phase if $1/h(q^{-1})$ is a stability polynomial.

(v) Stability by the Schur-Cohn Test
Given a polynomial

$$A_p(q^{-1}) = 1 + \sum_{s=1}^{p} a_s q^{-s}$$

the Schur-Cohn test consists of running the following backwards recursion:

Initialization:

$$a_i^{(p)} = a_i, \quad 1 \le i \le p; \qquad \sigma_p^2 = 1.$$

For: $k = p, p-1, \ldots, 1$

$$a_{k-1,j} = (a_{k,j} - a_{kk}a_{k,k-j})/(1 - a_{kk}^2), \quad 1 \le j \le k-1 \tag{C4.4}$$

$$\sigma_{k-1}^2 = \sigma_k^2/(1 - a_{k-1,k-1}^2).$$

Stability follows if $a_{kk}^2 < 1$, $1 \le j \le p$. If any $a_{kk}^2 > 1$ the algorithm then terminates.

In [V4] and [K1, p. 182] it is shown that the algorithm is a backward version of the Levinson algorithm for finding the partial autocorrelations or reflection coefficients (a_{kk}) of an AR(p) time series (whose parameters are $a_1 \cdots a_p$). The Levinson algorithm is itself a specialization of the Cholesky factorization algorithm when the matrix being factorized is the inverse of a Toeplitz matrix. From the above iteration we see that given a stable polynomial we can parameterize it in terms of the reflection coefficients a_{kk}, $1 \le k \le p$. Conversely, given a set of reflection coefficients a_{kk}, $1 \le k \le p$, $|a_{kk}| < 1$, we can reparameterize them by an AR polynomial from the following forward recursion:

For: $k = 2, 3, \ldots, p$

$$a_{k,j} = (a_{k-1,j} + a_{kk}a_{k-1,k-j}), \quad 1 \le j \le k-1. \tag{C4.5}$$

Then set

$$a_j = a_{p,j}, \quad 1 \le j \le p.$$

The reader is invited to show the equivalence of (C4.4) and (C4.5).

C5 INPUT/OUTPUT ENERGY BOUNDS

Energy Bound lemma: If $H(q^{-1})$ is a one-sided filter

$$H(q^{-1}) = \sum_{s=0}^{\infty} h_s q^{-s}$$

with bounded peak amplitude, i.e.,

$$\sup_{|\omega| \le \pi} \left| H(e^{-j\omega}) \right| \le C < \infty$$

and

$$y_k = H(q^{-1})x_k, \quad k \ge 1 \tag{C5.1}$$

then

$$\sum_{k=1}^{N} y_k^2 \le C^2 \sum_{k=1}^{N} x_k^2$$

and letting $N \to \infty$

$$\sum_{k=1}^{\infty} y_k^2 \le C^2 \sum_{k=1}^{\infty} x_k^2.$$

Proof. Introduce the "shut-down" or windowed input

$$x_{kN} = \begin{cases} x_k & 1 \le k \le N \\ 0 & \text{otherwise} \end{cases}$$

and corresponding output

$$y_{kN} = H(q^{-1})x_{kN}, \quad k \ge 1$$

so that

$$y_{kN} = y_k, \quad 1 \le k \le N.$$

Then

$$\sum_{k=1}^{N} y_k^2 = \sum_{k=1}^{N} y_{kN}^2 \le \sum_{k=1}^{\infty} y_{kN}^2$$
$$= \int_{-\pi}^{\pi} |y_N(e^{-j\omega})|^2 d\omega/2\pi$$

by Parseval's theorem [A6, Section 11.4] where

$$y_N(e^{-j\omega}) = \sum_{k=1}^{\infty} y_{kN} e^{-j\omega k}$$

is the Fourier series of $\{y_{kN}\}$. From (C5.1)

$$y_N(e^{-j\omega}) = H(e^{-j\omega})x_N(e^{-j\omega})$$
$$x_N(e^{-j\omega}) = \sum_{k=1}^{N} x_k e^{-j\omega k}$$

so that

$$\sum_{k=1}^{N} y_k^2 \le \int_{-\pi}^{\pi} \left|H(e^{-j\omega})\right|^2 \left|x_N(e^{-j\omega})\right|^2 d\omega/2\pi$$
$$\le C^2 \int_{-\pi}^{\pi} \left|x_N(e^{-j\omega})\right|^2 d\omega/2\pi$$
$$= C^2 \sum_{k=1}^{N} x_k^2$$

which is the required result. Of course if $\sum_{-\infty}^{\infty} |h_t| < \infty$ then $C \le \sum_{-\infty}^{\infty} |h_t|$.

C6 STABILITY OF LINEAR STATE SPACE SYSTEMS

(i) Exponential Stability
The linear state space system

$$\underline{x}_k = \underline{A}\underline{x}_{k-1}, \quad k \geq 1 \tag{C6.1}$$

is exponentially stable (ES) to 0 if there are constants $m > 0$, $0 < \theta < 1$ so that

$$\|\underline{x}_k\| \leq m\theta^{k-k_0}\|\underline{x}_{k_0}\|, \quad k \geq k_0. \tag{C6.2}$$

We also then call $\underline{A}$ a stability matrix.

(ii) Lyapunov Function
Stability also can be investigated by means of a Lyapunov function. This is typically system energy, which in the linear setting means a quadratic form in the state such as

$$V(\underline{x}) = \underline{x}^T\underline{P}\underline{x}, \ \underline{P} \text{ is positive definite} \tag{C6.3}$$

$$\Rightarrow \qquad 0 < \lambda_{min}(\underline{P})\|\underline{x}\|^2 \leq V(\underline{x}) \leq \lambda_{max}(\underline{P})\|\underline{x}\|^2. \tag{C6.4}$$

To prove stability the idea is to show $V_k = V(\underline{x}_k) \to 0$ as $k \to \infty$ and then use (C6.4) to conclude $\|\underline{x}_k\| \to 0$.

Now let $\underline{Q}$ be positive definite and suppose the following Lyapunov equation can be solved for the Lyapunov matrix $\underline{P}$

$$\underline{A}^T\underline{P}\underline{A} - \underline{P} = -\underline{Q}. \tag{C6.5}$$

Then we find $V(\underline{x})$ obeys the following property:

$$\begin{aligned} V(\underline{A}\underline{x}) &= \underline{x}^T\underline{A}^T\underline{P}\underline{A}\underline{x} \\ &= \underline{x}^T\underline{P}\underline{x} - \underline{x}^T\underline{Q}\underline{x} \\ &\leq V(\underline{x})(1-\rho) \end{aligned} \tag{C6.6}$$

where

$$\rho = \lambda_{min}(\underline{Q})/\lambda_{max}(\underline{P}).$$

The two properties (C6.4) and (C6.6) now yield ES. We have, from (C6.6)

$$\begin{aligned} V(\underline{x}_k) = V_k &\leq (1-\rho)^{k-k_0}V_{k_0} \\ \Rightarrow \qquad \|\underline{x}_k\| &\leq (1-\rho)^{k-k_0}m\|\underline{x}_{k_0}\|, \text{ by (C6.4)} \\ m &= \lambda_{max}(\underline{P})/\lambda_{min}(\underline{P}). \end{aligned}$$

So existence of a Lyapunov function (that is, a function satisfying (C6.4) and (C6.6)) implies ES. Remarkably the converse is also true.

(iii) Lyapunov Function and Exponential Stability

Suppose (C6.1) is ES, then given any positive definite $\underline{Q}$ let

$$\underline{P} = \sum_{s=0}^{\infty} \underline{A}^{Ts} \underline{Q} \underline{A}^{s}, \quad (\underline{A}^{Ts} = (A^T)^s).$$

First we check that the series converges. If $\underline{\alpha}$ is any fixed vector then

$$\begin{aligned} \underline{\alpha}^T \underline{P} \underline{\alpha} &= \sum_{s=0}^{\infty} \underline{\alpha}^T \underline{A}^{Ts} \underline{Q} \underline{A}^s \underline{\alpha} \\ &= \sum_{s=0}^{\infty} \underline{y}_s^T \underline{Q} \underline{y}_s \end{aligned}$$

where $y_s = \underline{A}^s \underline{\alpha}$ is the solution to (C6.1) with $\underline{x}_0 = \underline{\alpha}$. By the ES assumption

$$\|\underline{y}_s\| \leq \|\underline{\alpha}\| m \theta^s, \quad 0 < \theta < 1$$

so that

$$\underline{\alpha}^T \underline{P} \underline{\alpha} \leq m \lambda_{max}(\underline{Q}) \|\underline{\alpha}\|^2 (1 - \theta^2)^{-1}.$$

So indeed the series converges. We note that

$$\lambda_{max}(\underline{P}) \leq \lambda_{max}(\underline{Q}) m (1 - \theta^2)^{-1}.$$

On the other hand $\underline{P}$ is positive definite. This follows since

$$\underline{\alpha}^T \underline{P} \underline{\alpha} = \sum_{s=0}^{\infty} \underline{y}_s^T \underline{Q} \underline{y}_s \geq \underline{y}_0^T \underline{Q} \underline{y}_0 = \underline{\alpha}^T \underline{Q} \underline{\alpha} > 0, \quad \underline{\alpha} \neq \underline{0}.$$

The last inequality holds since $\underline{Q}$ is positive definite. We check that (C6.5) holds. We have

$$\underline{A}^T \underline{P} \underline{A} = \sum_{u=1}^{\infty} \underline{A}^{Tu} \underline{Q} \underline{A}^u = \underline{P} - \underline{Q}$$

as required. Thus, as already demonstrated, $V(\underline{x}) = \underline{x}^T \underline{P} \underline{x}$ obeys (C6.4) and (C6.6).

(iv) Exponential Stability and the Eigenvalues of $\underline{A}$

From (C6.1)

$$\underline{x}_k = \underline{A}^{k-k_0} \underline{x}_{k_0}, \quad k \geq k_0$$

and we see (why?) that ES follows if

$$\text{all eigenvalues of } \underline{A} \text{ have modulus } < 1. \tag{C6.7}$$

Equivalently we see that ES follows if

$$\text{all roots of } \det(qI - \underline{A}) \text{ have modulus } < 1. \tag{C6.8}$$

The converse result holds and is easily deduced from (C6.5). If $\underline{\xi}_u$ is a unit eigenvector of $\underline{A}$ with eigenvalue λ_u then

$$|\lambda_u|^2 \underline{\xi}_u^H \underline{P} \underline{\xi}_u - \underline{\xi}_u^H \underline{P} \underline{\xi}_u = -\underline{\xi}_u^H \underline{Q} \underline{\xi}_u$$

where the superscript H denotes complex conjugate transpose. Thus

$$\underline{\xi}_u^H \underline{Q} \underline{\xi}_u = (1 - |\lambda_u|^2) \underline{\xi}_u^H \underline{P} \underline{\xi}_u$$

and since both $\underline{Q}$, $\underline{P}$ are positive definite we must have

$$|\lambda_u|^2 < 1, \quad \text{for all } u.$$

This completes the proof that

$$\text{ES} \Rightarrow \text{(C6.7) or (C6.8).}$$

(v) Stability of Slowly Time Varying Linear Systems
Consider the time varying linear system

$$\underline{z}_k = \underline{A}_k \underline{z}_{k-1}. \tag{C6.9}$$

Suppose that

(C6A1) $\|\underline{A}_k\| \leq A < \infty$.

(C6A2) All eigenvalues of $\underline{A}_k$ have modulus $< \lambda < 1$.

One might hope that this would give a stability result for (C6.9) but more is needed. Suppose:

(C6A3) $\underline{A}_k$ is slowly time varying, i.e., $\|\underline{A}_k - \underline{A}_{k-1}\| \leq \epsilon$.

Then if ϵ is small enough, (C6.9) is ES to $\underline{0}$.

Proof. We reproduce part of the proof of [D2]. In [D2] it is shown that (C6A1), (C6A2) ensure

$$\|\underline{A}_k^r\| \leq m\lambda^r, \quad \text{for some } m > 0. \tag{C6.10}$$

Now introduce the Lyapunov matrix

$$\underline{P}_k = \sum_{r=0}^{\infty} \underline{A}_k^{Tr} \underline{A}_k^r \tag{C6.11}$$

so that

$$\underline{A}_k^T \underline{P}_k \underline{A}_k - \underline{P}_k = -I. \tag{C6.12}$$

Note that from (C6.12)

$$\underline{P}_k \geq I \tag{C6.13}$$

while from (C6.10) and (C6.11)

$$\|\underline{P}_k\| \leq m^2(1 - \lambda^2)^{-1} = P, \text{ say.} \tag{C6.14}$$

Then consider the following candidate Lyapunov function:

$$V(k, \underline{z}) = \underline{z}^T \underline{P}_k \underline{z}. \tag{C6.15}$$

Then from (C6.13) and (C6.14)

$$\|\underline{z}\|^2 \le V(k, \underline{z}) \le P\|\underline{z}\|^2 \tag{C6.16a}$$

$$\|V_z(k, \underline{z})\| \le 2P\|\underline{z}\|. \tag{C6.16b}$$

Next consider that

$$\begin{aligned} V(k, \underline{A}_k\underline{z}) &= \underline{z}^T \underline{A}_k^T \underline{P}_k \underline{A}_k \underline{z} \\ &= \underline{z}^T \underline{P}_k \underline{z} - \|\underline{z}\|^2 \\ &= V(k-1, \underline{z}) + \underline{z}^T(\underline{P}_k - \underline{P}_{k-1})\underline{z} - \|\underline{z}\|^2. \end{aligned} \tag{C6.17}$$

From (C6.12) we find

$$\begin{aligned} &\underline{A}_k^T \underline{P}_k \underline{A}_k - \underline{P}_k - (\underline{A}_{k-1}^T \underline{P}_{k-1} \underline{A}_{k-1} - \underline{P}_{k-1}) = \underline{0} \\ \Rightarrow \quad &\underline{A}_k^T(\underline{P}_k - \underline{P}_{k-1})\underline{A}_k - (\underline{P}_k - \underline{P}_{k-1}) = \underline{M}_k \end{aligned} \tag{C6.18}$$

where

$$\begin{aligned} \underline{M}_k &= \underline{A}_{k-1}^T \underline{P}_{k-1} \underline{A}_{k-1} - \underline{A}_k^T \underline{P}_{k-1} \underline{A}_k \\ &= -(\underline{A}_k - \underline{A}_{k-1})^T \underline{P}_{k-1} \underline{A}_k - \underline{A}_{k-1}^T \underline{P}_{k-1}(\underline{A}_k - \underline{A}_{k-1}). \end{aligned}$$

Thus from (C6.14), (C6A1), and (C6A3)

$$\|\underline{M}_k\| \le 2\epsilon PA. \tag{C6.19}$$

Next, from (C6.18)

$$\underline{P}_k - \underline{P}_{k-1} = -\sum_{r=0}^{\infty} \underline{A}_k^{Tr} \underline{M}_k \underline{A}_k^r$$

so that from (C6.10) and (C6.19)

$$\|\underline{P}_k - \underline{P}_{k-1}\| \le 2AP^2\epsilon. \tag{C6.20}$$

Returning to (C6.17) we find

$$V(k, \underline{A}_k\underline{z}) \le V(k-1, \underline{z}) - (1 - 2AP^2\epsilon)\|\underline{z}\|^2. \tag{C6.21}$$

Let us require ϵ to be small enough that

$$2AP^2\epsilon < 1. \tag{C6.22}$$

Then we have a Lyapunov function obeying the basic conditions (C6.16a), (C6.16b), (C6.21), and (C6.22).

To finish off the proof we note that, from (C6.16a)

$$V(k, \underline{A}_k\underline{z}) \le V(k-1, \underline{z})\left(1 - (1 - 2AP^2\epsilon)P^{-1}\right).$$

Thus setting $\underline{z} = \underline{z}_{k-1}$ we get

$$V(k, \underline{z}_k) \leq \lambda V(k-1, \underline{z}_{k-1})$$
$$0 < \lambda = 1 - (1 - 2AP^2\epsilon)/P < 1.$$

Thus

$$V(k, \underline{z}_k) \leq \lambda^{k-k_0} V(k_0, \underline{z}_{k_0})$$

and via (C6.16a) ES follows.

C7 STABILITY OF NONLINEAR STATE SPACE SYSTEMS

C7.1 Definitions

Here we consider nonlinear systems of the form (with $\underline{z}_k$ a p-vector)

$$\delta \underline{z}_k = \mu \underline{f}(k, \underline{z}_{k-1}), \quad k \geq 0. \tag{C7.1}$$

If $\underline{f}(\bullet, \bullet)$ does not depend on k, the system is said to be time invariant or autonomous.

Definition C7.1. Equilibrium Point. We say $\underline{z} = \underline{z}_e$ is an equilibrium point of (C7.1) if

$$\underline{f}(k, \underline{z}_e) = \underline{0} \quad \text{for all } k.$$

Definition C7.2. An equilibrium point $\underline{z}_e$ of (C7.1) is said to be stable if for every $\epsilon \geq 0$, $k_0 \geq 0$, there exists a $\delta(\epsilon, k_0) > 0$ such that

$$\|\underline{z}_0 - \underline{z}_e\| \leq \delta(\epsilon, k_0) \text{ implies } \|\underline{z}_k - \underline{z}_e\| < \epsilon \text{ for all } k \geq k_0.$$

This definition says roughly that if we start close to $\underline{z}_e$ we stay close to it if the system is stable.

Definition C7.3. An equilibrium point $\underline{z}_e$ of (C7.1) is said to be attractive if for some $\rho > 0$ and every η, $k_0 > 0$ there exists $T = T(\eta, k_0)$ such that

$$\|\underline{z}_0 - \underline{z}_e\| < \rho \text{ implies } \|\underline{z}_k - \underline{z}_e\| < \eta \text{ for all } k \geq k_0 + T.$$

Attractivity implies that all trajectories starting near $\underline{z}_e$ eventually approach $\underline{z}_e$. Stability and attractivity are independent properties (see [H1]).

Definition C7.4. An equilibrium point $\underline{z}_e$ of (C7.1) is said to be asymptotically stable if it is stable and attractive.

Definition C7.5. An equilibrium point $\underline{z}_e$ of (C7.1) is uniformly stable if $\delta(\epsilon, k_0)$ in Definition C7.2 does not depend on k_0.

Definition C7.6. An equilibrium point $\underline{z}_e$ of (C7.1) is uniformly asymptotically stable if it is uniformly stable and for some $\rho > 0$ and every $\eta > 0$ there exists $T = T(\eta)$ such that

$$\|\underline{z}_0 - \underline{z}_e\| < \rho \text{ implies } \|\underline{z}_k - \underline{z}_e\| < \eta \text{ for all } k \geq k_0 + T.$$

Definition C7.7. Exponential Stability. We say $\underline{z} = \underline{z}_e$ is an exponentially stable (ES) equilibrium point of (C7.1) if there are positive constants h, m, α, with $0 < \alpha < 1$ such that

$$\|\underline{z}_k - \underline{z}_e\| \leq m\alpha^{k-k_0}\|\underline{z}_{k_0} - \underline{z}_e\|, \quad k \geq k_0, \quad \|\underline{z}_{k_0} - \underline{z}_e\| \leq h. \tag{C7.2}$$

We call h the radius of convergence and we say $\underline{z}_{k_0}$ lies in the region of attraction of $\underline{z}_e$ or the stability region for $\underline{z}_e$. If Definition C7.7 holds for any finite h, then we say $\underline{z} = \underline{z}_e$ is a global ES equilibrium point. Often we loosely describe (C7.1) as being ES or exponentially convergent to $\underline{z}_e$. Without loss of generality we can always take $\underline{z}_e = \underline{0}$. Our concern throughout the book is mostly with ES. Note that $1-\alpha$ measures the rate at which $\|\underline{z}_k - \underline{z}_e\|$ decays to 0, so we call it the decay rate. It will usually be the case that decay rate $= 1 - \alpha = O(\mu)$.

Proving convergence or stability of a nonlinear difference equation such as (C7.1) is in general quite hard. It is often achieved by means of a Lyapunov function, which is typically a system energy that decreases with time.

C7.2 Lyapunov Functions and Exponential Stability

Exponential stability and Lyapunov functions are closely linked as seen in:

Theorem C7.1. Consider the system (C7.1) and suppose $\underline{f}(k, \underline{z})$ obeys a uniform Lipschitz condition

$$\|\underline{f}(k, \underline{z}) - \underline{f}(k, \underline{z}')\| \leq L\|\underline{z} - \underline{z}'\|, \quad \|\underline{z}p\|, \|\underline{z}'\| \leq h. \tag{C7.3}$$

Then the following two statements are equivalent:

(i) $\underline{z} = \underline{0}$ is a locally exponentially stable equilibrium point of (C7.1) of stability radius h and decay rate $O(\mu)$.

(ii) There is a Lyapunov function $V(k, \underline{z})$ such that for $\|\underline{z}\| \leq h$

(C7A1) $\alpha_1\|\underline{z}\|^2 \leq V(k, \underline{z}) \leq \alpha_2\|\underline{z}\|^2$

(C7A2) $V(k, \underline{z} + \mu\underline{f}(k, \underline{z})) - V(k-1, \underline{z}) \leq -\alpha_3\mu\|\underline{z}\|^2$

(C7A3) $\|V_z(k, \underline{z})\| \leq \alpha_4\|\underline{z}\|$, $V_z(k, \underline{z}) = \partial V(k, \underline{z})/\partial\underline{z}$

(C7A4) If in addition,

$$\sup_k \|\underline{f}_{zi}(k, \underline{z})\| \leq d, \quad \|\underline{z}\| \leq h, \quad 1 \leq i \leq p$$

where $\underline{f}_{zi}(k, \underline{z}) = \partial^2 \underline{f}(k, \underline{z})/\partial\underline{z}^T\partial z_i$, $1 \leq i \leq p$. Then

$$\|V_{zz}(k, \underline{z})\| \leq \alpha_5 + \alpha_6\|\underline{z}\|, \quad V_{zz}(k, \underline{z}) = \partial^2 V(k, \underline{z})/\partial\underline{z}\partial\underline{z}^T$$

where α_1–α_6 are positive and have positive limits as $\mu \to 0$.

Proof of Theorem C7.1.

(i) Suppose there is a Lyapunov function obeying (C7A1)–(C7A4). Then consider that

$$\begin{aligned} V_k &= V(k, \underline{z}_k) \\ &= V(k, \underline{z}_{k-1} + \mu \underline{f}(k, \underline{z}_{k-1})) \\ &\le V(k-1, \underline{z}_{k-1}) - \alpha_3 \mu \|\underline{z}_{k-1}\|^2, \quad \text{by (C7A2)} \\ &\le V_{k-1} - \alpha_3 \mu \alpha_2^{-1} V_{k-1}, \quad \text{by (C7A1)}. \end{aligned}$$

So provided μ is small enough that $\mu\alpha_3\alpha_2^{-1} < 1$ we have

$$V_k \le \alpha^{2(k-k_0)} V_{k_0}, \quad \alpha^2 = 1 - \mu\alpha_3\alpha_2^{-1}$$

$$\Rightarrow \quad \|\underline{z}_k\| \le m\alpha^{k-k_0} \|\underline{z}_{k_0}\|, \quad m^2 = \alpha_2/\alpha_1, \text{ by (C7A1)}$$

thus ES follows. Also from (C7A1) and (C7A2), $V(k, \mu \underline{f}(k, \underline{0})) = 0$, so that from (C7A1) $\underline{f}(k, \underline{0}) = \underline{0}$, i.e., $\underline{0}$ is an equilibrium point.

(ii) Suppose ES holds. The proof is in several steps:

(a) *Proof of (C7A1):*

Let $\underline{y}(\tau, \underline{z}, k)$ be the solution of (C7.1) at time τ having started at time k at $\underline{z}$. Then we introduce the following candidate Lyapunov function:

$$V(k, \underline{z}) = \mu \sum_{\tau=k}^{k+T} \|\underline{y}(\tau, \underline{z}, k)\|^2 \tag{C7.4}$$

where $T > 0$ is to be chosen. By ES we have

$$\|\underline{y}(\tau, \underline{z}, k)\| \le m\alpha^{\tau-k} \|\underline{z}\|$$

so that

$$\begin{aligned} V(k, \underline{z}) &\le \mu m^2 \|\underline{z}\|^2 \sum_{\tau=k}^{k+T} \alpha^{2(\tau-k)} \\ &= \mu m^2 \|\underline{z}\|^2 (1 - \alpha^{2(T+1)})/(1 - \alpha^2). \end{aligned}$$

So the upper bound in (C7A1) is satisfied. For the lower bound we see from (C7.3) that

$$\|\underline{y}(\tau, \underline{z}, k)\| \ge (1 - \mu L)\|\underline{y}(\tau - 1, \underline{z}, k)\| \ge (1 - \mu L)^{\tau-k} \|\underline{z}\|.$$

So

$$\begin{aligned} V(k, \underline{z}) &\ge \mu \sum_{\tau=k}^{k+T} (1 - \mu L)^{2(\tau-k)} \|\underline{z}\|^2 \\ &= \mu \frac{(1 - \mu L)^{2T+2} - 1}{(1 - \mu L)^2 - 1} \|\underline{z}\|^2 = \alpha_1 \|\underline{z}\|^2 \end{aligned}$$

as required.

(b) *Proof of (C7A2):*
Next consider that

$$V(k, \underline{z} + \mu \underline{f}(k, \underline{z})) = \mu \sum_{\tau=k}^{k+T} \|\underline{y}(\tau, \underline{z} + \mu \underline{f}(k, \underline{z}), k)\|^2$$
$$= \mu \sum_{\tau=k}^{k+T} \|\underline{y}(\tau, \underline{z}, k-1)\|^2$$
$$= \mu \sum_{k-1}^{k-1+T} \|\underline{y}(\tau, \underline{z}, k-1)\|^2$$
$$+ \mu\|\underline{y}(k+T, \underline{z}, k-1)\|^2 - \mu\|\underline{y}(k-1, \underline{z}, k-1)\|^2$$
$$\leq V(k-1, \underline{z}) + \mu m^2 \alpha^{2T} \|\underline{z}\|^2 - \mu\|\underline{z}\|^2.$$

Now take T so large that $m^2\alpha^{2T} < 1$. Then set

$$\alpha_3 = 1 - m^2\alpha^{2T}$$

so that (C7A2) is thus established.

(c) *Proof of (C7A3):*
We have from (C7.4)

$$V_z(k, \underline{z}) = 2\mu \sum_{\tau=k}^{k+T} \underline{y}_z^T(\tau, \underline{z}, k)\underline{y}(\tau, \underline{z}, k) \tag{C7.5}$$

where (see Appendix A for vector differentiation)

$$\underline{y}_z^T = \partial \underline{y}^T / \partial \underline{z}, \quad \underline{y}_z = \partial \underline{y} / \partial \underline{z}^T.$$

From (C7.1)

$$\delta \underline{y}(\tau, \underline{z}, k) = \mu \underline{f}(\tau, \underline{y}(\tau - 1, \underline{z}, k))$$
$$\Rightarrow \quad \delta \underline{y}_z(\tau, \underline{z}, k) = \mu \underline{f}_z^T \underline{y}_z(\tau - 1, \underline{z}, k) \tag{C7.6}$$
$$\Rightarrow \quad \|\delta \underline{y}_z(\tau, \underline{z}, k)\| \leq \mu L \|\underline{y}_z(\tau - 1, \underline{z}, k)\|, \quad \text{by (C7.3)}$$
$$\Rightarrow \quad \|\underline{y}_z(\tau, \underline{z}, k)\| \leq (1 + \mu L)\|\underline{y}_z(\tau - 1, \underline{z}, k)\|$$
$$\leq (1 + \mu L)^{\tau - k}\|\underline{y}_z(k, \underline{z}, k)\|$$
$$= (1 + \mu L)^{\tau - k}. \tag{C7.7}$$

Thus by the ES assumption

$$\|V_z(k, \underline{z})\| \le 2\mu \sum_{\tau=k}^{k+T} (1+\mu L)^{\tau-k} m\alpha^{\tau-k} \|\underline{z}\|$$

$$= \alpha_4 \|\underline{z}\|$$

$$\alpha_4 = 2\mu m \frac{\alpha^T (1+\mu L)^T - 1}{\alpha(1+\mu L) - 1}.$$

Thus (C7A3) is established.

(d) *Proof of (C7A4):*
From (C7.5)

$$V_{zz}(k, \underline{z}) = \underline{A}_k + [\underline{b}_{k1}, \ldots, \underline{b}_{kr}]$$

$$\underline{A}_k = 2\mu \sum_{\tau=k}^{k+T} \underline{y}_{z}^T(\tau, \underline{z}, k) \underline{y}_{z}(\tau, \underline{z}, k)$$

$$\underline{b}_{ki} = 2\mu \sum_{\tau=k}^{k+T} \underline{y}_{zi}^T(\tau, \underline{z}, k) \underline{y}(\tau, \underline{z}, k), \quad 1 \le i \le p$$

where

$$\underline{y}_{zi}(\tau, \underline{z}, k) = \frac{\partial}{\partial z_i} \frac{\partial}{\partial \underline{z}} \underline{y}(\tau, \underline{z}, k).$$

From (C7.7) we see that

$$\|\underline{A}_k\| \le 2\mu \sum_{\tau=k}^{k+T} (1+\mu L)^{2(\tau-k)} \tag{C7.8}$$

$$= 2\mu \sum_{s=0}^{T} (1+\mu L)^{2s}.$$

From (C7.6)

$$\delta \underline{y}_{zi}(\tau, \underline{z}, k) = \mu \underline{f}_z^T \underline{y}_{zi}(\tau-1, \underline{z}, k) + \mu \underline{f}_{zi}^T \underline{y}_z(\tau-1, \underline{z}, k)$$

$$\underline{f}_{zi} = \frac{\partial}{\partial z_i} \frac{\partial}{\partial \underline{z}} \underline{f}(k, \underline{z}).$$

Then we find using the regularity conditions and (C7.7)

$$\|\underline{y}_{zi}(\tau, \underline{z}, k)\| \le (1+\mu L) \|\underline{y}_{zi}(\tau-1, \underline{z}, k)\| + \mu d (1+\mu L)^{\tau-1-k}.$$

Iterating this gives

$$\|\underline{y}_{zi}(\tau, \underline{z}, k)\| \le (1+\mu L)^{\tau-k} \|\underline{y}_{zi}(k, \underline{z}, k)\| + \mu d (\tau - k)(1+\mu L)^{\tau-1-k}.$$

However $\underline{y}_{zi}(k, \underline{z}, k) = \underline{0}$, so that we find

$$\|\underline{b}_{ki}\| \le 2\mu \sum_{\tau=k}^{k+T} \mu d(\tau - k)(1+\mu L)^{\tau-k} m\alpha^{\tau-k}\|\underline{z}\|$$

$$= 2m\mu^2 d \sum_{s=0}^{T} s[(1+\mu L)\alpha]^s \|\underline{z}\|. \tag{C7.9}$$

Putting (C7.8) and (C7.9) together gives (C7A4).

(e) Finally we check the orders of α_1–α_6 as $\mu \to 0$. Since the decay rate is $O(\mu)$ we have

$$\alpha = \alpha(\mu) = 1 - \mu a(\mu), \qquad \lim_{\mu\to 0} a(\mu) = a_0 > 0.$$

(It is straightforward to show $a_0 < L$.) If we choose $T = T_0/\mu$ then

$$m\alpha^T = m\,(1 - \mu a(\mu))^{T_0/\mu}$$
$$\le me^{-a(\mu)T_0}.$$

So if we choose T_0 so that

$$\lim_{\mu\to 0} me^{-a(\mu)T_0} = m_0 e^{-a_0 T_0} < \frac{1}{4}$$

then for some μ_0 we have

$$me^{-aT_0} < \frac{3}{4}, \qquad \mu \le \mu_0.$$

Now we can investigate the α_is. We have

$$\lim_{\mu\to 0} \alpha_1 = \lim_{\mu\to 0} \mu \frac{(1-\mu L)^{2T+2} - 1}{(1-\mu L)^2 - 1}$$
$$= (2L)^{-1} \lim_{\mu\to 0} \left(-(1-\mu L)^{2T_0\mu^{-1}+2} + 1\right)$$
$$= (2L)^{-1}(-e^{-2LT_0} + 1) > 0.$$

Continuing:

$$\lim_{\mu\to 0} \alpha_2 = \lim_{\mu\to 0} \mu \frac{1-\alpha^{2T+2}}{1-\alpha^2}$$
$$= (2a)^{-1} \lim_{\mu\to 0} \left(1 - (1-\mu a)^{2T_0\mu^{-1}+2}\right)$$
$$= (2a)^{-1}(1 - e^{-2T_0 a_0}) > 0.$$

Next:

$$\lim_{\mu\to 0} \alpha_4 = \lim_{\mu\to 0} \mu m(\mu) \frac{\alpha^T(1+\mu L)^T - 1}{\alpha(1+\mu L) - 1}$$
$$= \frac{m(0)}{L-a}(e^{-aT_0}e^{LT_0} - 1) > 0$$

Calculations for α_5, α_6 are similar.

This completes the proof of Theorem C7.1.

For an autonomous system

$$\delta \underline{z}_k = \mu \underline{f}_{av}(\underline{z}_{k-1}) \tag{C7.10}$$

the result simplifies somewhat.

Theorem C7.2. Consider the system (C7.10) and suppose $\underline{f}_{av}(\underline{z})$ obeys a Lipschitz condition (for $\|\underline{z}\| \le h$). Then the following two statements are equivalent:

(i) $\underline{z} = \underline{0}$ is a locally exponentially stable equilibrium point of (C7.10) with stability radius h and decay rate $= O(\mu)$.

(ii) There is a Lyapunov function $V(\underline{z})$ such that for $\|\underline{z}\| \le h$

(C7a1) $\alpha_1\|\underline{z}\|^2 \le V(\underline{z}) \le \alpha_2\|\underline{z}\|^2$
(C7a2) $V(\underline{z} + \mu \underline{f}(\underline{z})) - V(\underline{z}) \le -\alpha_3\mu\|\underline{z}\|^2$
(C7a3) $\|V_z(\underline{z})\| \le \alpha_4\|\underline{z}\|$, $V_z = dV/d\underline{z}$
(C7a4) If also $\|\underline{f}_{av,zi}(\underline{z})\| \le d$, $\|\underline{z}\| \le h$ where $\underline{f}_{av,zi}(\underline{z}) = d^2\underline{f}_{av}(\underline{z})/d\underline{z}^T dz_i$, $1 \le i \le p$, then

$$\|V_{zz}(\underline{z})\| \le \alpha_5 + \alpha_6\|\underline{z}\|, \quad V_{zz} = d^2V/d\underline{z}d\underline{z}^T$$

where α_1–α_6 are positive constants and have positive limits as $\mu \to 0$.

Proof of Theorem C7.2. The proof technique is much the same as for Theorem C7.1. We need only note the time invariant nature of (C7.10). So if $\underline{y}(\tau, \underline{z}, k)$ is the solution of (C7.10) at time τ having started at $\underline{z}$ at time k, then

$$\underline{y}(\tau, \underline{z}, k) = \underline{y}(\tau - k, \underline{z}, 0).$$

Then the candidate Lyapunov function becomes

$$V(\underline{z}) = \mu \sum_{s=0}^{N} \|\underline{y}(s, \underline{z}, 0)\|^2$$

where N is to be chosen. One now proceeds much as for Theorem C7.1. Details are left to the reader.

C7.3 Continuity Property for Exponential Stability

Suppose the gradient of the system (C7.1) has a sort of time invariant continuity:

(C7A5) $\sup_{k\geq 0} \|\underline{f}_{\underline{z}}(k,\underline{\rho}) - \underline{f}_{\underline{z}}(k,\underline{0})\| = \psi(\underline{\rho}), \quad \|\underline{\rho}\| \leq h$ and $\psi(\underline{\rho}) \to 0$ as $\|\underline{\rho}\| \to 0$.

If we define $\psi(\delta) = \sup_{\underline{\rho}:\|\underline{\rho}\|\leq\delta} \psi(\underline{\rho})$, then $\psi(\underline{\rho}) \leq \psi(\|\underline{\rho}\|)$, and $\psi(\epsilon) \to 0$ as $\epsilon \to 0$. Note that $\psi(\epsilon)$ is non-increasing in ϵ.

Suppose also that $\underline{0}$ is a locally ES equilibrium point of stability radius h. Suppose $\underline{z}_k$, $\underline{z}'_k$ are two trajectories of (C7.1) starting at different initial conditions $\underline{z}_{k_0}$, $\underline{z}'_{k_0}$ with $\|\underline{z}_{k_0}\|, \|\underline{z}'_{k_0}\| \leq h/2$. Then for some m, α with $0 < \alpha < 1$,

$$\|\underline{z}_k - \underline{z}'_k\| \leq m\alpha^{k-k_0}\|\underline{z}_{k_0} - \underline{z}'_{k_0}\|.$$

Proof. Introduce $\underline{\Delta}_k = \underline{z}_k - \underline{z}'_k$, and then use (C7.1) to write

$$\delta\underline{\Delta}_k = \mu\underline{f}(k,\underline{\Delta}_{k-1}) + \mu\underline{g}(k,\underline{\Delta}_{k-1})$$

where

$$\underline{g}(k,\underline{\Delta}) = \underline{f}(k,\underline{z}'_{k-1}+\underline{\Delta}) - \underline{f}(k,\underline{z}'_{k-1}) - \underline{f}(k,\underline{\Delta}).$$

Now, since $\underline{f}(k,\underline{0}) = \underline{0}$, a Taylor series (see Appendix D, section D13) gives for some $\lambda^{(r)}_{k-1}, \lambda'^{(r)}_{k-1}$ with $0 < \lambda^{(r)}_{k-1}, \lambda'^{(r)}_{k-1} < 1$:

$$g_r(k,\underline{\Delta}_{k-1}) = f^T_{z,r}(\underline{z}'_{k-1} + \lambda'^{(r)}_{k-1}\underline{\Delta}_{k-1})\underline{\Delta}_{k-1} - f^T_{z,r}(\lambda^{(r)}_{k-1}\underline{\Delta}_{k-1})\underline{\Delta}_{k-1}$$

where an index r denotes the rth component, $1 \leq r \leq p$. Now apply (C7A5) to find

$$\|\underline{g}(k,\underline{\Delta}_{k-1})\| \leq \|\underline{\Delta}_{k-1}\|\left[\sum_{r=1}^{p}\psi(\underline{z}'_{k-1} + \lambda'^{(r)}_{k-1}\underline{\Delta}_{k-1}) + \sum_{r=1}^{p}\psi(\lambda^{(r)}_{k-1}\underline{\Delta}_{k-1})\right]$$
$$\leq \|\underline{\Delta}_{k-1}\|\rho'_{k-1}$$

where

$$\rho'_{k-1} = 2p\psi(\|\underline{z}'_{k-1}\| + \|\underline{\Delta}_{k-1}\|).$$

But by ES, for some m_0 and α_0,

$$\|\underline{z}'_k\| \leq m_0\alpha_0^{k-k_0}\|\underline{z}'_0\| \leq m_0\alpha_0^{k-k_0}h$$

and similarly for $\|\underline{z}_k\|$. Thus

$$\rho'_{k-1} \leq \rho_{k-1} = 2p\psi(3m_0h\alpha_0^{k-k_0-1}). \tag{C7.11}$$

By ES there is a Lyapunov function $V(k,\underline{z})$ obeying (C7A1)–(C7A3). So writing $\underline{f}_k = \underline{f}(k,\underline{\Delta}_{k-1})$, $\underline{g}_k = \underline{g}(k,\underline{\Delta}_{k-1})$ consider that, via a Taylor series expansion

$$V(k,\underline{\Delta}_k) = V(k,\underline{\Delta}_{k-1} + \mu\underline{f}_k + \mu\underline{g}_k)$$
$$= V(k,\underline{\Delta}_{k-1} + \mu\underline{f}_k) + \mu\underline{g}_k^T V_z(k,\underline{\Delta}_{k-1} + \mu\underline{f}_k + \mu\lambda''_k\underline{g}_k)$$

where $0 < \lambda''_k < 1$. In view of (C7A2), (C7A3) we find

$$\begin{aligned} V(k, \underline{\Delta}_k) &\le V(k-1, \underline{\Delta}_{k-1}) - \mu\alpha_3\|\underline{\Delta}_{k-1}\|^2 \\ &\quad + \mu\|\underline{g}_k\|\alpha_4\|\underline{\Delta}_{k-1} + \mu\underline{f}_k + \mu\lambda''_k\underline{g}_k\| \\ &\le V(k-1, \underline{\Delta}_{k-1}) - \mu\alpha_3\|\Delta_{k-1}\|^2 \\ &\quad + \mu\alpha_4\rho_{k-1}\|\underline{\Delta}_{k-1}\|(\|\underline{\Delta}_{k-1}\| + \mu L\|\underline{\Delta}_{k-1}\| + \mu\rho_{k-1}\|\underline{\Delta}_{k-1}\|) \\ &\le V(k-1, \underline{\Delta}_{k-1}) - \mu\alpha_3\|\underline{\Delta}_{k-1}\|^2 + \mu\alpha_4\rho_{k-1}C\|\underline{\Delta}_{k-1}\|^2 \end{aligned} \tag{C7.12}$$

for some constant C. Take k_1 large enough that $C\alpha_4\rho_{k-1} < \alpha_3/2$ for all $k \ge k_1$; then

$$\begin{aligned} V(k, \underline{\Delta}_k) &\le V(k-1, \underline{\Delta}_{k-1}) - \mu\alpha_3\|\underline{\Delta}_{k-1}\|^2/2 \\ &\le V(k-1, \underline{\Delta}_{k-1})(1 - \mu\alpha_3/2\alpha_2), \quad \text{by (C7A1).} \end{aligned}$$

Iterating this gives, with $\alpha^2 = 1 - \mu\alpha_3/2\alpha_2$:

$$V(k, \underline{\Delta}_k) \le \alpha^{2(k-k_1)} V(k_1 - 1, \Delta_{k_1-1}). \tag{C7.13}$$

Returning to (C7.12), let $\rho = \max_{k_0+1 \le k \le k_1} \rho_{k-1} = 2p\psi(3m_0h\alpha_0)$; then

$$\begin{aligned} V(k, \underline{\Delta}_k) &\le V(k-1, \underline{\Delta}_{k-1}) + \mu(\alpha_4 C\rho - \alpha_3)\|\underline{\Delta}_{k-1}\|^2, \quad k_0 + 1 \le k \le k_1 \\ &\le V(k-1, \underline{\Delta}_{k-1})(1 + \mu D), \quad \text{by (C7A1)} \end{aligned}$$

where $D = (\alpha_4 C\rho - \alpha_3)/\alpha_1$. Iterating this gives

$$V(k_1 - 1, \underline{\Delta}_{k_1-1}) \le V(k_0, \underline{\Delta}_0)(1 + \mu D)^{k_1-1-k_0}.$$

Using this in (C7.13) gives

$$V(k, \underline{\Delta}_k) \le \alpha^{2(k-k_1)}(1 + \mu D)^{k_1-1-k_0} V(k_0, \underline{\Delta}_{k_0}).$$

Now using (C7A1) again leads to the required result

$$\|\underline{\Delta}_k\| \le m\alpha^{k-k_0}\|\underline{\Delta}_{k_0}\|$$

where

$$m = \left[(1 + \mu D)^{k_1-1-k_0}\alpha_2/\alpha^{2(k_1-k_0)}\alpha_1\right]^{1/2}.$$

It is important to note here that in view of (C7.11), $k_1 - k_0$ does not depend on k_0. Indeed it is defined as the first integer r such that

$$2C\alpha_4 p\psi(3m_0\alpha_0^r h) < \alpha_3/2.$$

C7.4 Nonlinear Stability by Linearization

Sometimes it is possible to show ES of a system such as (C7.1) or (C7.10) by a linearization method. The basic result is:

Theorem C7.1. Considering the system (C7.1), suppose $\underline{0}$ is an equilibrium point and that there is a sequence of matrices $\underline{A}_k$ such that

$$\lim_{\|\underline{z}\|\to 0} \sup_{k\geq 0} \|\underline{f}(k,\underline{z}) - \underline{A}_k\underline{z}\|/\|\underline{z}\| = 0. \tag{C7.14}$$

If $\underline{0}$ is an exponentially stable equilibrium point of the linearized system

$$\delta \underline{z}_k = \mu \underline{A}_k \underline{z}_{k-1} \tag{C7.15}$$

then $\underline{0}$ is an exponentially stable equilibrium point of (C7.1).

Proof of Theorem C7.1. For the time varying linearized system (C7.15), let $\underline{B}_k = I + \mu\underline{A}_k$ and

$$\prod_k = \underline{B}_k \underline{B}_{k-1} \cdots \underline{B}_1$$

so that $\Phi_{\tau k} = \prod_\tau \prod_k^{-1}$ is the transition matrix for (C7.15). We firstly follow Theorem C7.1 to construct a Lyapunov function. Indeed (C7.4) becomes

$$V(k,\underline{z}) = \underline{z}^T \underline{M}_k \underline{z}$$

$$\underline{M}_k = \mu \sum_{\tau=k}^{k+T} \Phi_{\tau k}^T \Phi_{\tau k}$$

and for some positive α_1, α_2

$$\alpha_1 \|\underline{z}\|^2 \leq V(k,\underline{z}) \leq \alpha_2 \|\underline{z}\|^2. \tag{C7.16}$$

Now let us try to use this Lyapunov function on (C7.1). We have, putting $\underline{f}_k = \underline{f}(k, \underline{z}_{k-1})$:

$$\begin{aligned} V_k = V(k,\underline{z}_k) &= \underline{z}_k^T \underline{M}_k \underline{z}_k \\ &= (\underline{z}_{k-1} + \mu \underline{f}_k^T)\underline{M}_k(\underline{z}_{k-1} + \mu\underline{f}_k) \\ &= (\underline{z}_{k-1} + \mu\underline{A}_k\underline{z}_{k-1} + \mu\underline{e}_k)^T \underline{M}_k (\underline{z}_{k-1} + \mu\underline{A}_k\underline{z}_{k-1} + \mu\underline{e}_k) \end{aligned}$$

where we have introduced

$$\underline{e}_k = \underline{f}_k - \underline{A}_k\underline{z}_{k-1}.$$

Thus

$$V_k = \underline{z}_{k-1}^T(\underline{B}_k^T \underline{M}_k \underline{B}_k)\underline{z}_{k-\mathrm{i}} + 2\underline{z}_{k-1}^T \underline{B}_k^T \underline{M}_k \underline{e}_k + \underline{e}_k^T \underline{M}_k \underline{e}_k.$$

Now (C7A2) becomes

$$V(k, \underline{B}_k\underline{z}) - V(k-1,\underline{z}) \leq -\mu\alpha_3\|\underline{z}\|^2$$

or equivalently

$$\underline{B}_k^T \underline{M}_k \underline{B}_k - \underline{M}_{k-1} = -\mu\alpha_3 I.$$

Thus

$$V_k - V_{k-1} = -\mu\alpha_3\|\underline{z}_{k-1}\|^2 + 2\mu\underline{z}_{k-1}^T\underline{B}_k^T\underline{M}_k\underline{e}_k + \mu^2\underline{e}_k^T\underline{M}_k\underline{e}_k.$$

From (C7.14) we can choose h small enough that

$$\|\underline{e}_k\| \leq \epsilon(\alpha_1/\alpha_2)^{\frac{1}{2}}\|\underline{z}_{k-1}\|, \quad \text{for all } k$$

where ϵ is to be chosen. Thus

$$\underline{e}_k^T\underline{M}_k\underline{e}_k \leq \epsilon^2\alpha_1\|\underline{z}_{k-1}\|^2 \leq \epsilon^2 V_{k-1}$$

$$|2\underline{z}_{k-1}^T\underline{B}_k^T\underline{M}_k\underline{e}_k| \leq 2(\underline{z}_{k-1}^T\underline{B}_k^T\underline{M}_k\underline{B}_k\underline{z}_{k-1})^{\frac{1}{2}}(\underline{e}_k^T\underline{M}_k\underline{e}_k)^{\frac{1}{2}}$$

$$\leq 2V_{k-1}^{\frac{1}{2}}\epsilon V_{k-1}^{\frac{1}{2}}.$$

Thus

$$V_k \leq V_{k-1} - \mu\alpha_3\alpha_2^{-1}V_{k-1} + 2\mu\epsilon V_{k-1} + \mu^2\epsilon^2 V_{k-1}$$

$$= \lambda V_{k-1}$$

$$\lambda = (1+\mu\epsilon)^2 - \mu\alpha_2^{-1}\alpha_3.$$

We choose ϵ so small that $0 < \lambda < 1$ (e.g., as $\mu \to 0$, $\epsilon < \alpha_3\alpha_2^{-1}/2$ will do); then we have

$$V_k \leq \lambda^{k-k_0}V_{k_0}.$$

So that from (C7.16)

$$\|\underline{z}_k\|^2 \leq m\lambda^{k-k_0}\|\underline{z}_{k_0}\|^2$$

$$m = \alpha_2\alpha_1^{-1}$$

whereupon ES is established.

Further material on (continuous time) nonlinear system stability can be found in [V3].

C8 UNIFORM STATIONARITY

In Appendices C8 and C9 we outline a continuous time theory of uniform stationarity—largely because proofs are readily accessible. A discrete time theory may be similarly constructed.

A deterministic signal $x(t)$ is called uniformly stationary if the following limit exists uniformly in t:

$$\lim_{T\to\infty}\frac{1}{T}\int_t^{t+T} x(s)x(s+u)ds = R_x(u).$$

It then follows that the following spectral distribution function exists [W6, Section 23]:

$$G_x(\omega) = \lim_{T\to\infty}\int_{-T}^{T} R_x(u)\left[\frac{e^{-j\omega u}-1}{-ju}\right]du$$

And there is the following spectral representation [W6, Section 23]:

$$R_x(u) = \int_{-\infty}^{\infty} e^{-j\omega u} dG_x(\omega)/2\pi.$$

In particular

$$\lim_{T\to\infty} \frac{1}{T} \int_t^{t+T} x^2(s)ds = R_x(0) = \int_{-\infty}^{\infty} dG_x(\omega)/2\pi.$$

If $x(t)$ is passed through a stable linear filter $H(D)$

$$y(t) = H(D)x(t), \qquad D = d/dt$$

then $y(t)$ is uniformly stationary [W6, Section 29] and has spectral distribution function

$$dG_y(\omega) = |H(j\omega)|^2 dG_x(\omega).$$

Also there is a cross spectral distribution function

$$dG_{xy}(\omega) = H(j\omega)dG_x(\omega)$$

and

$$\lim_{T\to\infty} \frac{1}{T} \int_t^{t+T} x(s)y(s)ds = R_{xy}(0) = \int_{-\infty}^{\infty} H(j\omega)dG_x(\omega)/2\pi.$$

This result is useful for averaging calculations.

C9 ALMOST PERIODIC SIGNALS

Almost periodic signals provide the most straightforward class of signals that are uniformly stationary.

Definition C9.1 ([W6, Section 24]) A signal $x(t)$ is (uniformly) almost periodic if

(i) Given ϵ there is a pseudo period $T(\epsilon)$ with

$$|x(t + T(\epsilon)) - x(t)| \le \epsilon, \qquad \text{for all } t.$$

(ii) Given ϵ there is a value $L(\epsilon)$ such that every interval $[A, A + L(\epsilon)]$ contains at least one pseudo period.

A periodic signal of fundamental period T has periods that are integer multiples of T and so satisfies (i), (ii).

A useful characterization of almost periodic signals is the following:

$x(t)$ is almost periodic if and only if there is a sequence of complex amplitudes A_k and real frequencies ω_k such that given $\epsilon > 0$, there is $n(\epsilon)$ with

$$|x(t) - \sum_{k=1}^{n(\epsilon)} A_k e^{j\omega_k t}| \leq \epsilon, \quad \text{for all } t$$

(See [W6, Section 24].)

If $x(t)$ is almost periodic, then it is uniformly stationary and $R_x(u)$ is almost periodic ([W6, Section 24]). Also the spectral distribution function $G_x(\omega)$ consists solely of jumps, so that the spectral representation takes the form [W6, Section 25]

$$R_x(u) = \sum_{n=-\infty}^{\infty} G_n e^{j\omega_n u}$$

where $\{\omega_n\}$ is a sequence of discrete frequencies and G_n a sequence of positive numbers (the jumps).

D

Mathematical Analysis Review

The review material in this appendix is couched in terms of sequences indexed by integers (that is, countably infinite sequences), but all the results given are also valid for sequences indexed by a continuous variable.

D1 INTERVALS

Let a, b be real numbers with $a < b$. The open interval (a, b) is the set of all real numbers x with $a < x < b$. The closed interval $[a, b]$ is the set of all x with $a \leq x \leq b$.

D2 SUP AND INF

(i) Let S be a sequence of real numbers $\{x_n\}_{n=1}^{\infty}$. A real number b is called an upper bound for S if $x_n \leq b$ for every x_n in S. A sequence with no upper bound is called unbounded. Similarly define lower bound. The sequence can also be indexed by a continuous index in this definition (e.g., x_μ, $0 \leq \mu \leq \mu_0 < \infty$).

(ii) Let S be a sequence of real numbers. A real number b is a least upper bound for S if

(a) b is an upper bound for S;

(b) no number less than b is an upper bound for S.

It is easy to show that a sequence cannot have two least upper bounds. The least upper bound is called the supremum.

Similarly we can define a greatest lower bound or infimum.

Example D2.1.

$$x_n = (n+1)^{-1}, n = 0, 1, \ldots \Rightarrow \sup_n x_n = 1.$$

D3 LIMIT POINTS

(i) Let S be a sequence of real numbers or m-vectors; then p is called a limit point of S if every ϵ-ball centered on p contains at least one element of S. An ϵ-ball centered on p consists of all points whose distance from p is $\leq \epsilon$: $B_\epsilon(p) = \{x : \|x - p\| \leq \epsilon\}$.

Example D3.1.

$x_n = (n+1)^{-1}$; $n = 0, 1, \ldots$ has 0 as a limit point.

Example D3.2.

$x_n = \cos(2\pi n/10)$ has every point in the interval $[-1, 1]$ as a limit point.

Example D3.3.

$x_\mu = (\mu + 1)^{-1}$; $\mu \geq 0$ has 1 as a limit point for $\mu \to 0$.

(ii) If p is a limit point of S then every ϵ-ball centered on p has infinitely many points of S [A6, Section 3.6, p. 53].

D4 BOLZANO-WEIERSTRASS THEOREM

(i) A sequence S is called bounded if $\|x_n\| \leq a < \infty$ for all x_n in S (or $\|x_\mu\| \leq a < \infty$ for all x_μ in S).

(ii) A bounded sequence S containing infinitely many points has at least one limit point [A6, Section 3.8, p. 54].

D5 CONVERGENCE OF SEQUENCES

(i) A sequence S of real numbers or m-vectors $\{x_n\}$ converges if there is a limit point p such that

$$\text{given } \epsilon > 0 \text{ there is } N_\epsilon \text{ such that}$$
$$\|x_n - p\| \leq \epsilon \quad \text{for all } n \geq N_\epsilon.$$

We say $\{x_n\}$ converges to p and write $x_n \to p$.

(ii) Subsequence Characterization of Convergence

(a) $x_n \to p$ if and only if every subsequence $\{x_{k_n}\} \to p$ [A6, Section 4.2, p. 72].

(b) $x_n \to p$ if and only if every subsequence $\{x_{k_n}\}$ has a further subsequence converging to p.

Proof: See (D8) below.

(iii) Let S be a sequence of real numbers $\{x_n\}$. Define the superior sequence

$$\bar{x}_n = \sup_{k \geq n} x_k$$

and the inferior sequence

$$\underline{x}_n = \inf_{k \geq n} x_k.$$

Both $\{\bar{x}_n\}$ and $\{\underline{x}_n\}$ are monotonic and so must have limits, respectively, $\bar{x}$ called $\limsup x_n$ and $\underline{x}$ called $\liminf x_n$.

Example D5.1.

$x_n = \cos(2\pi n/10)$ has $\bar{x}_n = 1$ and $\underline{x}_n = -1$, so $\limsup x_n = 1$ and $\liminf x_n = -1$.

Example D5.2.

$x_n = \cos(2\pi n)/(n+1)$ has $\bar{x}_n = (n+1)^{-1}$ and $\underline{x}_n = -(n+1)^{-1}$, so $\limsup x_n = 0 = \liminf x_n$.

Note that $x_n \to p$ if and only if

$$\bar{x} = \limsup x_n = \liminf x_n = \underline{x} = p.$$

D6 CONVERGENCE OF SERIES

(i) Given a sequence $\{a_k\}$ the partial sum sequence is defined as

$$s_n = \sum_{k=1}^{n} a_k.$$

The pair $\{a_k\}$, $\{s_k\}$ is called an infinite series. The series is said to converge if s_k tends to a finite limit and diverge if s_k tends to $\pm\infty$. If $s_n \to s$ we call s the sum of the series and write

$$s = \sum_{k=1}^{\infty} a_k.$$

(ii) If $a_k \geq 0$ then $\sum_{k=0}^{\infty} a_k$ converges if and only if $\{s_n\}$ is bounded above [A6, Section 8.5].

(iii) Cauchy Criterion for Convergence
$\sum_{k=1}^{\infty} a_k$ converges if and only if for every $\epsilon > 0$ there is N_ϵ so that

$$\left| \sum_{k=n+1}^{n+r} a_k \right| < \epsilon \quad \text{for } n \geq N_\epsilon, r = 1, 2, \ldots$$

(see [A6, Section 8.5]).
Corollary: If $\sum_{k=1}^{\infty} a_k$ converges then $a_k \to 0$ as $k \to \infty$.

D7 TOEPLITZ LEMMA

Let $\{a_{ns}\}$ be a sequence of weights with the properties

(i) $\sum_{s=1}^{\infty} |a_{ns}| \leq c$
(ii) $a_{ns} \to 0$, as $n \to \infty$ for each s
(iii) $\sum_{s=1}^{\infty} a_{ns} \to A$, as $n \to \infty$

Suppose that

$$Z_s \to Z, \quad \text{as } s \to \infty.$$

Then as $n \to \infty$

$$\sum_{s=1}^{\infty} a_{ns} Z_s \to AZ.$$

Proof. Given $\epsilon > 0$, there exists n_0 so that for all $s \geq n_0$

$$|Z_s - Z| \leq \epsilon.$$

Then consider that

$$\sum_{s=1}^{\infty} a_{ns} Z_s - AZ = \sum_{s=1}^{\infty} a_{ns}(Z_s - Z) + (\sum_{s=1}^{\infty} a_{ns} - A)Z.$$

By (iii) the second term $\to 0$ as $n \to \infty$, so we need only consider the first. We have

$$|\sum_{s=1}^{\infty} a_{ns}(Z_s - Z)| \leq |\sum_{s=1}^{n_0-1} a_{ns}(Z_s - Z)| + |\sum_{s=n_0}^{\infty} a_{ns}(Z_s - Z)|.$$

Now the first term is bounded by

$$|\sum_{s=1}^{n_0} a_{ns}| \max_{1 \leq s \leq n_0} |Z_s - Z|$$

which goes to 0 by (ii). The second term is bounded by ϵc, but since ϵ is arbitrary, the result follows. Note the following: (i) we have indexed a_{ns} by an integer $n \to \infty$, but it could be indexed by a continuous variable, say μ (like n^{-1}) $\to 0$. (ii) It may be that $a_{ns} = 0$, $s > k_n$ where $k_n \to \infty$ as $n \to \infty$, so the sums may be finite.

Corollary: (Kronecker's lemma) B_n is a non-decreasing positive sequence with $B_n \to \infty$, then

$$B_n^{-1} \sum_{s=1}^{n} a_s \to 0 \quad \text{if} \quad \sum_{s=1}^{\infty} a_s / B_s < \infty.$$

Proof. Let

$$T_n = \sum_{s=1}^{n} a_s/B_s, \qquad T_0 = 0.$$

Then, setting $b_n = B_n - B_{n-1} \geq 0$ we have

$$\begin{aligned}\frac{1}{B_n}\sum_{s=1}^{n} a_s &= \frac{1}{B_n}\sum_{s=1}^{n} B_s(T_s - T_{s-1}) \\ &= \frac{1}{B_n}\sum_{s=1}^{n}(B_sT_s - B_{s-1}T_{s-1}) - \frac{1}{B_n}\sum_{s=1}^{n} b_sT_{s-1} \\ &= T_n - \frac{1}{B_n}\sum_{s=1}^{n} b_sT_{s-1}.\end{aligned}$$

By assumption $T_n \to T < \infty$; then the right hand side $\to 0$ by the Toeplitz lemma.

D8 METRIC SPACES [A6, SECTION 3.13, P. 60], [L13, CHAPTER 1]

A set S of "points" is a metric space if there is a distance function or metric $d(\bullet, \bullet)$ so that for any three points x, y, z in S

$$\begin{aligned} &d(x, y) = 0 \quad \text{if and only if } x = y \\ &d(x, y) = d(y, x) \\ &d(x, z) \leq d(x, y) + d(y, z). \end{aligned}$$

Example D8.1.

The space $C[0, T]$ of continuous functions on the interval $[0, T]$ with metric

$$d(x, y) = \sup_{0 \leq \tau \leq T} |x(\tau) - y(\tau)|$$

D9 LIMITS AND COMPLETENESS

Let S be a metric space with metric $d(\bullet, \bullet)$.

(i) $\{x_n\}$, a sequence in S, converges to p in S if

$$d(x_n, p) \to 0, \qquad \text{as } n \to \infty.$$

We write $x_n \to p$.

(ii) A sequence in a metric space can converge to at most one point in S [A6, Section 4.2, p. 71].

(iii) Cauchy Sequence

A sequence $\{x_n\}$ in a metric space S is called a Cauchy sequence if

$$\begin{aligned} &\text{given } \epsilon > 0 \text{ there is } N_\epsilon \text{ such that} \\ &d(x_n, x_m) \leq \epsilon \quad \text{for all } n, m \geq N_\epsilon. \end{aligned}$$

(iv) Every convergent sequence in a metric space is a Cauchy sequence [A6, Section 4.3, p. 72].

(v) A metric space is called complete if every Cauchy sequence in S has a limit in S.

Example D9.1.

$C[0, T]$ is complete. Let $\{x_n(t)\}$ be a Cauchy sequence in $C[0, T]$. For fixed t, $\{x_n(t)\}$ is a Cauchy sequence of real numbers and so has a limit $x_0(t)$: As t varies these limits trace out a function. However it is a standard result [A6, Section 9.4], [L13, Section 1.3] that the limit function of a uniformly convergent sequence of continuous functions is continuous.

D10 SUBSEQUENCE CHARACTERIZATIONS OF LIMITS

(i) In a metric space S a sequence $\{x_n\}$ converges to p if and only if every subsequence $\{x_{k_n}\}$ converges to p [A6, Section 4.2].

(ii) In a metric space S a sequence $\{x_n\}$ converges to p if and only if every subsequence has a further subsequence converging to p.

Proof. If $x_n \to p$, let $\{x_{k_n}\}$ be any subsequence and extract a further subsequence $\{x_{k_{nm}}\}$; by (i) it converges to p. Suppose the condition holds, yet $x_n \not\to p$; then there is a subsequence converging to $p_1 \neq p$. Now take a further subsequence. On the one hand it converges to p_1; on the other hand under the condition it must converge to p, so $p_1 = p$, which is a contradiction.

D11 COMPACTNESS

A set K lying in a metric space S is called compact if every sequence in K has a convergent subsequence.

Example D11.1.

Arzela-Ascoli theorem [L13, Section 5.2]. A set $K \subset C[0, T]$ is compact if and only if $x(\tau) \in K$ are uniformly bounded and equicontinuous, that is,

(i) $\sup_{0 \le \tau \le T} |x(\tau)| \le B < \infty$, for all $x(\tau)$ in K;

(ii) Equicontinuity

$$\text{given } \epsilon > 0 \text{ there is } \delta(\epsilon) \text{ so that}$$

$$\sup_{|\tau_2 - \tau_1| < \delta(\epsilon)} |x(\tau_2) - x(\tau_1)| \le \epsilon \quad \text{for all } x(t) \text{ in } K.$$

Note: We use sections D9(ii) and D10 together as follows: Suppose $\{x_\mu(\tau)\}$ are a sequence of continuous functions obeying the conditions D10(i) and D10(ii). Then take any subsequence $\{x_{\mu_t}(\tau)\}$; it has, by section D10, a further subsequence converging to a limit function $x_e(\tau)$. Suppose we show somehow that there is only one possible limit function, say $\bar{x}(\tau)$. Then by section D9(ii) the whole sequence converges to $\bar{x}(\tau)$.

D12 DIFFERENTIABILITY

(i) A vector function $\underline{x}(\tau)$ defined on $[0, T]$ is absolutely continuous if

$$\text{given } \epsilon > 0 \text{ there is } \delta(\epsilon) > 0 \text{ so that}$$
$$\sum_{k=1}^{n} \|\underline{x}(b_k) - \underline{x}(a_k)\| \leq \epsilon$$

where for every positive integer n, (a_k, b_k) $k = 1 \ldots n$ are disjoint open subintervals in $[0, T]$ having the property that $\sum_{k=1}^{n}(b_k - a_k) \leq \delta$.

(ii) $\underline{x}(\tau)$ defined on $[0, T]$ is absolutely continuous if and only if it is an indefinite integral [A7, Section 2.3]; i.e., for some integrable $\underline{g}(\tau)$,

$$\underline{x}(\tau + \delta) - \underline{x}(\tau) = \int_{\tau}^{\tau+\delta} \underline{g}(\sigma) d\sigma. \tag{D12.1}$$

(iii) If $\underline{x}(\tau)$ defined on $[0, T]$ is absolutely continuous with representation (D12.1) then $\underline{x}(\tau)$ is differentiable almost everywhere on $[0, T]$ and

$$\frac{d\underline{x}(\tau)}{d\tau} = \underline{g}(\tau).$$

The converse also holds [A7, Section 2.3].

D13 SPATIAL TAYLOR'S SERIES

If $g(\underline{z})$ is differentiable at each point of an open set S and $\underline{a}$, $\underline{a} + \underline{h}$ are points in S whose line segment (i.e., points of the form $\underline{a} + t\underline{h}$, $0 \leq t \leq 1$) lies in S, then there is a λ with $0 < \lambda < 1$ such that

$$g(\underline{a} + \underline{h}) = g(\underline{a}) + \underline{h}^T g_z(\underline{a} + \lambda \underline{h})$$

where $g_z(\underline{z}) = dg(\underline{z})/d\underline{z}$.
Proof: See [A7, Section 12.14].

D14 UNIFORM CONTINUITY

A function $\underline{g}(\underline{z})$ is said to be uniformly continuous in a set A if given ϵ there is a $\delta > 0$ (depending only on A) so that if $\underline{z}, \underline{z}' \in A$ then

$$\|\underline{g}(\underline{z}) - \underline{g}(\underline{z}')\| < \epsilon \quad \text{when } \|\underline{z} - \underline{z}'\| < \delta.$$

That is, one δ does for all $\underline{z}, \underline{z}' \in A$ (see [A7, Section 4.19]).

D15 INEQUALITIES

(i) Holder's Inequality

$$\sum |a_i||b_i| \le \left(\sum |a_i|^p\right)^{1/p} \left(\sum |b_i|^q\right)^{1/q}, \quad \frac{1}{p} + \frac{1}{q} = 1$$

To prove it, show that if ξ, η are positive

$$\xi\eta \le \frac{\xi^p}{p} + \frac{\eta^q}{q}, \quad \frac{1}{p} + \frac{1}{q} = 1$$

then sum the inequality with ξ replaced by

$$|a_i|/(\sum |a_i|^p)^{1/p}$$

and η replaced by

$$|b_i|/(\sum |b_i|^q)^{1/q}.$$

(ii) Minkowski's Inequality

$$\left(\sum (a_i + b_i)^p\right)^{1/p} \le \left(\sum |a_i|^p\right)^{\frac{1}{p}} + \left(\sum |b_i|^p\right)^{1/p}$$

To prove it, expand $(a_i + b_i)^p$ by the Binomial theorem and then apply Holder's inequality.

E

Probability Review

This appendix contains some definitions and results of probability theory used in the text. Our presentation is developed to some extent with an eye on "how to use" and so is somewhat more than a passive listing. For a very clear development of measure-theoretic probability the reader is referred to [B9].

E1 PROBABILITY SPACE

A sample space Ω is a nonempty set: a subset of Ω is called an event. A collection of events $\mathcal{F}$ is called a σ-field if it is closed under complements and countable unions, and contains the null set ϕ.

A probability measure is a function from Ω to the real line $\mathcal{R}$, having the properties:

(i) $P(A) \geq 0,\ A \subset \Omega$

(ii) $P(\Omega) = 1$

(iii) Countable additivity: if $A_i \in \mathcal{F}$, $A_i \cap A_j = \phi$, $i \neq j$, then

$$P\left(\bigcup_{i=1}^{\infty} A_i\right) = \sum_{i=1}^{\infty} P(A_i)$$

If $A \in \mathcal{F}$, a σ-field, we say A is $\mathcal{F}$-measurable. The triplet $(\Omega, \mathcal{F}, P)$ is called a probability space. The most significant axiom is countable additivity (see [B9]). Countable additivity is equivalent to finite additivity plus continuity, which says if $A_i \uparrow A$ then $P(A_i) \uparrow P(A)$ (where $\uparrow$ means "increases to").

If A_n is a sequence of subsets of Ω we define

$$\limsup_k A_k = \bigcap_{n=1}^{\infty} \bigcup_{k=n}^{\infty} A_k$$

$$\liminf_k A_k = \bigcup_{n=1}^{\infty} \bigcap_{k=n}^{\infty} A_k.$$

(cf. [C5, Section 4.2]). We see that

$\omega \in \limsup_k A_k$ if and only if

ω belongs to infinitely many of the A_k.

To put it another way

$\limsup_k A_k$ occurs if and only if

A_k occurs i.o. (infinitely often).

So we can say

$$\begin{aligned} P\left(\limsup_k A_k\right) &= P(A_k\ i.o.) \\ &= \lim_{n\to\infty} P\left(\bigcup_{k=n}^{\infty} A_k\right). \end{aligned}$$

E2 RANDOM VARIABLES

If $(\Omega, \mathcal{F}, P)$ is a probability space, then a random variable (RV) is a number (i.e., a real valued function) $X(\omega)$ associated with the event ω; $X(\omega)$ is $\mathcal{F}$-measurable, i.e., $\{\omega : X(\omega) \in \mathcal{R}\} = X^{-1}(\mathcal{R}) \in \mathcal{F}$ ($\mathcal{R} =$ the real line).

(i) *(Cumulative) Distribution Function (cdf).*
An RV induces a probability measure on Ω

$$P_X(A) = \mu_X(A) = P(\omega : X(\omega) \in A)$$

when A is a Borel measurable subset of $\mathcal{R}$. The measure may be equivalently expressed in terms of the cumulative distribution function (cdf) of X

$$\begin{aligned} F_X(x) &= P\{\omega : X(\omega) \le x\} \\ &= P_X(-\infty, x]. \end{aligned}$$

$F_X(x)$ is non-decreasing, right continuous, and $F_X(-\infty) = 0$, $F_X(\infty) = 1$. If $F_X(x)$ is absolutely continuous then X has a pdf $p_X(x) \ge 0$ with

$$F_X(x) = \int_{-\infty}^{x} p_X(x)dx.$$

A vector RV $\underline{X}$ is just a vector of RV's $\underline{X} = (X_1 \cdots X_n)^T$. It has a vector distribution function

$$F_{\underline{X}}(\underline{x}) = P(X_1 \le x_1, \cdots, X_n \le x_n)$$
$$= P(\omega : X_i(\omega) \le x_i, i = 1, \cdots, n).$$

(ii) *Expectation or Average.*
The expectation, expected value, or average of $X(\omega)$ is defined by the Lebesgue-Stieltjes integral (if it exists)

$$E(X) = \int X(\omega) dP_X(\omega)$$
$$= \int x dF_X(x)$$

where equality follows by a change of variables. The heuristic idea behind this definition is easily seen by rewriting the arithmetic average as follows:

$$\sum_{i=1}^{n} \frac{X_i}{n} = \sum_{u=1}^{m} a_u f_u$$

where f_u is the fraction of X_is taking the value a_u, $1 \le u \le m$. The Lebesgue-Stieltjes integral exists if and only if X is integrable, i.e., $E|X| < \infty$.

(iii) *Inequalities.*
Jensen's Inequality. If $J(x)$ is a convex function, i.e.,

$$J(\alpha x + (1-\alpha)y) \le \alpha J(x) + (1-\alpha)J(y)$$

$$\Rightarrow \qquad J\left(\sum_{i=1}^{n} \frac{x_i}{n}\right) \le \sum_{i=1}^{n} \frac{J(x_i)}{n}$$

which leads to

$$J(E(X)) \le E(J(X)).$$

Hölder's Inequality.

$$E|XY| \le \left(E|X|^p\right)^{1/p} \left(E|Y|^q\right)^{1/q}, \quad \frac{1}{p} + \frac{1}{q} = 1.$$

Minkowski's Inequality.

$$\left(E|X+Y|^m\right)^{1/m} \le \left(E|X|^m\right)^{1/m} + \left(E|Y|^m\right)^{1/m}, \quad m \ge 1.$$

The reader is invited to prove these inequalities.

E3 CONDITIONAL PROBABILITY [B9, SECTIONS 33, 34]

The conditional probability of an event A, given the event B is

$$P(A|B) = P(A \cap B)/P(B), \quad P(B) > 0.$$

Let $\mathcal{G}$ be a sub σ-field (of events) of $\mathcal{F}$. The conditional probability of A given $\mathcal{G}$ is an RV defined by two properties:

(i) $P(A|\mathcal{G})$ is $\mathcal{G}$-measurable and integrable

(ii) $\int_G P(A|\mathcal{G})dP = P(A \cap G),\ \ G \in \mathcal{G}$

A is independent of $\mathcal{G}$ if it is independent of every event G in $\mathcal{G}$, so from the second property

$$P(A|\mathcal{G}) = P(A)$$
$$\Leftrightarrow \qquad P(A \cap G) = P(A)P(G).$$

If X is an integrable RV then the conditional expectation of X given $\mathcal{G}$ is an RV defined by two properties:

(i) $E(X|\mathcal{G})$ is $\mathcal{G}$-measurable and integrable.

(ii) $E(X|\mathcal{G})$ obeys the orthogonality condition $E((X - E(X|\mathcal{G}))\Phi(\mathcal{G})) = 0$ where $\Phi(\mathcal{G})$ is a $\mathcal{G}$-measurable, integrable RV.

Of course, there is a conditional Jensen inequality: If $J(x)$ is convex then

$$J\left(E(X|\mathcal{G})\right) \le E\left(J(X)|\mathcal{G}\right).$$

The following iterated conditional expectation properties are useful. If X is integrable and $\mathcal{G}_1$ and $\mathcal{G}_2$ are two σ-fields with $\mathcal{G}_1 \subset \mathcal{G}_2$, then

$$E\left(E(X|\mathcal{G}_2)|\mathcal{G}_1\right) = E(X|\mathcal{G}_1) \quad \text{w.p.1.} \tag{E3.1}$$

If X is integrable and $\mathcal{F}_1$ and $\mathcal{F}_2$ are independent σ-fields then

$$E\left(E(X|\mathcal{F}_2)|\mathcal{F}_1\right) = E(X|\mathcal{F}_2) \quad \text{w.p.1.} \tag{E3.2}$$

E4 STOCHASTIC CONVERGENCE

If $\{X_n(\omega)\}$ is a sequence of RVs on a probability space $(\Omega, \mathcal{F}, P)$, the following modes of convergence are of interest.

(i) X_n converges in distribution (or weakly) to X, written as $X_n \Rightarrow X$ if

$$P(X_n \le \xi) \to P(X \le \xi), \text{ as } n \to \infty$$
at every continuity point of $F_X(\xi)$.

This also can be written

$$E\left(I(X_n \le \xi)\right) \to E\left(I(X \le \xi)\right)$$

where $I(\bullet)$ is an indicator function. The definition is equivalent to

$$E(f(X_n)) = \int f(x)dF_n(x) \quad \to \quad E(f(X)) = \int f(x)dF_X(x)$$

(where $F_n(x) = P(X_n \le x)$) for every bounded continuous function $f(\bullet)$.

(ii) X_n converges in probability to X, written as $X_n \xrightarrow{p} X$ if for every $\epsilon > 0$

$$P(|X_n - X| > \epsilon) \to 0, \quad \text{as } n \to \infty.$$

(iii) X_n converges w.p.1 (with probability 1) to X, written $X_n \to X$, w.p.1, if

$$P\left(\omega : \lim_{n\to\infty} X_n(\omega) = X(\omega)\right) = 1$$

or equivalently there is a set N with $P(N) = 0$ so that for $\omega \notin N$, given $\epsilon > 0$, there is an integer $n_\epsilon(\omega)$ with

$$|X_n(\omega) - X(\omega)| \le \epsilon, \text{ for all } n \ge n_\epsilon(\omega).$$

It has already been mentioned in Appendix B5 that

$$X_n \xrightarrow{p} X \quad \text{implies} \quad X_n \Rightarrow X.$$

Below we will see that

$$X_n \to X \text{ w.p.1} \quad \text{implies} \quad X_n \xrightarrow{p} X.$$

Definition (iii) is useless for proving w.p.1 convergence, so other criteria are needed.

Theorem E4.1 ([C5, Sections 4.1, 4.2])

(a) The following conditions are each equivalent to $X_n \to X$ w.p.1.
(i) For every $\epsilon > 0$, $\lim_{n\to\infty} P(|X_m - X| > \epsilon \text{ for some } m \ge n) = 0$.
(ii) For every $\epsilon > 0$, $\lim_{n\to\infty} P(\sup_{m\ge n} |X_m - X| > \epsilon) = 0$.
(iii) For every $\epsilon > 0$, $P(|X_n - X| > \epsilon, \ i.o.) = 0$.

(b) In case X is not known the following Cauchy-type conditions are equivalent to: X_n converges w.p.1:
(i) For every $\epsilon > 0$:

$$\lim_{n\to\infty} P(|X_n - X_{n+k}| > \epsilon, \text{ for some } k \ge 1) = 0.$$

(ii) For every $\epsilon > 0$:

$$\lim_{m\to\infty} \lim_{n\to\infty} P(|X_n - X_{n+k}| > \epsilon \text{ for some } k \text{ with } 1 \le k \le m) = 0.$$

Next note that (to get a usable criterion) b(ii) occurs if:

Theorem E4.2. X_n converges w.p.1 if

$$\lim_{m\to\infty} \lim_{n\to\infty} P\left(\max_{1\le k\le m} |X_n - X_{n+k}| > \epsilon\right) = 0. \tag{E4.1}$$

A somewhat less powerful but still useful result is:

Theorem E4.3. X_n converges to X w.p.1 if

$$\sum_{n=1}^{\infty} P(|X_n - X| > \epsilon) < \infty.$$

Proof. Follows from Theorem E4.1(ii)

$$P\left(\sup_{m\geq n}|X_m - X| > \epsilon\right) \leq \sum_{m=n}^{\infty} P(|X_m - X| > \epsilon) \to 0$$

as $n \to \infty$ under the stated condition.

E5 WEAK CONVERGENCE [B9, SECTION 25]

In determining the limit behavior of the cdfs of a sequence of RVs the following result is fundamental.

Theorem E5.1 (Helley's Theorem) If $F_n(x)$ is a sequence of cdf's, there is a subsequence $F_{n_k}(x)$ and a non-decreasing, right continuous function $F(x)$ with $F_{n_k}(x) \to F(x)$ at every continuity point of $F(x)$.

Although $0 \leq F(x) \leq 1$, $F(x)$ need not be a cdf. To ensure that limits can only be cdfs, it is necessary to constrain the sequence to prevent escape of probability mass.

Definition E5.1. A sequence of cdfs $\{F_n(x)\}$ (or probability measures $P_n(\bullet)$) is tight if given ϵ there is an A such that

$$P_n(-A, A] > 1 - \epsilon, \text{ for all } n;$$
$$\text{or } F_n(-A) < \epsilon/2,\ F_n(A) > 1 - \epsilon/2, \text{ for all } n$$

or equivalently, if

$$\lim_{A\to\infty} \sup_{n\geq 1} P(|X_n| > A) = 0.$$

With tightness we get:

Theorem E5.2 (Prohorov's Theorem) If $\{P_n\}$ (or $\{F_n(x)\}$) is a tight sequence of probability measures (or cdfs), then every subsequence P_{n_k} has a further subsequence $P_{n_{k_j}}$ and a probability measure P such that $P_{n_{k_j}} \to P$ as $j \to \infty$.

It is often possible to show the limit measure (also called the steady state or invariant measure) is unique, and then we have:

Corollary E5.1. If $\{P_n\}$ is tight and every convergent subsequence converges to the same P, then $P_n \Rightarrow P$.

E6 INTERCHANGE OF LIMIT AND EXPECTATION [C6, SECTIONS 3.2, 4.5]

(i) Monotone Convergence Theorem (MCT)
If $X_n \geq 0$, $X_n \uparrow X$ in probability, then $E(X_n) \uparrow E(X)$. An important application of this occurs when $Y_k \geq 0$: $\sum_{k=1}^{\infty} E(Y_k) < \infty \Rightarrow \sum_{k=1}^{\infty} Y_k$ converges w.p.1. (Prove it!)

(ii) Fatou's Lemma
If $X_n \geq 0$ then $E(\liminf_n X_n) \leq \liminf_n E(X_n)$.

(iii) Uniform Integrability.
$\{Y_n\}$ is called uniformly integrable if

$$\lim_{A\to\infty} \sup_{n\geq 1} E\left(I(|X_n| > A)|X_n|\right) = 0.$$

It is left to the reader to show that this occurs if and only if

(a) $\sup_n E|X_n| \leq \infty$.

(b) Given $\epsilon > 0$, there is $\delta(\epsilon) > 0$, so that for any $E \in \mathcal{F}$, $P(E) < \delta(\epsilon) \Rightarrow \sup_n \int_E |X_n| dP < \epsilon$.

If $X_n \xrightarrow{p} X$ and $\{X_n\}$ is uniformly integrable then $E(X_n) \to E(X)$.

Proof. For any $\epsilon > 0$

$$E|X_n - X| = \int_{|X_n - X| < \epsilon} |X_n - X| dP + \int_{|X_n - X| > \epsilon} |X_n - X| dP.$$

Given $\epsilon > 0$, take n so large that $P(|X_n - X| > \epsilon) < \delta(\epsilon)$. Then

$$E|X_n - X| < \epsilon P(|X_n - X| < \epsilon) + \epsilon < 2\epsilon$$

and since ϵ is arbitrary we get $E|X_n - X| \to 0$. Finally $||E(X_n)| - |E(X)|| \leq |E(X_n) - E(X)|$ so the result follows.

(iv) Dominated Convergence
If $X_n \xrightarrow{p} X$, $|X_n| \leq Y$, $E(Y) < \infty$, then $E(X_n) \to E(X)$.

Proof. For any E, $\int_E |X_n| dP \leq \int_E |Y| dP$, so that $\{X_n\}$ is uniformly integrable and the result follows.

(v) Strong Uniform Integrability
We say $\{X_n\}$ is strongly uniformly integrable if there is a RV Y with $E|Y| < \infty$ and $P(|X_n| \geq \alpha) \leq P(|Y| \geq \alpha)$. Note that then

$$\int_{X \geq \alpha} |X| dP = \alpha P(|X| \geq \alpha) + \int_{\alpha}^{\infty} P(|X| \geq t) dt$$

$$\Rightarrow \quad \int_{|X_n| \geq \alpha} |X_n| dP \leq \int_{|Y| \geq \alpha} Y dP$$

So that $\{X_n\}$ is uniformly integrable.

E7 MARTINGALES [B9, SECTION 35]

If $\{X_n\}$ is a sequence of RVs and $\{\mathcal{F}_n\}$ a sequence of sub σ-fields of $\mathcal{F}$ we say $\{X_n, \mathcal{F}_n\}$ is a martingale (MG) if

(i) $\mathcal{F}_n$ are increasing, i.e., $\mathcal{F}_{n+1} \supset \mathcal{F}_n$
(ii) X_n is $\mathcal{F}_n$-measurable (we say X_n is adapted to $\mathcal{F}_n$)
(iii) $\sup_n E|X_n| < \infty$
(iv) $E(X_{n+1}|\mathcal{F}_n) = X_n$ w.p.1.

If $X_n \geq 0$ and in (iv), "=" is replaced by "$\geq$", we call X_n a submartingale. If $X_n \geq 0$ and in (iv), "=" is replaced by "$\leq$", we call X_n a super martingale.

If $(X_n, \mathcal{F}_n)$ is an MG, it is then an MG with respect to its own history, i.e., we can replace $\mathcal{F}_n$ by $\mathcal{G}_n = \sigma$-field generated by $X_1 \cdots X_n$. We can think of $\mathcal{G}_n$ as the "history" of X_n. In any case (iv) then becomes

$$E(X_{n+1}|X_1 \cdots X_n) = X_n \quad \text{w.p.1.}$$

We do not always want $\mathcal{F}_n = \mathcal{G}_n$, because it is often important that $\mathcal{F}_n$ contains other "histories" than that of $\{X_n\}$. One of the reasons for the power of MG methods is the Martingale Convergence theorem (MGCT).

Theorem E7.1 (Doob's MGCT) If $\{X_n, \mathcal{F}_n\}$ is an MG, and $\sup_n E|X_n| < \infty$, then X_n converges w.p.1, and the limit RV, call it X, has $E|X| < \infty$.

The proof is rather long: However there is a short proof, if finite second moments are assumed. This proof is based on checking the criterion of Theorem E4.2. To that end we need the following result.

Theorem E7.2 (MG maximal inequality) If $X_n \geq 0$ is a sub MG then for any $\epsilon > 0$:

$$P\left(\max_{1 \leq k \leq n} X_k \geq \epsilon\right) \leq E(X_n)/\epsilon.$$

Proof. Let

$$\begin{aligned} A_1 &= \{\omega : X_1(\omega) \geq \epsilon\} \\ A_k &= \{\omega : \max_{1 \leq i < k} X_i < \epsilon \leq X_k\}, \quad k = 2 \cdots n \\ A &= \bigcup_{k=1}^{n} A_k = \{\omega : \max_{1 \leq i \leq n} Y_i \geq \epsilon\}. \end{aligned}$$

Note that $A_k \in \mathcal{F}_k = \sigma(X_1 \cdots X_k)$ and the A_k are disjoint. Thus if I_A is the indicator for the event A, then

$$\begin{aligned}
E(I_A X_n) &= E(X_n \sum_{k=1}^{n} I_{A_k}) \\
&= \sum_{k=1}^{n} E(X_n I_{A_k}) \\
&= \sum_{k=1}^{n} E\left(E(X_n I_{A_k}|\mathcal{F}_k)\right) \\
&\geq \sum_{k=1}^{n} E(I_{A_k} X_k), \quad \text{by sub MG property.}
\end{aligned}$$

But on A_k, $X_k \geq \epsilon$ so

$$E(I_A X_n) \geq \epsilon \sum_{k=1}^{n} E(I_{A_k}) = \epsilon E(I_A) = \epsilon P(A)$$

$$\Rightarrow \qquad \epsilon P(A) \leq E(I_A X_n) \leq E(X_n)$$

which is the required result.

Now we use this to obtain

Theorem E7.3 (Square Integrable MGCT) Suppose $\{X_n, \mathcal{F}_n\}$ is an MG with property $\sup_n E(X_n^2) < \infty$. Then X_n converges w.p.1.

Proof. We have to show (E4.1). However for fixed n, $Y_{n,k} = |X_n - X_{n+k}|$ is a sub MG since

$$\begin{aligned}
E(Y_{n,k}|\mathcal{F}_{k+n-1}) &= E\left(|X_n - X_{n+k}||\mathcal{F}_{k+n-1}\right) \\
&\geq |E(X_n - X_{n+k}|\mathcal{F}_{k+n-1})|, \quad \text{by Jensen's inequality} \\
&= |X_n - X_{n+k-1}| = Y_{n,k-1}.
\end{aligned}$$

So by the MG maximal inequality

$$\begin{aligned}
P\left(\max_{1\leq k\leq m} |X_n - X_{n+k}| > \epsilon\right) &\leq E|X_n - X_{n+m}|/\epsilon \\
&\leq E^{\frac{1}{2}}(X_n - X_{n+m})^2/\epsilon.
\end{aligned}$$

But by the MG property

$$E(X_n - X_{n+r})^2 = E(X_{k+r}^2) - E(X_n^2).$$

Setting $r = 1$ shows that $E(X_t^2)$ is a non-decreasing sequence, and since it is bounded it must have a finite limit. But then we see that

$$\lim_{m\to\infty} \lim_{n\to\infty} E(X_{n+m} - X_n)^2 = 0$$

so that (E4.1) is established.

Another MGCT of great use is:

Theorem E7.4 (Super MGCT) If $\{V_{n-1}, \alpha_n, \beta_n\}$ are sequences of non-negative RVs adapted to a sequence of increasing σ-fields $\mathcal{F}_n$ and

$$E(V_n|\mathcal{F}_n) \le V_{n-1} - \alpha_n + \beta_n$$

then if $\sum_{n=1}^{\infty} \beta_n < \infty$ w.p.1, then

$$V_n \text{ converges w.p.1}$$

$$\textstyle\sum_{n=1}^{\infty} \alpha_n < \infty \text{ w.p.1.}$$

Proof. There are some subtleties in the proof (see [N4]), so we prove a slightly weaker (albeit much less useful) result instead. We suppose $\sum_{n=1}^{\infty} E(\beta_n) < \infty$. Then we have

$$0 \le E(V_n) \le E(V_{n-1}) - E(\alpha_n) + E(\beta_n).$$

Summing up gives

$$\sum_{u=1}^{n} E(\alpha_u) + E(V_n) \le E(V_0) + \sum_{u=1}^{n} E(\beta_u).$$

Thus we deduce $\sum_{u=1}^{\infty} E(\alpha_u) < \infty \Rightarrow \sum_{u=1}^{\infty} \alpha_u < \infty$ w.p.1 by MCT. Next let $W_n = -V_n + \sum_{u=1}^{n} \beta_u$ so that

$$E(W_n|\mathcal{F}_{n-1}) \ge W_{n-1}$$

while

$$E|W_n| \le E(V_n) + \sum_{u=1}^{n} E(\beta_u) < E(V_0) + 2\sum_{u=1}^{\infty} E(\beta_u) < \infty.$$

So by the sub MGCT, $W_n \to W < \infty$ w.p.1 and since $\sum_{u=1}^{\infty} \beta_u < \infty$ w.p.1 we deduce V_n converges w.p.1.

A further useful MGCT is [H4, Theorem 2.17].

Theorem E7.5 (Chow's Conditional MGCT) If $S_n = \sum_{i=1}^{n} X_i$ is an MG (with respect to increasing σ-fields $\mathcal{F}_n$) and for some p with $1 \le p \le 2$

$$\sum_{i=1}^{\infty} E(|X_i|^p|\mathcal{F}_{i-1}) < \infty, \quad \text{w.p.1}$$

then S_n converges w.p.1.

E8 STATIONARY PROCESSES

A sequence of RVs $\{X_n\}$ is strictly stationary if $\{X_{n_1} \cdots X_{n_m}\}$ has the same joint distribution as $\{X_{n_1+k} \cdots X_{n_m+k}\}$ for each $k \ge 1$ and any collection of integers $\{n_1 \cdots n_m\}$.

Discussion of autocovariances and spectra associated with stationary processes has already been given in Appendix B.

Theorem E8.1 (Wold Decomposition [C1, Section 1.3]) Given a zero mean stationary process $\{X_t\}$ with absolutely continuous spectrum $F(\omega)$ the Wold decomposition asserts that under certain mild regularity conditions $\{X_t\}$ can be represented as a filtered white noise

$$X_t = h(q^{-1})\epsilon_t$$

where $\{\epsilon_t\}$ are uncorrelated, of variance

$$\sigma^2 = \frac{1}{2\pi}\int_{-\pi}^{\pi} \log_e F(\omega)d\omega$$

and $h(q^{-1})$ is a square summable one-sided transfer function

$$h(q^{-1}) = \sum_{u=0}^{\infty} h_u q^{-u}, \quad \sum_{u=0}^{\infty} h_u^2 < \infty.$$

The connection of $h(q^{-1})$ to $F(\omega)$ is, of course,

$$F(\omega) = |h(e^{-j\omega})|^2\sigma^2.$$

Perhaps the fundamental limit theorem for stationary processes is the Ergodic theorem, which asserts w.p.1 convergence of averages. To state it we need, however, to develop a little background.

A transformation T from Ω to Ω is called measure or probability preserving if $P(T^{-1}(A)) = P(A)$, $A \in \mathcal{F}$. Every probability preserving measure generates a strictly stationary process by the relation $X_n(\omega) = X(T^n\omega)$. An event $A \in \mathcal{F}$ is called invariant if $T^{-1}(A) = A$. The collection of invariant events form a σ-field, call it $\mathcal{T}$. T is called ergodic if for $A \in \mathcal{T}$, $P(A) = 0$ or 1. Ergodicity can be a little better understood by noting it is implied by mixing. T is mixing if for $A, B \in \mathcal{F}$

$$P(A \cap T^{-n}B) - P(A)P(B) \to 0, \text{ as } n \to \infty.$$

This says that events, long distant in time, are nearly independent. We can now state the theorem [S20, Section 3.5]:

Theorem E8.2 (Pointwise Ergodic Theorem) If T is probability preserving and ergodic and $E|X_0| < \infty$ then

$$\frac{1}{n}\sum_{k=1}^{n} X_k \to E(X_0) \quad \text{w.p.1}.$$

E9 MIXING [E1, SECTION 7.2]

To allow the development of limit theorems for correlated signals (i.e., sequences of correlated RVs) it is necessary to have constraints on the dependence between events that are widely separated in time. This is accomplished by various "mixing" measures. Let $\{X_n\}$ be a sequence of RVs and let $\mathcal{F}_{-\infty}^{k}$ be an increasing sequence of σ-fields containing

$$\sigma(X_k, X_{k-1}, \ldots) = \sigma\text{-fields generated by the "history" of } X_k$$

and let $\mathcal{F}_k^{\infty}$ be a decreasing sequence of σ-fields containing

$$\sigma(X_k, X_{k+1}, \ldots) = \sigma\text{-fields generated by the "future" of } X_k.$$

Definition E9.1. A sequence of RVs $\{X_n\}$ is said to be strong mixing if, uniformly in k

$$|P(A \cap B) - P(A)P(B)| \leq \alpha_n \to 0, \text{ as } n \to \infty$$

for $A \in \mathcal{F}_{-\infty}^k$, $B \in \mathcal{F}_{n+k}^{\infty}$.

Definition E9.2. A sequence of RVs $\{X_n\}$ is said to be uniform mixing if, uniformly in k

$$|P(B|A) - P(B)| \leq \phi_n \to 0, \text{ as } n \to \infty$$

for $A \in \mathcal{F}_{-\infty}^k$, $B \in \mathcal{F}_{n+k}^{\infty}$.

The following inequalities are basic tools by which these mixing measures are used to prove limit results. Suppose Y is $\mathcal{F}_{-\infty}^k$ measurable, X is $\mathcal{F}_{k+n}^{\infty}$ measurable. Then

(i) If $p^{-1} + q^{-1} < 1$ and $\{X_n\}$ is strong mixing,

$$|E(XY) - E(X)E(Y)| \leq 8\alpha_n^{1-p^{-1}-q^{-1}} \|X\|_p \|Y\|_q$$

where $\|X\|_p = (E|X|^p)^{1/p}$ and $\|Y\|_q = (E|Y|^q)^{1/q}$.
If $p^{-1} + q^{-1} = 1$

$$E\left(|E(X|\mathcal{F}_k) - E(X)|\right) \leq 8 \times 2^{1/p} \alpha_n^{1/p} \|X\|_q.$$

(ii) If $p^{-1} + q^{-1} = 1$ and $\{X_n\}$ is uniform mixing,

$$|E(XY) - E(X)E(Y)| \leq 2\phi_n^{q^{-1}} \|X\|_p \|Y\|_q$$

$$|E(X|\mathcal{F}_k) - E(X)| \leq 2^{q^{-1}} \phi_n^{q^{-1}} \left(E(|X|^p|\mathcal{F}_k) + E|X|^p\right)^{p^{-1}}$$

$$\|E(X|\mathcal{F}_k) - E(X)\|_v \leq 2\phi_n^{q^{-1}} \|E(|X|^p|\mathcal{F}_k)\|_v^{p^{-1}}.$$

E10 STRONG LAW OF LARGE NUMBERS (SLLN)

Perhaps the most celebrated SLLN is:

Theorem E10.1 (Kolmogorov's SLLN) If $\{\epsilon_k\}$ is i.i.d. (independent identically distributed) with mean zero and $E|\epsilon_k| < \infty$. Then

$$\frac{1}{n} \sum_{k=1}^{n} \epsilon_k \to 0 \quad \text{w.p.1}, \qquad \text{as } n \to \infty.$$

Proof: See [H4, Theorem 2.19].

Here, for illustrative purposes, we prove a weaker result.

Theorem E10.2. If $\{\epsilon_k\}$ are i.i.d., zero mean, $E(\epsilon_k^2) = \sigma^2 < \infty$ then

$$\frac{1}{n}\sum_{k=1}^{n} \epsilon_k \to 0 \quad \text{w.p.1.}$$

Proof. By Kronecker's lemma, the result occurs if $\sum_{s=1}^{\infty} \epsilon_s/s < \infty$ w.p.1. So introduce $T_n = \sum_{s=1}^{n} \epsilon_s/s$, which is an MG, and $E(T_n^2) = \sigma^2 \sum_{s=1}^{n} s^{-2} \leq \sigma^2 \sum_{s=1}^{\infty} s^{-2} < \infty$. So by the MGCT, T_n converges w.p.1, and so the result is established.

We have already mentioned the ergodic theorem. Now we give an SLLN for correlated sequences using mixing conditions.

Theorem E10.3. If $\{X_k\}$ are zero mean with $\sup_k E(X_k^4) < \mu_4 < \infty$ and if $\{X_k\}$ is strong mixing with $\sum_{r=1}^{\infty} \alpha_r^{1/4} < \infty$ or uniform mixing with $\sum_{r=1}^{\infty} \phi_r^{1/2} < \infty$, then

$$\frac{1}{n}\sum_{k=1}^{n} X_k \to 0 \quad \text{w.p.1.}$$

Proof. We use the MG decomposition of (E12.3). It is valid under our conditions. Summing (E12.3) gives

$$\frac{1}{n}\sum_{k=1}^{n} X_k = \frac{1}{n}\sum_{k=1}^{n} \nu_k + \frac{v_0 - v_n}{n}.$$

We need $v_N/N \to 0$ w.p.1 and an SLLN for ν_k. However

$$\sum_{N=1}^{\infty} P(v_N/N > \epsilon) < \sum_{N=1}^{\infty} E(v_N^2)/N^2\epsilon^2 < \infty$$

if $\sup_N E(v_N^2) < \infty$, and this follows from the lemma in Appendix E12.

The argument based on uniform mixing is similar. An SLLN for ν_k follows from Kronecker's lemma if $\sum_{k=1}^{n} \nu_k/k$ converges w.p.1. This is an MG and converges w.p.1 if $\sum_k E(\nu_k^2) < \infty$, which follows since $\sup_k E(X_k^2) < \infty$ and $\sup_k E(v_k^2) < \infty$.

E11 CENTRAL LIMIT THEOREM (CLT)

CLTs can be developed under a variety of conditions. The simplest general CLT is:

Theorem E11.1. If $\{X_k\}$ are i.i.d., zero mean, unit variance, then

$$\frac{1}{\sqrt{n}}\sum_{k=1}^{n} X_k \Rightarrow N(0, 1)$$

where $N(0, 1)$ denotes an RV with Gaussian pdf and zero mean, unit variance.

Proof. A very simple proof is given in [B9, Section 27].

More powerful CLTs are based on MG results. The following result follows from [H4, p. 58].

Theorem E11.2. Suppose $\{X_n, \mathcal{F}_n\}$ is an MG difference, i.e., X_n is adapted to $\mathcal{F}_n$, $\mathcal{F}_n$ are increasing σ-fields, $E(X_n|\mathcal{F}_{n-1}) = 0$. Introduce:

$$\text{partial sum: } S_n = \sum_{u=1}^{n} X_u$$

$$\text{partial variance: } V_n^2 = \sum_{u=1}^{n} E(X_u^2|\mathcal{F}_{n-1})$$

$$\text{variance: } s_n^2 = E(V_n^2) = E(S_n^2)$$

Then (a, b) or (a, b') imply $S_n/s_n \Rightarrow N(0, 1)$, as $n \to \infty$.

(a) $s_n^{-2}V_n^2 \xrightarrow{p} 1$, as $n \to \infty$.

(b) $\sum_{u=1}^{n} E(X_u^2 I(|X_u| > \epsilon s_n))/s_n^2 \to 0$, as $n \to \infty$, for each $\epsilon > 0$.

(b)' $\max_{1 \le u \le n} |X_u|/s_n \xrightarrow{p} 0$.

E12 MARTINGALE APPROXIMATION

For many calculations involving SLLNs, CLTs, and weak convergence in general, a remarkable MG approximation result (due to [G10]) has found wide use. The idea can be clearly illustrated firstly in a special case.

Suppose $\{X_k\}$ is zero mean stationary with Wold decomposition

$$X_k = \sum_{s=0}^{\infty} h_s \epsilon_{k-s} = h(q^{-1})\epsilon_k.$$

But suppose in fact $\{\epsilon_k\}$ are independently distributed: Then $\{X_k\}$ is called a linear process. Now consider the following algebraic decomposition:

$$h(q^{-1}) = \sum_{s=0}^{\infty} h_s q^{-s}$$

$$\begin{aligned}
\Rightarrow \qquad h(1) - h(q^{-1}) &= \sum_{s=0}^{\infty} h_s(1 - q^{-s}) \\
&= (1 - q^{-1}) \sum_{s=0}^{\infty} h_s \sum_{u=0}^{s-1} q^{-u} \\
&= (1 - q^{-1}) \sum_{u=0}^{\infty} q^{-u} \sum_{s=u+1}^{\infty} h_s \\
&= (1 - q^{-1}) \sum_{u=0}^{\infty} q^{-u} h_u^*
\end{aligned}$$

where

$$h_u^* = \sum_{s=u+1}^{\infty} h_s.$$

Then we can write

$$\begin{aligned} X_k &= \left(h(1) - (1 - q^{-1})h^*(q^{-1})\right) \epsilon_k \\ &= h(1)\epsilon_k - (v_k - v_{k-1}) \end{aligned} \tag{E12.1}$$

where we have introduced the stationary process

$$v_k = h^*(q^{-1})\epsilon_k.$$

For $\{v_k\}$ to exist in mean square it is necessary that $\sum_{s=0}^{\infty}(h_s^*)^2 < \infty$ and this can be shown to hold if $\sum_{s=0}^{\infty} s^2 h_s^2 < \infty$ (see [P2]).

The decomposition (E12.1) provides an approximation of X_k by a white noise. To see its utility we show how an SLLN, CLT can be rapidly obtained for X_k.

(i) SLLN
Sum (E12.1) to find

$$\frac{1}{N}\sum_{k=1}^{N} X_k = \frac{h(1)}{N}\sum_{k=1}^{N}\epsilon_k - \frac{v_N - v_0}{N}.$$

By Kolmogorov's SLLN

$$\frac{1}{N}\sum_{k=1}^{N}\epsilon_k \to 0 \quad \text{w.p.1.}$$

So if we show $v_N/N \to 0$ w.p.1, we have an SLLN for X_k. Consider that since $E(v_0^2) < \infty$

$$\sum_{N=1}^{\infty} P(v_N/N > \epsilon) < \sum_{N=1}^{\infty} E(v_N^2)/(N^2\epsilon^2) < \infty$$

so that by Theorem E4.3, $v_N/N \to 0$ w.p.1.

(ii) CLT
Now we have

$$\frac{1}{\sqrt{N}}\sum_{k=1}^{N} X_k = \frac{h(1)}{\sqrt{N}}\sum_{k=1}^{N}\epsilon_k - \frac{v_N - v_0}{\sqrt{N}}.$$

So from the CLT for ϵ_k

$$\frac{1}{\sqrt{N}}\sum_{k=1}^{N}\epsilon_k \Rightarrow N(0, \sigma^2)$$

we obtain a CLT for X_k if we can show

$$\frac{v_N}{\sqrt{N}} \xrightarrow{p} 0.$$

However

$$P\left(|v_N|/\sqrt{N} > \epsilon\right) < E(v_N^2)/N\epsilon^2 \to 0.$$

Further development of these ideas in this special case is in [P2].

Now we look for a decomposition like (E12.1) in a more general setting. Given $\{X_k\}$ (not necessarily stationary) and an increasing sequence of σ-fields $\mathcal{F}_k$ containing $\sigma(X_k, X_{k-1}, \ldots)$, introduce the MG difference (or MG version of white noise)

$$\nu_k = \sum_{r=0}^{\infty} \left(E(X_{k+r}|\mathcal{F}_k) - E(X_{k+r}|\mathcal{F}_{k-1})\right).$$

It is quickly checked that $E(\nu_k|\mathcal{F}_{k-1}) = 0$. We also introduce the residual sequence

$$v_{k-1} = \sum_{r=0}^{\infty} E(X_{k+r}|\mathcal{F}_{k-1}) \tag{E12.2}$$

= accumulated $r+1$ step-ahead forecasts of X from the "history" $\mathcal{F}_k$.

Then we find

$$\begin{aligned}
v_{k-1} - v_k &= \sum_{r=0}^{\infty} E(X_{k+r}|\mathcal{F}_{k-1}) - \sum_{r=0}^{\infty} E(X_{k+1+r}|\mathcal{F}_k) \\
&= \sum_{r=0}^{\infty} E(X_{k+r}|\mathcal{F}_{k-1}) - \sum_{r=1}^{\infty} E(X_{k+r}|\mathcal{F}_k) \\
&= X_k - \left(\sum_{r=0}^{\infty}(E(X_{k+r}|\mathcal{F}_k) - E(X_{k+r}|\mathcal{F}_{k-1}))\right) \\
&= X_k - \nu_k
\end{aligned}$$

and so we obtain the general decomposition

$$X_k = \nu_k + v_{k-1} - v_k. \tag{E12.3}$$

Finally we need to ensure v_k, ν_k are well defined. Since $E|\nu_k| \leq E|X_k| + 2E|v_k|$, we need only check that $\sup_k E|v_k| < \infty$. We have:

(i) If $\{X_k\}$ is strong mixing, then from Appendix E9

$$E\left(|E(X_{k+r}|\mathcal{F}_{k-1})|\right) \leq 8\sqrt{2}\alpha_{r+1}^{1/2} E^{1/2}(X_{k+r}^2)$$

so $\sup_k E|v_k| < \infty$ if

$$\sum_{r=1}^{\infty} \alpha_r^{1/2} < \infty, \quad \sup_k E(X_k^2) < \infty.$$

(ii) If $\{X_k\}$ is uniform mixing, Appendix E9 gives

$$E\left(|E(X_{k+r}|\mathcal{F}_{k-1})|\right) \le 2\phi_r^{1/2}\left(E[E(|X_{k+r}|^2|\mathcal{F}_k)]\right)^{\frac{1}{2}}$$

$$\le 2\phi_r^{1/2}\left(E(X_{k+r}^2)\right)^{\frac{1}{2}}, \quad \text{by (E3.1)}$$

so $\sup_k E|v_k| < \infty$ if

$$\sum_{r=1}^{\infty}\phi_r^{1/2} < \infty, \qquad \sup_k E(X_k^2) < \infty.$$

More generally, the following lemma is useful

Lemma E12.1. If $\{X_k\}$ is strong mixing

$$E(|v_{k-1}|^m) \le 8^m \sup_t \left(E(|X_t|^{2m})\right)^{\frac{1}{2}}\left(\sum_{n=0}^{\infty}\alpha_n^{1/(2m)}\right)^m.$$

If $\{X_k\}$ is uniform mixing

$$E(|v_{k-1}|^m) \le 2^m \sup_t E(|X_t|^m)\left(\sum_{n=0}^{\infty}\phi_n^{1-1/m}\right)^m.$$

Proof.

$$E(|v_{k-1}|^m) \le \left|\sum_{r=0}^{\infty} E\left(|v_{k-1}|^{m-1}E(X_{k+r}|\mathcal{F}_{k-1})\mathrm{sgn}(v_{k-1})\right)\right|$$

$$= \left|\sum_{r=0}^{\infty} E\left(E(X_{k+r}|v_{k-1}|^{m-1}\mathrm{sgn}(v_{k-1})|\mathcal{F}_{k-1})\right)\right|$$

$$= \left|\sum_{r=0}^{\infty} E\left(X_{k+r}|v_{k-1}|^{m-1}\mathrm{sgn}(v_{k-1})\right)\right|$$

$$\le \sum_{r=0}^{\infty} 8\alpha_r^{1/(2m)}\left(E(|X_{k+r}|^{2m})\right)^{\frac{1}{2m}}\left(E(|v_{k-1}|^{(m-1)m/(m-1)})\right)^{\frac{m-1}{m}}$$

$$\le 8\sup_t\left(E(|X_t|^{2m})\right)^{\frac{1}{2m}}\left(E(|v_{k-1}|^m)\right)^{1-1/m}\left(\sum_{r=0}^{\infty}\alpha_r^{1/(2m)}\right)^m.$$

From which the first result follows. The second follows in a similar way.

Bibliography

[A1] Albert, A.E., and Gardner, L.A. (1967). *Stochastic Approximation and Nonlinear Regression*. MIT Press, Cambridge, MA.

[A2] Alexander, S.T. (1986). *Adaptive Signal Processing: Theory and Applications*. Springer-Verlag, New York.

[A3] Anderson, B.D.O., and Moore, J.B. (1979). *Optimal Filtering*. Prentice-Hall, Englewood Cliffs, NJ.

[A4] Anderson, B.D.O., and Johnson, Jr., C.R. (1982). "Exponential convergence of adaptive identification and control algorithms." *Automatica* 18, pp. 1–13.

[A5] Anderson, B.D.O. (1977). "Exponential stability of linear equations arising in adaptive identification." *IEEE Trans. Automatic Control* 22, pp. 83–88.

[A6] Apostol, T. (1974). *Mathematical Analysis*. Addison-Wesley, New York.

[A7] Ash, R.B. (1972). *Real Analysis and Probability*. Academic Press, New York.

[B1] Bai, E., Fu, L., and Sastry, S. (1988). "Averaging analysis for discrete time and sampled data adaptive systems." *IEEE Trans. Circuits and Systems* 35, pp. 137–148.

[B2] Balachandra, M., and Sethna, P.R. (1975). "A generalization of the method of averaging for systems with two-time scales." *Archive for Rat. Mech. Anal.* 58, pp. 261–283.

[B3] Bauer, F.L. (1955). "Ein directes interationsverfahren zur Hurwitz-Zerlegung eines polynomes." *Elek, Ubertr.* 9, pp. 290–295.

[B4] Bellanger, M.G. (1986). "New applications of digital signal processing in communications." *IEEE ASSP Magazine* 3, pp. 6–11.

[B5] Bellanger, P. (1987). *Adaptive Digital Filters and Signal Analysis*. Marcel Dekker, New York.

[B6] Benveniste, A. (1987). "Design of adaptive algorithms for the tracking of time-varying systems." *Int. Journal of Adaptive Control and Signal Processing* 1, pp. 3–29.

[B7] Benveniste, A., Metivier, M., and Priouret, P. (1990). *Adaptive Algorithms and Stochastic Approximations*. Springer-Verlag, New York.

[B8] Bershad, N.J. (1986). "Analysis of the normalized LMS algorithm with Gaussian inputs." *IEEE Trans. Acoustics, Speech, Signal Processing* 34, pp. 793–806.

[B9] Billingsley, P. (1986). *Probability and Measure*, 2nd ed. John Wiley, New York.

[B10] Bitmead, R.R., and Johnson, C.R.J., Jr. (1987). "Discrete averaging principles and robust adaptive identification." In *Control and Dynamics Systems: Advances in Theory and Applications* XXV, edited by C.T. Leondes, pp. 237–271. Academic Press, New York.

[B11] Bitmead, R.R., and Anderson, B.D.O. (1980). "Lyapunov techniques for the exponential stability of linear difference equations with random coefficients." *IEEE Trans. Automatic Control* 25, pp. 782–787.

[B12] Blankenship, G., and Papanicoloau, G.C. (1978). "Stability and control of stochastic systems with wide-band noise disturbances." *SIAM Journal of Applied Mathematics* 34, pp. 437–476.

[B13] Bogoliuboff, N.N., and Mitropolskii, Y.A. (1961). *Asymptotic Methods in the Theory of Non-linear Oscillations*. Gordon & Breach, New York.

[B14] Brillinger, D. (1981). *Time Series: Data Analysis and Theory*. Holden-Day, San Francisco.

[C1] Caines, P.E. (1988). *Linear Stochastic Systems*. Prentice-Hall, Englewood Cliffs, NJ.

[C2] Caraiscos, C., and Liu, B. (1984). "A roundoff error analysis of the LMS adaptive algorithm." *IEEE Trans. Acoustics, Speech, Signal Processing* 32, pp. 34–41.

[C3] Carter, G. (1987). "Coherence and time delay estimation." *IEEE Proc.* 75, pp. 236–255.

[C4] Chen, H.F., and Guo, L. (1991). *Identification and Stochastic Adaptive Control*. Birkhäuser Boston, Cambridge, MA.

[C5] Chung, K.L. (1974). *A Course in Probability Theory*, 2nd ed. Academic Press, New York.

[C6] Cioffi, J.M. (1985). "When do I use an RLS adaptive filter?" *Proc. 19th Asilomar Conf. Circ. System Computers*, Pacific Grove, CA.

[C7] Cowan, C.F.N., and Grant, P.M. (1985). *Adaptive Filters*. Prentice-Hall, Englewood Cliffs, NJ.

[D1] D'Azzo, J.J., and Houpis, C.C. (1966). *Feedback Control System Analysis and Synthesis*, 2nd ed. McGraw Hill, New York.

[D2] Desoer, C.A. (1970). "Slowly varying discrete system $X_{i+1} = A_i X_i$." *Electronics Letters* 6, pp. 339–340.

[D3] Deveritskii, D.P., and Fradkov, A.L. (1974). "Two models for analyzing the dynamics of adaptive algorithms." *Automatikha i Telemekhanica* 1, pp. 66–75.

[D4] Ding, Z., Kennedy, R.A., Anderson, B.D.O., and Johnson, C.R., Jr. (1991). "Ill-convergence of Godard blind equalizers in data communication systems." *IEEE Trans. Communications* 39, pp. 1313–1326.

[D5] Ding, Z., Johnson, C.R., Jr., and Kennedy, R.A. (1992). "On the (non)existence of undesirable equilibria of Godard blind equalizers." *IEEE Trans. Signal Processing* 40, pp. 2425–2431.

[D6] Donoho, D.L. (1981). "On minimum entropy deconvolution." In *Applied Time Series Analysis II*, edited by D.F. Findley. Academic Press, New York.

[E1] Ethier, S.N., and Kurtz, T.G. (1986). *Markov Processes: Characterization and Convergence*. John Wiley, New York.

[E2] Etter, D., and Stearns, S.M. (1981). "Adaptive estimation of time delays in sampled data systems." *IEEE Trans. Acoustics, Speech, Signal Processing* 29, pp. 582–587.

[F1] Farden, D.C. (1981). "Tracking properties of adaptive signal processing algorithms." *IEEE Trans. Acoustics, Speech, Signal Processing* 29, pp. 439–446.

[F2] Feuer, A., and Weinstein, E. (1985). "Convergence analysis of LMS filters with uncorrelated Gaussian data." *IEEE Trans. Acoustics, Speech, Signal Processing* 33, pp. 222–230.

[F3] Foschini, G.J. (1985). "Equalizing without altering or detecting data." *AT&T Technical Journal* 6, pp. 1885–1911.

[F4] Freidlin, M.I., and Wentzell, A.D. (1984). *Random Perturbations of Dynamical Systems.* Springer-Verlag, New York.

[G1] Gardner, W.A. (1987). "Nonstationary learning characteristics of the LMS algorithm." *IEEE Trans. Circuits and Systems* 34, pp. 1199–1207.

[G2] Gear, C.W. (1971). *Numerical Initial Value Problems in Ordinary Differential Equations.* Prentice-Hall, Englewood Cliffs, NJ.

[G3] Gersh, W., and Kitagawa, G. (1985). "A smoothness priors time-varying AR coefficient modeling of non-stationary covariance time series." *IEEE Trans. Automatic Control* 30, pp. 48–57.

[G4] Gertler, I., and Banyasz, G. (1974). "A recursive (on-line) maximum likelihood identification method." *IEEE Trans. Automatic Control* 19, pp. 816–820.

[G5] Gibson, J.D. (1984). "Adaptive prediction for speech encoding." *IEEE ASSP Magazine* 1, pp. 12–26.

[G6] Gikhman, I.I. (1952). "Poporodu odnoi Teoremy N.I. Bogolybova." *Ukrain. Math. Zh.* 4, pp. 215–218.

[G7] Godard, D. (1974). "Channel equalization using a Kalman filter for fast data transmission." *IBM Journal of Research and Development* 18, pp. 267–273.

[G8] Godard, D. (1980). "Self-recovering equalization and carrier tracking in two-dimensional data communication systems." *IEEE Trans. Communications* 28, pp. 1867–1875.

[G9] Goodwin, G.C., and Sin, K.S. (1984). *Adaptive Filtering, Prediction and Control.* Prentice-Hall, Englewood Cliffs, NJ.

[G10] Gordin, M.I. (1969). "The central limit theorem for stationary processes." *Soviet Math. Dokl.* 10, pp. 1174–1176.

[G11] Grenander, U., and Szego, G. (1958). *Toeplitz Forms and Their Applications.* University of California Press, Berkeley.

[G12] Gritton, C.W.K., and Lin, D.W. (1984). "Echo cancellation algorithms." *IEEE ASSP Magazine* 1, pp. 30–38.

[G13] Gupta, S.C., and Hasdorff, L. (1970). *Fundamentals of Automatic Control.* John Wiley, New York.

[G14] Guo, L. (1990). "Estimating time-varying parameters by the Kalman filter based algorithm: Stability and convergence." *IEEE Trans. Automatic Control* 35, pp. 141–147.

[H1] Hahn, W. (1967). *Stability of Motion*, Springer-Verlag, New York.

[H2] Hajivandi, M., and Gardner, W.A. (1990). "Measures of tracking performance for the LMS algorithm." *IEEE Trans. Acoustics, Speech, Signal Processing* 38, pp. 1953–1958.

[H3] Hale, J.K. (1980). *Ordinary Differential Equations.* Krieger, Melbourne, FL.

[H4] Hall, P., and Heyde, C.C. (1980). *Martingale Limit Theory and its Application.* Academic Press, New York.

[H5] Hamming, R.W. (1989). *Digital Filters*, 3d ed. Prentice-Hall, Englewood Cliffs, NJ.

[H6] Hastings-James, R., and Sage, M.W. (1969). "Recursive generalized least squares procedure for on-line identification of process parameters." *Proc. IEE* 116, pp. 2057–2062.

[H7] Haykin, S. (1986). *Adaptive Filter Theory.* Prentice-Hall, Englewood Cliffs, NJ.

[H8] Hewer, G.A. (1973). "Analysis of a discrete matrix Ricatti equation of linear control and Kalman filtering." *Journal of Math. Anal. and Appl.* 42, pp. 226–236.

[H9] Hewer, G.A. (1971). "An iterative technique for the computation of the steady state gains for the discrete optimal regulator." *IEEE Trans. Automatic Control* 16, pp. 382–383.

[H10] Honig, M.L., and Messerschmidt, D.G. (1984). *Adaptive Filters: Structures, Algorithms and Applications*. Kluwer Academic Publishers, Norwell, MA.

[H11] Hwang, S.Y. (1978). "Solution of complex integrals using the Laurent expansion." *IEEE Trans. Acoustics, Speech, Signal Processing* 26, pp. 263–265.

[J1] Jaszwinski, A.H. (1970). *Stochastic Processes and Filtering Theory*. Academic Press, New York.

[J2] Jayant, N.S., and Noll, P. (1984). *Digital Coding of Waveforms*. Prentice-Hall, Englewood Cliffs, NJ.

[J3] Johnson, C.R., Jr. (1988). *Lectures on Adaptive Parameter Estimation*. Prentice-Hall, Englewood Cliffs, NJ.

[J4] Johnson, C.R., Jr.; Dasgupta, S.; and Sethares, W. (1988). "Averaging analysis of local stability of a real constant modulus algorithm adaptive filter." *IEEE Trans. Acoustics, Speech, Signal Processing* 36, pp. 900–910.

[J5] Johnson, C.R., Jr. (1991). "Admissibility in blind adaptive channel equalization." *IEEE Control Systems Magazine* 11, pp. 3–15.

[K1] Kailath, T. (1980). *Linear Systems*. Prentice-Hall, Englewood Cliffs, NJ.

[K2] Khasminski, R.Z. (1966). "On stochastic processes defined by differential equations with a small parameter." *Theory of Probability Appl.* 11, pp. 211–228.

[K3] Korostelev, A.P. (1981). "Multistep procedures of stochastic optimization." *Avtomatikha i Telemekhanica* 5, pp. 82–90.

[K4] Kushner, H.J. (1984). *Approximation and Weak Convergence Methods for Random Processes, with Applications to Stochastic System Theory*. MIT Press, Cambridge, MA.

[K5] Kushner, H.J., and Huang, H. (1981). "Asymptotic properties of stochastic approximations with constant coefficients." *SIAM Journal of Control* 19, pp. 87–105.

[L1] Landau, I.D. (1976). "Unbiased recursive identification using model reference adaptive techniques." *IEEE Trans. Automatic Control* 21, pp. 194–202.

[L2] Landau, I.D. (1979). *Adaptive Control: The Model Reference Approach*. Marcel Dekker, New York.

[L3] Larimore, M.G., Treichler, J.R., Johnson, C.R., Jr. (1980). "SHARF: An algorithm for adapting IIR digital filters." *IEEE Trans. Acoustics, Speech, Signal Processing* 28, pp. 428–440.

[L4] Lindsey, W.C., and Simon, M.K. (1973). *Telecommunication System Engineering*. Prentice-Hall, Engelwood Cliffs, NJ.

[L5] Ljung, L. (1977). "On positive real functions and the convergence of some recursive schemes." *IEEE Trans. Automatic Control* 22, pp. 539–551.

[L6] Ljung, L. (1977). "Analysis of recursive stochastic algorithms." *IEEE Trans. Automatic Control* 22, pp. 551–575.

[L7] Ljung, L. (1979). "Asymptotic behavior of the extended Kalman filter as a parameter estimator for linear systems." *IEEE Trans. Automatic Control* 24, pp. 36–50.

[L8] Ljung, L., and Soderstrom, T. (1983). *Theory and Practice of Recursive Identification*. MIT Press, Cambridge, MA.

[L9] Ljung, L. (1985). "Asymptotic variance expressions for identified black-box transfer function models." *IEEE Trans. Automatic Control* 30, pp. 834–844.

[L10] Ljung, L. (1987). *System Identification: Theory for the User*. Prentice-Hall, Englewood Cliffs, NJ.

[L11] Ljung, L., and Gunnarrson, S. (1990). "Adaptation and tracking in system identification: A survey." *Automatica* 26, pp. 7–21.

[L12] Ljung, L., and Priouret, P. (1991). "A result on the mean square error obtained using general tracking algorithms." *Int. Journal of Adaptive Control and Signal Processing* 5, pp. 231–250.

[L13] Lusternik, L.A., and Sobolev, V.J. (1974). *Elements of Functional Analysis*, 3rd ed. Gordon & Breach, New York.

[M1] Macchi, O. (1986). "Optimization of adaptive identification for time varying filters." *IEEE Trans. Automatic Control* 31, pp. 283–287.

[M2] McGarty, T.P. (1974). *Stochastic Systems and State Estimation*. John Wiley, New York.

[M3] Mayne, D.Q. (1963). "Optimal nonstationary estimation of the parameters of a linear system with Gaussian inputs." *Journal of Electronics Control* 14, p. 101.

[M4] Meerkov, S.M. (1973). "Averaging of trajectories of slow dynamic systems." (in Russian), *Differentsial'nye Uravneniya* 9, pp. 1609–1617.

[M5] Metivier, M., and Priouret, P. (1984). "Applications of a Kushner and Clark lemma to general cases of stochastic algorithms." *IEEE Trans. Information Theory* 30, pp. 140–151.

[M6] Morgan, A.P., and Narendra, K.S. (1977). "On the uniform asymptotic stability of certain linear nonautonomous differential equations." *SIAM Journal of Control* 15, pp. 5–24.

[M7] Mulgrew, B., and Cowan, C. (1988). *Adaptive Filters and Equalisers*. Kluwer Academic, Norwell, MA.

[N1] Nagumo, J.I., and Noda, A. (1967). "A learning method for system identification." *IEEE Trans. Automatic Control* 12, pp. 282–287.

[N2] Narendra, K.S., and Annaswamy, A.M. (1989). *Stable Adaptive Systems*. Prentice-Hall, Englewood Cliffs, NJ.

[N3] Nehorai, A., and Starer, D. (1990). "Adaptive pole estimation." *IEEE Trans. Acoustics, Speech, Signal Processing* 38, pp. 825–838.

[N4] Neveu, J. (1975). *Discrete Parameter Martingales*. North Holland, Amsterdam.

[P1] Panuska, V. (1968). "A stochastic approximation method for identification of linear systems using adaptive filtering." *Joint Automatic Control Conference*, Ann Arbor, MI.

[P2] Phillips, P.C.B., and Solo, V. (1992). "Asymptotics for linear processes." *Ann. Stat.* 20, pp. 971–1001.

[P3] Pomet, J.B., and Praly, L. (1989). "Adaptive nonlinear regulation: Equation error from the Lyapunov equation." *Proc. of 28th IEEE Conf. on Decision and Control*, IEEE Press, Tampa, FL, pp. 1008–1013.

[P4] Proakis, J.G. (1974). "Channel identification for high speed digital communications." *IEEE Trans. Automatic Control* 19, pp. 916–922.

[R1] Riedle, B., and Kokotovic, P. (1986). "Integral manifolds of slow adaptation." *IEEE Trans. Automatic Control* 31, pp. 316–324.

[R2] Roy, S., and Shynk, J.J. (1990). "Analysis of the momentum LMS algorithm." *IEEE Trans. Acoustics, Speech, Signal Processing* 38, pp. 2088–2098.

[R3] Roseau, M. (1969). "Sur une classe de systemes dynamiques soumis a des excitations periodiques de longue periode." *C.R. Acad. Sci. Paris* 258, pp. 409–412.

[R4] Rudin, W. (1966). *Real and Complex Analysis*. McGraw Hill, New York.

[S1] Sato, Y. (1975). "A method of self-recovering equalization for multilevel amplitude modulation systems." *IEEE Trans. Communications* 23, pp. 679–682.

[S2] Sanders, J.A., and Verhulst, F. (1985). *Averaging Methods in Nonlinear Dynamical Systems*. Springer-Verlag, New York.

[S3] Sastry, S., and Bodson, M. (1989). *Adaptive Control, Stability, Convergence, and Robustness*. Prentice-Hall, Englewood Cliffs, NJ.

[S4] Sethares, W.A., Kennedy, R.A., and Gu, Z. (1991). "An approach to blind equalization of non-minimum phase systems." *Proc. 30th IEEE Conf. Decision and Control*, IEEE Press, Brighton, UK, pp. 1529–1532.

[S5] Shalvi, O., and Weinstein, E. (1990). "New criteria for blind deconvolution of nonminimum phase systems (channels)." *IEEE Trans. Information Theory* 36, pp. 312–321.

[S6] Shilman, S.V., and Yastrebov, A.I. (1976). "Convergence of a class of multistep stochastic adaptation algorithms." *Avtomatikha i Telemekhanika* 8, pp. 111–118.

[S7] Shilman, S.V., and Yastrebov, A.I. (1978). "Properties of a class of multistep gradient and pseudo-gradient algorithms of adaptation and learning." *Avtomatikha i Telemekhanika* 4, pp. 95–104.

[S8] Shynk, J.J, (1989). "Adaptive IIR filtering." *IEEE ASSP Magazine* 6, pp. 4–21.

[S9] Soderstrom, T. (1973). "An on-line algorithm for approximate maximum likelihood identification of linear dynamic systems." *Report 7308*, Dept. of Automatic Control, Lund Inst. of Technology, Lund, Sweden.

[S10] Soderstrom, T., and Stoica, P. (1983). *The Instrumental Variable Approach to System Identification*. Springer-Verlag, New York.

[S11] Solo, V. (1978). *Time Series Recursions and Stochastic Approximation*. Ph.D dissertation, The Australian National University, Canberra, Australia.

[S12] Solo, V. (1979). "The convergence of AML." *IEEE Trans. Automatic Control* 24, pp. 958–962.

[S13] Solo, V. (1980). "Some aspects of recursive parameter estimation." *Int. Journal of Control* 32, pp. 395–410.

[S14] Solo, V. (1989). "Adaptive spectral factorization." *IEEE Trans. Automatic Control* 34, pp. 1047–1051.

[S15] Solo, V. (1989). "The limiting behaviour of LMS." *IEEE Trans. Acoustics, Speech, Signal Processing* 31, pp. 1909–1922.

[S16] Solo, V. (1992). "The error variance of LMS." *IEEE Trans. Signal Processing* 40, pp. 803–813.

[S17] Solo, V. (1992). "Deterministic adaptive control with slowly varying parameters: an averaging analysis." *Technical Report*, Dept. ECE, The Johns Hopkins University.

[S18] Starer, D., and Nehorai, A. (1991). "Adaptive polynomial factorization by coefficient matching." *IEEE Trans. Signal Processing* 39, pp. 527–530.

[S19] Stearns, S.D. (1981). "Error surface of recursive adaptive filters." *IEEE Trans. Circuits and Systems* 28, pp. 603–606.

[S20] Stout, W.F. (1974). *Almost Sure Convergence*. Academic Press, New York.

[T1] Treichler, J.R., and Larimore, M.G. (1985). "New processing techniques based on the constant modulus adaptive algorithm." *IEEE Trans. Acoustics, Speech, Signal Processing* 33, pp. 420–431.

[T2] Tsypkin, Ya.Z. (1971). *Adaptation and Learning in Automatic Systems*. Academic Press, New York.

[V1] Ventsel, A.D., and Freidlin, M.I. (1983). *Fluctuations in Dynamical Systems Caused by Small Random Perturbations*. Springer-Verlag, New York.

[V2] Verdu, S., Anderson, B.D.O., and Kennedy, R.A. (1991). "Anchored blind equalization." *Proc. 25th Conf. Information Sciences and Systems*, The Johns Hopkins University, March 1991, pp. 774–779.

[V3] Vidyasagar, M. (1978). *Nonlinear Systems Analysis*. Prentice-Hall, Englewood Cliffs, NJ.

[V4] Vieira, A., and Kailath, T. (1977). "On another approach to the Schur-Cohn criterion." *IEEE Trans. Circuits and Systems* 24, pp. 218–220.

[V5] Volosov, V.M. (1961). "The method of averaging." *Soviet Math. Dokl.* 2, pp. 221–224.

[W1] White, S.A. (1975). "An adaptive recursive filter." *Proc. of 9th Asilomar Conf. Circ. System Comp.*, p. 21.

[W2] Widrow, B., and Hoff, M.E. (1960). "Adaptive switching circuits." *IRE Wescon Convention Record Part IV*. pp. 96–104.

[W3] Widrow, B., Glover, J.R., et al. (1975). "Adaptive noise cancelling: Principles and applications." *IEEE Proc.* 63, pp. 1692–1716.

[W4] Widrow, B., McCool, J.M., Larimore, M.G., Johnson, C.R., Jr. (1976). "Stationary and nonstationary learning characteristics of the LMS adaptive filter." *IEEE Proc.* 64, pp. 1151–1162.

[W5] Widrow, B., and Stearns, S.D. (1985). *Adaptive Signal Processing*. Prentice-Hall, Englewood Cliffs, NJ.

[W6] Wiener, N. (1958). *The Fourier Integral and Certain of its Applications*. Dover, Mineola, NY.

[W7] Wilson, G.T. (1969). "Factorization of the covariance generating function of a pure moving average process." *SIAM Journal on Numerical Anal.* 6, pp. 1–8.

[Y1] Young, P.C. (1971). "An instrumental variable method for real time identification of a noisy process." *Automatica* 6, pp. 271–287.

[Y2] Young, P.C. (1984). *Recursive Estimation and Time-Series Analysis*. Springer-Verlag, New York.

Index

D

E

F

G

H

I